Protein Science: Architecture and Function

Protein Science: Architecture and Function

Edited by Anton Torres

MURPHY & MOORE
www.murphy-moorepublishing.com

Published by Murphy & Moore Publishing,
1 Rockefeller Plaza,
New York City, NY 10020, USA
www.murphy-moorepublishing.com

Protein Science: Architecture and Function
Edited by Anton Torres

International Standard Book Number: 978-1-63987-464-4 (Hardback)

Cataloging-in-Publication Data

Protein science : architecture and function / edited by Anton Torres.
p. cm.
Includes bibliographical references and index.
ISBN 978-1-63987-464-4
1. Proteins. 2. Proteins--Structure. 3. Proteomics. 4. Protein engineering.
I. Torres, Anton.
QD431 .P76 2022
572.6--dc23

TABLE OF CONTENTS

PREFACE

The world is advancing at a fast pace like never before. Therefore, the need is to keep up with the latest developments. This book was an idea that came to fruition when the specialists in the area realized the need to coordinate together and document essential themes in the subject. That's when I was requested to be the editor. Editing this book has been an honour as it brings together diverse authors researching on different streams of the field. The book collates essential materials contributed by veterans in the area which can be utilized by students and researchers alike.

The large macromolecules which consist of long chains of amino acid residues are known as proteins. Peptide bonds are responsible for keeping individual amino acid residues together. Different proteins have different sequences of amino acids, which are governed by the sequence of nucleotide of their genes. Protein science refers to the application of various methods which are used to analyze the function, structure, and biochemical significance of proteins. This is achieved using different techniques such as protein purification, cellular localization, proteomics, structure determination, bioinformatics and chemical analysis. Proteins are responsible for carrying out a large variety of functions like transportation of molecules, catalysis of metabolic reactions, responding to stimuli and DNA replication. This book elucidates the concepts and innovative models around prospective developments with respect to protein science, their architecture and their function. It aims to present researches that have transformed this discipline and aided its advancement. This book aims to serve as a resource guide for students and experts alike and contribute to the growth of the discipline.

Each chapter is a sole-standing publication that reflects each author´s interpretation. Thus, the book displays a multi-facetted picture of our current understanding of application, resources and aspects of the field. I would like to thank the contributors of this book and my family for their endless support.

Editor

1

SARS coronavirus papain-like protease inhibits the type I interferon signaling pathway through interaction with the STING-TRAF3-TBK1 complex

Xiaojuan Chen[1], Xingxing Yang[1,2], Yang Zheng[1], Yudong Yang[1], Yaling Xing[1], Zhongbin Chen[1,2]✉

[1] Division of Infection and Immunity, Department of Electromagnetic and Laser Biology, Beijing Institute of Radiation Medicine, Beijing 100850, China
[2] Anhui Medical University, Hefei 230032, China
✉ Correspondence: chenzb@bmi.ac.cn (Z. Chen)

ABSTRACT

SARS coronavirus (SARS-CoV) develops an antagonistic mechanism by which to evade the antiviral activities of interferon (IFN). Previous studies suggested that SARS-CoV papain-like protease (PLpro) inhibits activation of the IRF3 pathway, which would normally elicit a robust IFN response, but the mechanism(s) used by SARS PLpro to inhibit activation of the IRF3 pathway is not fully known. In this study, we uncovered a novel mechanism that may explain how SARS PLpro efficiently inhibits activation of the IRF3 pathway. We found that expression of the membrane-anchored PLpro domain (PLpro-TM) from SARS-CoV inhibits STING/TBK1/IKKε-mediated activation of type I IFNs and disrupts the phosphorylation and dimerization of IRF3, which are activated by STING and TBK1. Meanwhile, we showed that PLpro-TM physically interacts with TRAF3, TBK1, IKKε, STING, and IRF3, the key components that assemble the STING-TRAF3-TBK1 complex for activation of IFN expression. However, the interaction between the components in STING-TRAF3-TBK1 complex is disrupted by PLpro-TM. Furthermore, SARS PLpro-TM reduces the levels of ubiquitinated forms of RIG-I, STING, TRAF3, TBK1, and IRF3 in the STING-TRAF3-TBK1 complex. These results collectively point to a new mechanism used by SARS-CoV through which PLpro negatively regulates IRF3 activation by interaction with STING-TRAF3-TBK1 complex, yielding a SARS-CoV countermeasure against host innate immunity.

Xiaojuan Chen, Xingxing Yang, and Yang Zheng contributed equally to this work.

KEYWORDS SARS coronavirus, papain-like protease, interferon, deubiquitinase, STING-TRAF3-TBK1 complex

INTRODUCTION

The innate immune system represents the first line of defense and initiates counteractive responses that protect the host from viral infection through evolutionarily conserved pattern recognition receptors (PRR) (Kawai and Akira, 2007). PRRs include the membrane-bound Toll-like receptors (TLRs) and cytosolic sensors, such as retinoic acid-inducible gene-I (RIG-I)-like receptors (RLRs), which sense RNA viruses (Kawai and Akira, 2007). RIG-I specifically detects the intracellular double-stranded viral RNA bearing 5′ triphosphate and panhandle structures to activate antiviral signaling (Hornung et al., 2006; Pichlmair et al., 2006). Once a host is invaded by a virus, PRRs transmit signals to the downstream kinases that activate transcription factors, including IFN regulatory factor-3 (IRF3), nuclear factor κB (NF-κB), and ATF-2/c-jun, with the help of different adaptor molecules (MAVS/IPS-1/VISA/Cardif for RIG-I, TRIF for TLR3, and MyD88 for TLR7/8/9) to activate IFN production (Barral et al., 2009; Yoneyama and Fujita, 2009; Chen and Jiang, 2013).

A previous study suggested that TRAF family members are involved in the regulation of inflammation and antiviral responses (Saha and Cheng, 2006). TRAF proteins play significant roles in signal transduction of antiviral innate immune responses. For example, TRAF3 uniquely regulates the type I IFN response and specifically contributes to TBK1-dependent IRF3 activation (Oganesyan et al., 2006; He et al., 2007). TRAF3 stimulates the non-canonical IKK-related kinase TBK1, which induces phosphorylation and dimerization of IRF3, resulting in nuclear translocation and activation of IRF3 (Fitzgerald et al., 2003). TRAF3 also interacts with several key signaling molecules, such as STING (stimulator of interferon genes, also known as MITA/ERIS/MYPS), which resides at the ER/mitochondrial membrane and forms dimers (Zhong et al., 2008; Sun et al., 2009). TRAF3, STING, and TBK1 form a signaling complex that transmits upstream sensory responses to downstream effectors and then phosphorylates IRF3 to activate type I IFN production.

Viruses have evolved elaborate mechanisms to evade or inactivate the innate immune signaling pathway for their replication (Schindler et al., 2007). Severe acute respiratory syndrome (SARS) is a highly contagious respiratory disease that appeared first in China in 2002 and has infected more than 8000 people worldwide and killed about 800 of those infected. SARS coronavirus (SARS-CoV) has a single-stranded, positive sense RNA genome of approximately 29.7 kb (Marra et al., 2003; Rota et al., 2003). Numerous studies have revealed that SARS-CoV develops an antagonistic mechanism to evade the antiviral activities of IFN (Devaraj et al., 2007; Thiel and Weber, 2008; Perlman and Netland, 2009; Zielecki et al., 2013). Several studies have shown that SARS-CoV papain-like protease (PLpro-TM) has deubiquitination (DUB) activity. PLpro-TM serves as an IFN antagonist (Sulea et al., 2005; Barretto et al., 2006; Frieman et al., 2009; Clementz et al., 2010). Previous reports have also indicated that PLpro-TM inhibits the phosphorylation and nuclear translocation of IRF3 (Devaraj et al., 2007; Frieman et al., 2009). However, there are some discrepancies in how PLpro-TM negatively regulates IRF3 activation. One study reported that PLpro-TM interacts with IRF3 (Devaraj et al., 2007). Another study rejected the finding that PLpro-TM has any influence on the phosphorylation of IRF3 (Frieman et al., 2009). A recent study has reported that PLP2 of mouse hepatitis virus A59 targets TBK1 to negatively regulate type I IFN by its DUB activity (Wang et al., 2011). It has also been reported that M protein of SARS-CoV impedes the formation of TRAF3-TBK1-IKKε complex and inhibits the production of type I IFN (Siu et al., 2009). The mechanism or mechanisms used by PLpro-TM to inhibit activation of the IRF3 pathway are not fully understood.

In the current study, we found that SARS-CoV PLpro-TM inhibits STING/TBK1/IKKε-mediated activation of type I IFN, and disrupts the phosphorylation and dimerization of IRF3. Furthermore, PLpro-TM interacts with STING, TRAF3, and TBK1, and disrupts the interaction between the components in STING-TRAF3-TBK1 complex that is required for the activation of IRF3. PLpro-TM reduces the levels of ubiquitinated forms of RIG-I, STING, TRAF3, TBK1, and IRF3, and contributes to disruption of the signaling required for the induction of IFN. Our work reveals a novel mechanism that explains how SARS PLpro efficiently inhibits activation of the IRF3 pathway, resulting in the SARS-CoV countermeasure against host innate immunity.

RESULTS

PLpro-TM inhibits STING/TBK1/IKKε-mediated activation of Type I interferon

Subsequent to our previous report that SARS coronavirus PLpro negatively regulates IRF3-dependent innate immunity (Devaraj et al., 2007), we determined to address whether SARS PLpro-TM blocks STING/TBK1/IKKε-mediated activation of Type I IFN. In this study, we assessed the level of STING/TBK1/IKKε-activated IFNβ- and IRF3- promoter activities in the presence of SARS PLpro-TM. HEK293T cells were co-transfected with PLpro-TM and a combination of plasmids encoding firefly luciferase under the control of IFN-β promoter (Fig. 1A, 1C and 1E) or IRF3 promoter of PRD(III-I)4 (Fig. 1B, 1D and 1F), and STING, TBK1, or IKKε as an activator of the IFNβ signaling pathway. We found that the stimulation of HEK-293T cells with STING alone increased the activity of IFNβ- and IRF3- promoter more than 20-fold. Co-expression of STING with SARS PLpro-TM resulted in a significant decrease in IFNβ- and IRF3- promoter activities, indicating that SARS PLpro-TM can antagonize STING-mediated activation of IFNβ- and IRF3- promoter activities (Fig. 1A and 1B). SARS PLpro-TM significantly inhibited IFNβ- and IRF3- promoter activities, which are activated by TBK1 (Fig. 1C and 1D) and IKKε (Fig. 1E and 1F). These results indicate that SARS-CoV PLpro-TM antagonizes STING/TBK1/IKKε-mediated IFN-β transcription by interfering with the activation of IRF3 pathway.

PLpro-TM antagonizes STING and TBK1-induced IRF3 activation

Since SARS-CoV PLpro-TM antagonizes STING/TBK1/IKKε-mediated IFN-β transcription at the level of IRF3 activation, we examined whether PLpro-TM affects STING/TBK1/IKKε induced dimerization and phosphorylation of IRF3. HEK293T cells were co-transfected with HA-STING, Flag-TBK1, or Flag-IKKε in the presence or absence of V5-tagged PLpro-TM. We found that stimulation of HEK293T cells with STING alone triggered the phosphorylation and dimerization of IRF3 (Fig. 2A, lane 2). When co-transfected with PLpro-TM, IRF3 dimer and phosphorylated IRF3 were eliminated from the cells (Fig. 2A, lane 3). Overexpression of TBK1 induced dimerization and phosphorylation of IRF3 (Fig. 2B, lane 2), but PLpro-TM significantly reduced the production of phosphorylated IRF3 and completely abolished the dimerization of IRF3 (Fig. 2B, lane 3). SARS-CoV PLpro-TM did not change the amount of IRF3 dimer or phosphorylated IRF3, each of which was activated by IKKε

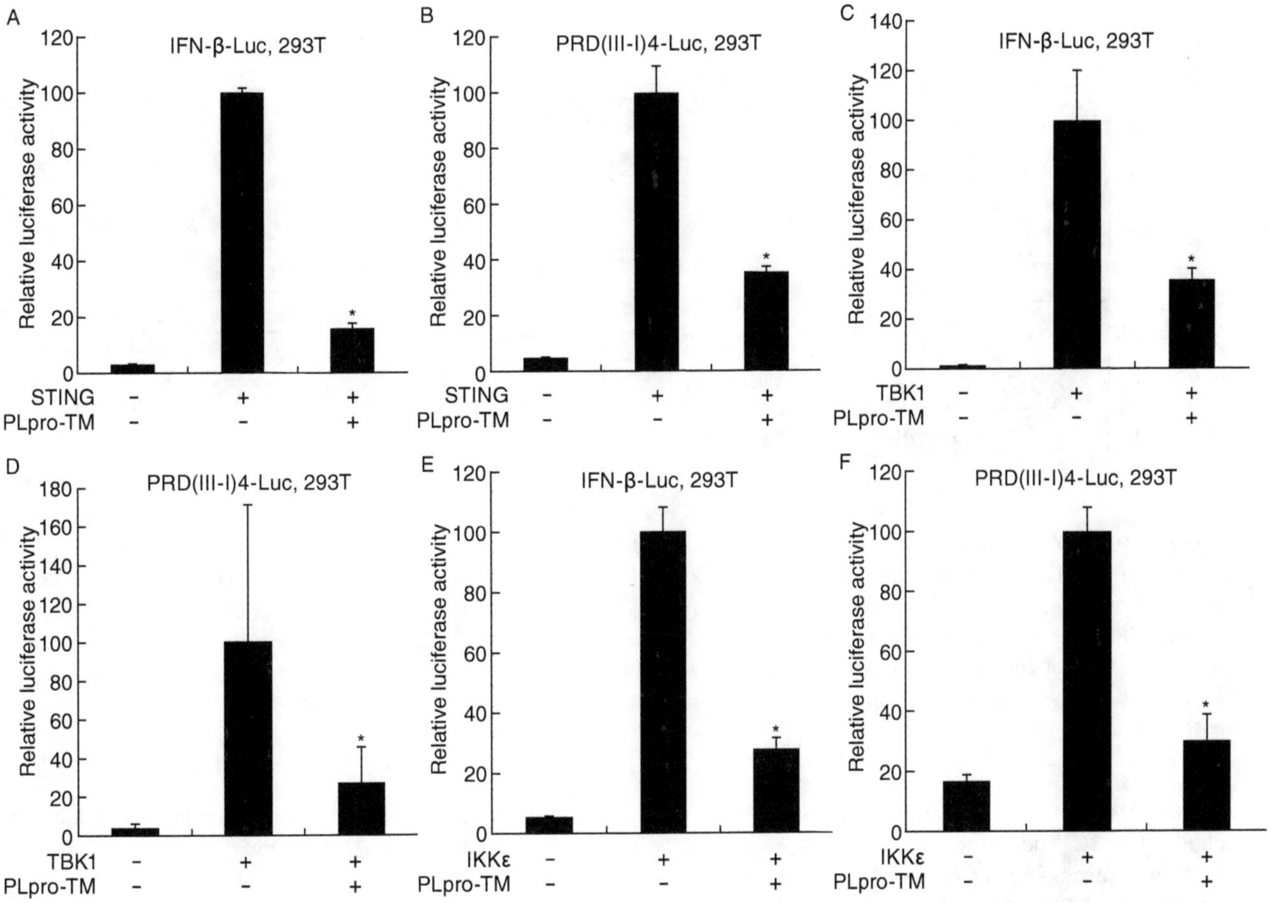

Figure 1. SARS coronavirus PLpro-TM is a potent IFN antagonist. HEK293T cells were co-transfected with either IFNβ-Luc (A, C, E) or PRD (III-I) 4-Luc (B, D, F). The plasmids expressing HA-STING (A and B), Flag-TBK1 (C and D), or Flag-IKKε (E and F) were applied to activate the IFN expression pathway and IRF3-dependent IFN expression pathway, in the presence or absence of PLpro TM-V5. pRL-TK encoding Rellina luciferase was applied for normalization of transfection efficiency. HEK293T cells were transfected by Lipofectamine 2000 according to the manufacturer's instructions and incubated for 24 h. The luciferase activities were assayed with the Dual Luciferase Reporter Assay. The results were expressed as mean relative luciferase activity (firefly luciferase activity divided by Renilla luciferase activity) including the standard deviation from repeated experiments carried out in triplicate. Asterisks indicate statistical significance ($P < 0.05$).

(Fig. 2C, lanes 2 and 3). Thus, SARS-CoV PLpro-TM inhibited STING- and TBK1-mediated dimerization and phosphorylation of IRF3.

PLpro-TM interacts with STING-TRAF3-TBK1 complex

Because SARS PLpro-TM inhibited STING- and TBK1-dependent activation of the IRF3 pathway, PLpro-TM might associate with STING-TRAF3-TBK1 complex, either directly or as part of the multiprotein complex. To confirm this hypothesis, we examined whether PLpro-TM interacted with the key signaling components in the complex. Twenty-four hours after transfection of HEK293T cells with PLpro-TM and Flag-STING, TRAF3, TBK1, or IKKε, the interactions between PLpro-TM and these regulating proteins were assessed via co-immunoprecipitation and Western blotting. The interactions between PLpro-TM and STING, TRAF3, TBK1, and IKKε were confirmed (Fig. 3A). The association between PLpro-TM and IRF3 was also identified (Fig. 3B). These results indicate that PLpro-TM interacts with the STING-TRAF3-TBK1 complex.

Both PLpro-TM and STING are membrane-bound proteins that are anchored to the endoplasmic reticulum (ER) membrane with the transmembrane (TM) domain (Huang et al., 2012; Ouyang et al., 2012; Shang et al., 2012; Shu et al., 2012; Yin et al., 2012), and the TM domains at the N-terminus of STING were found to be indispensable for IRF3 activation *in vitro* (Tanaka and Chen, 2012). Our recent study demonstrated that coronavirus PLpros negatively regulate antiviral innate immune response by disruption of STING-mediated signaling (Sun et al., 2012). In this study, we further addressed which STING domains are required for the interaction between STING and PLpro-TM. We

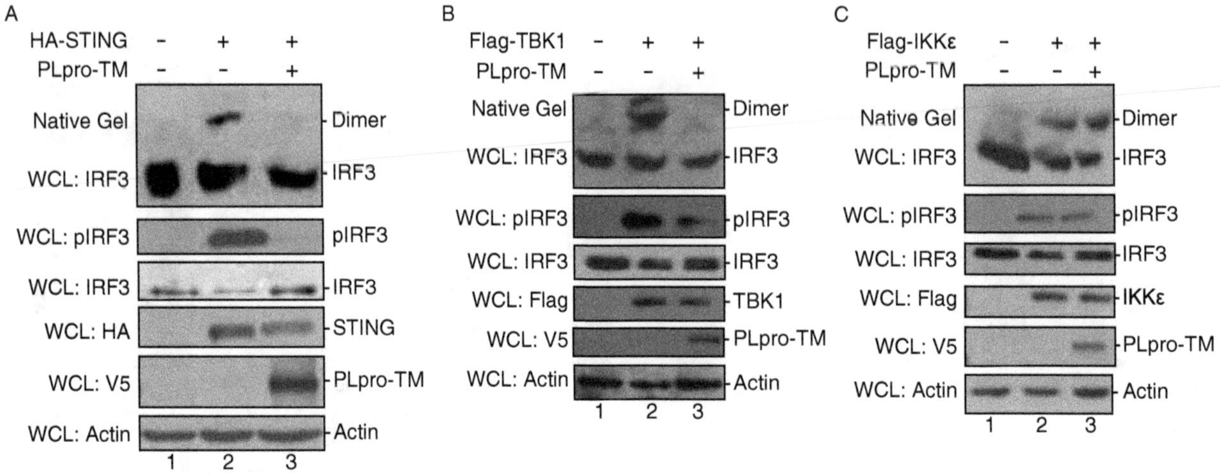

Figure 2. PLpro-TM antagonizes STING and TBK1-induced IRF3 activation. HEK293T cells were mock transfected or transfected PLpro-TM with HA-tagged STING (A), Flag-tagged TBK1 (B), or Flag-tagged IKKε (C) which was applied to activate IRF3 pathway. Proteins were extracted 24 h after transfection, the cell lysate was separated by Native PAGE followed by Western blotting to detect IRF3 monomer and dimer forms (top of Fig. 2A–C). The expression of HA-STING, Flag-TBK1, Flag-IKKε, pIRF3, IRF3, and PLpro-TM was analyzed by Western blotting performed with anti-HA, anti-Flag, anti-pIRF3, anti-IRF3, and anti-V5 antibodies, respectively. Actin was detected from whole cell lysate (WCL) as a loading control (bottom of Fig. 2A–C). Each experiment was repeated at least 3 times.

constructed a series of deletion mutants of STING, as described in Fig. 3C. The association of STING mutants with PLpro-TM was assessed by co-transfection of PLpro-TM and each of the STING mutants into HEK293T cells. We found that the association between STING and PLpro-TM was completely abolished when 4 TM domains at the N-terminus of STING were deleted (Fig. 3D, lanes 5 and 8). Partial deletion of TM domains or the C-terminal helicase domain did not affect the interaction between STING and PLpro-TM (Fig. 3D, lanes 4, 6, and 7). These findings indicate that PLpro-TM interacts with STING through the TM domains.

PLpro-TM disrupts STING-TRAF3-TBK1 interaction

Because PLpro-TM interacts with the components of the STING-TRAF3-TBK1 complex, we hypothesized that PLpro-TM would impede the formation of functional STING-TRAF3-TBK1 tripartite complex. To test this hypothesis, we assessed the impacts of PLpro-TM on the assembly of the STING-TRAF3-TBK1 complex. DNA plasmids expressing TRAF3, MAVS, STING, and TBK1 were co-transfected into HEK293T cells in the absence and in the presence of PLpro-TM. The cell lysate was co-immunoprecipitated (Fig. 4). We observed that co-immunoprecipitation of TRAF3 with MAVS or TBK1 was disrupted in the presence of PLpro-TM (Fig. 4A, lane 4 and Fig. 4D, lane 3). The assembly of STING with MAVS was obviously disrupted by PLpro-TM (Fig. 4B, lane 3). We also found that PLpro-TM disrupted the co-immunoprecipitation of STING with IRF3 (Fig. 4F, lane 3). However, the expression of PLpro-TM had no effect on co-immunoprecipitation of TRAF3 and STING (Fig. 4C, lane 3), and the interaction of STING with TBK1 (Fig. 4E, lane 3). Thus, SARS-CoV PLpro-TM disrupted the interaction between the key components of STING-TRAF3-TBK1 complex, which explains how PLpro-TM suppresses the activation and nuclear translocation of IRF3.

PLpro-TM blocks ubiquitination of STING-TRAF3-TBK1 complex

Multiple regulatory molecules located upstream of IRF3 in the IFN pathway require ubiquitination, especially K63-linked ubiquitination, and deubiquitination, which play critical roles in the activation of IFN responses (Bibeau-Poirier and Servant, 2008; Bhoj and Chen, 2009; Isaacson and Ploegh, 2009; Zhong et al., 2010). Our team and others have previously reported that SARS PLpro-TM has DUB activity, which acts as a negative regulator of the innate immune response (Barretto et al., 2005; Lindner et al., 2005; Sulea et al., 2005; Barretto et al., 2006; Ratia et al., 2006; Devaraj et al., 2007; Chen et al., 2009; Frieman et al., 2009; Clementz et al., 2010; Sun et al., 2012). In the current study, we examined whether SARS-CoV PLpro-TM could recognize and deubiquitinate the key signaling molecules in the IFN signaling pathway. HEK293T cells were co-transfected with HA-Ub-K63 and plasmids which encode RIG-I, TRAF3, STING, TBK1, or IRF3. The cell lysate was immunoprecipitated to verify the ubiquitination status of those immunoprecipitated proteins. We found reductions at the levels of

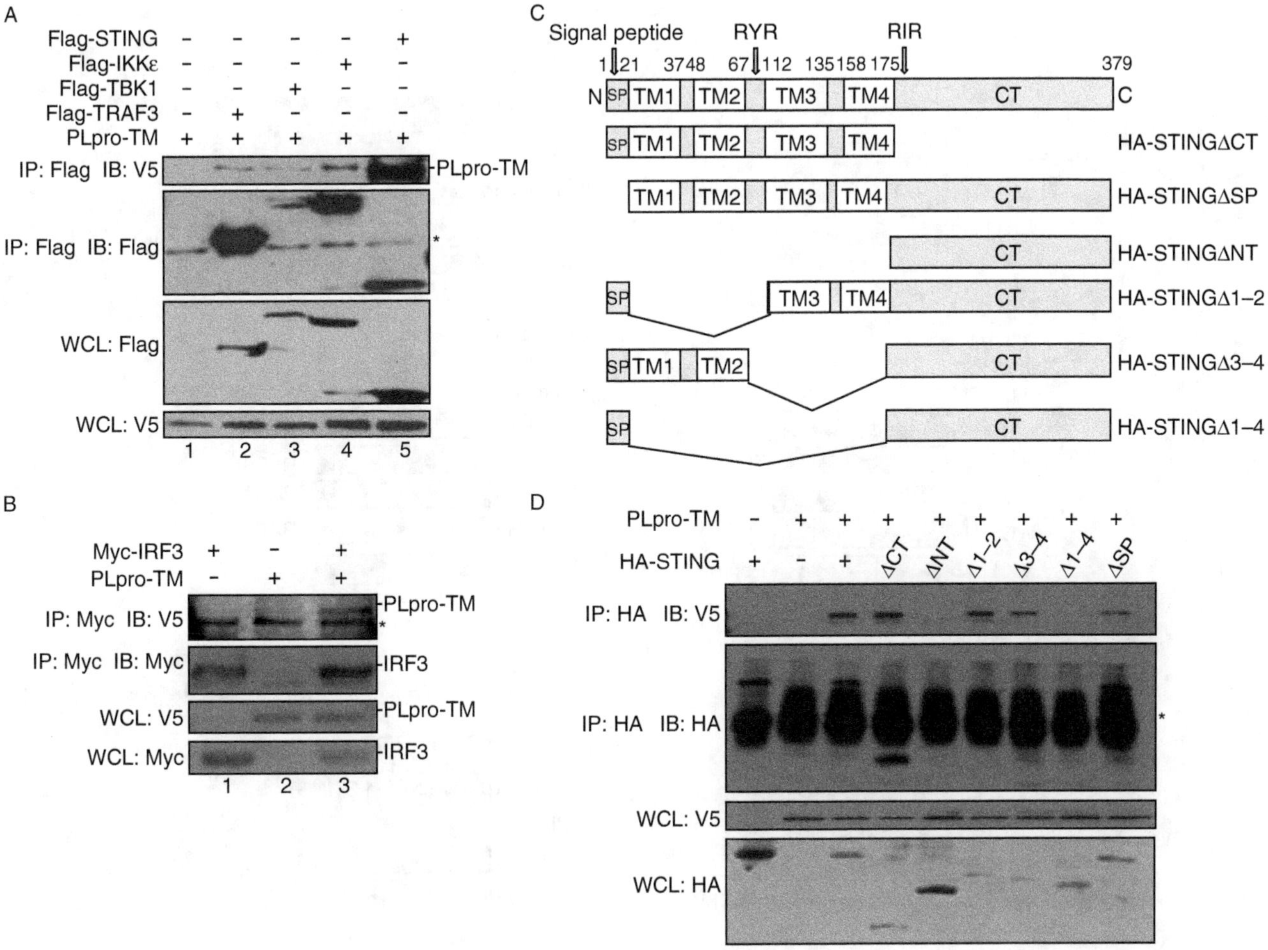

Figure 3. PLpro-TM interacts with STING-TRAF3-TBK1 complex. (A) HEK293T cells were co-transfected with expression plasmids for Flag-tagged STING, TBK1, IKKε, and TRAF3 both with and without PLpro-TM. The cell lysate was extracted 24 h after transfection. Co-immunoprecipitation experiments were performed with the indicated antibodies. The co-immunoprecipitation product was blotted by using anti-V5 (top panel), and re-probed by using anti-Flag (second panel) to detect TRAF3, TBK1, IKKε, and STING. The input lysate was blotted by using anti-Flag to detect TRAF3, TBK1, IKKε, STING (third panel), and anti-V5 to detect PLpro-TM (bottom panel). The asterisk indicates the nonspecific band. (B) HEK293T cells were transfected with Myc-tagged IRF3 in the absence and in the presence of PLpro-TM. The cell lysate was extracted 24 h after transfection. Co-immunoprecipitation was performed with the indicated antibodies. The co-immunoprecipitation product was blotted by using anti-V5 (top panel), and re-probed by using anti-Myc (second panel) to detect IRF3. The input lysate was blotted using anti-V5 to detect PLpro-TM (third panel), and anti-Myc to detect IRF3 (bottom panel). The asterisk indicates the nonspecific band. (C) Diagrams of full-length and truncated STING proteins used in the study. TM, transmembrane domain. CT, C-terminus domain. (D) The association of PLpro-TM with HA-STING or its truncated mutants. HEK293T cells were co-transfected with plasmids expressing HA-tagged STING or its truncated mutants, with or without PLpro-TM. The cell lysate was extracted 24 h after transfection. Co-immunoprecipitation was performed with the indicated antibodies. The co-immunoprecipitation product was blotted by using anti-V5 (top panel), and re-probed using anti-HA (second panel) to detect STING and its truncated mutants. The input lysate was blotted by using anti-V5 to detect PLpro-TM (third panel), and anti-HA to detect STING and its truncated mutants (bottom panel). The asterisk indicates the nonspecific band. Every experiment was repeated for at least 3 times.

poly-ubiquitinated RIG-I (Fig. 5A), TRAF3 (Fig. 5B), STING (Fig. 5C), TBK1 (Fig. 5D), and IRF3 (Fig. 5E) in cells expressing PLpro-TM. These findings indicated that PLpro-TM inhibited the ubiquitination of RIG-I, TRAF3, STING, TBK1, and IRF3 through its DUB activity. As ubiquitination of these signaling regulators is required in the formation of STING-TRAF3-TBK1, this result is in agreement with the finding that SARS-CoV PLpro-TM blocks activation of IFN by disrupting the formation of STING-TRAF3-TBK1 mediated complex.

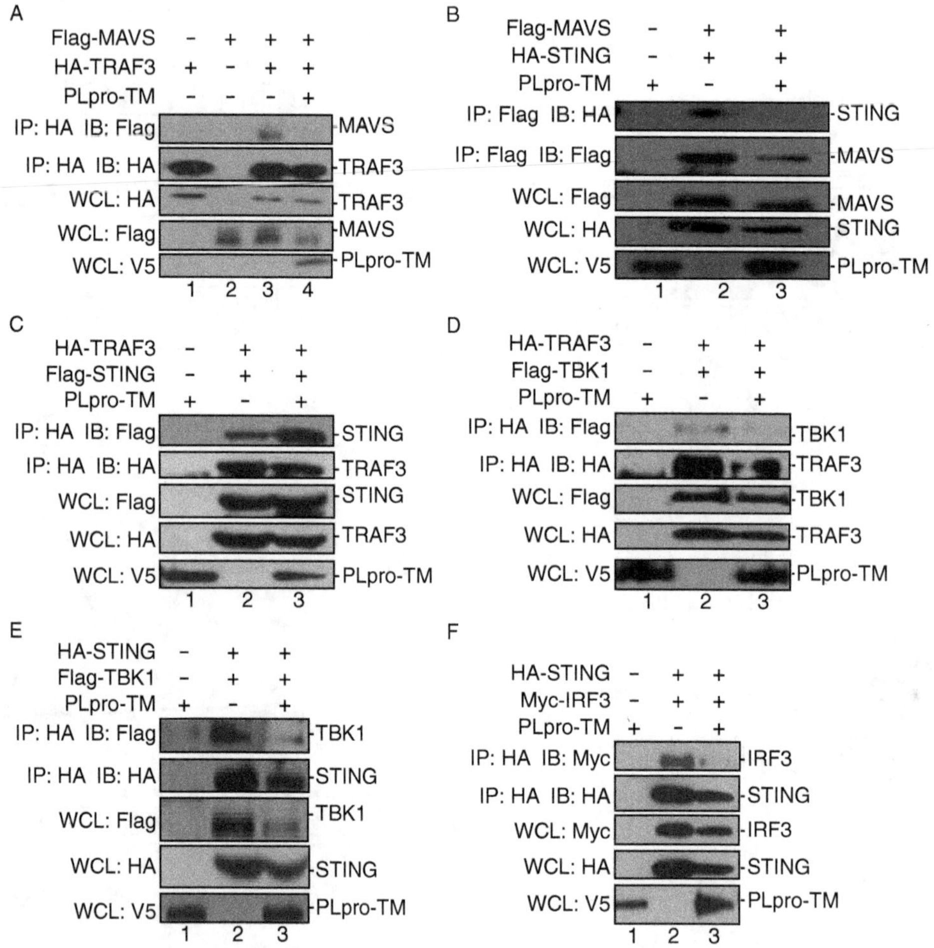

Figure 4. PLpro-TM disrupts the formation of STING-TRAF3-TBK1 complex. (A, B) HEK293T cells were transfected with Flag-MAVS together with either HA-tagged TRAF3 (A) or HA-tagged STING (B) with or without PLpro-TM. Twenty-four hours after transfection, the cell lysate was collected, and subjected to immunoprecipitation (IP) and immunoblotting (IB) using the indicated antibodies. Co-immunoprecipitation was blotted using anti-Flag or anti-HA (Fig. 4A and 4B, top panels), and re-probed using anti-HA or anti-Flag (Fig. 4A and 4B, second panels) for HA-TRAF3 (Fig. 4A) or Flag-MAVS (Fig. 4B). The input lysate was blotted using anti-Flag or anti-HA to detect MAVS (Fig. 4A, third panel) or STING (Fig. 4B, third panel), and anti-V5 to detect PLpro-TM (Fig. 4A and 4B, bottom panels). (C and D) HEK293T cells were co-transfected with HA-TRAF3 together with either Flag-tagged STING (C) or Flag-tagged TBK1 (D) and PLpro-TM. Twenty-four hours after transfection, the cell lysate was prepared and subjected to immunoprecipitation (IP) and immunoblotting (IB) with the indicated antibodies. (E and F) HA-STING and Flag-tagged TBK1 (E) or Myc-tagged IRF3 (F) were co-transfected into HEK293T cells with or without PLpro-TM. Twenty-four hours after transfection, the cell lysate was collected and subjected to immunoprecipitation (IP) and immunoblotting (IB) with the indicated antibodies. Each experiment was repeated at least 3 times.

DISCUSSION

SARS-CoV PLpro (PLpro-TM) acts as an IFN antagonist that reduces the production of IFN by inhibiting activation of the IRF3 pathway (Devaraj et al., 2007; Frieman et al., 2009; Clementz et al., 2010; Sun et al., 2012), but the underlying mechanisms used by SARS PLpro-TM to inhibit activation of IRF3 pathway are not fully understood. In this study, we uncovered a novel mechanism used by SARS PLpro-TM to inhibit activation of the IRF3 pathway. We found that (1) SARS-CoV PLpro-TM inhibits the activation of transcription factor IRF3 and negatively modulates the IFNβ signaling pathways, which results from PLpro-TM impeding STING- and TBK1-activated phosphorylation and dimerization of IRF3. (2) PLpro-TM interacts with TRAF3, TBK1, IKKε, STING, and IRF3, the key components that assemble a regulator complex for activation of IFN expression. However, the interaction between the components in STING-TRAF3-TBK1 complex is disrupted by PLpro-TM. (3) SARS PLpro-TM reduces the levels of ubiquitinated forms of RIG-I, STING, TRAF3, TBK1, and IRF3, the key components in the

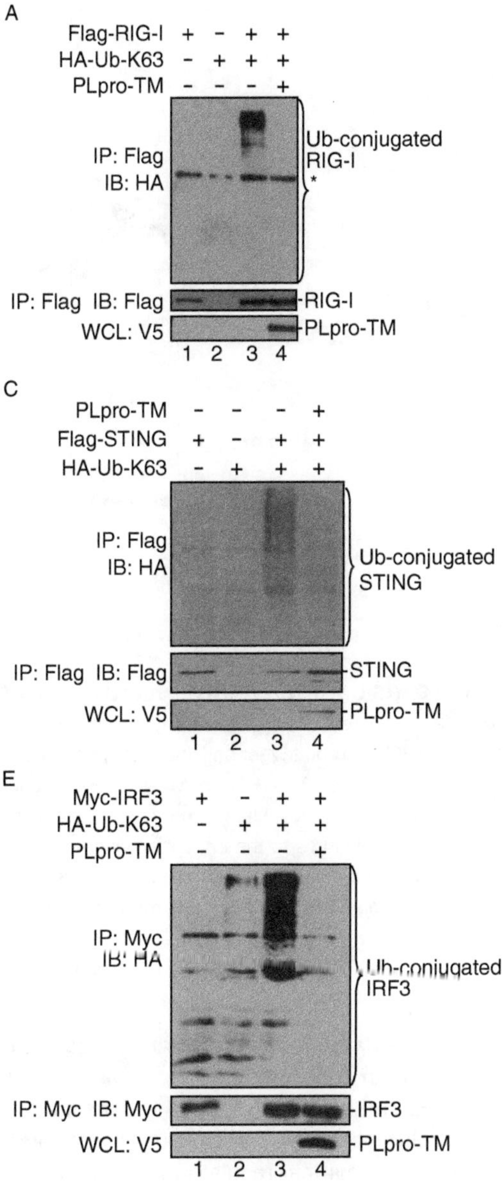

Figure 5. PLpro-TM reduces the ubiquitination levels of signaling molecules. The plasmids expressing Flag-tagged RIG-I (A), TRAF3 (B), STING (C), TBK1 (D), and Myc-tagged IRF3 (E) together with HA-Ub-K63 were co-transfected into HEK293T cells in the presence and in the absence of V5-tagged PLpro-TM. Cells were incubated for 24 h after transfection and treated with 25 μmol/L MG132 for 4 h prior to the harvest of the cell lysate. The cell lysate was immunoprecipitated with the indicated antibodies. The products were immunoblotted using anti-HA to evaluate K63-linked ubiquitinated proteins (upper panels). The whole cell lysate (WCL) was blotted to evaluate the expression of PLpro-TM (bottom panels). Each experiment was repeated at least 3 times. The asterisk indicates the nonspecific band from cellular proteins with high affinity with Agarose.

STING-TRAF3-TBK1 complex. These results revealed a novel mechanism used by SARS coronavirus to negatively regulate activation of innate immunity (Fig. 6).

The most striking finding of this study is that PLpro-TM disrupts the STING-TRAF3-TBK1 complex, which leads to inhibiting IRF3 activation. STING is important in regulating the production of IFN (Ishikawa and Barber, 2008). STING binds to both TBK1 and IRF3 and is necessary for the activation of IRF3 (Tanaka and Chen, 2012). TRAF3 locates at ER-to-Golgi transport compartments. Co-localization of TRAF3 and STING allows them to be involved in the induction of type I IFN innate immunity (van Zuylen et al., 2012). TRAF3 binds to TBK1 and forms a complex with TBK1, linking upstream signaling responses of RIG-I/Mda5-MAVS-STING to the TBK1-directed phosphorylation of IRF3 and IFNβ transcription (Fig. 6A) (Fitzgerald et al., 2003;

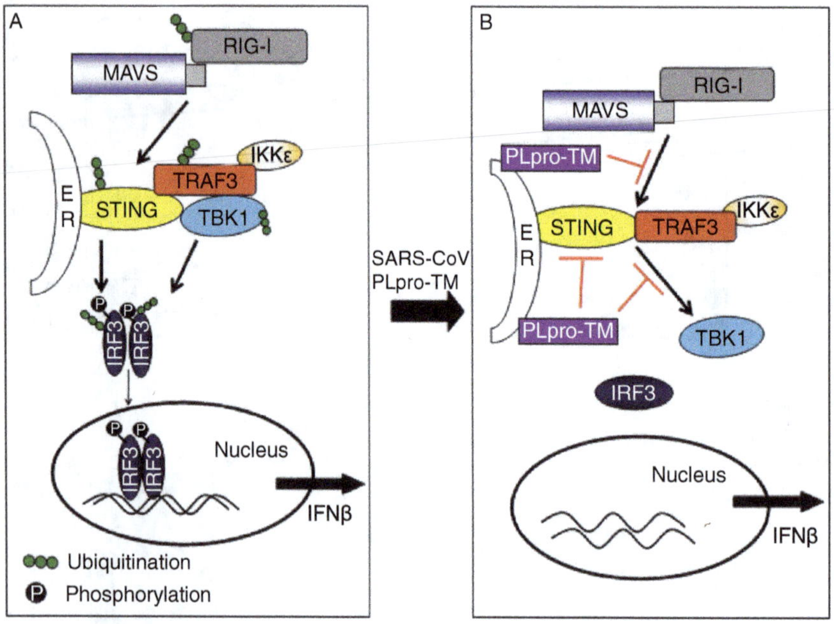

Figure 6. A hypothetical model depicting the mechanisms used by SARS-CoV PLpro-TM to disrupt the STING-TRAF3-TBK1 complex required for activation of the IFN-β production pathway. (A) Viral infections are sensed by pattern-recognition receptors (PRRs), which in turn recognize pathogen-associated molecular patterns (PAMPs) and trigger antiviral responses. RIG-I identifies the 5′-3p RNA, and then interacts with MAVS to activate the IFN responses factor 3 (IRF3). In MAMs (mitochondrial-associated membranes), MAVS also interacts with STING that locates at the ER (endoplasmic reticulum), and induces the ubiquitination and dimerization of STING. The activated STING recruits TBK1 and IRF3 and contributes to the phosphorylation of IRF3 mediated by TBK1. STING recruits TBK1 and IKKε and forms the TBK1-IKKε complex via the association with TRAF3. The TBK1 complex induces the phosphorylation, dimerization, and nuclear translocation of IRF3. (B) When SARS-CoV invades the host, these processes are disrupted by SARS-CoV PLpro-TM. PLpro-TM disrupts activation of the IRF3 pathway induced by STING-TRAF3-TBK1 by blocking the assembly of the complex, and it deubiquitinates the signaling molecules RIG-I, TRAF3, STING, TBK1, and IRF3.

Oganesyan et al., 2006). In this study, we showed that SARS-CoV PLpro-TM interacts with the components of STING-TRAF3-TBK1 complex, and that PLpro-TM inhibits formation of the STING-TRAF3-TBK1 complex that is necessary for IFNβ transcription. PLpro-TM also disrupts the interaction between these key signaling proteins and prevents STING-TRAF3-TBK1 complex from activating IRF3 (Fig. 6B). We confirm that SARS PLpro-TM interacts with STING, a recently identified regulator that facilitates the recruitment of TBK1. This is in agreement with our recent finding that NL63 CoV PLP2 disrupts the dimerization of STING and deubiquitinates STING for modulation of IFN expression (Sun et al., 2012). Along with our previous findings, this study revealed a new mechanism that PLpro-TM interacts with STING by disruption of the STING-TRAF3-TBK1 complex, which leads to inhibiting IRF3 activation. Our findings suggest that SARS coronavirus PLpro-TM interacts with STING through the TM domain at the N-terminus of STING. STING forms a membrane trafficking system that mediates the dynamic translocation of STING and leads to the effective induction of innate immune responses (Saitoh et al., 2010), but the interaction between PLpro-TM and STING prohibits the membrane trafficking of STING, which results in inhibition of phosphorylation and dimerization of IRF3 and blocks IFN signaling transduction (Fig. 6B).

Ubiquitination and deubiquitination emerge as the key mechanisms that regulate the virus-induced type I IFN signaling pathways (Bibeau-Poirier and Servant, 2008; Bhoj and Chen, 2009; Isaacson and Ploegh, 2009; Zhong et al., 2010). Ubiquitination, especially K63-linked ubiquitination, is critically involved in regulation of the activation of related receptors, such as RIG-I, and transduction of cell signaling pathways in innate immune responses (Zeng et al., 2010). In contrast, deubiquitination inhibits the activation of signaling molecules, and cellular and viral DUBs play important roles in negative regulation of host innate immunity. A20, named after its cDNA clone No. and also referred to as tumor necrosis factor α-induced protein (TNFAIP) 3, has DUB activity that is a central gatekeeper in inflammation and immunity. It removes K-63 linked poly-ubiquitin chains from RIP1, TRAF6, RIP2, and NEMO and negatively regulates the innate immune responses (Coornaert et al., 2009). Cellular protein DUBA specifically deubiquitinates TRAF3 that is critical for activation of IRF3, resulting in inhibiting phosphorylation of IRF3 (Kayagaki et al., 2007; Wang et al., 2011). Recent studies revealed that coronaviruses, including human CoV NL63, MHV, and SARS-CoV, have evolved to encode DUBs for modulating the innate immunity (Barretto et al., 2005; Sulea et al., 2005; Chen et al., 2007; Frieman et al., 2009; Clementz et al., 2010; Wang et al., 2011; Sun et al., 2012). We have reported that human CoV NL63 reduces the ubiquitinated forms of STING, RIG-I, TBK1, and IRF3 to negatively regulate IFN signaling (Sun et al., 2012). MHV-A59 PLP2 distinctively targets and deubiquitinates TBK1 to negatively regulate cellular type I IFN signaling pathway (Wang et al., 2011). DUBs have been reported in other viral proteins, such as arterivirus EAV PLP2 (van Kasteren et al., 2013), herpes simplex virus type 1 (HSV-1) UL36 (Kattenhorn et al., 2005), and porcine reproductive and respiratory syndrome virus (PRRSV) nsp2 (Chen et al., 2010; Sun et al., 2010). These viral proteins act as IFN antagonists through the DUB activity.

Multiple SARS coronaviral proteins, such as ORF3b, ORF6, N, M, and nsp1, have been suggested to inhibit IFN production (Kamitani et al., 2006; Devaraj et al., 2007; Frieman et al., 2007; Kopecky-Bromberg et al., 2007; Wathelet et al., 2007; Zust et al., 2007; Narayanan et al., 2008; Siu et al., 2009). SARS M protein is associated with TRAF3, which is crucial for the IFN responses and impairs the function of TRAF3-TBK1-IKKε (Siu et al., 2009). The formation of TRAF3-TBK1 complex is an essential step in the activation of IRF3 (Siu et al., 2009). We found in thestudy that when TRAF3 and TBK1 were co-expressed in HEK293T cells, the interaction between TRAF3 and TBK1 was observed in cells. However, the interaction between TRAF3 and TBK1 was completely disrupted in the presence of PLpro-TM. Furthermore, the interaction between STING and TBK1 was significantly decreased when coexpressed with PLpro-TM. These data collectively indicate that PLpro-TM impeded the interactions between the components of STING-TRAF3-TBK1 complex, which results in inhibition of the activation of IRF3. There are differences between M protein and PLpro-TM in the regulation of IRF3 activation. SARS M protein inhibits the activation of IRF3 mediated by both TBK1 and IKKε (Siu et al., 2009). When IKKε and PLpro-TM were co-expressed in HEK293T cells, we found that SARS-CoV PLpro-TM did not inhibit IKKε-mediated phosphorylation and dimerization of IRF3. However, IKKε activation of the IRF3-dependent IFNβ-Luc and PRD (III-I)4-Luc reporters is obviously inhibited by PLpro-TM. This discrepancy may be due to one or more other mechanisms explaining how PLpro blocks IKKε-mediated ubiquitination and translocation of IRF3. Additionally, the Gn cytoplasmic tail of Hantavirus forms a complex with TRAF3 that disrupts the formation of TBK1-TRAF3 complexes and their downstream signaling responses required for IFNβ transcription (Alff et al., 2008; Matthys and Mackow, 2012). This may be a common mechanism used by human viruses to regulate antiviral innate immunity by preventing formation of the STING-TRAF3-TBK1 complex.

In summary, our findings uncover a new mechanism used by SARS-CoV through which PLpro-TM negatively regulates IFN activation, as depicted in Fig. 6. SARS PLpro-TM blocks: 1) STING and TBK1-medaited phosphorylation and dimerization of IRF3; 2) the STING-TRAF3-TBK1 complex formation; and 3) the ubiquitination of the key signaling molecules in the IFNβ expression pathway, such as STING, TRAF3, TBK1, and IRF3. Our study has yielded strong evidence of the mechanisms underlying SARS-CoV countermeasures against host innate immunity.

MATERIALS AND METHODS

Cell culture, transfection reagent, and antibodies

HEK 293T cells were cultured in Dulbecco's Modified Eagle's Medium (GIBCO) supplemented with 10% heat-inactivated fetal bovine serum (FBS) and maintained at 37°C in a humidified 5% CO_2 incubator. Transient transfection reagent Lipofectamine 2000 was purchased from Invitrogen. Anti-HA (MBL), anti-Flag (MBL), anti-V5 (MBL), anti-Myc (MBL), anti-β-Actin (Beyotime), anti-IRF3 (Abcam), and anti-pIRF3 (Abcam) antibodies were used in the study.

Plasmid DNAs

The wild-type SARS-CoV PLpro-TM, which containing PLpro core domain (aa 1541 to 1855 and the TM domain (aa 2139–2425), and plasmids of IFN-β-Luc, PRD (III-I) 4-Luc, and HA-tagged Ub-K63 were previously described (Devaraj et al., 2007; Clementz et al., 2010). Constructs of Flag-MAVS, Myc-IRF3, Flag-TBK1, Flag-RIG-I, and Flag-RIG-IN were kindly provided by Dr. Himanshu Kuma and Shizuo Akira (Immunology Frontier Research Center, Osaka University, Osaka, Japan). Flag-STING, HA-STING, and Flag-IKKε were kindly provided by Dr. Zhengfan Jiang (School of Life Sciences, Peking University, Beijing, China) and Dr. Xuemin Zhang (National Biomedicine Center, Beijing). HA-TRAF3 and Flag-TRAF3 plasmids were kindly supplied by Dr. Kui Li (University of Tennessee, Memphis, Tennessee, USA).

Construction of STING deletion mutants

A series of STING mutants were constructed as shown in Fig. 3C. Briefly, STING mutants were amplified by PCR or overlap PCR from HA-STING construct, and inserted into pCMV-HA vector with a HA tag at the N-terminus of each STING mutant. The forward (P_1) primer and reverse primer (P_2) for each mutant were as follows: STINGΔCT P_1: 5′-CGGAATTCCGCCCCACTCCAGCCTGCATCCATCCC-3′, STINGΔCT P_2: 5′-CCGCTCAGAGTCAGAGCTCTGGCAGGATCAGCCG-3′. STINGΔSP P_1: 5′-CGGAATTCCGGCAGCCTTGGTTCTGCTGAGTGCC-3′, STINGΔSP P_2: 5′-CCGCTCGAGTCAAGAGAAATCCGTGCGGAGAGG-3′. STINGΔNT P_1: 5′-CGGAATTCCGCAGGCCCGGATTCGAACTTACATTC-3′, STINGΔNT P_2: 5′-CCGCTCGAGTCAAGAGAAATCCGTGCGGAGAGG-3′. STINGΔ1–2 P_1: 5′-GTGAAGGGCGGGCCGACCGCCTTCTGGGCCCCGTGACCCC-3′, STINGΔ1–2 P_2: 5′-GCGGTCGGCCCGCCCTTCAC-3′.

STINGΔ3–4 P_1: 5′-TAAGTTCGAATCCGGGCCTGAGCCAGGCTG-CAGACCCCGTTTAA C-3′, STINGΔ3–4 P_2: 5′-CAGGCCCGGATTC GAACTTAC-3′. STINGΔ1–4 P_1: 5′-TAAGTTCGAATCCGGGCCTGC TTCTGGGCCCCGTGACCCC-3′, Δ1–4 P_2: 5′-CAGGCCCGGATTC GAACTTAC-3′. PCRs were performed with LA Taq (TaKaRa) for the initial denaturation at 94°C for 5 min, then 30 cycles of 94°C for 30 s, 55°C for 30 s, and 72°C for 60 s, and extension at 72°C for 60 s. For some of the STING mutants, overlap PCR was performed. During the first-round PCR, 2 small fragments were amplified with the use of HA-STING as a template for the initial denaturation at 94°C for 5 min, then 30 cycles of 94°C for 30 s, 55°C for 30 s, and 72°C for 60 s, and extensions at 72°C for 10 min. The amplicons were purified and fused together by 5 cycles of 94°C for 30 s, 55°C for 30 s, and 72°C for 60 s, and extensions at 72°C for 10 min, followed by 25 cycles of 94°C for 30 s, 55°C for 30 s, and 72°C for 60 s, and extensions at 72°C for 10 min. All the STING deletion mutants were cloned into pCMV-HA vector and confirmed by endonuclease digestion analysis and DNA sequencing.

Luciferase assay

HEK293T cells were transfected with the reporter plasmid DNAs (pRL-TK, IFN-β-Luc, or PRD (III-I) 4-Luc, the IRF3 promoter controlled Luciferase reporter, and SARS PLpro-TM) using Lipofectamine 2000 according to the manufacturer's protocol and incubated for 24 h. Firefly luciferase and Renilla luciferase activities were assayed using the Dual Luciferase Reporter Assay Kit (Promega). The results were expressed as mean values of the relative luciferase activity (firefly luciferase activity divided by Renilla luciferase activity) including the standard deviation from repeated experiments carried out in triplicate. For statistical analysis, differences between the values of regulators and regulators co-transfected with PLpro-TM were subjected to unpaired, 2-tailed Student's *t* test performed with the use of Microsoft SPSS 12.0 software. A *P* value of <0.05 was considered statistically significant (Vaux et al., 2012).

Western blotting analysis

HEK 293T cells were seeded in 60-mm-diameter dishes and incubated at 37°C for 18 h. The cells were subsequently transfected with SARS PLpro-TM or empty vector using Lipofectamine 2000 reagent according to the manufacturer's instruction. Twenty-four hours after transfection, cells were lysed in RIPA buffer, containing protease inhibitor cocktail at 4°C for 30 min, spun down at 12,000 rpm for 10 min. Protein extracts in 40 µL of 2× SDS-PAGE sample buffer were boiled for 10 min. Samples were separated on SDS-PAGE gel and transferred to NC membrane. Blots were incubated with indicated primary antibodies. After 3 washings in 1× TBS-T buffer, blots were incubated with HRP-conjugated secondary antibodies (Beyotime, China). Antibody-antigen reactions were detected with the use of Western Lighting Plus-ECL chemiluminescence reagents (Biomed, China). Each experiment was repeated at least 3 times.

IRF3 dimer detection

To assess IRF3 dimerization, HEK293T cells were transfected with HA-STING or Flag-TBK1/IKKε (1 µg) in the presence and in the absence of V5-tagged PLpro-TM (1 µg). Twenty-four hours after transfection, the cell lysate was fractionated by 8% Native Gel in running buffer containing 1% sodium deoxycholate at 4°C, as previously described (Frieman et al., 2009). After electrophoresis, proteins were transferred and analyzed by Western blotting with the indicated antibodies. Each experiment was repeated at least 3 times.

Co-immunoprecipitation (Co-IP) analysis

The effects of SARS PLpro-TM in disrupting the interaction between STING and associated proteins in cultured cells were assessed as previously described (Gack et al., 2007). Briefly, HEK293T cells were co-transfected with Flag/HA-STING (2 µg), Flag-TRAF3 (2 µg), Flag-TBK1/IKKε/MAVS (2 µg), and Myc-IRF3 (2 µg) in the presence and in the absence of V5-tagged PLpro-TM (2 µg) with the use of Lipofectamine 2000. Twenty-four hours later, cells were lysed in RIPA buffer (50 mmol/L Tris-HCl pH 7.4, 150 mmol/L NaCl, 2 mmol/L EDTA, 1% NP-40) containing protease inhibitor cocktail (1 mmol/L, Roche), 0.1% SDS, and 10 µmol/L NEM at 4°C for 30 min. The cell extracts were then spun down at 12,000 rpm for 10 min at 4°C. The soluble lysate was immunoprecipitated with anti-Flag/HA antibody for at least 12 h, the lysate was then precleared by the addition of 40 µL of protein A + G agrose (Beyotime) and rocked at 4°C for 6 h. This was followed by spinning down the beads. The bead-antibody-protein complex was spun down and washed 4 times with 1 mL RIPA buffer or PBS. The protein was eluted from the beads in 40 µL of 2× SDS-PAGE sample buffer subjected to boiling for 10 min. The proteins were separated by SDS-PAGE and transferred to NC membrane for Western blotting. Each experiment was repeated at least 3 times.

Assessing ubiquitination of signaling molecules

The effect of SARS PLpro-TM on ubiquitinated proteins in cultured cells was assessed as previously described (Evans et al., 2004; Clementz et al., 2010). Briefly, Flag-tagged RIG-I, TRAF3, STING, TBK1, IKKε, and Myc-tagged IRF3 (1.0 µg each) were co-transfected into HEK293T cells cultured in 60-mm dishes together with pcDNA3.1-HA-Ub-K63. Transfection was performed with Lipofectamine 2000 according to the manufacturer's instructions. Empty vector pcDNA3.1/V5-HisB was used to standardize the total amount of DNA used for transfection. After 24 h, cells were incubated with 25 µmol/L MG132 for 4 h, then harvested by the addition of 300 µL RIPA buffer (50 mmol/L Tris-HCl pH 7.4, 150 mmol/L NaCl, 2 mmol/L EDTA, 1% NP-40) containing protease inhibitor cocktail (1 mmol/L, Roche), 0.1% SDS, and 10 µmol/L NEM at 4°C for 30 min, spun down at 12,000 rpm for 10 min. Non-covalently bound proteins were dissociated by boiling the lysate in 1% SDS; samples were diluted 1:10 in lysis buffer (50 mmol/L Tris-HCl pH 7.4, 150 mmol/L NaCl, 2 mmol/L EDTA, 1% NP-40) containing protease inhibitor cocktail and 10 µmol/L NEM. The soluble lysate was then immunoprecipitated with anti-Flag/HA antibody for at least 12 h, the lysate was then precleared by adding 40 µL protein A + G agrose (Beyotime) and rocked at 4°C for 6 h, followed by spinning down the beads. The beads-antibody-protein complex was spun down and washed 4 times with 1 mL RIPA buffer or PBS. The protein was eluted from the beads in 40 µL of 2× SDS-PAGE sample buffer subjected to boiling for 10 min. The samples were separated by SDS-PAGE and transferred to NC membrane for Western blotting. Each experiment was repeated at least 3 times.

ACKNOWLEDGEMENTS

This work was supported by grants received from the National Natural Science Foundation of China (Grant Nos. 81172799 and 81273231 to Z.C., and Grant No.81102478 to Y. X.).

ABBREVIATIONS

DUB, deubiquitination; EAV, equine arteritis virus; ER, endoplasmic reticulum; HSV-1, herpes simplex virus type 1; IFN, interferon; IKKε, IkappaB kinase; IRF3, IFN regulatory factor-3; MHV, mouse hepatitis virus; MAVS, mitochondria antiviral signaling protein; NEMO, NF-κB essential modulator; NF-κB, nuclear factor κB; ORF, open reading frames; PLpro, papain-like protease; PRR, pattern recognition receptors; PRRSV, porcine reproductive and respiratory syndrome virus; RIP, receptor interacting protein; RLRs, retinoic acid-inducible gene-I (RIG-I)-like receptors; SARS-CoV, severe acute respiratory syndrome coronavirus; STING, stimulator of interferon genes; TBK1, TANK-binding kinase 1; TLRs, Toll-like receptors; TM, transmembrane; PLpro-TM, the membrane-anchored PLpro domain; TRAF, tumor necrosis factor receptor-associated factor; PRD, positive regulatory domain.

COMPLIANCE WITH ETHICS GUIDELINES

Xiaojuan Chen, Xingxing Yang, Yang Zheng, Yudong Yang, Yaling Xing, and Zhongbin Chen declare that they have no conflict of interest.

This article does not contain any studies with human or animal subjects performed by the any of the authors.

REFERENCES

Alff PJ, Sen N, Gorbunova E, Gavrilovskaya IN, Mackow ER (2008) The NY-1 hantavirus Gn cytoplasmic tail coprecipitates TRAF3 and inhibits cellular interferon responses by disrupting TBK1-TRAF3 complex formation. J Virol 82:9115–9122

Barral PM, Sarkar D, Su ZZ, Barber GN, DeSalle R, Racaniello VR, Fisher PB (2009) Functions of the cytoplasmic RNA sensors RIG-I and MDA-5: key regulators of innate immunity. Pharmacol Ther 124:219–234

Barretto N, Jukneliene D, Ratia K, Chen Z, Mesecar AD, Baker SC (2005) The papain-like protease of severe acute respiratory syndrome coronavirus has deubiquitinating activity. J Virol 79:15189–15198

Barretto N, Jukneliene D, Ratia K, Chen Z, Mesecar AD, Baker SC (2006) Deubiquitinating activity of the SARS-CoV papain-like protease. Adv Exp Med Biol 581:37–41

Bhoj VG, Chen ZJ (2009) Ubiquitylation in innate and adaptive immunity. Nature 458:430–437

Bibeau-Poirier A, Servant MJ (2008) Roles of ubiquitination in pattern-recognition receptors and type I interferon receptor signaling. Cytokine 43:359–367

Chen H, Jiang Z (2013) The essential adaptors of innate immune signaling. Protein Cell 4:27–39

Chen Z, Wang Y, Ratia K, Mesecar AD, Wilkinson KD, Baker SC (2007) Proteolytic processing and deubiquitinating activity of papain-like proteases of human coronavirus NL63. J Virol 81:6007–6018

Chen X, Chou CY, Chang GG (2009) Thiopurine analogue inhibitors of severe acute respiratory syndrome-coronavirus papain-like protease, a deubiquitinating and deISGylating enzyme. Antivir Chem Chemother 19:151–156

Chen Z, Zhou X, Lunney JK, Lawson S, Sun Z, Brown E, Christopher-Hennings J, Knudsen D, Nelson E, Fang Y (2010) Immunodominant epitopes in nsp2 of porcine reproductive and respiratory syndrome virus are dispensable for replication, but play an important role in modulation of the host immune response. J Gen Virol 91:1047–1057

Clementz MA, Chen Z, Banach BS, Wang Y, Sun L, Ratia K, Baez-Santos YM, Wang J, Takayama J, Ghosh AK et al (2010) Deubiquitinating and interferon antagonism activities of coronavirus papain-like proteases. J Virol 84:4619–4629

Coornaert B, Carpentier I, Beyaert R (2009) A20: central gatekeeper in inflammation and immunity. J Biol Chem 284:8217–8221

Devaraj SG, Wang N, Chen Z, Chen Z, Tseng M, Barretto N, Lin R, Peters CJ, Tseng CT, Baker SC et al (2007) Regulation of IRF-3-dependent innate immunity by the papain-like protease domain of the severe acute respiratory syndrome coronavirus. J Biol Chem 282:32208–32221

Evans PC, Ovaa H, Hamon M, Kilshaw PJ, Hamm S, Bauer S, Ploegh HL, Smith TS (2004) Zinc-finger protein A20, a regulator of inflammation and cell survival, has de-ubiquitinating activity. Biochem J 378:727–734

Fitzgerald KA, McWhirter SM, Faia KL, Rowe DC, Latz E, Golenbock DT, Coyle AJ, Liao SM, Maniatis T (2003) IKKepsilon and TBK1 are essential components of the IRF3 signaling pathway. Nat Immunol 4:491–496

Frieman M, Yount B, Heise M, Kopecky-Bromberg SA, Palese P, Baric RS (2007) Severe acute respiratory syndrome coronavirus ORF6 antagonizes STAT1 function by sequestering nuclear import factors on the rough endoplasmic reticulum/Golgi membrane. J Virol 81:9812–9824

Frieman M, Ratia K, Johnston RE, Mesecar AD, Baric RS (2009) Severe acute respiratory syndrome coronavirus papain-like protease ubiquitin-like domain and catalytic domain regulate antagonism of IRF3 and NF-kappaB signaling. J Virol 83:6689–6705

Gack MU, Shin YC, Joo CH, Urano T, Liang C, Sun L, Takeuchi O, Akira S, Chen Z, Inoue S et al (2007) TRIM25 RING-finger E3 ubiquitin ligase is essential for RIG-I-mediated antiviral activity. Nature 446:916–920

He JQ, Oganesyan G, Saha SK, Zarnegar B, Cheng G (2007) TRAF3 and its biological function. Adv Exp Med Biol 597:48–59

Hornung V, Ellegast J, Kim S, Brzozka K, Jung A, Kato H, Poeck H, Akira S, Conzelmann KK, Schlee M et al (2006) 5′-Triphosphate RNA is the ligand for RIG-I. Science 314:994–997

Huang YH, Liu XY, Du XX, Jiang ZF, Su XD (2012) The structural basis for the sensing and binding of cyclic di-GMP by STING. Nat Struct Mol Biol 19:728–730

Isaacson MK, Ploegh HL (2009) Ubiquitination, ubiquitin-like modifiers, and deubiquitination in viral infection. Cell Host Microbe 5:559–570

Ishikawa H, Barber GN (2008) STING is an endoplasmic reticulum adaptor that facilitates innate immune signalling. Nature 455:674–678

Kamitani W, Narayanan K, Huang C, Lokugamage K, Ikegami T, Ito N, Kubo H, Makino S (2006) Severe acute respiratory syndrome coronavirus nsp1 protein suppresses host gene expression by promoting host mRNA degradation. Proc Natl Acad Sci USA 103:12885–12890

Kattenhorn LM, Korbel GA, Kessler BM, Spooner E, Ploegh HL (2005) A deubiquitinating enzyme encoded by HSV-1 belongs to a family of cysteine proteases that is conserved across the family Herpesviridae. Mol Cell 19:547–557

Kawai T, Akira S (2007) Antiviral signaling through pattern recognition receptors. J Biochem 141:137–145

Kayagaki N, Phung Q, Chan S, Chaudhari R, Quan C, O'Rourke KM, Eby M, Pietras E, Cheng G, Bazan JF et al (2007) DUBA: a deubiquitinase that regulates type I interferon production. Science 318:1628–1632

Kopecky-Bromberg SA, Martinez-Sobrido L, Frieman M, Baric RA, Palese P (2007) Severe acute respiratory syndrome coronavirus open reading frame (ORF) 3b, ORF 6, and nucleocapsid proteins function as interferon antagonists. J Virol 81:548–557

Lindner HA, Fotouhi-Ardakani N, Lytvyn V, Lachance P, Sulea T, Menard R (2005) The papain-like protease from the severe acute respiratory syndrome coronavirus is a deubiquitinating enzyme. J Virol 79:15199–15208

Marra MA, Jones SJ, Astell CR, Holt RA, Brooks-Wilson A, Butterfield YS, Khattra J, Asano JK, Barber SA, Chan SY et al (2003) The genome sequence of the SARS-associated coronavirus. Science 300:1399–1404

Matthys V, Mackow ER (2012) Hantavirus regulation of type I interferon responses. Adv Virol 2012:524024

Narayanan K, Huang C, Lokugamage K, Kamitani W, Ikegami T, Tseng CT, Makino S (2008) Severe acute respiratory syndrome coronavirus nsp1 suppresses host gene expression, including that of type I interferon, in infected cells. J Virol 82:4471–4479

Oganesyan G, Saha SK, Guo B, He JQ, Shahangian A, Zarnegar B, Perry A, Cheng G (2006) Critical role of TRAF3 in the Toll-like receptor-dependent and -independent antiviral response. Nature 439:208–211

Ouyang S, Song X, Wang Y, Ru H, Shaw N, Jiang Y, Niu F, Zhu Y, Qiu W, Parvatiyar K et al (2012) Structural analysis of the STING adaptor protein reveals a hydrophobic dimer interface and mode of cyclic di-GMP binding. Immunity 36:1073–1086

Perlman S, Netland J (2009) Coronaviruses post-SARS: update on replication and pathogenesis. Nat Rev Microbiol 7:439–450

Pichlmair A, Schulz O, Tan CP, Naslund TI, Liljestrom P, Weber F, Reis e Sousa C (2006) RIG-I-mediated antiviral responses to single-stranded RNA bearing 5′-phosphates. Science 314:997–1001

Ratia K, Saikatendu KS, Santarsiero BD, Barretto N, Baker SC, Stevens RC, Mesecar AD (2006) Severe acute respiratory syndrome coronavirus papain-like protease: structure of a viral deubiquitinating enzyme. Proc Natl Acad Sci USA 103:5717–5722

Rota PA, Oberste MS, Monroe SS, Nix WA, Campagnoli R, Icenogle JP, Penaranda S, Bankamp B, Maher K, Chen MH et al (2003) Characterization of a novel coronavirus associated with severe acute respiratory syndrome. Science 300:1394–1399

Saha SK, Cheng G (2006) TRAF3: a new regulator of type I interferons. Cell Cycle 5:804–807

Saitoh T, Fujita N, Yoshimori T, Akira S (2010) Regulation of dsDNA-induced innate immune responses by membrane trafficking. Autophagy 6:430–432

Schindler C, Levy DE, Decker T (2007) JAK-STAT signaling: from interferons to cytokines. J Biol Chem 282:20059–20063

Shang G, Zhu D, Li N, Zhang J, Zhu C, Lu D, Liu C, Yu Q, Zhao Y, Xu S et al (2012) Crystal structures of STING protein reveal basis for recognition of cyclic di-GMP. Nat Struct Mol Biol 19:725–727

Shu C, Yi G, Watts T, Kao CC, Li P (2012) Structure of STING bound to cyclic di-GMP reveals the mechanism of cyclic dinucleotide recognition by the immune system. Nat Struct Mol Biol 19:722–724

Siu KL, Kok KH, Ng MH, Poon VK, Yuen KY, Zheng BJ, Jin DY (2009) Severe acute respiratory syndrome coronavirus M protein inhibits type I interferon production by impeding the formation of TRAF3. TANK.TBK1/IKKepsilon complex. J Biol Chem 284:16202–16209

Sulea T, Lindner HA, Purisima EO, Menard R (2005) Deubiquitination, a new function of the severe acute respiratory syndrome coronavirus papain-like protease? J Virol 79:4550–4551

Sun W, Li Y, Chen L, Chen H, You F, Zhou X, Zhai Z, Chen D et al (2009) ERIS, an endoplasmic reticulum IFN stimulator, activiates innate immune signaling through dimerization. Pro Natl Acad Sci USA 106:8653–8658

Sun Z, Chen Z, Lawson SR, Fang Y (2010) The cysteine protease domain of porcine reproductive and respiratory syndrome virus nonstructural protein 2 possesses deubiquitinating and interferon antagonism functions. J Virol 84:7832–7846

Sun L, Xing Y, Chen X, Zheng Y, Yang Y, Nichols DB, Clementz MA, Banach BS, Li K, Baker SC et al (2012) Coronavirus papain-like proteases negatively regulate antiviral innate immune response through disruption of STING-mediated signaling. PLoS One 7: e30802

Tanaka Y, Chen ZJ (2012) STING specifies IRF3 phosphorylation by TBK1 in the cytosolic DNA signaling pathway. Sci Signal 5:ra20

Thiel V, Weber F (2008) Interferon and cytokine responses to SARS-coronavirus infection. Cytokine Growth Factor Rev 19:121–132

van Kasteren PB, Bailey-Elkin BA, James TW, Ninaber DK, Beugeling C, Khajehpour M, Snijder EJ, Mark BL, Kikkert M (2013) Deubiquitinase function of arterivirus papain-like protease 2 suppresses the innate immune response in infected host cells. Proc Natl Acad Sci USA 110:E838–E847

van Zuylen WJ, Doyon P, Clement JF, Khan KA, D'Ambrosio LM, Do F, St-Amant-Verret M, Wissanji T, Emery G, Gingras AC et al (2012) Proteomic profiling of the TRAF3 interactome network reveals a new role for the ER-to-Golgi transport compartments in innate immunity. PLoS Pathog 8:e1002747

Vaux DL, Fidler F, Cumming G (2012) Replicates and repeats—what is the difference and is it significant? A brief discussion of statistics and experimental design. EMBO Rep 13:291–296

Wang G, Chen G, Zheng D, Cheng G, Tang H (2011) PLP2 of mouse hepatitis virus A59 (MHV-A59) targets TBK1 to negatively regulate cellular type I interferon signaling pathway. PLoS One 6:e17192

Wathelet MG, Orr M, Frieman MB, Baric RS (2007) Severe acute respiratory syndrome coronavirus evades antiviral signaling: role of nsp1 and rational design of an attenuated strain. J Virol 81:11620–11633

Yin Q, Tian Y, Kabaleeswaran V, Jiang X, Tu D, Eck MJ, Chen ZJ, Wu H (2012) Cyclic di-GMP sensing via the innate immune signaling protein STING. Mol Cell 46:735–745

Yoneyama M, Fujita T (2009) RNA recognition and signal transduction by RIG-I-like receptors. Immunol Rev 227:54–65

Zeng W, Sun L, Jiang X, Chen X, Hou F, Adhikari A, Xu M, Chen ZJ (2010) Reconstitution of the RIG-I pathway reveals a signaling role of unanchored polyubiquitin chains in innate immunity. Cell 141:315–330

hong B, Yang Y, Li S, Wang YY, Li Y, Diao F, Lei C, He X, Zhang L, Tien P et al (2008) The adaptor protein MITA links virus-sensing receptors to IRF3 transcription factor activation. Immunity 29:538–550

Zhong B, Zhang Y, Tan B, Liu TT, Wang YY, Shu HB (2010) The E3 ubiquitin ligase RNF5 targets virus-induced signaling adaptor for ubiquitination and degradation. J Immunol 184: 6249–6255

Zielecki F, Weber M, Eickmann M, Spiegelberg L, Zaki AM, Matrosovich M, Becker S, Weber F (2013) Human cell tropism and innate immune system interactions of human respiratory coronavirus EMC compared to those of severe acute respiratory syndrome coronavirus. J Virol 87:5300–5304

Zust R, Cervantes-Barragan L, Kuri T, Blakqori G, Weber F, Ludewig B, Thiel V (2007) Coronavirus non-structural protein 1 is a major pathogenicity factor: implications for the rational design of coronavirus vaccines. PLoS Pathog 3:e109

2

The emerging roles of the DDX41 protein in immunity and diseases

Yan Jiang[1], Yanping Zhu[1], Zhi-Jie Liu[1,2], Songying Ouyang[1,3]✉

[1] National Laboratory of Biomacromolecules, Institute of Biophysics, Chinese Academy of Sciences, Beijing 100101, China
[2] iHuman Institute, Shanghai Tech University, Shanghai 201210, China
[3] Department of Microbiology, Immunology and Molecular Genetics, University of California Los Angeles, Los Angeles, CA 90095, USA
✉ Correspondence: ouyangsy@ibp.ac.cn (S. Ouyang)

ABSTRACT

RNA helicases are involved in almost every aspect of RNA, from transcription to RNA decay. DExD/H-box helicases comprise the largest SF2 helicase superfamily, which are characterized by two conserved RecA-like domains. In recent years, an increasing number of unexpected functions of these proteins have been discovered. They play important roles not only in innate immune response but also in diseases like cancers and chronic hepatitis C. In this review, we summarize the recent literatures on one member of the SF2 superfamily, the DEAD-box protein DDX41. After bacterial or viral infection, DNA or cyclic-di-GMP is released to cells. After phosphorylation of Tyr414 by BTK kinase, DDX41 will act as a sensor to recognize the invaders, followed by induction of type I interferons (IFN). After the immune response, DDX41 is degraded by the E3 ligase TRIM21, using Lys9 and Lys115 of DDX41 as the ubiquitination sites. Besides the roles in innate immunity, DDX41 is also related to diseases. An increasing number of both inherited and acquired mutations in *DDX41* gene are identified from myelodysplastic syndrome and/or acute myeloid leukemia (MDS/AML) patients. The review focuses on DDX41, as well as its homolog Abstrakt in Drosophila, which is important for survival at all stages throughout the life cycle of the fly.

KEYWORDS DDX41, innate immunity, RNA helicases, myelodysplastic syndrome, acute myeloid leukemia

INTRODUCTION

RNA helicases are enzymes that utilize the energy derived from NTP hydrolysis to unwind double-stranded RNA (dsRNA) molecules (Luking et al., 1998) or disrupt RNA-protein interactions (Jankowsky et al., 2001). DExD/H-box helicases comprise the largest SF2 helicase superfamily. Within the DExD/H-box family, the proteins are further classified as DEAD, DEAH, DExH, and DExD helicases based on the amino acid sequence of the conserved motif II (Fullam and Schroder, 2013). DEAD-box proteins, which are named after the strictly conserved sequence Asp-Glu-Ala-Asp (D-E-A-D), are widely found in organisms from bacteria to humans (Linder et al., 1989). They are involved in many aspects of RNA metabolism, such as transcription, pre-mRNA splicing, transport, translation, mRNA decay, and ribosome biogenesis (Rocak and Linder, 2004). The core of DEAD-box proteins consists of two tandem repeats of RecA-like domains, with motifs I, Ia, Ib, II, and III in the N-terminal domain and motifs IV, V, and VI in the C-terminal domain. The DEAD sequence is located in motif II (Caruthers and McKay, 2002). Subsequent studies identified three additional motifs that are characteristic of the DEAD-box proteins, namely the Q-motif (Cordin et al., 2004), GG motif (Schmid and Linder, 1992), and QxxR motif (Caruthers et al., 2000).

Pathogen-associated molecular patterns (PAMPs) of pathogens can be detected by cells of the innate immune response with the help of germline-encoded pattern recognition receptors (PRRs) to induce type I interferons (IFN) (Medzhitov and Janeway, 2000). In recent years it has been reported that several DExD/H-box helicases contribute to antiviral immunity, either by acting as sensors for viral nucleic acids or by facilitating downstream signaling events. The RIG-like helicases (RLHs), including RIG-I, MDA5, and LGP2 are recognized as one of the most important groups of anti-viral PRRs (Schmidt et al., 2012). Another DEAD box helicase, DDX3 is reported to act as a sensor for viral RNA in conjunction with RIG-I and MDA5. The authors proposed that DDX3 can sensitize the RLH system for dsRNA ligands at early stages of infection when levels of RIG-I are still low (Oshiumi et al., 2010). DHX9 is identified as a sensor for dsRNA in myeloid cells (Zhang et al., 2011c), and as a sensor for CpG DNA in plasmacytoid dendritic cells (pDCs) (Kim et al., 2010). DDX1 can directly bind to poly(I:C) (Zhang et al., 2011a). DDX21 and DHX36 are located downstream of DDX1. Both DDX21 and DHX36 interact with the downstream protein TIR-domain-containing-adapter-inducing interferon-β (TRIF). This suggests that DDX1 senses dsRNA and then triggers signaling via DDX21 and DHX36 to TRIF (Zhang et al., 2011a). DDX60 is proved to act in conjunction with RIG-I or MDA5 to mediate responses to viral dsRNA (Miyashita et al., 2011). Besides the involvement in antiviral immune response, DEAD-box proteins also play important roles in virus replication. DDX1 can bind to Hepatitis C virus (HCV) 3′ (+) UTR as well as its reverse complementary 5′ (−) UTR (Tingting et al., 2006), suggesting a possible role in the initiation of HCV RNA replication. DDX1 has also been reported to be important for the human immunodeficiency virus type 1 (HIV-1) replication as it binds to and serves as a cofactor of the HIV-1 Rev protein (Fang et al., 2004). DDX3 is required for HCV RNA replication, wherein the HCV core binds to the C-terminus of DDX3 (Owsianka and Patel, 1999). DDX5 was found to interact with the HCV NS5B protein, and to be important for HCV RNA replication (Goh et al., 2004). Taken together, the DExD/H-box helicases may have a much broader role in innate immunity and cancers than previously appreciated. In this review, we will focus on the functions of DDX41 and its Drosophila homolog Abstrakt (Abs).

FUNCTIONS OF DDX41

Functions of DDX41 in innate immunity

Yong-Jun Liu's group carried out an siRNA screen with 59 members of the DExD/H-box helicase family to address potential involvement of other DExD/H-box helicases in innate immunity (Zhang et al., 2011b). They identified DDX41, a member of the DEAD-box proteins, as an intracellular DNA sensor in myeloid dendritic cells. Knockdown of DDX41 expression blocked activation of the mitogen-activated protein kinase TBK1 (TANK-binding kinase 1) and the transcription factors NF-κB and IRF3 (interferon regulatory factor) by B-form DNA but not RNA. The stimulator of interferon genes (STING) functions as a key scaffolding and adaptor protein to facilitate the signal transduction initiated from upstream cytosolic dsDNA receptors to downstream effectors TBK1 and IRF3, leading to the expression of type I IFNs (Liu and Wang, 2016). DDX41 functions through the STING-TBK1-IRF3 pathway. It directly binds DNA and STING via its DEAD domain (Fig. 1). The authors also showed that the Walker A and B motifs are required for the binding (Zhang et al., 2011b). During bacterial infection, certain bacterial species can release secondary messengers cyclic di-GMP (c-di-GMP) or cyclic di-AMP (c-di-AMP) to act as PAMP to trigger host type I interferon innate immune response (Hengge, 2009). In 2011, STING was reported to be an innate immune sensor of cyclic dinucleotides (Burdette et al., 2011). In 2012, our group determined the crystal structure of STING C-terminal domain complexed with c-di-GMP in a unique binding mode (Ouyang et al., 2012). Interestingly, Parvatiyar et al. showed that DDX41 specifically and directly interacts with c-di-GMP. Knockdown of DDX41 via short hairpin RNA inhibited immune response to c-di-GMP and resulted in defective activation of the STING-TBK1-IRF3 pathway. They proposed a three molecule synergistic binding mechanism wherein the detection of c-di-GMP by DDX41 enhances the DDX41-STING interaction, which in turn increases the binding affinity of STING and c-di-GMP; the molecular interactions eventually lead to activating the interferon response. They also demonstrated that the DEAD domain plays key roles in DDX41 detecting c-di-GMP (Fig. 1) (Parvatiyar et al., 2012). Uncontrolled sensing of DNA or RNA and excessive production of type I interferon could induce autoimmune diseases, so DDX41 must be degraded or inactivated after immune response. In 2013, Yong-Jun Liu's group identified the E3 ligase TRIM21 as a DDX41-interacting protein. Overexpression of TRIM21 results in more degradation of DDX41 and less production of type I interferon in response to intracellular dsDNA. The SPRY-PRY domain of TRIM21 interacts with the DEAD domain of DDX41. Lys9 and Lys115 of DDX41 are the ubiquitination sites of TRIM21 (Fig. 1) (Zhang et al., 2013). However, how DDX41 functions as a DNA sensor is not completely understood. Reports in literature about DDX41 appeared very infrequently after that until 2015, when Lee et al. demonstrated that BTK-deficient cells have impaired IFN-β production and TBK1/IRF3 activation when stimulated with pathogens. Tyr364 and Tyr414 of DDX41 are critical for its recognition of DNA and binding to STING. Tyr414 is identified as the BTK phosphorylation site (Fig. 1). Besides, BTK's kinase domain can bind the DEAD domain of DDX41 (Lee et al., 2015). Recently, the porcine DDX41 is reported involved in the dsDNA- and dsDNA-virus-mediated type I IFN signaling pathway in porcine kidney cells (Zhu et al., 2014). In addition, the function of DDX41 in innate immune regulation was also studied in teleosts. Quynh et al. cloned and characterized the DDX41-encoding gene from the olive flounder and found that olive flounder DDX41 is also a cytosolic viral DNA sensor. It showed antiviral function similar to DDX41 in mammals (Quynh et al., 2015).

Functions of DDX41 in human diseases

Besides the roles in innate immunity, DDX41 is also related to diseases. So far, 10 genes (RUNX1, CEBPA, TERC, TERT, GATA2, SRP72, ANKRD26, ACD, ETV6, and DDX41) germline heterozygous mutations have been reported associate with familial myelodysplasia syndrome (MDS)/ acute myeloid leukemia (AML) (Cardoso et al., 2016). In

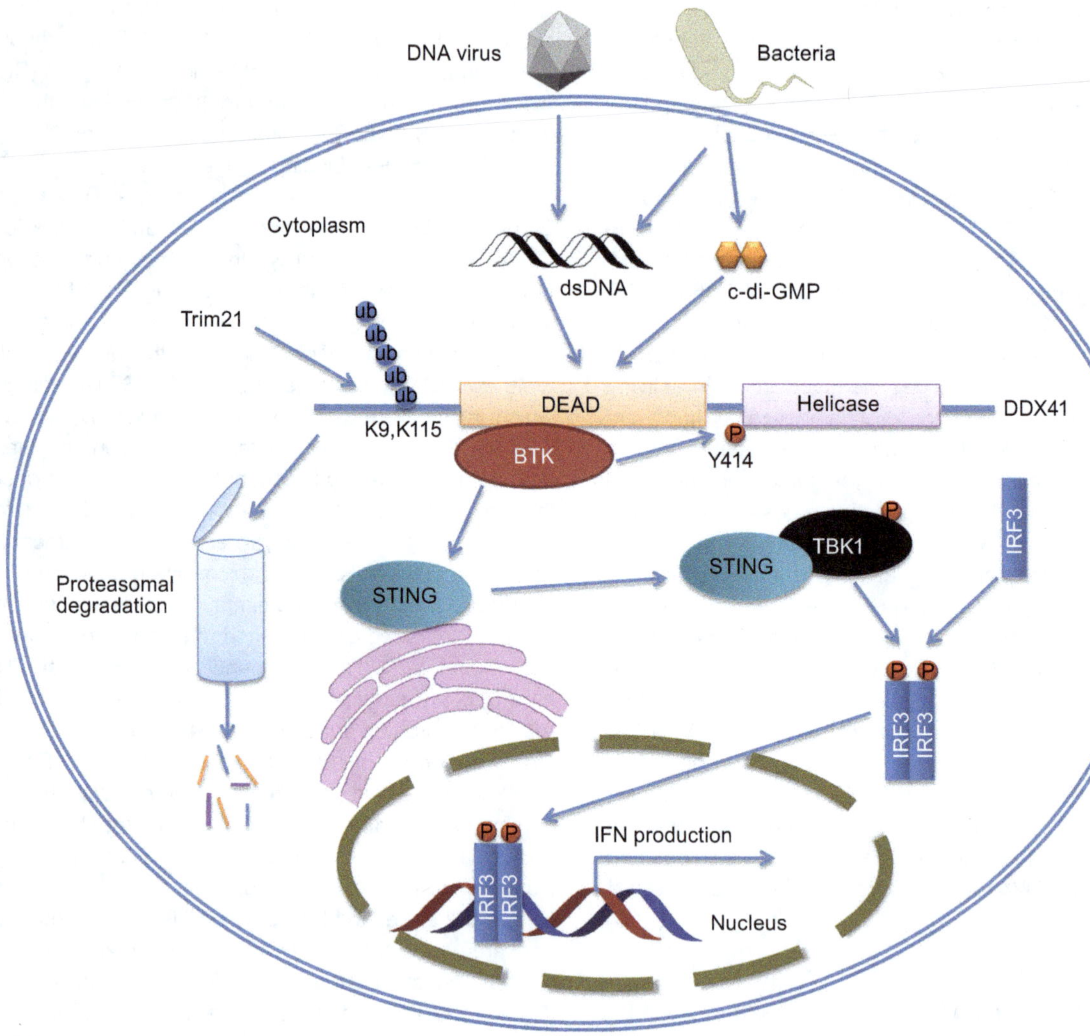

Figure 1. The signaling pathway of DDX41 in innate immunity. After infection, virus and bacteria release dsDNA or c-di-GMP to cells. DDX41 is then phosphorylated by BTK kinase at Y414 site and activated. The activated DDX41 can detect foreign PAMPs with DEAD domain, and then activate STING. Activated STING translocates from the endoplasmic reticulum (ER) to Golgi apparatus and interacts with TBK1 (Ouyang et al., 2012). The STING-TBK1 complex is required for the activation of TBK1 and the subsequent phosphorylation and nuclear translocation of IRF3, ultimately leading to expression of type I interferons. After immune response, DDX41 is ubiquitinated by Trim 21 at Lys9 and Lys115 sites, leading to degradation in proteasome.

2012, Somatic DDX41 mutations were reported in the study of sporadic acute myeloid leukemia (AML) syndrome by Ding et al. (2012). In 2015, Polprasert et al. identified a familial AML syndrome characterized by long latency and germline mutations in *DDX41* gene. Most of these mutations are located in DEAD domain, suggesting importance of this domain. In xenograft experiments with cell lines in which DDX41 was knocked down, they observed accelerated tumor growth compared to mock transduced cells, implying DDX41 functioning as a tumor suppressor. To further elucidate the roles of DDX41, they performed a pull down assay in HEK293 cells to identify proteins associated with DDX41. Their finding indicates that spliceosomal proteins constitute the top functional group associated with DDX41 and mutations in DDX41 (R525H) alter the native DDX41 interactome especially for major components in the U2 and U5 spliceosomes. It was concluded that DDX41 defects lead to loss of tumor suppressor function due to altered pre-mRNA splicing and RNA processing (Polprasert et al., 2015). The R525H mutation not only causes a defect in pre-rRNA processing but also inhibits cell cycle progression through the MDM2-RB-E2F axis (Kadono et al., 2016). Lewinsohn et al.

identified five novel mutations including missense mutations within important functional domains, and splicing mutations predicted to result in truncated proteins from a family with MDS/AML (Lewinsohn et al., 2016). Two novel articles about germline DDX41 mutations and variants in families with MDS/AML are also published recently (Cardoso et al., 2016; Li et al., 2016). We made a structure prediction for the human DDX41 (aa 150–564) by I-TASSER server (Roy et al., 2010; Roy et al., 2012; Zhang, 2008) and analyzed amino acid conservation by the ConSurf Server (http://consurf.tau.ac.il/). We found that mutations related to diseases are conserved except M155I mutation (Fig. 2). With an increasing number of both inherited and acquired mutations in *DDX41* gene being identified, further study of how DDX41 mutations lead to hematologic malignancies is needed. Thus, DDX41 and other members of the DEAD/H-box family gene mutations are identified functionally relevant as a novel family of mutations with possible implications for prognosis and treatment of myeloid malignancies, which may lead to approaches of therapeutic schedule (Antony-Debre and Steidl, 2015).

FUNCTIONS OF ABSTRAKT

Abstrakt in Drosophila is the homolog of DDX41, which shares 66% identity and 80% similarity at the amino acid level. The gene was initially identified by a very specific phenotype, the failure of the Bolwig nerve to fasciculate and project normally (Schmucker et al., 1997). Abstrakt is essential for survival at all stages throughout the life cycle of the fly. Mutants show specific defects in many developmental processes, including cell-shape changes, localization of RNA, and apoptosis. Mutations of E236K and V426M show a temperature-sensitive behavior, leading to lethality at 29°C and above (Irion and Leptin, 1999). In 2004, Irion et al. found that Abstrakt has a role in controlling cell polarity and asymmetric cell division in multiple cell types. The Abs protein interacts with *insc* RNA. Abs mutants show loss of Insc protein levels but no change of *insc* RNA levels. Although Abs is predicted to participate in RNA metabolism, they demonstrated a novel role for Abs in the post-transcriptional regulation of *insc* expression (Irion et al., 2004). An unexpected interaction of Abs and Sorting Nexin-2 (SNX2) was found by Johnny K. Ngsee' group in 2005 (Abdul-Ghani

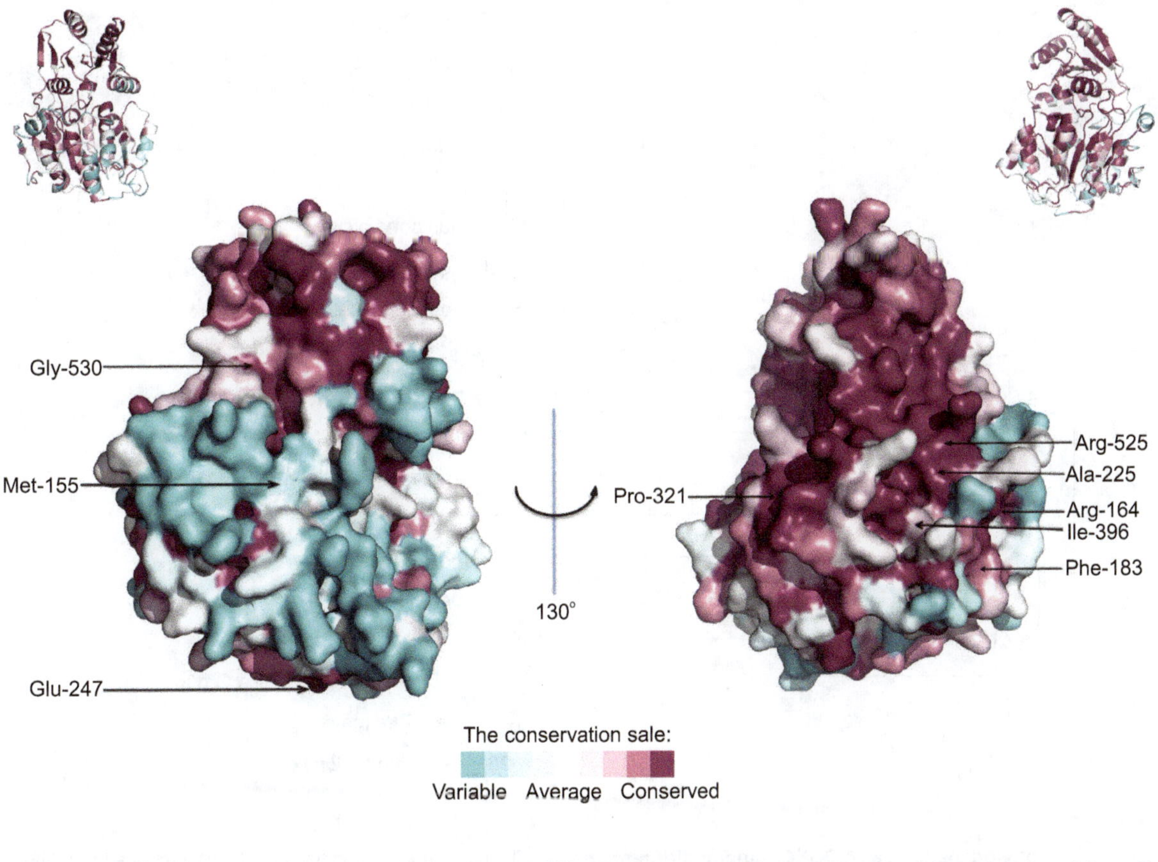

Figure 2. The amino acids conservation of human DDX41 DEAD and helicase domains. Graphical representation of the amino acids conservation using the predicted structure of human DDX41 is shown. Mutations related with MDS/AML are labeled. The color scale is based upon the level of conservation as determined by the ConSurf server. Category 9 corresponds to fully conserved residues. Small ribbon diagrams are shown at the top.

et al., 2005). They found that the N-terminal domain of Abs interacts with the phox homology (PX) domain of SNX2, suggesting that Abs also participates in protein sorting. They also found that Abs might shuttle between the nucleus and cytoplasm with a bias towards the nucleus, and the N-terminal domain is responsible for its nuclear localization. Deletion of this domain In Abs (aa 194–622) resulted in distinct punctate cytoplasmic distribution and loss of nuclear localization (Abdul-Ghani et al., 2005).

CONCLUSIONS

An increasing number of important roles of DExD/H-box helicases are discovered. In addition to the involvement in RNA metabolism, they are related to immune response, viral replication, and cell growth. The studies discussed in this review illustrate the progress being made in our understanding of the emerging roles of DDX41. DDX41 is widely distributed in eukaryotes, especially in animals. The phylogenetic tree shows that the amino acid sequence is conserved among species (Fig. 3). Although several papers showed that DDX41 is important for innate immunity, the mechanism of how it functions is still unknown. Questions remain, for example, how can it distinguish self-DNA and foreign-DNA? Furthermore, structural investigation of the DEAD domain is urgent to address a few important questions. First, is there structural difference between different DEAD domains as amino acid sequences are very similar? Second, there is a long disordered region at N-terminal except the DEAD domain and Helicase domain. Until now there has been no reports about the roles of this long disordered region. In Drosophila, the N-terminal region of the DEAD-box protein Abstrakt mediates its localization into nucleus. Is this the function of DDX41 as well? Third, Mutations in *DDX41* gene have been identified from MDS/AML patients. DDX41 defects lead to loss of tumor suppressor function due to altered pre-mRNA splicing and RNA processing. Can DDX41 be a drug target for the diseases?

In conclusion, although many new roles of DDX41 have been discovered, there are still many problems to be solved, where structural elucidation can have a significant impact. Added importance of structural investigation comes from the indication that DDX41 may function as a tumor suppressor, and could be a good therapeutic target for disease treatment.

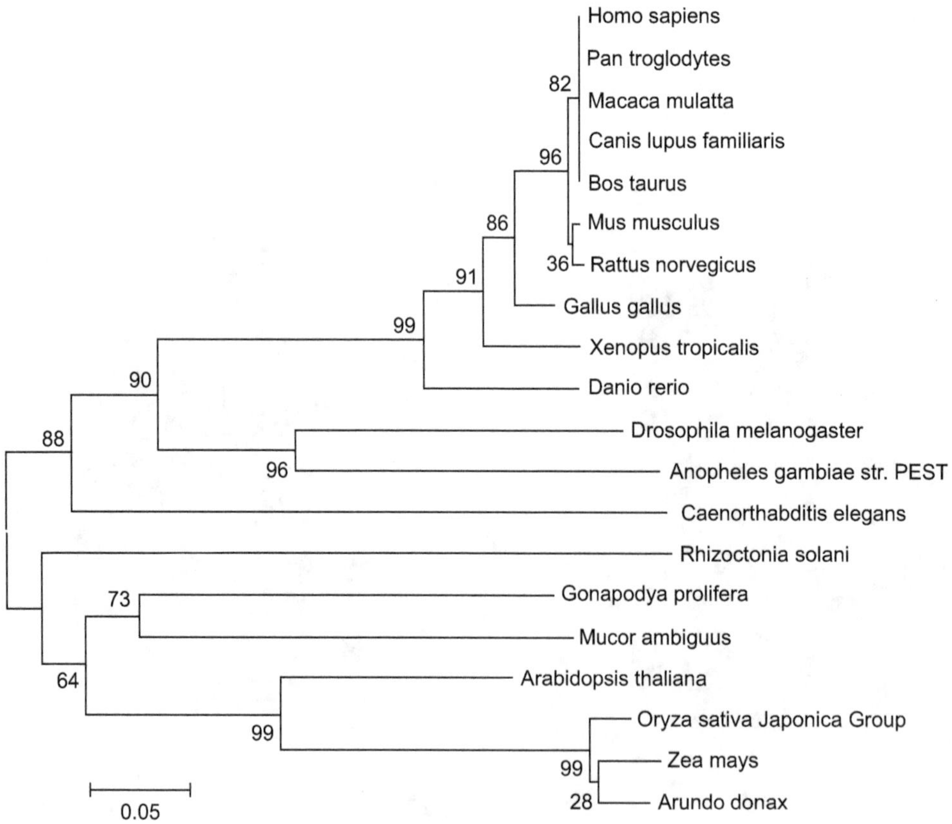

Figure 3. The phylogenetic tree of DDX41 among different species. The sequences of indicated DDX41 proteins from different species are downloaded from GenBank. The tree is generated using MEGA5 software (http://www.megasoftware.net/) with Neighbor Joining method. Length of branch represents divergence and numbers on branches are indicated as percentage of bootstrap values in 1000 sampling replicates.

ACKNOWLEDGMENTS

This work was supported by the National Basic Research Program (973 Program) (No. 2014CB910400 and 2013CB911103), the National Natural Science Foundation of China (Grants No. 31570875, 31330019, 81590761, 31560727 and 81501353), the Beijing Nova Program (Grant No. Z141102001814020) to S.O., Youth Innovation Promotion Association CAS (Grant No. 2013065) to S.O., and the special project of Ebola virus research from the president foundation of Chinese Academy of Sciences.

ABBREVIATIONS

MDS, myelodysplastic syndrome; AML, acute myeloid leukemia; BTK, Bruton's tyrosine kinase; D-E-A-D, Asp-Glu-Ala-Asp; HCV, hepatitis C virus; IFN, interferon; IRF3, interferon regulatory factor 3; mDC, myeloid dendritic cells; PAMP, pathogen-associated molecular pattern; PRR, pattern recognition receptor; STING, stimulator of interferon genes; TBK1, TANK-binding kinase 1; TRIF, TIR-domain-containing-adapter-inducing interferon-β

COMPLIANCE WITH ETHICS GUIDELINES

Yan Jiang, Zhi-Jie Liu, and Songying Ouyang declare that they have no conflict of interest. This article does not contain any studies with human or animal subjects performed by the any of the authors.

REFERENCES

Abdul-Ghani M, Hartman KL, Ngsee JK (2005) Abstrakt interacts with and regulates the expression of sorting nexin-2. J Cell Physiol 204:210–218

Antony-Debre I, Steidl U (2015) Functionally relevant RNA helicase mutations in familial and sporadic myeloid malignancies. Cancer Cell 27:609–611

Burdette DL, Monroe KM, Sotelo-Troha K, Iwig JS, Eckert B, Hyodo M, Hayakawa Y, Vance RE (2011) STING is a direct innate immune sensor of cyclic di-GMP. Nature 478:515–518

Cardoso SR, Ryan G, Walne AJ, Ellison A, Lowe R, Tummala H, Rio-Machin A, Collopy L, Al Seraihi A, Wallis Y et al (2016) Germline heterozygous DDX41 variants in a subset of familial myelodysplasia and acute myeloid leukemia. Leukemia. doi:10.1038/leu.2016.124

Caruthers J, McKay D (2002) Helicase structure and mechanism. Curr Opin Struct Biol 12:123–133

Caruthers JM, Johnson ER, McKay DB (2000) Crystal structure of yeast initiation factor 4A, a DEAD-box RNA helicase. Proc Natl Acad Sci USA 97:13080–13085

Cordin O, Tanner NK, Doère M, Linder P, Banroques J (2004) The newly discovered Q motif of DEAD-box RNA helicases regulates RNA-binding and helicase activity. EMBO J 23:2478–2487

Ding L, Ley TJ, Larson DE, Miller CA, Koboldt DC, Welch JS, Ritchey JK, Young MA, Lamprecht T, McLellan MD et al (2012) Clonal evolution in relapsed acute myeloid leukaemia revealed by whole-genome sequencing. Nature 481:506–510

Fang J, Kubota S, Yang B, Zhou N, Zhang H, Godbout R, Pomerantz RJ (2004) A DEAD box protein facilitates HIV-1 replication as a cellular co-factor of Rev. Virology 330:471–480

Fullam A, Schroder M (2013) DExD/H-box RNA helicases as mediators of anti-viral innate immunity and essential host factors for viral replication. Biochim Biophys Acta 1829:854–865

Goh PY, Tan YJ, Lim SP, Tan YH, Lim SG, Fuller-Pace F, Hong W (2004) Cellular RNA helicase p68 relocalization and interaction with the hepatitis C virus (HCV) NS5B protein and the potential role of p68 in HCV RNA replication. J Virol 78: 5288–5298

Hengge R (2009) Principles of c-di-GMP signalling in bacteria. Nat Rev Microbiol 7:263–273

Irion U, Leptin M, Siller K, Fuerstenberg S, Cai Y, Doe CQ, Chia W, Yang X (2004) Abstrakt, a DEAD box protein, regulates Insc levels and asymmetric division of neural and mesodermal progenitors. Curr Biol 14:138–144

Irion U, Leptin M (1999) Developmental and cell biological functions of the *Drosophila* DEAD-box protein Abstrakt. Curr Biol 9:1373–1381

Jankowsky E, Gross C, Shuman S, Pyle A (2001) Active disruption of an RNA–protein interaction by a DExH: D RNA helicase. Science 291:121–125

Kadono M, Kanai A, Nagamachi A, Shinriki S, Kawata J, Iwato K, Kyo T, Oshima K, Yokoyama A, Kawamura T et al (2016) Biological implications of somatic DDX41 p.R525H mutation in acute myeloid leukemia. Exp Hematol 44:745–754

Kim T, Pazhoor S, Bao M, Zhang Z, Hanabuchi S, Facchinetti V, Bover L, Plumas J, Chaperot L, Qin J, Liu Y (2010) Aspartate–glutamate–alanine–histidine box motif (DEAH)/RNA helicase A helicases sense microbial DNA in human plasmacytoid dendritic cells. Proc Natl Acad Sci USA 107:15181–15186

Lee KG, Kim SS, Kui L, Voon DC, Mauduit M, Bist P, Bi X, Pereira NA, Liu C, Sukumaran B et al (2015) Bruton's tyrosine kinase phosphorylates DDX41 and activates its binding of dsDNA and STING to initiate type 1 interferon response. Cell Rep 10:1055–1065

Lewinsohn M, Brown AL, Weinel LM, Phung C, Rafidi G, Lee MK, Schreiber AW, Feng J, Babic M, Chong CE et al (2016) Novel germ line DDX41 mutations define families with a lower age of MDS/AML onset and lymphoid malignancies. Blood 127:1017–1023

Li R, Sobreira N, Witmer PD, Pratz KW, Braunstein EM (2016) Two novel germline DDX41 mutations in a family with inherited myelodysplasia/acute myeloid leukemia. Haematologica 101: e228–e231

Linder P, Lasko PF, Ashburner M, Leroy P, Nielsen PJ, Nishi K, Schnier J, Slonimski PP (1989) Birth of the D-E-A-D box. Nature 337:121–122

Liu X, Wang C (2016) The emerging roles of the STING adaptor protein in immunity and diseases. Immunology 147:285–291

Luking A, Stahl U, Schmidt U (1998) The protein family of RNA helicases. Crit Rev Biochem Mol Biol 33:259–296

Medzhitov R, Janeway CJ (2000) Innate immune recognition: mechanisms and pathways. Immunol Rev 173:89–97

Miyashita M, Oshiumi H, Matsumoto M, Seya T (2011) DDX60, a DEXD/H box helicase, is a novel antiviral factor promoting RIG-I-like receptor-mediated signaling. Mol Cell Biol 31:3802–3819

Oshiumi H, Sakai K, Matsumoto M, Seya T (2010) DEAD/H BOX 3 (DDX3) helicase binds the RIG-I adaptor IPS-1 to up-regulate IFN-beta-inducing potential. Eur J Immunol 40:940–948

Ouyang S, Song X, Wang Y, Ru H, Shaw N, Jiang Y, Niu F, Zhu Y, Qiu W, Parvatiyar K et al (2012) Structural analysis of the STING adaptor protein reveals a hydrophobic dimer interface and mode of cyclic di-GMP binding. Immunity 36:1073–1086

Owsianka AM, Patel AH (1999) Hepatitis C virus core protein interacts with a human DEAD box protein DDX3. Virology 257:330–340

Parvatiyar K, Zhang Z, Teles RM, Ouyang S, Jiang Y, Iyer SS, Zaver SA, Schenk M, Zeng S, Zhong W et al (2012) The helicase DDX41 recognizes the bacterial secondary messengers cyclic di-GMP and cyclic di-AMP to activate a type I interferon immune response. Nat Immunol 13:1155–1161

Polprasert C, Schulze I, Sekeres MA, Makishima H, Przychodzen B, Hosono N, Singh J, Padgett RA, Gu X, Phillips JG et al (2015) Inherited and somatic defects in DDX41 in myeloid neoplasms. Cancer Cell 27:658–670

Quynh NT, Hikima J, Kim YR, Fagutao FF, Kim MS, Aoki T, Jung TS (2015) The cytosolic sensor, DDX41, activates antiviral and inflammatory immunity in response to stimulation with double-stranded DNA adherent cells of the olive flounder, *Paralichthys olivaceus*. Fish Shellfish Immunol 44:576–583

Rocak S, Linder P (2004) DEAD-box proteins: the driving forces behind RNA metabolism. Nat Rev Mol Cell Biol 5:232–241

Roy A, Kucukural A, Zhang Y (2010) I-TASSER: a unified platform for automated protein structure and function prediction. Nat Protoc 5:725–738

Roy A, Yang J, Zhang Y (2012) COFACTOR: an accurate comparative algorithm for structure-based protein function annotation. Nucleic Acids Res 40:W471–W477

Schmid S, Linder P (1992) D-E-A-D protein family of putative RNA helicases. Mol Microbiol 6:283–291

Schmidt A, Rothenfusser S, Hopfner KP (2012) Sensing of viral nucleic acids by RIG-I: from translocation to translation. Eur J Cell Biol 91:78–85

Schmucker D, Jackle H, Gaul U (1997) Genetic analysis of the larval optic nerve projection in *Drosophila*. Development 124:937–948

Tingting P, Caiyun F, Zhigang Y, Pengyuan Y, Zhenghong Y (2006) Subproteomic analysis of the cellular proteins associated with the 3′ untranslated region of the hepatitis C virus genome in human liver cells. Biochem Biophys Res Commun 347:683–691

Zhang Y (2008) I-TASSER server for protein 3D structure prediction. BMC Bioinform 9:40

Zhang Z, Kim T, Bao M, Facchinetti V, Jung SY, Ghaffari AA, Qin J, Cheng G, Liu YJ (2011a) DDX1, DDX21, and DHX36 helicases form a complex with the adaptor molecule TRIF to sense dsRNA in dendritic cells. Immunity 34:866–878

Zhang Z, Yuan B, Bao M, Lu N, Kim T, Liu YJ (2011b) The helicase DDX41 senses intracellular DNA mediated by the adaptor STING in dendritic cells. Nat Immunol 12:959–965

Zhang Z, Yuan B, Lu N, Facchinetti V, Liu YJ (2011c) DHX9 pairs with IPS-1 to sense double-stranded RNA in myeloid dendritic cells. J Immunol 187:4501–4508

Zhang Z, Bao M, Lu N, Weng L, Yuan B, Liu YJ (2013) The E3 ubiquitin ligase TRIM21 negatively regulates the innate immune response to intracellular double-stranded DNA. Nat Immunol 14:172–178

Zhu X, Wang D, Zhang H, Zhou Y, Luo R, Chen H, Xiao S, Fang L (2014) Molecular cloning and functional characterization of porcine DEAD (Asp-Glu-Ala-Asp) box polypeptide 41 (DDX41). Dev Comp Immunol 47:191–196

3

Mitochondrial calcium uniporter protein MCU is involved in oxidative stress-induced cell death

Yajin Liao[1,2], Yumin Hao[1], Hong Chen[1], Qing He[1,2], Zengqiang Yuan[1]✉, Jinbo Cheng[1]✉

[1] State Key Laboratory of Brain and Cognitive Sciences, Institute of Biophysics, Chinese Academy of Sciences, Beijing 100101, China
[2] College of Life Sciences, University of Chinese Academy of Sciences, Beijing 100049, China
✉ Correspondence: zqyuan@ibp.ac.cn (Z. Yuan), cheng_jinbo@126.com (J. Cheng)

ABSTRACT

Mitochondrial calcium uniporter (MCU) is a conserved Ca^{2+} transporter at mitochondrial in eukaryotic cells. However, the role of MCU protein in oxidative stress-induced cell death remains unclear. Here, we showed that ectopically expressed MCU is mitochondrial localized in both HeLa and primary cerebellar granule neurons (CGNs). Knockdown of endogenous MCU decreases mitochondrial Ca^{2+} uptake following histamine stimulation and attenuates cell death induced by oxidative stress in both HeLa cells and CGNs. We also found MCU interacts with VDAC1 and mediates VDAC1 overexpression-induced cell death in CGNs. This finding demonstrates that MCU-VDAC1 complex regulates mitochondrial Ca^{2+} uptake and oxidative stress-induced apoptosis, which might represent therapeutic targets for oxidative stress related diseases.

KEYWORDS MCU, VDAC1, oxidative stress, calcium uptake, cell death

INTRODUCTION

As the major source of ATP, mitochondrion plays an essential role in cellular physiology and metabolism in eukaryotic cells. On the one hand, low mitochondrial Ca^{2+} concentration fails to effectively activate pyruvate dehydrogenase as to produce enough ATP (Robb-Gaspers et al., 1998); on the other hand, overloaded mitochondrial Ca^{2+} reduces mitochondrial membrane potential (Δψ) and triggers cell death. Therefore, mitochondrial Ca^{2+} homeostasis is essential for physiological function and survival of the cells.

Some proteins involved in mitochondrial Ca^{2+} uptake have been discovered in recent years. VDAC (voltage-dependent anion channel) is the first identified Ca^{2+} transport channel, which locates in the outer membrane of mitochondria (OMM) and highly permeable to Ca^{2+} (Xu et al., 1999). VDAC1 will lose its Ca^{2+} transport capacity when the Ca^{2+} binding sites are blocked by Ruthenium Red (RuR) (Gincel et al., 2001). MCU is another identified Ca^{2+} uniporter protein, which is localized in the inner membrane of mitochondrion (IMM) (Baughman et al., 2011). Numbers of proteins have been shown to regulate the activity of MCU, such as essential mitochondrial calcium uptake1/2 (MICU1/2), MCU regulator (EMRE), MCUb and MCU regulator 1 (MCUR1) (Ahuja and Muallem, 2014; Alam et al., 2012; Mallilankaraman et al., 2012a; Mallilankaraman et al., 2012b; Raffaello et al., 2013; Sancak et al., 2013). When cells undergo severe damage, mitochondrial permeability transition pore (PTP) will open and release pro-apoptotic factors, especially cytochrome c, from mitochondria to the cytoplasm (Kinnally et al., 2011). Overexpression of VDAC1 renders cells sensitive to oxidative stress inducers, such as thapsigargin (TG) and arsenic trioxide (As2O3) (Ben-Hail and Shoshan-Barmatz, 2012; Shoshan-Barmatz et al., 2010; Shoshan-Barmatz et al., 2009). Ectopically expressed VDAC1 increases the concentration of Ca^{2+} in the inner-membrane space of mitochondrion, leading to mitochondrial Ca^{2+} overload and cytochrome c release (Brustovetsky et al., 2002; Madesh and Hajnoczky, 2001; Naranmandura et al., 2012; Rapizzi et al., 2002). Cytochrome c interacts with Apaf1 and initiates activation of caspase-3 (Chu et al., 2001). As a subunit of the PTP, VDAC1 is involved in cytochrome c release and apoptosis (Zheng et al., 2004). Whether MCU, the inner membrane Ca^{2+} uniporter, has the similar function as

VDAC1 protein in neurons is poorly studied. In our study, we found that MCU plays an important role in oxidative stress-induced apoptosis. In addition, we showed that MCU interacts with VDAC1 and is involved in the VDAC1-mediated cell death in CGNs.

RESULTS

MCU regulates the mitochondrial Ca^{2+} uptake in primary CGNs

First, we found that overexpressed MCU was mainly localized in the mitochondria in HeLa cells and CGNs (Fig. 1A and 1B). To identify whether MCU functions as mitochondrial Ca^{2+} uniporter in HeLa cells, two shRNAs against MCU were designed. As shown in Fig. 2A, shRNA#1 had a better knockdown efficiency compared with shRNA#2. In order to monitor mitochondrial Ca^{2+} levels, we transfected cells with a mitochondrial targeted Ca^{2+} indicator, Mito-GCaMP3. Knockdown of MCU dramatically decreased histamine-induced mitochondrial Ca^{2+} level, which includes both the peak Ca^{2+} level and the recovery time to baseline (Fig. 2B–D). Furthermore,

Figure 2. MCU is required for mitochondrial calcium uniporter in HeLa cells and CGNs. (A) Two shRNA against MCU were designed to knockdown MCU in HeLa cells and the knockdown efficiency of each was detected by Western blot. (B) Knocking down MCU decreased mitochondrial Ca^{2+} uptake by assaying Mito-GCaMP3. Control or MCU knockdown HeLa cells were transfected with Mito-GCaMP3. GCaMP3 fluorescence was measured before and during exposure to histamine stimulation (100 μmol/L). (C) Representative traces Ca^{2+} response to histamine stimulation in control or MCU knockdown HeLa cells. (D) Graph shows quantification of peak Ca^{2+} intensities. (E) Representative traces of Ca^{2+} uptake in digitonin-permeabilized HeLa cells. The level of Ca^{2+} was indicated by the calcium Green-5N. (F) Graph shows relative Ca^{2+} uptake. (G) CGNs were transfected with shMCU or shvector together with Mito-GCaMP3. GCaMP3 fluorescence was measured before and after exposure to histamine (100 μmol/L). (H) Representative traces showing Ca^{2+} response after histamine stimulation over time in MCU wild type or knockdown cells were. (I) Graph shows quantification of peak Ca^{2+} intensities.

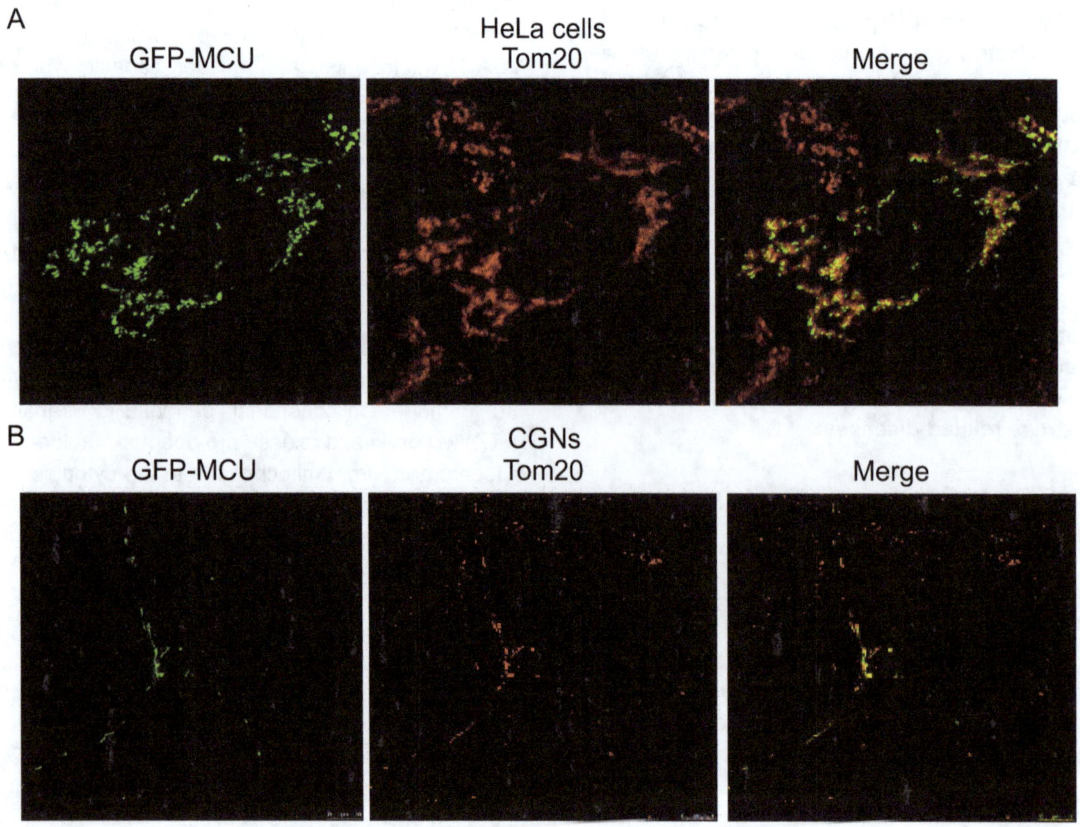

Figure 1. The sublocation of MCU in HeLa cells and CGN. (A) GFP-MCU (green) was tranfected into HeLa cells. Twenty-four hours post transfection, cells were fixed and stained with Tom20 (Red). (B) GFP-MCU (green) and mito-Red (red) were co-tranfected into CGNs. Twenty-four hours post transfection, cells were fixed and imaged by laser confocal microscope using a 40× objective lens.

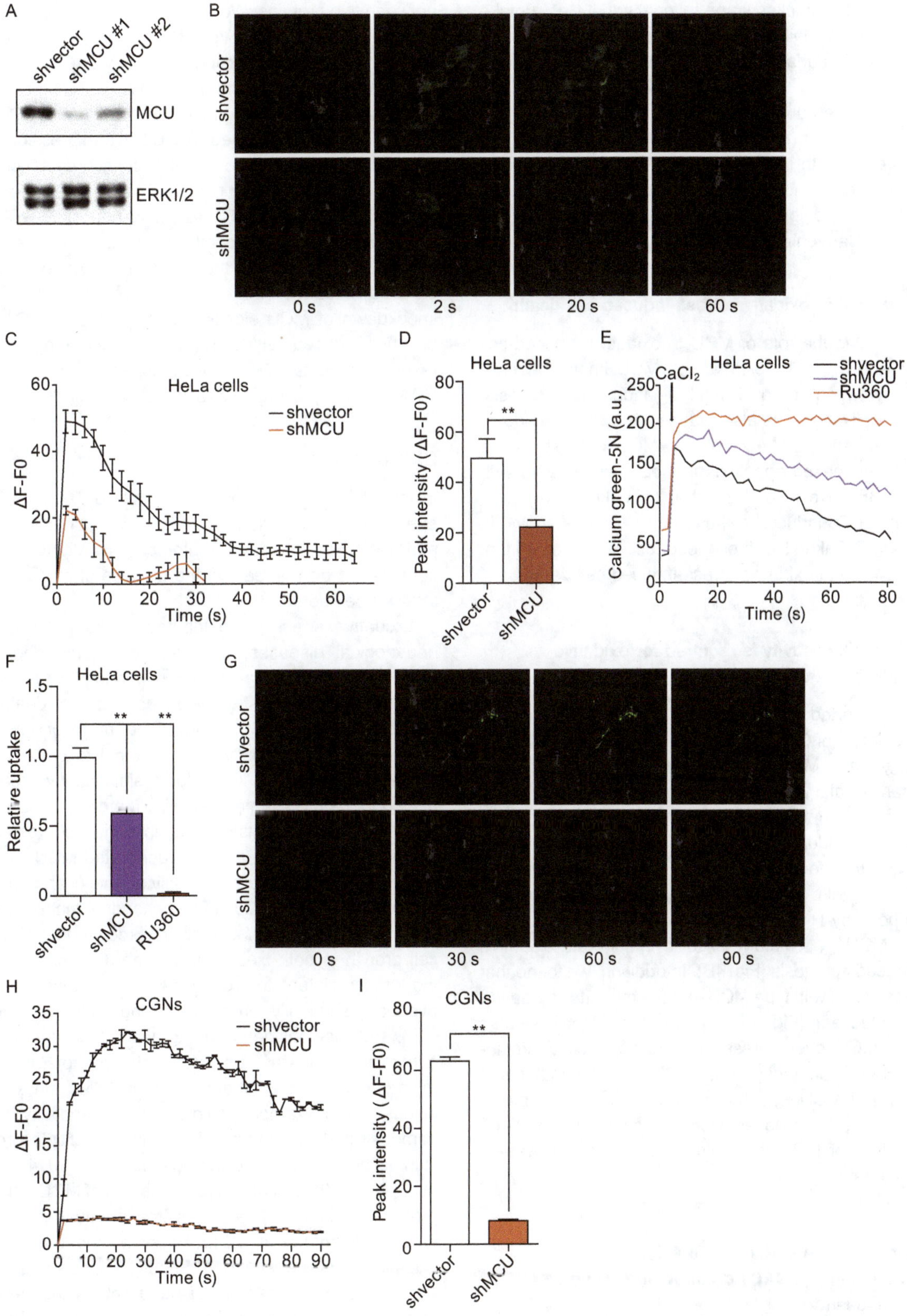
A
shvector
shMCU #1
shMCU #2
MCU
ERK1/2
B
shvector
shMCU
0 s
2 s
20 s
60 s
C
HeLa cells
shvector
shMCU
ΔF-F0
Time (s)
D
HeLa cells
Peak intensity (ΔF-F0)
**
shvector
shMCU
E
CaCl2
HeLa cells
shvector
shMCU
Ru360
Calcium green-5N (a.u.)
Time (s)
F
HeLa cells
Relative uptake
**
**
shvector
shMCU
RU360
G
shvector
shMCU
0 s
30 s
60 s
90 s
H
CGNs
shvector
shMCU
ΔF-F0
Time (s)
I
CGNs
Peak intensity (ΔF-F0)
**
shvector
shMCU

the effect of MCU knockdown on mitochondrial Ca^{2+} uptake was examined in digitonin-permeabilized cells. In normal condition, mitochondria could rapidly clear most of externally added 50 μmol/L Ca^{2+}. However, MCU knockdown remarkably inhibited external Ca^{2+} clearance, similar to Ru360, a specific MCU blocker (Fig. 2E and 2F). Consistently, in CGN cells transfected with Mito-GCaMP3, MCU knockdown dramatically blocked mitochondrial Ca^{2+} uptake (Fig. 2G–I). Together, these findings implicate MCU functions as a mitochondrial Ca^{2+} uniporter in both HeLa cells and CGNs.

MCU is involved in oxidative stress-induced cell death

Next, we studied the role of MCU in oxidative induced-cell death. As shown in Fig. 3A and 3B, MCU knockdown decreased hydrogen peroxide (H_2O_2)-induced apoptosis. Caspase-3 activation was also reduced in MCU knockdown cells (Fig. 3C and 3D). Accordingly, MCU overexpression significantly increased apoptosis in HeLa cells treated with H_2O_2 (Fig. 3E). We also found that MCU knockdown decreased the H_2O_2-induced cell death in CGNs as shown in Fig. 3F and 3G. Taken together, these results suggested that MCU is involved in oxidative stress-induced cell death.

MCU Ca^{2+} uptake activity is required for oxidative stress-induced cell death

It has been reported that D260 and E263 of MCU are critical for the regulation of Ca^{2+} uptake (Fig. 4A) and the mutations of MCU_{D260A} and MCU_{E263A} lose the Ca^{2+} uptake activity (Baughman et al., 2011). To further confirm whether the function of Ca^{2+} uptake activity of MCU was required for oxidative stress-induced cell death, we constructed the MCU mutants and transfected into cells. We observed that overexpression of wild type MCU significantly increased cell death induced by H_2O_2. However, expression of MCU_{D260A} mutant or MCU_{E263A} mutant failed to exaggerate oxidative stress-induced apoptosis (Fig. 4B). In addition, we found that overexpression of wild type MCU, not the mutants, increased caspase-3 cleavage (Fig. 4C and 4D). Consistently, we also found that MCU overexpression increased cell death induced by H_2O_2 in primary cultured CGNs, and the mutants of MCU_{D260A} and MCU_{E263A} had no effect on the apoptosis (Fig. 4E). Thus, these observations indicated that the Ca^{2+} uptake activity of MCU is required for oxidative stress-induced cell death.

MCU interacts with VDAC1 and functions as a downstream of VDAC1 during oxidative stress-induced cell death

It has been shown that VDAC1 overexpression made cells sensitive to oxidative stress (Rapizzi et al., 2002). Since both VDAC1 and MCU are involved in mitochondrial calcium uptake, we then ask whether there is functional interaction between two proteins. We firstly observed there is a physical interaction of MCU and VDAC1 (Fig. 5A). Next, we investigated the biological function of MCU-VDAC1 interaction in cells. We found that overexpression of VDAC1 increased the apoptosis induced by H_2O_2 in HeLa cells and MCU knockdown significantly inhibited VDAC1 overexpression-induced cell death, which suggests MCU functions as the downstream of VDAC1 during oxidative stress-induced apoptosis (Fig. 5B and 5C). VDAC1 has also been shown to be involved in neuronal cell death (Fernandez-Echevarria et al., 2014). Here, we found that overexpression of VDAC1 increased oxidative stress-induced apoptosis in CGNs and knockdown of MCU significantly mitigated VDAC1 overexpression-induced cell death (Fig. 5D). Taken together, these results suggested MCU functions as the downstream of VDAC1 during oxidative stress-induced cell death.

DISCUSSION

In this study, we demonstrate the role of MCU in oxidative stress-induced cell death by loss- and gain-of-function experiments. We found MCU interacts with VDAC1 and functions as the downstream in the process of oxidative stress-induced cell death.

Oxidative stress is contributed to the pathogenesis of neurological diseases, such as stroke and degenerative diseases. Mitochondrial calcium overload plays an important role in the oxidative stress-induced neuronal death. In our study, we found MCU is localized at mitochondria and functioned as a crucially Ca^{2+} channel in both HeLa and primary CGNs. MCU knockdown significantly blocked the mitochondrial Ca^{2+} uptake activity. Moreover, our functional studies indicate MCU is involved in oxidative stress-induced cell death. Recently, Pan et al. generated $MCU^{-/-}$ mice and found that $MCU^{-/-}$ mice are grossly normal (Pan et al., 2013), but a significant reduction of mitochondrial matrix calcium. In this study, we found that MCU knockdown did not affect the cell growth in both HeLa and primary CGNs, but rendered the cells resistant to oxidative stress. Mitochondrial Ca^{2+} overload is usually observed during ischemia/reperfusion (I/R), and it is considered to aggravate I/R injury. As MCU is a mitochondrial Ca^{2+} uniporter, inhibition of MCU activity might be a therapy strategy for oxidative stress-induced diseases. For example, inhibition of MCU activity has been demonstrated to attenuate I/R injury in multiple organs including brain, heart and liver (Dong et al., 2014; Schwartz et al., 2013; Zhao et al., 2013). Inhibition of MCU activity also protects brain and heart from iron overload-induced dysfunction (Kumfu et al., 2012; Sripetchwandee et al., 2013a; Sripetchwandee et al., 2014; Sripetchwandee et al., 2013b). Accordingly, enhancement of MCU activity increases Ca^{2+} level in mitochondria and promotes oxidative stress-induced apoptosis (De Stefani et al., 2011). Interestingly, recent studies showed that the role of MCU in apoptosis is dependent on cell type. Knockdown of MCU protect HeLa cells, not in MEF cells, against oxidative stress induced apoptosis

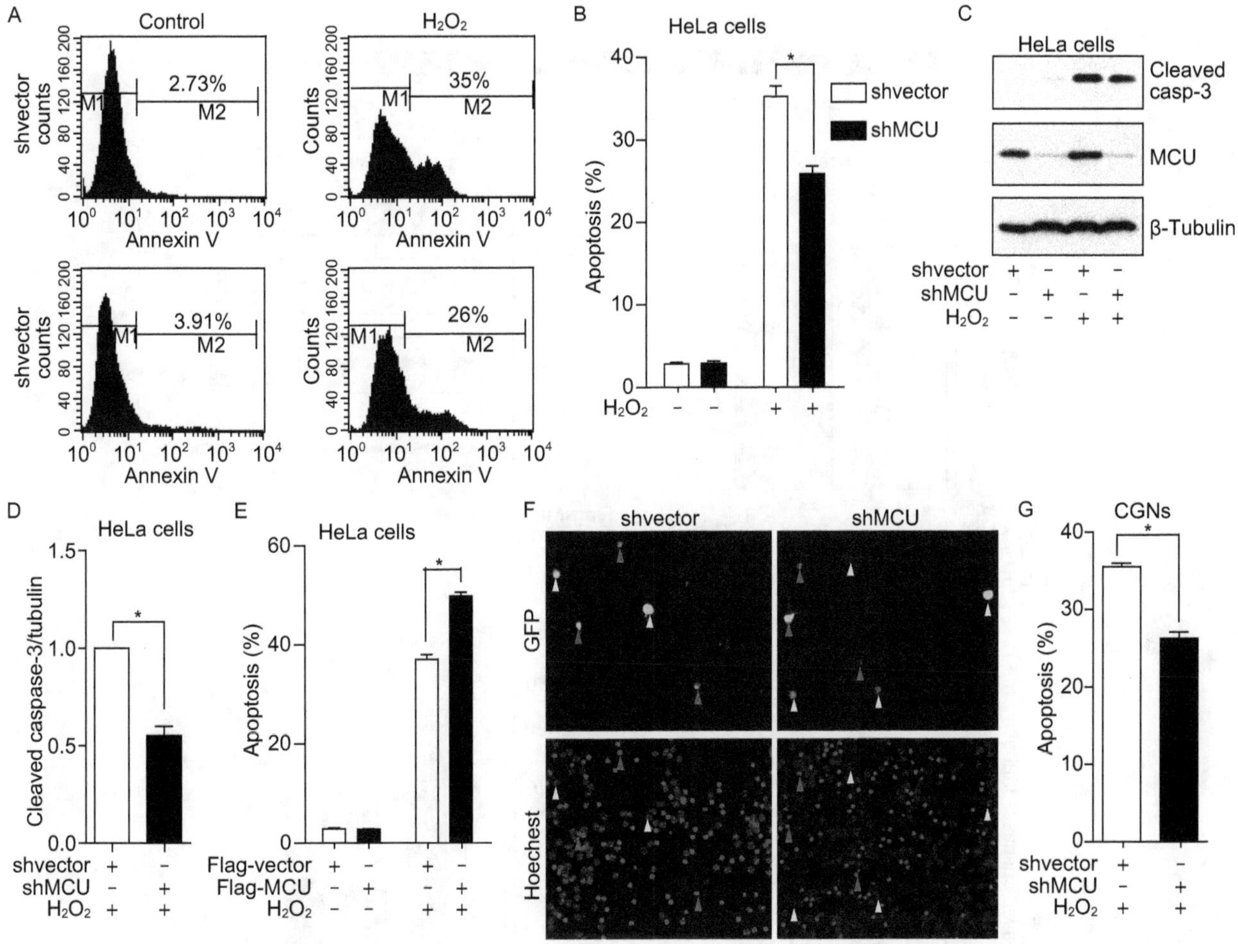

Figure 3. MCU is involved in oxidative stress-induced cell death. (A and B) MCU silenced HeLa cells or control cells were treated with H_2O_2 for 24 h, then, cells were collected and stained with FITC labeled annexin V and finally analysis by flow cytometer. (C) MCU silenced HeLa cells or control cells were treated with H_2O_2 for 24 h, and cells were collected and the cleaved caspase-3 was detected. (D) Graph shows normalized level of cleaved caspase-3. (E) HeLa cells tranfected with Flag-vector or Flag-MCU were exposed to H_2O_2 for 24 h and apoptosis was analyzed by FACS. (F and G) CGNs transfected with GFP vector together with shMCU or shvector were treated with 70 μmol/L H_2O_2 for 24 h. Green arrowhead stands for the healthy cells and red arrowhead indicates apoptotic cells.

(De Stefani et al., 2011; Hall et al., 2014; Pan et al., 2013). In this study, we found that MCU knockdown attenuates oxidative stress-induced apoptosis in both HeLa and primary CGNs.

PTP is regarded as the gatekeeper of apoptosis, and its opening is regulated by several proteins, such as VDAC1, adenosine nucleotide translocase (ANT) and cyclophilin D (Halestrap et al., 1997; Pestana et al., 2010). Overexpression of these proteins makes cells sensitive to apoptotic-inducing stimulus. Overexpressed VDAC1 in HeLa cells enhances cell death upon treatment with H_2O_2, staurosporine (STS), TG or As_2O_3 (Keinan et al., 2013). In this study, we found MCU knockdown remarkably inhibited VDAC1 overexpression induced-cell death, suggesting MCU functions as a downstream of VDAC1 during oxidative stress-induced cell death.

In summary, we showed that VDAC1 (outer member) and MCU (inner member) form complex and mediate mitochondrial calcium uptake and stress-induced cell death. Furthermore, we showed that MCU is involved in oxidative stress-induced apoptosis as a downstream regulator of VDAC1. Therefore, the inhibition of the activity of MCU or disruption of VDAC1-MCU interaction might be a strategy to stroke and degenerative diseases.

MATERIALS AND METHODS

Materials

H_2O_2, anti-Flag M2 mAb and Anti-MCU pAb were purchased from Sigma-Aldrich. Anti-active caspase-3 pAb was purchased from Millipore. Anti-GFP pAb, anti-GAPDH mAb, anti-β-tubulin mAb were

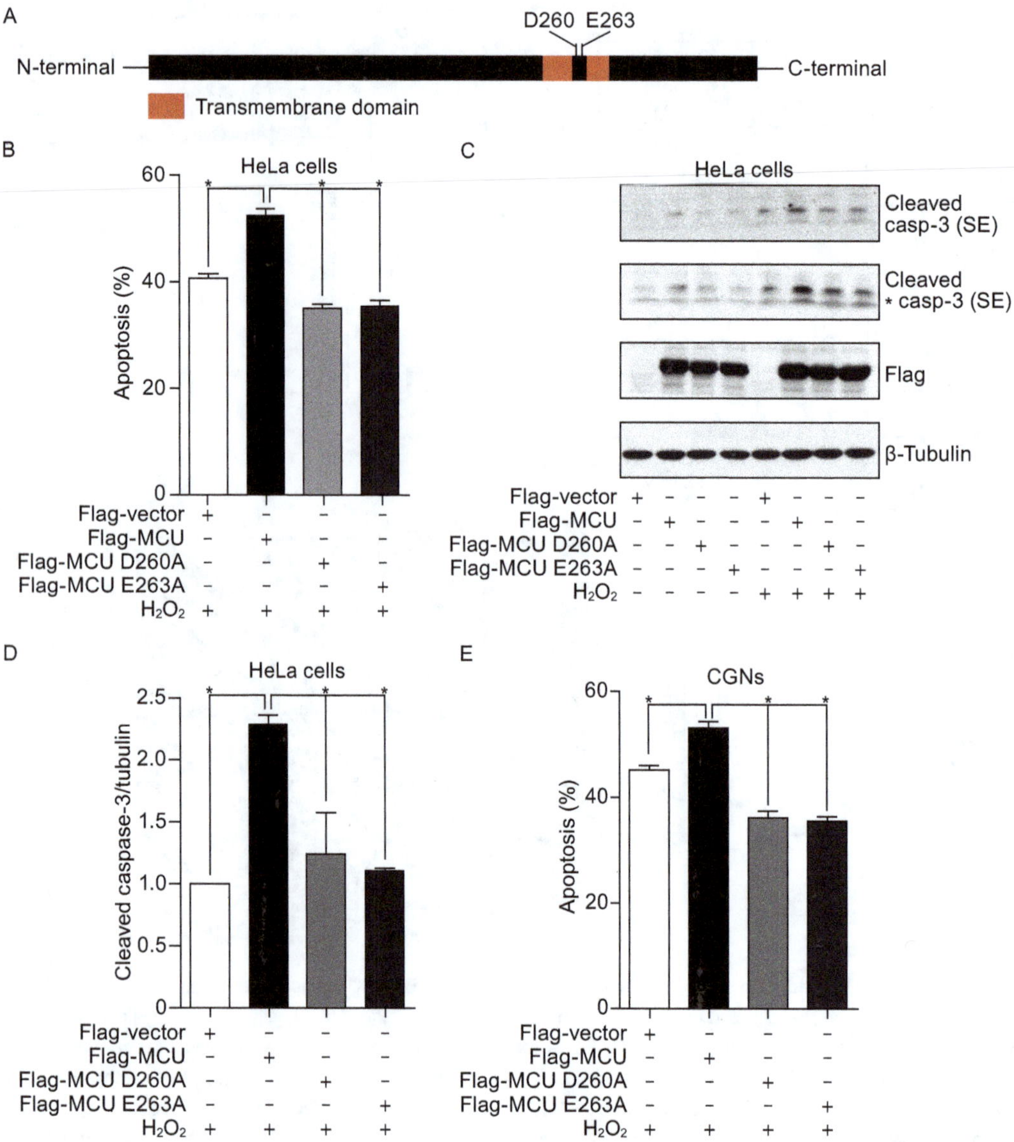

Figure 4. The calcium uptake activity of MCU is required for oxidative stress-induced cell death. (A) The critical site for calcium uptake activity of MCU. (B) HeLa cells were overexpressed with wild type MCU, MCU_{D260A} or MCU_{E263A}, and then cells were treated with H_2O_2 for 24 h. The apoptosis was finally analyzed by FACS. (C) HeLa cells transfected with wild type MCU, MCU_{D260A} or MCU_{E263A} as indicated were treated with 200 μmol/L H_2O_2 for 24 h. Cleaved caspase-3 were analyzed ("*" means non-specific band). (D) Graph shows normalized level of cleaved caspase-3. (E) CGNs transfected with GFP vector plus wild type MCU, MCU_{D260A} mutant, MCU_{E263A} mutant or empty vector were treated with 70 μmol/L H_2O_2 for 24 h.

purchased from CWbiotech. Primers for shRNA were synthesized by Invitrogen and cloned into pLKO.1 vector.

Cell culture and transfection

HeLa cells were maintained in DMEM supplied with 10% fetal bovine serum at 37°C in a humidified atmosphere with 5% CO_2. Mouse primary cerebellar granule neurons (CGNs) were isolated from 10 days old mouse as previously described (Xie et al., 2012). Isolated CGNs were cultured in BME supplemented with 10 μmol/L cytosine arabinoside (AraC) and 25 mmol/L glucose. Lipofectamine 2000 reagent was applied for transfection in HeLa cells according to the manufacturer's instructions. For transfection of CGNs, the classical calcium phosphate coprecipitation technique was used.

Imaging of mitochondrial calcium

Mitochondrial Ca^{2+} uptake in intact cells was detected as previously described (Qiu et al., 2013). Briefly, cells plated on 35-mm glass-bottom dishes were transfected with Mito-GCaMP3 to monitor the concentration of Ca^{2+} in mitochondrial. 24 h post transfection, time-lapse confocal microscopy was started at 1-sec intervals using a

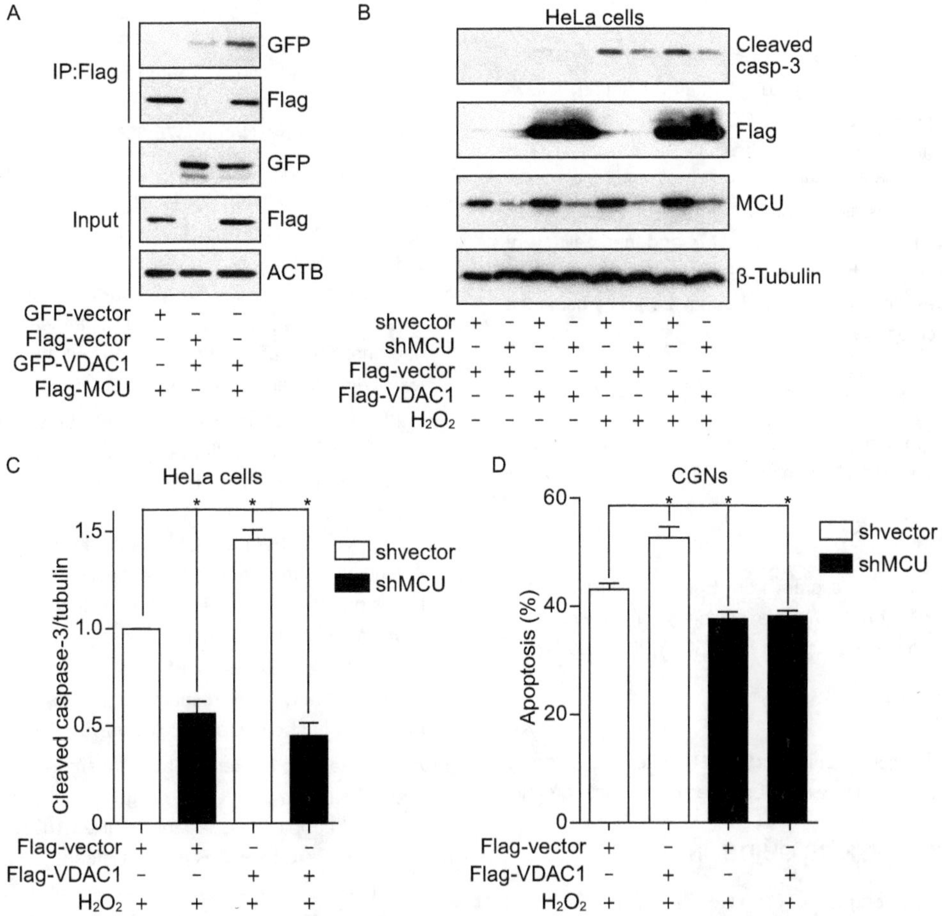

Figure 5. MCU interacts with VDAC1 and functions as a downstream of VDAC1. (A) Flag-MCU was transfected into cells together with GFP-VDAC1 or empty vector. Twenty-four hours later, cells were collected and immunoprecipitated using anti-Flag M2 beads. (B) Normal or MCU knockdown HeLa cells were transfected with Flag-VDAC1 or empty vector. Twenty-four hours later, cells were treated with 200 μmol/L H_2O_2 for another 24 h and cleaved caspase-3 was analyzed. (C) Graph shows normalized level of cleaved caspase-3. (D) CGNs transfected with shMCU or shvector, Flag-VDAC1 or empty vector as indicated were treated with 70 μmol/L H_2O_2 for 24 h.

40× objective lens. Images were obtained using laser scanning confocal microscope. GCaMP3 was excited using the 488 nm line of an argon laser and detected at 530–550 nm. Ten to fifteen cells were randomly selected in each scan by drawing regions around individual cells, and the green fluorescence intensity was monitored sequentially. The results are representative of at least three independent experiments, and we have confirmed the reproducibility of these findings.

Calcium uptake in permeabilized HeLa cells

Mitochondrial Ca^{2+} uptake in permeabilized cells was tested as previously described (Sancak et al., 2013). Briefly, HeLa cells grown in 10 cm tissue culture plates were trypsinized and resuspended in 10 mL of DMEM. One million of each cell lines were transferred to microcentrifuge tubes, followed by spining down for 3 min at 800 ×*g* at room temperature. Then cells were washed with PBS once and resuspended in KCl buffer (125 mmol/L KCl, 2 mmol/L K_2HPO_4, 1 mmol/L $MgCl_2$, 20 mmol/L HEPES, pH 7.2), supplemented with 5 mmol/L glutamate and malate, 0.01% digitonin and 0.8 μmol/L Green-5N. Fluorescence was monitored every 0.2 s before and after addition of 50 μmol/L final concentration of Ca^{2+} at 27°C using a Thermo Scientific Varioskan Flash, filter sets (Ex506/Em532).

Immunoblot analysis

Cells were lysed in RIPA buffer (strong) (biyuntian, China). The concentration of total proteins was determined using BCA protein concentration detection kit (biyuntian, China). The same amount of proteins were loaded on SDS-PAGE and resolved by electrophoresis, followed by transferring onto NC membrane. Then, the membranes were blocked with blocking buffer (5% fat-free dry milk in TBST buffer) and incubated with primary antibodies and HRP labeled secondary antibodies, respectively. Lastly, specific proteins were visualized with enhanced ECL plus Western blotting substrate according the manufacturer's instructions (Thermo).

Apoptosis analysis

For HeLa cells, apoptotic cells were quantified by flow cytometry using an Annexin-V-FITC apoptosis detection kit (biyuntian) following the manufacturer's instructions. For CGNs, apoptotic cells were detected by nuclei staining using Hoechest 33342. Primary CGNs were isolated and seeded on glass piece. Seven days later, cells were transfected with plasmid or shRNA. Forty-eight hours later, cells were treated with 70 μmol/L H_2O_2 for 24 h, and then cells were fixed and stained with Hoechest 33342. Neuronal apoptosis assay was performed as described (Konishi and Bonni, 2003) by using the Zeiss imager D1 microscope.

Statistical analysis of data

Statistical data are presented as mean ± S.D. Significance was calculated by Student's *t*-test (* means $P < 0.05$, ** means $P < 0.01$).

ACKNOWLEDGEMENTS

This work was supported by a grant from Beijing Nature Science Foundation (Grant No. 7132147). We sincerely thank Professor Guangju Ji for kindly providing plasmid Mito-GCaMP3 for us.

ABBREVIATIONS

MCU, mitochondrial calcium uniporter; CGN, cerebellar granule neurons; VDAC, voltage-dependent anion-selective channel protein.

COMPLIANCE WITH ETHICS GUIDELINES

Yajin Liao, Yumin Hao, Hong Chen, Qing He, Zengqiang Yuan and Jinbo Cheng declare that they have no conflict of interest. All institutional and national guidelines for the care and use of laboratory animals were followed.

REFERENCES

Ahuja M, Muallem S (2014) The gatekeepers of mitochondrial calcium influx: MICU1 and MICU2. EMBO Rep 15:205–206

Alam MR, Groschner LN, Parichatikanond W, Kuo L, Bondarenko AI, Rost R, Waldeck-Weiermair M, Malli R, Graier WF (2012) Mitochondrial Ca^{2+} uptake 1 (MICU1) and mitochondrial Ca^{2+} uniporter (MCU) contribute to metabolism-secretion coupling in clonal pancreatic beta-cells. J Biol Chem 287:34445–34454

Baughman JM, Perocchi F, Girgis HS, Plovanich M, Belcher-Timme CA, Sancak Y, Bao XR, Strittmatter L, Goldberger O, Bogorad RL, Koteliansky V, Mootha VK (2011) Integrative genomics identifies MCU as an essential component of the mitochondrial calcium uniporter. Nature 476:341–345

Ben-Hail D, Shoshan-Barmatz V (2012) Anion-transport blockers inhibit VDAC1 channel conductance, oligomerization and apoptosis. Febs J 279:429–429

Brustovetsky N, Brustovetsky T, Jemmerson R, Dubinsky JM (2002) Calcium-induced cytochrome c release from CNS mitochondria is associated with the permeability transition and rupture of the outer membrane. J Neurochem 80:207–218

Chu ZL, Pio F, Xie Z, Welsh K, Krajewska M, Krajewski S, Godzik A, Reed JC (2001) A novel enhancer of the Apaf1 apoptosome involved in cytochrome c-dependent caspase activation and apoptosis. J Biol Chem 276:9239–9245

De Stefani D, Raffaello A, Teardo E, Szabo I, Rizzuto R (2011) A forty-kilodalton protein of the inner membrane is the mitochondrial calcium uniporter. Nature 476:336–340

Dong HL, Wang SL, Zhang ZW, Yu AL, Liu Z (2014) The effect of mitochondrial calcium uniporter opener spermine on diazoxide against focal cerebral ischemia–reperfusion injury in rats. J Stroke Cerebrovasc 23:303–309

Fernandez-Echevarria C, Diaz M, Ferrer I, Canerina-Amaro A, Marin R (2014) Abeta promotes VDAC1 channel dephosphorylation in neuronal lipid rafts. Relevance to the mechanisms of neurotoxicity in Alzheimer's disease. Neuroscience 278:354–366

Gincel D, Zaid H, Shoshan-Barmatz V (2001) Calcium binding and translocation by the voltage-dependent anion channel: a possible regulatory mechanism in mitochondrial function. Biochem J 358:147–155

Halestrap AP, Connern CP, Griffiths EJ, Kerr PM (1997) Cyclosporin A binding to mitochondrial cyclophilin inhibits the permeability transition pore and protects hearts from ischaemia/reperfusion injury. Mol Cell Biochem 174:167–172

Hall DD, Wu YJ, Domann FE, Spitz DR, Anderson ME (2014) Mitochondrial calcium uniporter activity is dispensable for MDA-MB-231 breast carcinoma cell survival. PLoS ONE 9:e96866

Keinan N, Pahima H, Ben-Hail D, Shoshan-Barmatz V (2013) The role of calcium in VDAC1 oligomerization and mitochondria-mediated apoptosis. Bba-Mol Cell Res 1833:1745–1754

Kinnally KW, Peixoto PM, Ryu SY, Dejean LM (2011) Is mPTP the gatekeeper for necrosis, apoptosis, or both? Biochimica et biophysica acta 1813:616–622

Konishi Y, Bonni A (2003) The E2F–Cdc2 cell-cycle pathway specifically mediates activity deprivation-induced apoptosis of postmitotic neurons. J Neurosci 23:1649–1658

Kumfu S, Chattipakorn S, Fucharoen S, Chattipakorn N (2012) Mitochondrial calcium uniporter blocker prevents cardiac mitochondrial dysfunction induced by iron overload in thalassemic mice. Biometals 25:1167–1175

Madesh M, Hajnoczky G (2001) VDAC-dependent permeabilization of the outer mitochondrial membrane by superoxide induces rapid and massive cytochrome *c* release. J Cell Biol 155:1003–1015

Mallilankaraman K, Cardenas C, Doonan PJ, Chandramoorthy HC, Irrinki KM, Golenar T, Csordas G, Madireddi P, Yang J, Muller M, Miller R, Kolesar JE, Molgo J, Kaufman B, Hajnoczky G, Foskett JK, Madesh M (2012a) MCUR1 is an essential component of mitochondrial Ca^{2+} uptake that regulates cellular metabolism. Nature Cell Biol 14:1336–1343

Mallilankaraman K, Doonan P, Cardenas C, Chandramoorthy HC, Muller M, Miller R, Hoffman NE, Gandhirajan RK, Molgo J, Birnbaum MJ, Rothberg BS, Mak DO, Foskett JK, Madesh M (2012b) MICU1 is an essential gatekeeper for MCU-mediated mitochondrial Ca(2+) uptake that regulates cell survival. Cell 151:630–644

Naranmandura H, Chen X, Tanaka M, Wang WW, Rehman K, Xu S, Chen Z, Chen SQ, Suzuki N (2012) Release of apoptotic cytochrome C from mitochondria by dimethylarsinous acid occurs through interaction with voltage-dependent anion channel in vitro. Toxicol Sci 128:137–146

Pan X, Liu J, Nguyen T, Liu CY, Sun JH, Teng YJ, Fergusson MM, Rovira II, Allen M, Springer DA, Aponte AM, Gucek M, Balaban RS, Murphy E, Finkel T (2013) The physiological role of mitochondrial calcium revealed by mice lacking the mitochondrial calcium uniporter. Nature Cell Biol 15:1464–1472

Pestana CR, Silva CH, Uyemura SA, Santos AC, Curti C (2010) Impact of adenosine nucleotide translocase (ANT) proline isomerization on Ca^{2+}-induced cysteine relative mobility/mitochondrial permeability transition pore. J Bioenerg Biomembr 42:329–335

Qiu J, Tan YW, Hagenston AM, Martel MA, Kneisel N, Skehel PA, Wyllie DJ, Bading H, Hardingham GE (2013) Mitochondrial calcium uniporter MCU controls excitotoxicity and is transcriptionally repressed by neuroprotective nuclear calcium signals. Nat Commun 4:2034

Raffaello A, De Stefani D, Sabbadin D, Teardo E, Merli G, Picard A, Checchetto V, Moro S, Szabo I, Rizzuto R (2013) The mitochondrial calcium uniporter is a multimer that can include a dominant-negative pore-forming subunit. EMBO J 32:2362–2376

Rapizzi E, Pinton P, Szabadkai G, Wieckowski MR, Vandecasteele G, Baird G, Tuft RA, Fogarty KE, Rizzuto R (2002) Recombinant expression of the voltage-dependent anion channel enhances the transfer of Ca2+ microdomains to mitochondria. J Cell Biol 159:613–624

Robb-Gaspers LD, Burnett P, Rutter GA, Denton RM, Rizzuto R, Thomas AP (1998) Integrating cytosolic calcium signals into mitochondrial metabolic responses. Embo J 17:4987–5000

Sancak Y, Markhard AL, Kitami T, Kovacs-Bogdan E, Kamer KJ, Udeshi ND, Carr SA, Chaudhuri D, Clapham DE, Li AA, Calvo SE, Goldberger O, Mootha VK (2013) EMRE is an essential component of the mitochondrial calcium uniporter complex. Science 342:1379–1382

Schwartz J, Holmuhamedov E, Zhang X, Lovelace GL, Smith CD, Lemasters JJ (2013) Minocycline and doxycycline, but not other tetracycline-derived compounds, protect liver cells from chemical hypoxia and ischemia/reperfusion injury by inhibition of the mitochondrial calcium uniporter. Toxicol Appl Pharmacol 273:172–179

Shoshan-Barmatz V, Zakar M, Rosenthal K, Abu-Hamad S (2009) Key regions of VDAC1 functioning in apoptosis induction and regulation by hexokinase. Biochimica et biophysica acta 1787:421–430

Shoshan-Barmatz V, Keinan N, Abu-Hamad S, Tyomkin D, Aram L (2010) Apoptosis is regulated by the VDAC1 N-terminal region and by VDAC oligomerization: release of cytochrome *c*, AIF and Smac/Diablo. Bba-Bioenergetics 1797:1281–1291

Sripetchwandee J, Kenknight S, Sanit J, Chattipakorn S, Chattipakorn N (2013a) Blocking mitochondrial calcium uniporter completely prevents cardiac mitochondrial dysfunction caused by iron overload. J Am Coll Cardiol 61:E705–E705

Sripetchwandee J, Sanit J, Chattipakorn N, Chattipakorn SC (2013b) Mitochondrial calcium uniporter blocker effectively prevents brain mitochondrial dysfunction caused by iron overload. Life Sci 92:298–304

Sripetchwandee J, KenKnight SB, Sanit J, Chattipakorn S, Chattipakorn N (2014) Blockade of mitochondrial calcium uniporter prevents cardiac mitochondrial dysfunction caused by iron overload. Acta Physiol 210:330–341

Xie Q, Hao Y, Tao L, Peng S, Rao C, Chen H, You H, Dong MQ, Yuan Z (2012) Lysine methylation of FOXO3 regulates oxidative stress-induced neuronal cell death. EMBO reports 13:371–377

Xu X, Decker W, Sampson MJ, Craigen WJ, Colombini M (1999) Mouse VDAC isoforms expressed in yeast: channel properties and their roles in mitochondrial outer membrane permeability. J Membr Biol 170:89–102

Zhao Q, Wang SL, Li Y, Wang P, Li SH, Guo YL, Yao RY (2013) The role of the mitochondrial calcium uniporter in cerebral ischemia/reperfusion injury in rats involves regulation of mitochondrial energy metabolism. Mol Med Rep 7:1073–1080

Zheng Y, Shi Y, Tian C, Jiang C, Jin H, Chen J, Almasan A, Tang H, Chen Q (2004) Essential role of the voltage-dependent anion channel (VDAC) in mitochondrial permeability transition pore opening and cytochrome *c* release induced by arsenic trioxide. Oncogene 23:1239–1247

4

A new unconventional HLA-A2-restricted epitope from HBV core protein elicits antiviral cytotoxic T lymphocytes

Lu Sun[1], Yu Zhang[1], Bao Zhao[1], Mengmeng Deng[1], Jun Liu[1], Xin Li[1], Junwei Hou[1], Mingming Gui[3], Shuijun Zhang[1], Xiaodong Li[2], George F. Gao[1✉], Songdong Meng[1✉]

[1] CAS Key Laboratory of Pathogenic Microbiology and Immunology, Institute of Microbiology, Chinese Academy of Sciences (CAS), Beijing 100101, China
[2] Beijing Institute of Infectious Diseases, Beijing 302 Hospital, Beijing 100039, China
[3] Xinjiang Agricultural University, Ürümqi 830052, China
✉ Correspondence: gaof@im.ac.cn (G. F. Gao), mengsd@im.ac.cn (S. Meng)

ABSTRACT

Cytotoxic T cells (CTLs) play a key role in the control of Hepatitis B virus (HBV) infection and viral clearance. However, most of identified CTL epitopes are derived from HBV of genotypes A and D, and few have been defined in virus of genotypes B and C which are more prevalent in Asia. As HBV core protein (HBc) is the most conservative and immunogenic component, in this study we used an overlapping 9-mer peptide pool covering HBc to screen and identify specific CTL epitopes. An unconventional HLA-A2-restricted epitope HBc141 149 was discovered and structurally characterized by crystallization analysis. The immunogenicity and anti-HBV activity were further determined in HBV and HLA-A2 transgenic mice. Finally, we show that mutations in HBc141–149 epitope are associated with viral parameters and disease progression in HBV infected patients. Our data therefore provide insights into the structure characteristics of this unconventional epitope binding to MHC-I molecules, as well as epitope specific CTL activity that orchestrate T cell response and immune evasion in HBV infected patients.

KEYWORDS chronic hepatitis B, HLA-A2, HBc peptide, CTL response, antiviral cytotoxity

Lu Sun and Yu Zhang contributed equally to this paper.

INTRODUCTION

Around 350 million people worldwide are chronically infected with Hepatitis B virus (HBV). Chronic HBV infection is a major cause of cirrhosis, liver failure, and hepatocellular carcinoma (HCC) (Lavanchy, 2005; Neuveut et al., 2010). HBV-specific $CD8^+$ T lymphocytes (CTL)-mediated immune response is multi-specific, polyclonal, and vigorous during acute hepatitis B (AHB), which plays a vital role in viral control and viral clearance, as well as disease pathogenesis (Yukihiro, 2012; Westover and Hughes, 2007; Tan et al., 2008). Whereas HBV-specific CTL response is minimal or undetectable in chronic hepatitis B (CHB) with viral persistence and immune tolerance, indicating the key role of HBV-specific T-cell response in determination of disease progression and outcome (Bertoletti and Gehring, 2006; Zhang et al., 2013).

The HBV genome of ~3.2 kb in length efficiently encodes several overlapping viral proteins, including the polymerase, core, HBe, envelope (Pre-S1, S2, S), and X proteins. Analysis of CTLs specific for viral epitopes within core (Sendi et al., 2009; Liu et al., 2012;), envelope (Liu et al., 2008), polymerase (Rehermann et al., 1995), and X (Hwang et al., 2002) proteins showed that the highly conserved HBV core protein (HBc) elicits the strongest CTL responses than other viral proteins, suggesting that HBc-specific T cell response may play a leading role in viral control and clearance. To identify immune-dominant HBV-specific CTL epitopes, especially epitopes from HBc protein, is therefore necessary for monitoring T cell responses during disease progression, as well as for developing epitope-based therapeutic vaccines against CHB (Inchauspe and Michel, 2007; Gordon et al., 2013; Liu et al., 2013a, b).

So far 60 HBV-specific human leukocyte antigen (HLA) class I restricted and 32 HBV-specific HLA class II restricted epitopes have been identified in all 8 HBV genotypes (Desmond et al., 2008; Liu et al., 2008; Guo et al., 2011; Zhang et al., 2013; Chen et al., 2013; Tan et al., 2013). However, most of these identified epitopes are derived from HBV of genotypes A and D, few HLA class I restricted epitopes have been defined in virus of genotypes B and C, which are more prevalent in Asia. Meanwhile, currently the mostly used methods to predict and identify T cell epitopes are either by computer algorithms based on the mode of peptide binding to MHC molecules, or measurement of T cell responses of PBMCs simulated with panels of overlapping peptides (Liu et al., 2011). These epitope identification strategies may ignore unconventional T cell epitopes as computational analysis is largely based on the characteristic of anchor residues.

In this study, an overlapping 9-mer peptide pool was used to screen and identify HBV genotypes B- and C-derived T cell epitopes. A new unconventional $CD8^+$ T cell epitope HBc141–149 derived from viral core protein, which shares partial sequence identity with previously reported HBc141–151 (Bertoni et al., 1997) and HBc139–148 (Lee et al., 1997), was discovered. Its immunological function and clinical relevance were further assessed.

RESULTS

Identification of a new HLA-A*0201-restricted epitope from HBV core protein

HBV core protein is the most conservative and immunogenic component of HBV proteins. We used an overlapping 9-mer peptide pool covering the whole length of core protein (aa 1–150) and its variants (totally 191 peptides) for T2 binding assay to screen genotype B- and C virus-derived HBcAg-specific T cell epitopes (Zhang et al., 2013), as shown in Fig. 1A. Several peptides were found to have high affinity for binding to HLA-A*0201 molecules, as evidenced by the FI (0.72 for HBc183–191, 2.41 for HBc141–149, 2.64 for HBc60–68 (V60), 2.47 for HBc18–27, and 0.04 for HBc82–90), as shown in Fig. 1B. HBc141–149 spanning from HBc aa 141 to 149 was chosen as the focus of this study due to its high binding affinity among the top hits from screening. To determine the minimal sequence length of the epitope, panels of N- or C-terminally truncated or extended peptides of HBc141–149 were synthesized. The FIs of truncated peptide HBc141–148 (STLPETTV) and HBc142–149 (TLPETTVV) were only 0.124 and 0.151, respectively (Fig. 1C). The FIs of decapeptides HBc140–149 (LSTLPETTVV) and HBc141–150 (STLPETTVVR) were only 0.033 and 0.005, respectively (Fig. 1D). All HBc141–149 truncated and extended peptides displayed little binding to HLA-A*0201, which indicates that the 9-mer peptide HBc141–149 is the optimized epitope in length. In addition to T2 cell binding assay, the capability of HBc141–149 to bind to HLA-A*0201 was observed in the refolding assay (Fig. 1E).

The structure of HLA-A*0201/HBc141–149 complex

Next, the complex of HLA-A2 and HBc141–149 was prepared for crystallization to characterize the binding features of HBc141–149. The crystal structure of the HLA complex was determined to 2.3 Å resolution (Table 1). The structure of HLA-A*0201/HBc141–149 shows that HBc141–149 possesses a typical conformation of an HLA-A2-restricted 9-mer epitope (Fig. 2A and 2B). The unambiguous electron densities of HBc141–149 clearly show that position 2 (T2) and position 9 (V9) are buried in pockets B and F, respectively (Fig. 2C). Compared to the typical HLA-A2-restricted epitopes which have an anchor residue Leu or Met at position 2, HBc141–149 has Thr at position 2. The side chain OH of Thr does not disrupt its inserting into the hydrophobic pocket B properly (Fig. 2D). Instead, the side chain OH of Thr can form a strong hydrogen bond interaction with H atom of Glu on the α1 domain of the heavy chain, which helps HBc141–149 binding to the HLA-A2 heavy chain and stabilizes the entire complex. The side chains of amino acids P4, E5, and V8 protrude out from the HLA-A2 surface and may be involved in T cell receptor (TCR) attachment and recognition. To the best of our knowledge, this is the first structure showing the binding features of HLA-A2 to HBc141–149 with an unconventional P2 anchor Thr.

HBc141–149 peptide generates specific CTL response in HLA-A2.1/Kb transgenic mice

Then, the immunogenicity of HBc141–149 epitope was determined *in vivo*. Female HLA-A2.1/Kb transgenic mice were immunized with HBc DNA-prime/HBc141–149 peptide boost regimen three times, using heat shock protein gp96 as adjuvant (Li et al., 2011). HBc141–149-specific CTL was detected by ELISPOT assay 1 week after the last immunization. As can be seen in Fig. 3A, similar to the HBc18–27 peptide-immunized mice (positive control), a strong CTL response was observed in splenocytes from HBc141–149 peptide-immunized mice (SFC, 145.4 ± 58.6). No peptide-specific $CD8^+$ T cell response was detected from HBc82–90 peptide-immunized mice (negative control). Similar results were observed in the killing assay using HBV plasmid-transfected 293T cells (Fig. 3B) or HBc141–149 peptide-pulsed T2 cells (Fig. 3C) as target cells.

HBc141–149 is a naturally processed epitope in AHB

To further determine the epitope-specific CTLs, fresh PBMCs from $HLA\text{-}A2^+$ AHB patients were stimulated with HBc141–149 peptide and detected by *ex vivo* IFN-γ ELISPOT assays. As shown in Fig. 3D, a much higher peptide-specific CTL

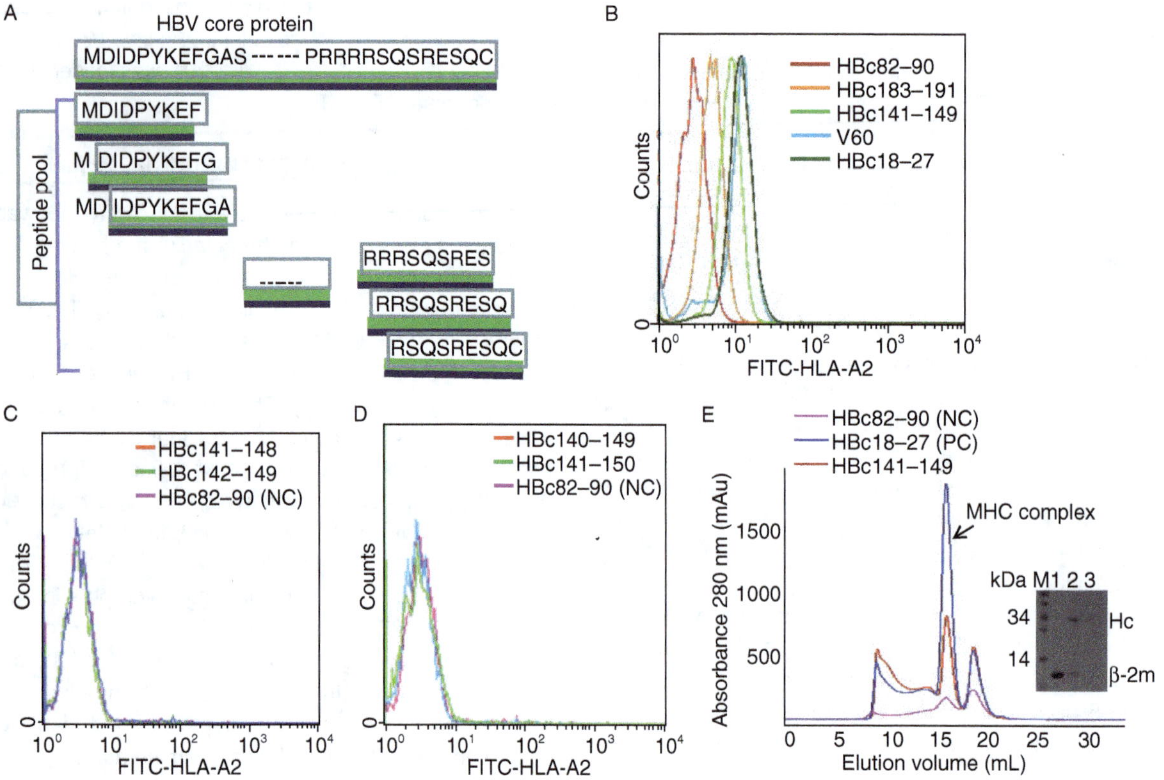

Figure 1. Binding affinity of HBc141–149 peptide to HLA-A*0201. (A) The overlapping method of 8 aa overlap was used to synthesize a total of 191 nonapeptides covering HBc aa 1–150. (B) MHC stabilization assay with T2 cells was used to quantify peptide binding affinity of HBc18–27, HBc82–90, HBc183–191, HBc141–149, and HBc60–68 (V60) to HLA-A*0201 molecules by flow cytometry, respectively. HBc18–27 and HBc82–90 peptides served as positive and negative controls, respectively. The results shown are representative of three independent experiments. (C and D) Measurement of the binding affinity of HBc141–149N- or C-terminally truncated (C) or extended (D) peptides to HLA-A*0201. HBc82-90 served as a negative control (NC). (E) *In vitro* refolding of HBC141–149 peptide with HLA-A*0201 heavy chain and β2m. Gel filtration chromatography was used to analyze the refolded complexes on a Superdex200 16/60 column. The HLA complex with the expected molecular mass of 45 kDa eluted at the volume of 15.9 mL. The HLA complex (peak 2) was analyzed by SDS-PAGE electrophoresis and coomassie blue staining (line 2).

response was observed in PBMCs stimulated with HBc141–149 or the immunodominant peptide HBc18–27 as the positive control than negative control peptide HBc82–90 (119.68 ± 76.66 for HBc141–149 or 164.58 ± 112.41 for HBc18–27 vs. 20.94 ± 17.57 for HBc82–90, both $P < 0.01$). The high standard deviations observed in ELISPOT assay may reflect the random between-patient variation. Taken together, these results indicate that HBc141–149 is an HLA-A2-restricted CD8$^+$ T cell epitope and is naturally processed in patients with HBV infection.

HBc141–149 epitope elicits antiviral T cell immunity in HLA-A2.1/HBV transgenic mice

Next, we examined whether HBc141–149 epitope was able to induce anti-HBV T cell response using F1 hybrids of HBV transgenic BALB/C mice and HLA-A2.1/kb transgenic mice as the experimental model, which are HBV immunotolerant. HLA-A2.1/HBV transgenic mice were immunized with an HBc DNA prime/ HBc141–149 peptide boost formulation. As shown in Fig. 4A, compared to mice immunized with negative control peptide, the number of SFCs increased by around 6-fold in HBc141–149 peptide-immunized mice. Similar result was obtained in cytotoxicity assay using HBV plasmid-transfected 293T cells as target cells. We then evaluated HBc141–149 peptide-induced T cell response could lead to inhibition of HBV. Immunization with HBc141–149 led to a 35.5% decrease in serum HBsAg levels ($P < 0.05$) (Fig. 4C), and a significant reduction of viral DNA levels ($P < 0.05$) (Fig. 4D) at week 8 compared to the negative control. Meanwhile, significant lower levels of serum HBsAg and viral DNA were observed after immunization (at week 8) than those before immunization (at week 0) in HBc141–149-treated mice but not in control peptide-treated mice. And these results indicate that the epitope-specific CTL response induced by the HBc141–149 could significantly inhibit HBV replication in the transgenic mice.

Table 1. X-ray diffraction data processing and refinement statistics

	HLA-A2/HBc141–149
Data processing	
Space group	$P2_1$
Cell parameters (Å)	
a (Å)	52.980
b (Å)	80.787
c (Å)	56.488
α (°)	90.000
β (°)	112.596
γ (°)	90.000
Resolution range (Å)	50.00–2.30
Total reflections	96418
Unique reflections	19518
Completeness (%)	99.8 (99.2)
Redundancy	4.9 (4.3)
R_{merge} (%)	0.123 (0.435)
I/σ	12.868 (2.939)
Refinement	
R_{work} (%)	0.2142
R_{free} (%)	0.2651
r.m.s. deviation	
Bond lengths (Å)	0.002
Bond angles (°)	0.627

Values in parentheses are given for the highest resolution shell.

HBV accumulates mutations within HBc141–149 epitope in HBV-infected patients

Finally, to address the clinical relevance of the HBc141–149 epitope in CHB, mutations within this epitope were analyzed in 197 CHB and 64 ACLF patients (Table 2). In CHB, compared to patients infected with wild-type isolates (HBc141–149, STLPETTVV), patients infected with HBc141–149 mutants had much higher ALT levels (247.7 ± 18.62 vs. 545.2 ± 137.7, $P < 0.05$) (Fig. 5A) and aspartate aminotransferase (AST) levels (187.6 ± 13.78 vs. 1668 ± 700.5, $P < 0.05$) (Fig. 5B). Notably, compared to CHB patients infected with HBc141–149 wild-type viruses, patients with HBc141–149 mutants had much higher (approximately 3.3-fold) HBV DNA loads ($P < 0.05$) (Fig. 5C). There were no statistical differences in sex and age between patients infected with the wild-type and mutant isolates. The prevalence of the HBc141–149 mutations increased with the disease progression in HBV-infected patients (Fig. 5D). To investigate the impact of these sequence variations within HBc141–149 epitope on its immunogenicity, three 9-mer peptides containing main epitope variations of HBcV149I, HBcT147A, and HBcT147C in CHB patients were synthesized, respectively, for HLA-A2.1/Kb transgenic mice immunization. As seen in Fig. 5E, compared to the wild-type HBc141–149 peptide-immunized mice (SFC, 96.7 ± 6.02), CTL responses by ELISPOT assay were significantly decreased in HBcV149I mutant peptide- (SFC, 66.3 ± 5.69) or HBcT147A mutant peptide- immunized mice (SFC, 73 ± 7.81). Taken together, these results suggest that the HBc141–149 mutations associated with necroinflammation and higher HBV levels, and it may also be associated with poor prognosis, which may be due to viral immune evasion.

DISCUSSION

In this study, we identified a new HLA-A2-restricted $CD8^+$ T cell epitope HBc141–149 by screening an overlapping 9-mer peptide pool covering HBV core protein. This unconventional HLA-A2 restricted epitope was further determined and structurally characterized by HLA-A2 transgenic mouse model and crystallographic analysis. Moreover, we demonstrated that the HBc141–149 epitope exhibits antiviral capability in HLA-A2.1/HBV transgenic mice. Finally, our results show that mutations in HBc141–149 epitope correlate with clinically relevant parameters in CHB. Our work may therefore provide a comprehensive evaluation of the impact of the newly defined epitope on viral specific T cell response and suggests a possible immune evasion for maintenance of viral persistence in patients with HBV infection.

The identification of HBV-specific T cell epitopes is mostly based on the binding mode of the peptides to MHC molecules in silico prediction or assessing T cell responses with panels of overlapping peptides in HBV-infected patients. These methods are effective and rapid to purposefully identify epitopes, however, a considerable number of atypical or unconventional epitopes may be ignored or missed by these methods due to the limitation of anchor residues analysis-based computation (Liu et al., 2011). In this study, an overlapping 9-mer peptide pool was used to screen B and C genotype-derived HBc-specific CTL epitopes, and HBc141–149 was identified as a new 9-mer HLA-A2-restricted T cell epitope. Importantly, similar numbers of HBc141–149-specific and immunodominant epitope HBc18–27-specific $CD8^+$ T cells were observed in AHB patients (Fig. 3C), indicating that HBc141–149 is an immunodominant epitope in HBV-infected patients. However, the newly defined epitope HBc141–149 (STLPETTVV) in this study possesses Thr at position 2. Different T cell epitope prediction programs, including SYFPEITHI (http://www.syfpeithi.de/Scripts/MHCServer.dll/Epitope-Prediction.htm), NET-MHC (http://www.cbs.dtu.dk/services/NetMHC/), and BIMAS (http: //www-bimas.cit.nih.gov) show the binding score of the epitope HBc141–149 to HLA-A2 is only 0, 1.465, and 0.414, respectively. As HBc141–149 does not get a high score, it could be omitted in conventional screening. Interestingly, as for HBc141–149 (STLPETTVV), the side chain OH of Thr2 can form a stronger hydrogen bond interaction with Glu on α1 domain of the heavy chain, which helps its inserting into

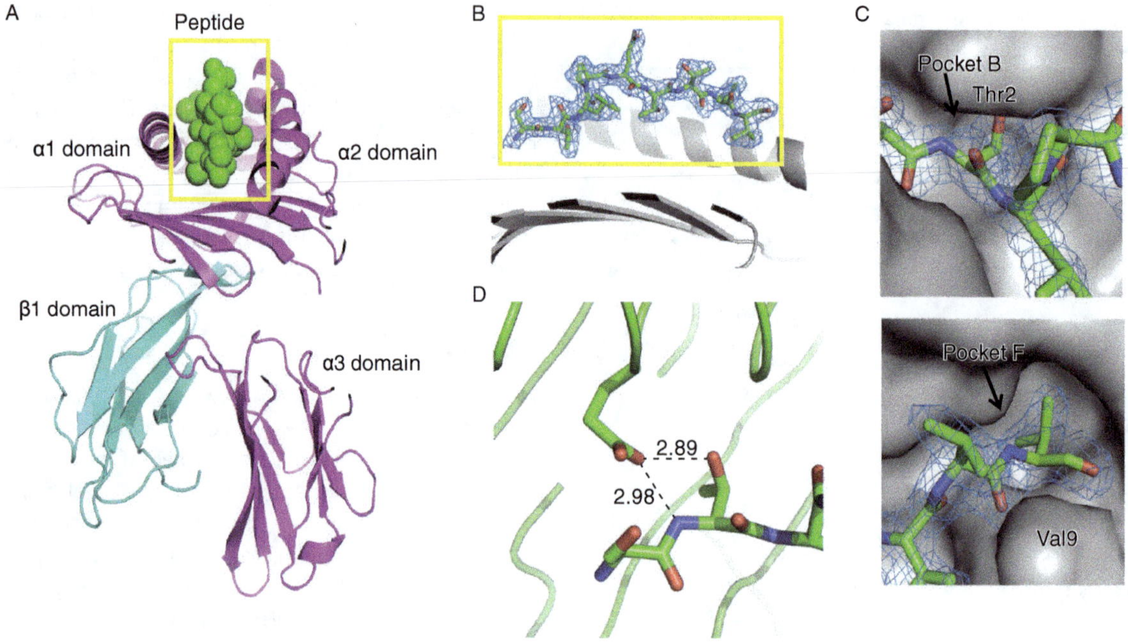

Figure 2. Structure of HLA-A*0201 complexed with HBc141–149. (A) Overview of the structure of HBc141–149 peptide presented by HLA-A*0201; HBc141–149 is presented in the peptide-binding cleft. (B) HBc141–149 peptide is presented in the peptide-binding clefts with electron density at the 1σ contour level. (C) Residues Thr2 and Val9 (green sticks) are located in pocket B and pocket F of HLA-A*0201. The heavy chain of HLA-A*0201 is show in gray surfaces. (D) The side chain OH of Thr forms hydrogen bond interaction with H atom of Glu on the alpha 1 domain of the heavy chain.

pocket B (Fig. 2D) of HLA-A*0201 and stabilize the entire complex. Therefore, our data indicated that Thr may also act as a dominant role as a P2 anchor for HLA-A2-binding peptides, which may be taken into consideration in the future designation of these online prediction programs.

CD8$^+$ T cells are the main effector cells responsible for viral clearance as well as disease pathogenesis during HBV infection (Thimme et al., 2003; Harari et al., 2006; Ouyang et al., 2013). As a consequence of long-term interaction between HBV and infected patients, the virus evolves mutations to reduce epitopes for evading immune detection and clearance, especially escape from T cell recognition (Maman et al., 2011; Westover and Hughes, 2007). Indeed, nonsynonymous mutations in HBV epitope have been found to be associated with disease progression in CHB (Kim et al., 2011; Frelin et al., 2009). Consistent to these studies, we found that mutations in the newly defined epitope are positively related to viral parameters and pathogenesis of liver disease. The effect of mutations within HBc141–149 on viral replication capability, as well as the potential impact of the epitope-specific CTLs on the interplay of HBV and chronically infected patients remains to be addressed.

In summary, this study indicates that HBc141–149 is an immunodominant HLA-A2- restricted CTL epitope with atypical binding characteristics to HLA-A2 molecules, has potent anti-HBV immune activity and the clinical significance in patients with HBV infection. We further demonstrated that mutations within this epitope may affect disease progression in CHB. Given the key role of CD8$^+$ T cells in viral clearance, our work provides valuable insight for the functional implications of HBc141–149 epitope-specific T cell response in HBV infection. Understanding the epitope-specific CTL function in the complex regulation networks that orchestrate T cell response, viral persistence and immunoevasion in CHB will allow to predict disease progression and develop immunotherapeutic approaches against HBV infection.

MATERIALS AND METHODS

Patients and blood samples

The database of HBV infected patients' records built by the 302 Hospital (Beijing, China) (http://www.hbvdb.com) was used to perform the Clinical information analyses of HBc141–149. In this study we investigated the associations between HBc141–149 and the progress of liver disease or patients' clinical indicators. The criteria for diagnosis of acute hepatitis B (AHB), chronic hepatitis B (CHB), acute on chronic liver failure (ACLF) complied with those of the Management Scheme of Diagnostic and Therapy of Viral Hepatitis and the Diagnostic and Treatment Guideline for Liver Failure, was issued by the Chinese Society of Infectious Diseases and Parasitology and the Chinese Society of Hepatology. CHB is further divided into two groups: mild chronic hepatitis B (CHB-M) and severe chronic hepatitis B (CHB-S). The criteria for CHB-M patients: mild-to-moderate liver disease activities which do not attain the criteria of

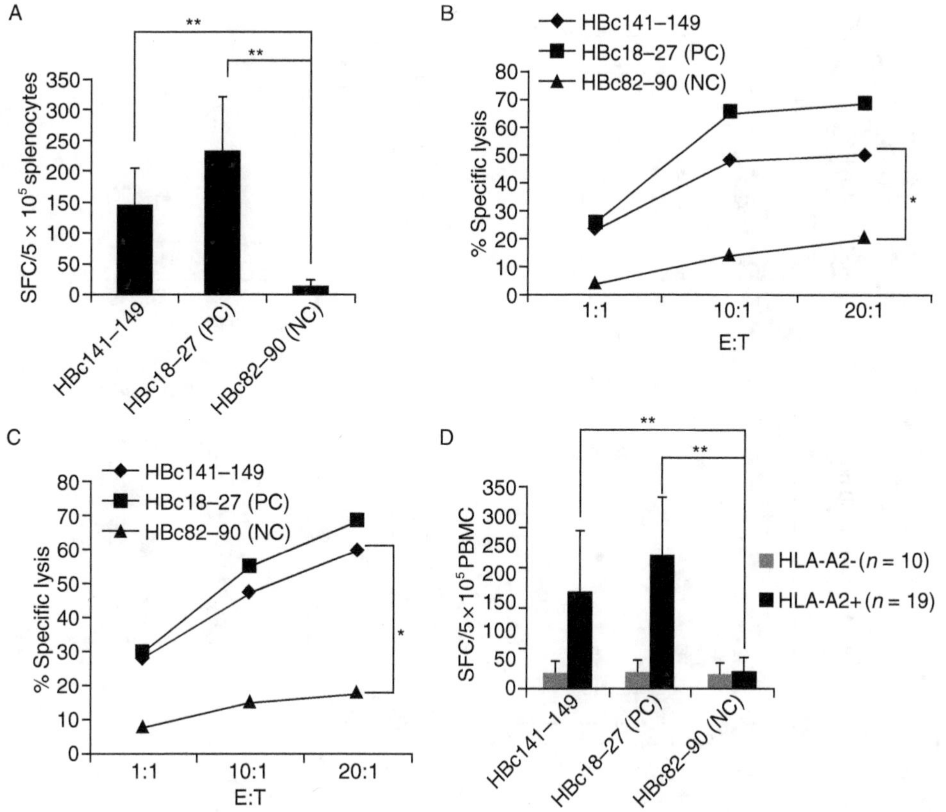

Figure 3. Detection of HBc141–149 peptide-specific CD8$^+$ T cell response in HLA-A2.1/Kb mice and AHB patients. Female HLA-A2.1/Kb transgenic mice were immunized with an HBV DNA prime/peptide boost regimen at weeks 1, 3, and 4. HBc82–90 and HBc18–27 peptides served as negative and positive controls in immunization, respectively. (A) Fresh splenocytes (5×10^5) isolated from immunized mice were stimulated with HBc141–149, HBc18–27 or HBc82–90 peptide, and peptide-specific CTLs were detected by IFN γ ELISPOT assay, (B and C) 293T cells labeled with CFSE were transfected with pHBV1.3 as target cells (B), and T2 cells labeled with CFSE were loaded with 20 μg/mL peptide at 37°C for 1 h as target cells (C). The target cells were then mixed with HBc141–149-stimulated splenocytes at different ratios: 1:1, 1:10, and 1:20, or HBc18–27, HBc82–90-stimulated splenocytes served as positive or negative controls. After 4 h, the mixed samples were stained with PI, and the killing of target cells were analyzed by FACS. The data shown are the mean ± SD of five mice. (D) AHB patients (*n* = 29) were divided into HLA-A2-positive (*n* = 19) and HLA-A2-negative (*n* = 10) groups. PBMCs (2×10^5/well) from these patients were stimulated with the indicated peptides for detection of peptide-specific CTLs by ELISPOT assay. HBc82–90 peptide served as negative control for background evaluation in ELISPOT assay. PBMCs from HLA-A2- patients were used for specificity evaluation of CTLs. Paired samples *t*-tests were used for ELISPOT assay in AHB patients. *$P < 0.05$ and **$P < 0.01$ by *t*-test. Data are representative of two independent experiments.

CHB-S. The criteria for CHB-S patients: severe liver disease with evident clinical manifestation and significant alteration of biochemical parameters (serum albumin level ≤32 g/L, serum total bilirubin (TBIL) >85.5 μmol/L, plasma prothrombin activity (PTA) 40%–60%, or serum cholinesterase <4500 IU/L (Xu et al., 2011). The criteria for ACLF: a history of CHB with symptoms of extreme fatigue and severe digestive problems with biochemical parameters of TBIL >10 times normal level (171 μmol/L) and PTA ≤40% (Yang et al., 2002). All the enrolled patients' clinical characteristics are described in Table 2.

A total of 29 AHB patients were enrolled for blood collection, which were divided into two groups: HLA-A2-positive (*n* = 19) and HLA-A2-negative (*n* = 10). All patients were negative for HCV, HDV, and HIV-1 infection. 10 mL of blood samples were collected from each patient. All patients were hospitalized in Beijing 302 Hospital from September 2010 to September 2011. All study participants have written informed consent and the study was approved by the Ethics Committee of Beijing 302 Hospital.

Mice

HBV transgenic BALB/c mice were purchased from Transgenic Engineering Lab, Infectious Disease Center (Guangzhou, China), which were generated with a viral DNA construct, pHBV1.3, containing 1.3 copies of the HBV genome. Serum HBV s antigen (HBsAg) and viral DNA, as well as HBc expression in hepatocytes in mice's liver, were tested positive for all transgenic mice. The HLA-A2.1/Kb transgenic mice (Zhang et al., 2007) were kindly provided

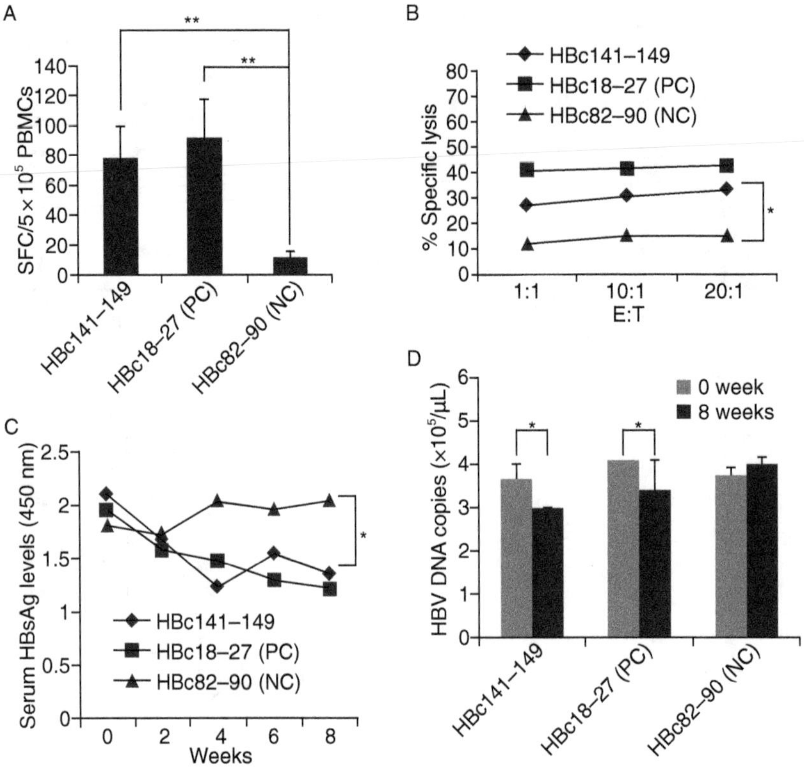

Figure 4. Immunization with HBc141–149 peptide induces specific CTL responses inhibits viral replication in HLA-A2^{+}/HBV transgenic mice. F1 hybrids of HLA-A2.1/kb transgenic mice and HBV transgenic BALB/c mice (n = 5 in each group) were immunized three times as in Fig. 3. (A) ELISPOT assay. Splenocytes from immunized mice were stimulated with HBc141–149 peptide, or with HBc82–90 or HBc18–27 as negative or positive controls. (B) 293T cells labeled with CFSE were transfected with pHBV1.3 as target cells, and mixed with HBc141–149-stimulated splenocytes at different ratios: 1:1, 1:10, and 1:20, or with HBc18–27 or HBc82–90-stimulated splenocytes served as positive or negative controls . After 4 h, the mixed samples were stained with PI and analyzed by FACS. (C) Serum secretion of HBsAg was detected by ELISA every other week. (D) HBV DNA levels in serum were quantified by real-time PCR at weeks 0 and 8. The data shown are the mean ± SD of five mice. Paired t-tests were used. *P < 0.05 and **P < 0.01. Data are representative of two independent experiments.

by Professor Huang WL (IMCAS, Beijing, China). The HLA-A2.1/Kb mice and HBV transgenic mice were crossed to gain the F1 hybrids of HLA-A2.1/HBV transgenic mice. All F1 hybrids were screened for serum HBsAg by ELISA, viral DNA by real-time PCR, and HLA-A2 by PCR-SSP (PROTRANS, Deutschland) before experimental manipulations.

Plasmid constructs

The wild-type HBc gene was cloned into pcDNA3.1 (Invitrogen) and the recombinant plasmid was designated pcDNA3.1-HBc. pHBV1.3 containing 1.3 copies of the full-length HBV genomic sequence was maintained in the lab.

Peptide synthesis

The HBc sequences of genotypes B and C were attained using the protein database from NCBI and a total of 171 HBc sequences of genotype B and 159 HBc sequences of genotype C included. The sequences were compared and served as a basis on peptide synthesis. If the variation rate of a certain amino acid (aa) was more than 10%, a series of peptides associated with this variation would be synthesized. We adopted the overlapping method (8-aa overlap) to synthesize a total of 191 nonapeptides (9-mers) covering HBc1–150 aa (Zhang et al., 2013). All of these peptides were synthesized at Jier Biological (Shanghai, China), and their purity was determined as >95%. The HBc18–27 (FLPSDFFPSV) and HBc82–90 (RELVVSYVN) peptides were used as positive and negative controls, respectively.

Cell culture and transfection

The 293T cell line was obtained from the ATCC (Manassas, VA, USA) and cultured in Roswell's Park Memorial Institute medium supplemented with 10% fetal bovine serum. The T2 cell line was kindly provided by Professor BinGao (IMCAS, Beijing, China) and cultured in RPMI 1640 with 10% fetal bovine serum (Gibco BRL, Paisley, UK). The transfection reagent Lipo2000 (Invitrogen, USA) were used in cell transfection according to the instruction provided

Table 2. Clinical and virologic characteristic of CHB-M, CHB-S, and ACLF patients

	CHB-M	CHB-S	ACLF	*P*-Value
Numbers	95	102	64	
Age (mean ± SD)	42.51 ± 13.7	43.37 ± 14.57	45.89 ± 12.09	$P < 0.01$
Gender (M/F)	12/83	18/84	13/42	$P > 0.05$
HBV genotype				
B (%)	17 (17.9%)	17 (16.7%)	5 (7.8%)	$P > 0.05$
C (%)	78 (82.1%)	85 (83.3%)	59 (92.2%)	$P < 0.01$
HBc141–149 variation	3 (3.1%)	12 (11.8%)	9 (14.1%)	$P < 0.05$
HBe positive (%)	72 (86.3%)	49 (47.1%)	6 (9.3%)	$P < 0.01$
HBV DNA copies/Ml (mean)	2.53×10^6	8.26×10^6	1.31×10^6	$P > 0.05$
(range)	($2354–3.5 \times 10^7$)	($500–2.64 \times 10^8$)	($5980–1.32 \times 10^7$)	
ALT(IU/L) (mean ± SD)	140.2 ± 186.7	284.5 ± 283.6	211.8 ± 244.1	$P < 0.01$
AST(IU/L) (mean ± SD)	87.7 ± 91.0	322.3 ± 932.2	238 ± 257.9	$P < 0.01$

HBV, hepatitis B virus; CHB-M, mild chronic hepatitis B; CHB-S, severe chronic hepatitis B; ACLF, acute-on-chronic liver failure; M, male; F, female; HBeAg, hepatitis B e antigen; ALT, alanine aminotransferase; AST, aspartate aminotransferase.

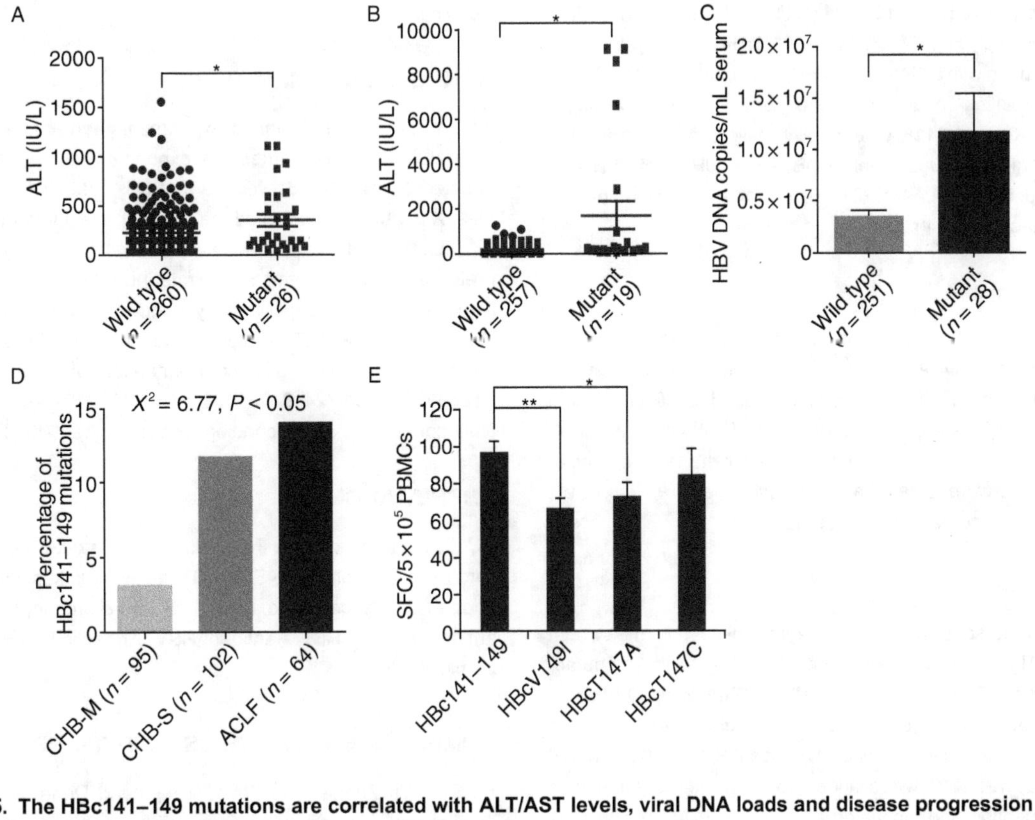

Figure 5. The HBc141–149 mutations are correlated with ALT/AST levels, viral DNA loads and disease progression in HBV-infected patients. (A–C) Distribution of serum ALT (A) and AST levels (B), and HBV DNA loads (C) in CHB patients infected with HBV isolates containing wild-type (STLPETTVV) or mutant HBc141–149. Student's *t*-test was used to determine *P*-values. $^*P < 0.05$. (D) Distribution of the HBc141–149 mutation rate in CHB and ACLF patients. CHB-M, mild chronic hepatitis B; CHB-S, severe chronic hepatitis B; ACLF, acute-on-chronic liver failure. Pearson's χ test was used to determine *P*-values. (E) Female HLA-A2.1/Kb transgenic mice were immunized with the wild-type HBc141–149 or mutant (HBcV149I, HBcT147A or HBcT147C) peptide as in Fig. 3. Fresh splenocytes (5×10^5) isolated from immunized mice were stimulated with wild-type or mutant peptide, respectively. Peptide-specific CTLs were detected by IFN-γ ELISPOT assay.

by the manufacturer. Cells and supernatants were harvested at 24 h, 48 h, and 72 h after transfection, respectively.

T2 binding assay

T2 cells were used to perform MHC stabilization assays as previously described (Zhou et al., 2006). The binding activity of each peptide was calculated as the fluorescent index (FI), and the FI was determined by: (mean FITC fluorescence with the given peptide − mean FITC fluorescence without peptide) (mean FITC fluorescence without peptide). Peptides regarded as epitopes with high affinity should meet the following criteria: FI ≥1.

Refolding, protein crystallography, and structure determination

Recombinant proteins of HLA-A*0201 heavy chain and β2m were expressed in *Escherichia coli* (Garboczi et al., 1992.) and the gradual dilution method was performed during refolding process (Liu et al., 2012). Then, Superdex 200 10/300 GL gel filtration chromatography followed by Resource-Q anion-exchanged chromatography (GE Healthcare) was used for the concentration and purification of the soluble portion of the refolded complex. The hanging-drop vapor diffusion method was performed at 18°C to crystallize the purified complexes. At a final concentration of 10 mg/mL in 0.1 mol/L Bis-Tris (pH 6.5) and 25% (*w*/*v*) polyethylene glycol 3350, HLA-A0201/HBc141–149 crystals were obtained. Equipped with an R-AXIS V||++ image-plate detector, Rigaku MicroMax007 rotating-anode X-ray generator was operated at 40 kV and 20 mA (Cu Kα; λ = 1.5418 Å) to collect crystallographic data at 100 K in house. With Protein Data Bank (PDB) entry 1JF1 as the search model, the structure of HLA-A*0201/HBc141–149 was determined by molecular replacement with the program MolRep.

Immunization of mice

Mice (6–8 weeks old) were immunized i.m. with 50 μg of plasmid pcDNA3.1-HBc or pcDNA3.1 (control) at week 1 and subcutaneously with 50 μg of HBc141–149 peptide bound to 30 μg of heat shock protein gp96 as adjuvant (Li et al., 2011) at weeks 3 and 4, respectively. Mice were sacrificed 1 week after the last immunization. Splenocytes were isolated as previously described (Liu et al., 2009). Each group contained five to seven mice.

IFN-γ ELISPOT assay

To detect epitope-specific T cells, enzyme-linked immunosorbent spot (ELISPOT) assay was performed according to the manufacturer's instruction. Briefly, ninety-six well PVDF plates (BD-Pharmigen, San Diego, CA) were precoated overnight at 4°C with the coating Ab and blocked for 1 h at 37°C. Patient PBMCs (2×10^5) or murine splenocytes (10^6) were added to each well together with 50 μg/mL of peptide and incubated at 37°C for 24–48 h with phytohemagglutinin (PHA)-stimulated T cells as positive controls. Each test was performed at least in triplicate. The spots were counted and analyzed using an ELISPOT Reader (Cellular Technology Ltd, USA).

Cytotoxicity assay

293T cells were labeled with 2 μmol/L CFSE as target cells after transfection with pHBV1.3 and seeded into a 96-well plate. Then CTLs were added at different ratios: 1:1, 10:1, and 20:1. Plates were incubated for 4–6 h at 37°C, and CFSE positive target cells were stained with propidium iodide (PI) using a Vybrant Apoptosis Assay Kit (Invitrogen, USA). In addition, T2 cells were loaded with 20 μg/mL HBc141–149 peptide at 37°C for 1 h as target cells, and seeded into a 96-well plate. The cytotoxicity assay was performed as described above. Each assay was performed in triplicate.

Detection of HBsAg and HBeAg by ELISA, and HBV DNA by real-time PCR

ELISAs and real-time PCR for detection of serum HBsAg, HBeAg and viral DNA copies were performed as described (Fan et al., 2013).

Statistical analysis

Differences between groups were determined using Student's *t*-test. Pearson's χ test was used to detect the correlation between variation rate and disease progress in HBV infected patients. Clinical statistical analyses were performed with SPSS version 16.0 software (SPSS Inc., Chicago, Illinois). *P* values <0.05 were considered significant.

ACKNOWLEDGEMENTS

L.S., Y.Z. and S.M. designed experiments and analyzed data; L.S., Y.Z., M.D. and J.H. carried out experiments; X.L. and M.G. constructed pcDNA3.1-HBc vectors and collected clinical patient samples. X.L. and D.X. analyzed the clinical patient information. L.S., Y. Z., J.L., S.Z., B.Z. and S.M. discussed the hypothesis and interpreted the data. L.S. and S.M. wrote the manuscript.

This work was supported by a grant from the National Basic Research Program (973 Program) (No. 2014CB542602); grants from the National Natural Science Foundation of China (Grant Nos. 31230026, 91029724, 81021003, and 81102018); Key Projects in the National Science and Technology Program Grant 2013ZX10002001.

ABBREVIATIONS

ACLF, acute on chronic liver failure; AHB, acute hepatitis B; CTL, cytotoxic T lymphocyte; ELISPOT, enzyme-linked immunosorbent spot; FI, fluorescent index; HBc, HBV core protein; HBeAg, HBV e antigen; HBsAg, HBV s antigen; HBV, Hepatitis B virus; SFC, spot-forming cells.

COMPLIANCE WITH ETHICS GUIDELINES

Lu Sun, Yu Zhang, Bao Zhao, Mengmeng Deng, Jun Liu, Xin Li, Junwei Hou, Mingming Gui, Shuijun Zhang, Xiaodong Li, George F. Gao, and Songdong Meng declare that they have no conflict of interest.

For studies with human subjects in this article: All patients were hospitalized in Beijing 302 Hospital from September 2010 to September 2011. Written informed consent was provided by all study participants. The study was approved by the Ethics Committee of Beijing 302 Hospital.

study participants. The study was approved by the Ethics Committee of Beijing 302 Hospital.

For studies with animals in this article: Animals received humane care, and the study of mice was in strict accordance with "The Regulation of the Institute of Microbiology, Chinese Academy of Sciences of Research Ethics Committee".

REFERENCES

Bertoletti A, Gehring AJ (2006) The immune response during hepatitis B virus infection. J Gen Virol 87:1439–1449

Bertoni R, Sidney J, Fowler P, Chesnut RW, Chisari FV, Sette A (1997) Human histocompatibility leukocyte antigen-binding supermotifs predict broadly cross-reactive cytotoxic T lymphocyte responses in patients with acute hepatitis. J Clin Invest 100:503–513

Chen X, Wang W, Wang S, Meng G, Zhang M, Ni B, Wu Y, Wang L (2013) An immunodominant HLA-A*1101-restricted $CD8^+$ T cell response targeting hepatitis B surface antigen in chronic hepatitis B patients. J Gen Virol 94:2717–2723

Desmond CP, Bartholomeusz A, Gaudieri S, Revill PA, Lewin SR (2008) A systematic review of T-cell epitopes in hepatitis B virus: identification, genotypic variation and relevance to antiviral therapeutics. Antivir Ther 13:161–175

Fan H, Yan X, Zhang Y, Zhang X, Gao Y, Xu Y, Wang F, Meng S (2013) Increased expression of Gp96 by HBx-induced NF-κB activation feedback enhances hepatitis B virus production. PLoS ONE 8:e65588

Frelin L, Wahlstrom T, Tucker AE, Jones J, Hughes J, Lee BO, Billaud JN, Peters C, Whitacre D, Peterson D, Milich DR (2009) A mechanism to explain the selection of the hepatitis e antigen-negative mutant during chronic hepatitis B virus infection. J Virol 83:1379–1392

Garboczi DN, Hung DT, Wiley DC (1992) HLA-A2-peptide complexes: refolding and crystallization of molecules expressed in *Escherichia coli* and complexed with single antigenic peptides. Proc Natl Acad Sci USA 89:3429–3433

Gordon SC et al (2013) Antiviral Therapy for Chronic HBV Infection and Development of Hepatocellular Carcinoma in a U.S. Population. Clin Gastroenterol Hepatol S1542–3565:01511–01515

Guo YJ, Zhu Y, Sun SH (2011) Identification and functional studies of HLA-A0201 restricted CTL epitopes in the X protein of hepatitis B virus. Acta Virol 55:107–115

Harari A, Dutoit V, Cellerai C, Bart PA, Du Pasquier RA, Pantaleo G (2006) Functional signatures of protective antiviral T-cell immunity in human virus infections. Immunol Rev 211:236–254

Hwang YK, Kim NK, Park JM, Lee Ky, Han WK, Kim HI, Cheong HS (2002) HLA-A2 1 restricted peptides from the HBx antigen induce specific CTL responses in vitro and in vivo. Vaccine 20:3770–3777

Inchauspe G, Michel ML (2007) Vaccines and immunotherapies against hepatitis B and hepatitis C viruses. J Viral Hepat 14 (Suppl. 1):97–103

Kim D, Lyoo KS, Smith D, Hur W, Hong SW, Sung PS, Yoon SK, Mehta S (2011) Number of mutations within CTL-defined epitopes of the hepatitis B virus (HBV) core region is associated with HBV disease progression. J Med Virol 83:2082–2087

Lavanchy D (2005) Worldwide epidemiology of HBV infection, disease burden, and vaccine prevention. J Clin Virol Suppl 1:S1–S3

Lee HG, Lim JS, Lee KY, Choi YK, Choe IS, Chung TW, Kim K (1997) Peptide-specific CTL induction in HBV-seropositive PBMC by stimulation with peptides in vitro: novel epitopes identified from chronic carriers. Virus Res 50:185–194

Li Y, Song H, Li J, Wang Y, Yan X, Zhao B, Zhang X, Wang S, Chen L, Qiu B, Meng S (2011) Hansenula polymorpha expressed heat shock protein gp96 exerts potent T cell activation activity as an adjuvant. J Biotechnol 151:343–349

Liu HG, Fan ZP, Chen WW, Yang HY, Liu QF, Zhang H, Tien P, Wang FS (2008) A mutant HBs antigen (HBsAg)183–191 epitope elicits specific cytotoxic T lymphocytes in acute hepatitis B patients. Clin Exp Immunol 151:441–447

Liu Z, Li XH, Qiu LP, Zhang XJ, Chen LZ, Cao S, Wang FS, Meng SD (2009) Treg suppress CTL responses upon immunization with HSP gp96. Eur J Immunol 39:3110–3120

Liu J, Zhang SH, Tan SG, Zheng BW, Gao GF (2011) Revival of the identification of cytotoxic T-lymphocyte epitopes for immunological diagnosis. Exp Biol Med (Maywood) 236:253–267

Liu Q, Zheng Y, Yu Y, Tan Q, Huang X (2012) Identification of HLA-A*0201-restricted $CD8^+$ T-cell epitope C64–72 from hepatitis B virus core protein. Int Immunopharmacol 13:141–147

Liu J et al (2013a) Conserved epitopes dominate cross-CD8+ T-cell responses against influenza A H1N1 virus among Asian populations. Eur J Immunol 43:2055–2069

Liu J et al (2013b) Cross-allele cytotoxic T lymphocyte responses against 2009 pandemic H1N1 influenza A virus among HLA-A24 and HLA-A3 supertype-positive individuals. J Virol 86:13281–13294

Maman Y, Blancher A, Benichou J, Yablonka A, Efroni S, Louzoun Y (2011) Immune-induced evolutionary selection focused on a single reading frame in overlapping hepatitis B virus proteins J Virol 85:4558–4566

Neuveut C, Wei Y, Buendia MA (2010) Mechanisms of HBV-related hepatocarcinogenesis. J Hepatol 52:594–604

Ouyang L, Li X, Liang Z, Yang D, Gong F, Shen G, Weng X, Wu X (2013) CD8(low) T-cell subpopulation is increased in patients with chronic hepatitis B virus infection. Mol Immunol 56:698–704

Rehermann B, Fowler P, Sidney J, Person J, Redeker A, Brown M, Moss B, Sette A, Chisari FV (1995) The cytotoxic T lymphocyte response to multiple hepatitis B virus polymerase epitopes during and after acute viral hepatitis. J Exp Med 181:1047–1058

Sendi H, Mehrab-Mohseni M, Shahraz S, Norder H, Alavian SM, Noorinayer B, Zali MR, Pumpens P, Bonkovsky HL, Magnius LO (2009) CTL escape mutations of core protein are more frequent in strains of HBeAg negative patients with low levels of HBV DNA. J Clin Virol 46:259–264

Tan AT et al (2008) Host ethnicity and virus genotype shape the hepatitis B virus-specific T-cell repertoire. J Virol 82:10986–10997

Tan AT, Sodsai P, Chia A, Moreau E, Chng MH, Tham CY, Ho ZZ, Banu N, Hirankarn N, Bertoletti A (2013) Immunoprevalence and immunodominance of HLA-Cw0801 restricted T cell response targeting the HBV envelope transmembrane region. J Virol 88:1332–1341

Thimme R, Wieland S, Steiger C, Ghrayeb J, Reimann KA, Purcell RH, Chisari FV (2003) CD8(+) T cells mediate viral clearance and disease pathogenesis during acute hepatitis B virus infection. J Virol 77:68–76

Westover KM, Hughes AL (2007) Evolution of cytotoxic T-lymphocyte epitopes in hepatitis B virus. Infect Genet Evol 7: 254–262

Xu Z et al (2011) Association of hepatitis B virus mutations in basal core promoter and precore regions with severity of liver disease: an investigation of 793 Chinese patients with mild and severe chronic hepatitis B and acute-on-chronic liver failure. J Gastroenterol 46:391–400

Yang PL, Althage A, Chung J, Chisari FV (2002) Hydrodynamic injection of viral DNA: a mouse model of acute hepatitis B virus infection. Proc Natl Acad Sci USA 99:13825–13830

Yukihiro S (2012) T cell immunopathogenesis and immunotherapeutic strategies for chronic hepatitis B virus infection. World J Gastroenterol 18:2443–2451

Zhang Y, Li S, Shan M, Pan X, Zhuang K, He L, Gould K, Tien P (2007) Hepatitis B virus core antigen epitopes presented by HLA-A2 single-chain trimers induce functional epitope-specific $CD8^+$ T-cell responses in HLA-A2.1/Kb transgenic mice. Immunology 121:105–112

Zhang Y, Ren Y, Wu Y, Zhao B, Qiu L, Li X, Xu D, Liu J, Gao GF, Meng S (2013) The L60V variation in hepatitis B virus core protein elicits new epitope-specific cytotoxic T lymphocytes and enhances viral replication. J Virol 87:8075–8084

Zhou M et al (2006) Screening and identification of severe acute respiratory syndrome-associated coronavirus-specific CTL epitopes. J Immunol 177:2138–2145

5

Insight into the antifungal mechanism of *Neosartorya fischeri* antifungal protein

Máté Virágh[1], Annamária Marton[2], Csaba Vizler[2], Liliána Tóth[1], Csaba Vágvölgyi[1], Florentine Marx[3], László Galgóczy[1]✉

[1] Department of Microbiology, Faculty of Science and Informatics, University of Szeged, Szeged, Hungary
[2] Institute of Biochemistry, Biological Research Centre, Hungarian Academy of Sciences, Szeged, Hungary
[3] Division of Molecular Biology, Biocenter, Innsbruck Medical University, Innsbruck, Austria
✉ Correspondence: galgoczi@gmail.com (L. Galgóczy)

ABSTRACT

Small, cysteine-rich, highly stable antifungal proteins secreted by filamentous Ascomycetes have great potential for the development of novel antifungal strategies. However, their practical application is still limited due to their not fully clarified mode of action. The aim of this work was to provide a deep insight into the antifungal mechanism of *Neosartorya fischeri* antifungal protein (NFAP), a novel representative of this protein group. Within a short exposure time to NFAP, reduced cellular metabolism, apoptosis induction, changes in the actin distribution and chitin deposition at the hyphal tip were observed in NFAP-sensitive *Aspergillus nidulans*. NFAP did show neither a direct membrane disrupting-effect nor uptake by endocytosis. Investigation of *A. nidulans* signalling mutants revealed that NFAP activates the cAMP/protein kinase A pathway via G-protein signalling which leads to apoptosis and inhibition of polar growth. In contrast, NFAP does not have any influence on the cell wall integrity pathway, but an unknown cell wall integrity pathway-independent mitogen activated protein kinase A-activated target is assumed to be involved in the cell death induction. Taken together, it was concluded that NFAP shows similarities, but also differences in its mode of antifungal action compared to two most investigated NFAP-related proteins from *Aspergillus giganteus* and *Penicillium chrysogenum*.

KEYWORDS *Neosartorya fischeri* antifungal protein, *Aspergillus nidulans*, cAMP/Pka signalling, Pkc/Mpk signalling, antifungal mechanism

INTRODUCTION

The increased incidence of fungal infections and the fast development of drug resistant filamentous fungi causing mycoses, plant infections or damage to cultural heritages strongly demand for the development of new antifungal strategies. Antifungal peptides have great potential in these fields, and also have significant commercial potential on the global market for antifungals set to be worth $12.2 billion by 2016 (Montesinos, 2007; Duncan and O'Neil, 2013).

In this respect, cysteine-rich antifungal proteins secreted by filamentous Ascomycetes are promising candidates. The *Neosartorya fischeri* antifungal protein (NFAP) secreted by *N. fischeri* NRRL181 is a novel representative of this protein group (Kovács et al., 2011). NFAP effectively inhibits the growth of numerous filamentous Ascomycetes including potential human and plant pathogens (Kovács et al., 2011; Virágh et al., 2014). In our previous work we demonstrated that NFAP interferes with the organization of the cell wall, destroys the chitin filaments and triggers apoptotic/necrotic events through reactive oxygen species accumulation in the NFAP-sensitive *Aspergillus nidulans* (Galgóczy et al., 2013). We also reported that *Pichia pastoris* KM71H produces heterologous NFAP in an antifungally active and folded state, and the antifungal effect is comparable to the native protein (Virágh et al., 2014). In spite of these promising results, a detailed insight into the antifungal mechanism of NFAP is still missing. This, however, represents essential prerequisite for the practical application of NFAP. Therefore, the antifungal mechanism of NFAP on *A. nidulans* was examined in more detail in the present study.

The following objectives were addressed in this work: investigation of (i) the cell viability and apoptotic/necrotic processes, (ii) the membrane permeabilization activity, and (iii) the actin and chitin distribution in the presence of NFAP. Furthermore, we examined the (iv) the localization of NFAP; and (v) the signalling pathways involved in the mode of action of NFAP.

RESULTS

Cell viability and apoptotic/necrotic events in the presence of NFAP

The two-colour fluorescent FUN1 stain passively diffuses into the fungal cells and stains the cytoplasm and metabolically inactive vacuoles green, while the metabolically active vacuoles red. After short-time exposure to NFAP (30 min) reduced cellular metabolism was detected in *A. nidulans* FGSC A4 hyphae indicated by the presence of only green fluorescent vacuoles compared to the untreated control, which contained red fluorescent vacuoles too (Fig. 1A). This phenomenon was also

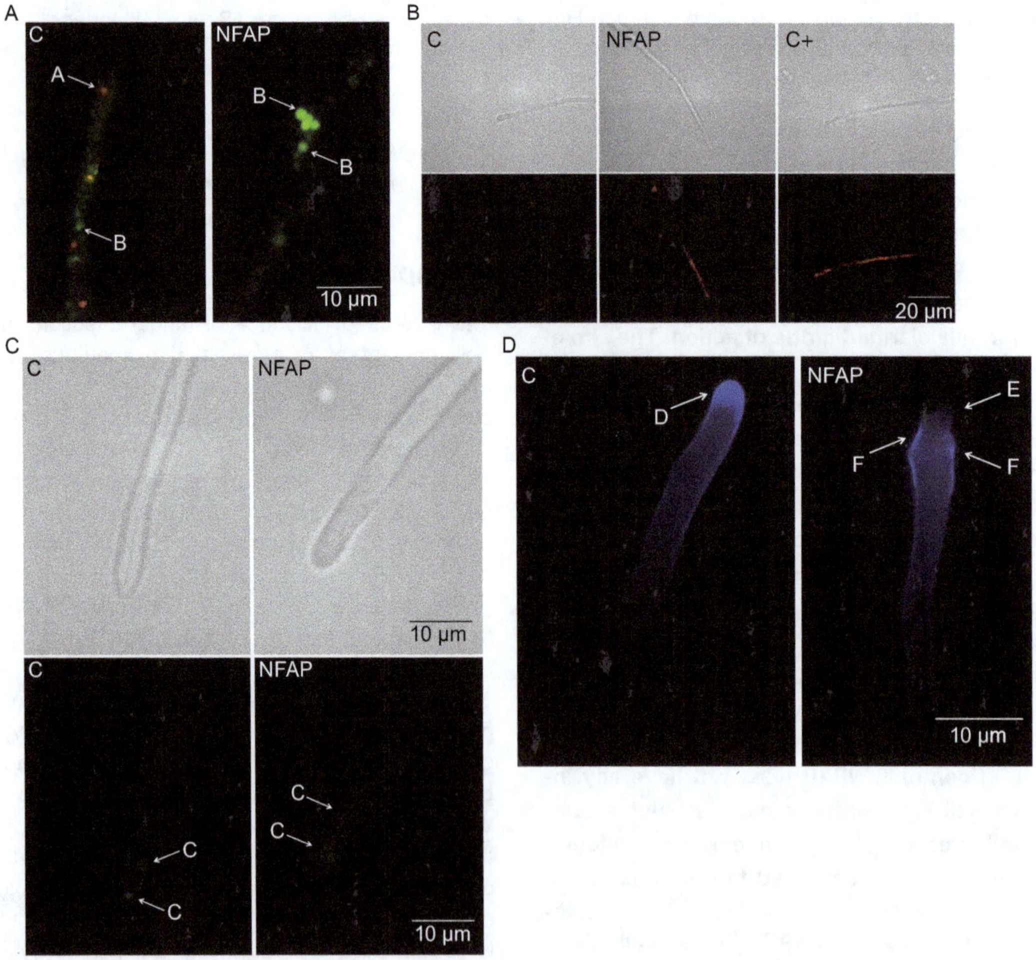

Figure 1. Physiological changes in *Aspergillus nidulans* in the presence of *Neosartorya fischeri* antifungal protein (NFAP). (A) Viability staining of *Aspergillus nidulans* FGSC A4 hyphae with FUN-1 dye after NFAP treatment for 30 min at 37°C. Red vacuoles A indicates metabolic activity, while green vacuoles B metabolic inactivity. (B) Propidium iodide (PI) staining of *A. nidulans* FGSC A4 hyphae after NFAP treatment for 16 h at 37°C. Intracellular red fluorescence indicates membrane disruption. (C) Actin distribution at *A. nidulans* Actin-GFP hyphal tips in response to NFAP treatment for 30 min at 30°C. C: actin patch. (D) Calcofluor white (CFW) staining of *A. nidulans* FGSC A4 hyphae after NFAP treatment for 30 min at 37°C. D: cap-like CFW fluorescence - site of the chitin assembly, E: lack of the cap-like CFW fluorescence, F: delocalized chitin deposition. C: untreated control, C + : positive PI staining control-hypha was treated with 70% Et-OH for 1 h at 4°C, NFAP: NFAP-treated (25 µg/mL) hyphae. Upper images, light microscopy; lower images, fluorescence microscopy of PI staining (B) and of actin distribution (C).

observed after 60 min and 16 h of NFAP-treatment (data not shown).

Annexin V-FITC Apoptosis Detection Kit dyes the apoptotic cells green, while the necrotic cells are counterstained red. Living cells do not show any fluorescence. After 30 and 60 min a significant increase in the number of apoptotic 16 h-old germlings was observed in the NFAP-treated samples (26.1 ± 7.5% and 31.8 ± 8.8%, respectively; $P < 0.0001$) compared the untreated control where only a few percent of the cells (2.2 ± 1.3%) showed green fluorescence. Only few necrotic cells were detected in the untreated (5.1 ± 1.9%) and NFAP-treated samples after 30 min (4.8 ± 3.7%) and 60 min (5.2 ± 0.9%). These results suggest that NFAP exerts its antifungal effect through induction of apoptosis. After 16 h, however, almost all germlings (98.9 ± 1.1%) showed red fluorescence which indicated the death of the cells.

NFAP does not cause direct membrane disruption

The membrane disrupting activity of NFAP was investigated by applying the membrane impermeant, red-fluorescent nuclear and chromosome stain propidium iodide (PI). This dye can penetrate only into the dead cells where the membrane integrity is damaged. Most of the *A. nidulans* FGSC A4 hyphae did not show any red fluorescence after a 30- and 60-min-long incubation in the presence of NFAP (data not shown). This indicates that the plasma membrane was still intact at these time points, similarly to the untreated control. After 16 h almost only red fluorescent hyphae appeared in the NFAP-treated sample pointing towards a massive membrane disruption as a consequence of the long-time cell killing effect of NFAP (Fig. 1B).

Changes in the actin distribution and chitin deposition at the hyphal tip

To study the morphological aberrations of NFAP-exposed *A. nidulans* hyphae we analysed the actin distribution and the deposition of chitin at the hyphal tips. Actin-GFP expressing *A. nidulans* showed typical actin patch distribution at the hyphal tips clustered near the apical region and scattered behind the tips (Taheri-Talesh et al., 2008) (Fig. 1C). In contrast, actin patches were disturbed in hyphae that were treated with NFAP for 30 min only (Fig. 1C). Calcofluor white (CFW) staining revealed delocalized chitin deposition at hyphal tips of *A. nidulans* FGSC A4 after NFAP treatment for 30 min. In contrast, the untreated control sample exhibited a characteristic cap-like CFW fluorescence (Fig. 1D). These effects in actin and chitin delocalization were also observed after 60 min and 16 h of incubation with NFAP (data not shown). These results indicate that in the presence of NFAP the normal actin polarization/localization and chitin distribution are disturbed in *A. nidulans*.

NFAP shows no active internalization

A produced polyclonal NFAP-antiserum reacted specifically with NFAP, and no signals were obtained with the control serum that was collected before the first injection (data not shown). An indirect immunofluorescence staining method was applied to study the localization of NFAP in *A. nidulans* FGSC A4. NFAP did not enter the fungal cell in detectable concentration levels after 30 and 60 min of exposure (data not shown), but after 16 h NFAP-specific fluorescence signals accumulated at hyphal fractures, twists and in cell-wall bubbles (Fig. 2A). To clarify whether the NFAP internalization

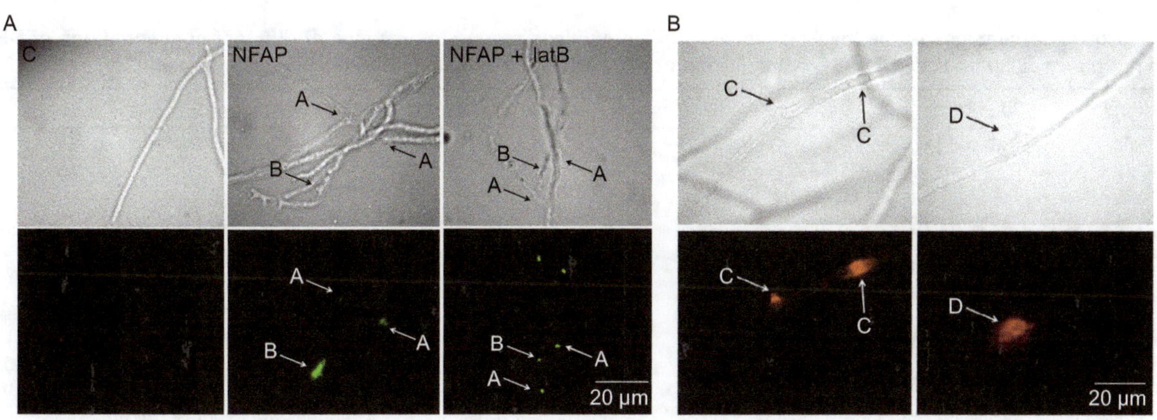

Figure 2. Localization of *Neosartorya fischeri* antifungal protein (NFAP) in *Aspergillus nidulans*. (A) Indirect immunofluorescence staining of *Aspergillus nidulans* FGSC A4 hyphae with rat anti-NFAP serum and FITC-conjugated swine anti-rat IgG, after NFAP treatment for 16 h at 37°C. NFAP accumulation in A: hyphal fracture and twist; B: cell-wall bubble. C: untreated control, NFAP: NFAP-treated (25 μg/mL), NFAP + latB: NFAP- (25 μg/mL) and latrunculin B- (5 μg/mL) treated hyphae. (B) Propidium iodide (PI) staining of *Aspergillus nidulans* FGSC A4 hyphae after 25 μg/mL NFAP treatment for 90 min at 37°C. Red fluorescence at hyphal bubble C and around hyphal fracture D indicates membrane disruption and outflow of the hyphal/cell content. Upper images, light microscopy; lower images, fluorescence microscopy of indirect immunofluorescence staining (A) and of PI staining (B).

was the consequence of an endocytic mechanism, the indirect immunofluorescence staining was repeated in the presence of 5 μg/mL latrunculin B (latB). LatB selectively inhibits the actin polymerization and therefore disturbs endocytosis. If the NFAP accumulation in the hyphae after 16 h is a consequence of an endocytotic mechanism no fluorescence signal should be observable in the cell. NFAP-specific fluorescence signals appeared in hyphae only after 16 h of incubation with NFAP and latB, but not after the exposure for 30 or 60 min (Fig. 2A). Based on these results, we conclude that NFAP is presumably not internalized by endocytosis. Instead, the accumulation of NFAP after 16 h is possibly a consequence of a passive diffusion at disrupted sites of the cell wall and plasma membrane.

NFAP interferes with the G-protein signal transduction and cell wall integrity pathways in *Aspergillus nidulans*

NFAP activates the cAMP/Pka signalling cascade through heterotrimeric G-protein

The *A. nidulans* FGSC1035 strain carries a dominant negative mutation in the guanidine nucleotide binding domain of heterotrimeric G-protein α-subunit (*fadA*). As a consequence of this mutation the ability of the G_α subunit to dissociate from $G_{\beta\gamma}$ is inhibited causing constitutive inactivation of heterotrimeric G-protein signalling (Yu et al., 1996). Growth inhibition assays revealed that FGSC1035 strain was less susceptible to NFAP than its isogenic recipient FGSC116 strain indicating that G-protein signalling is necessary for NFAP toxicity (Table 1). In *A. nidulans* the cAMP/protein kinase A (Pka) pathway is activated by heterotrimeric G-protein signalling, and it is involved in the regulation of morphological development (Shimizu and Keller, 2001), and in the induction of apoptosis (Semighini et al., 2006). To investigate if cAMP/Pka signalling plays a role in the mechanism of action of NFAP, we tested the susceptibility of an *A. nidulans* strain with depleted Pka activity (Shimizu and Keller, 2001). The *pkaA* deletion mutant exhibited reduced sensitivity towards NFAP compared to its recipient RKIS1 strain suggesting that NFAP activates the cAMP/Pka signalling cascade and possibly promotes programmed cell death (PCD) (Table 1). Our assumption was further supported by susceptibility tests applying NFAP in combination with 8-bromoadenosine 3',5'-cyclic monophosphate (8-Br-cAMP) and caffeine on strain FGSC A4. 8-Br-cAMP is an activator of cAMP/Pka signalling in fungi (Gorovits and Yarden, 2003), while caffeine represses the Pka signalling by reduction of cAMP levels (Kuranda et al., 2006). NFAP aggravated the toxicity of 8-Br-cAMP indicating a synergistic interaction between the two compounds and an activating effect of NFAP in cAMP/Pka signalling (Table 2). In contrast, caffeine ameliorated the growth of FGSC A4 strain exposed to NFAP (Table 2). The Δ*pkaA* strain, however, showed hypersensitivity (63% ± 6.5% growth) to 20 mmol/L caffeine compared to its recipient RKIS1 strain (91% ± 6.8% growth). This growth reduction induced by caffeine was cured by the addition of NFAP and overgrowth was detected compared to the untreated control (Table 2). These results suggest that NFAP and caffeine act in an opposite manner on Pka signalling. The overgrowth effect may be explained by the response of the fungus that activates rescue mechanisms against the co-administration of two antifungals, which have different cellular targets (Ouedraogo et al., 2011).

It is worth to mention here that FGSC A4 and Actin-GFP strains showed the same changes in chitin deposition and the

Table 1. Growth percentages of the investigated *Aspergillus nidulans* strains in presence of different concentrations of NFAP in *in vitro* broth microdilution test after 48 h of incubation at 30°C or 37°C (depending on the investigated strain)

NFAP/Strain	200 μg/mL	100 μg/mL	50 μg/mL	Type
RhoA^{G14V}	48 % ± 15.2%^{ns}	57% ± 4.5%^{ns}	86% ± 12.5%^{ns}	Mutant
Δ*mpkA*	108% ± 9.1%***	105% ± 6.7%***	103% ± 0.9%**	Mutant
GR5	47% ± 7.5%	58% ± 5.4%	83% ± 11.9%	Isogenic recipient of RhoAG^{14V} and Δ*mpkA*
alcA-PkcA[a]	48% ± 10.6%***	57% ± 0.6%***	64% ± 12.5%***	Mutant
alcA-PkcA[b]	74% ± 7.6%^{ns}	77% ± 0.5%^{ns}	99% ± 0.2%^{ns}	Mutant
R153	72% ± 2.8%	83% ± 4.4%	102% ± 1.8%	Isogenic recipient of *alcA*-Pkca
Δ*pkaA*	56% ± 6.1%***	69% ± 5.3%***	83% ± 11.5%**	Mutant
RKIS 1	40% ± 2.7%	48% ± 1.6%	66% ± 4.0%	Isogenic recipient of Δ*pkaA*
FGSC 1035	52% ± 1.9%*	78% ± 0.2%***	101% ± 0.4%***	Mutant
FGSC 116	42% ± 6.7%	59% ± 1.9%	75% ± 2.3%	Isogenic recipient of FGSC 1035

The untreated control is taken as 100% of growth. The mean growth percentages and their standard deviations from three replicates (n = 3) are indicated in the cells. Significant differences (P-values) were determined based on the comparison with the growth percentages of parental strain. ***: P < 0.0001; **: P < 0.005; *: P < 0.05; ^{ns}: no significant differences. NFAP: *Neosartorya fischeri* antifungal protein.

[a] *A. nidulans alcA*-PkcA in the presence of glucose.

[b] *A. nidulans alcA*-PkcA in the presence of glycerol.

Table 2. Growth percentages of *Aspergillus nidulans* strains in presence of NFAP and NFAP-8 Br-cAMP/caffeine combinations in *in vitro* broth microdilution test after 48 h of incubation at 37°C

NFAP/other compounds	0 μg/mL	50 μg/mL	100 μg/mL	200 μg/mL
FGSC A4				
NFAP	100%	91% ± 6.0%	67% ± 6.7%	52% ± 3.2%
NFAP + 5 mmol/L 8-Br-cAMP	70% ± 1.6%	58% ± 4.2%***	57% ± 3.8%***	41% ± 3.2%***
NFAP + 20 mmol/L caffeine	71% ± 2.1%	74% ± 3.7%^{ns}	61% ± 2.6%***	108% ± 1.5%***
Δ*pkaA*				
NFAP	100%	83% ± 11.5%	69% ± 5.3%	56% ± 6.1%
NFAP + 20 mmol/L caffeine	63% ± 6.5%	123% ± 7.9%***	195% ± 30.6%***	196% ± 3.0%***

The untreated control is taken as 100% of growth. The mean growth percentages and their standard deviations from three replicates (n = 3) are indicated in the cells. Significant differences (P-values) were determined based on the comparison with the growth percentages in the presence of 8-Br-cAMP or caffeine alone. ***: $P < 0.0001$; ^{ns}: no significant differences. 8-Br-cAMP: 8-bromoadenosine 3',5'-cyclic monophosphate, NFAP: *Neosartorya fischeri* antifungal protein.

actin distribution after 60 min of exposure to 5 mmol/L 8-Br-cAMP (Fig S1a and S1b) as it was observed in the presence of NFAP (Fig. 1D and 1C). In contrast to the FGSC A4 strain (Fig. 1D), chitin delocalization was not observed in the *fadA* and *ΔpkaA* mutants when they were treated with 25 μg/mL NFAP for 30 min (Fig S2a and S2b). These observations further strengthen the role of the NFAP in the inhibition of polar growth and chitin assembly via cAMP/Pka signalling.

NFAP acts on an unknown cell wall integrity pathway-independent mitogen-activated protein kinase A-downstream target

RhoA is an essential protein for polar growth, branching, and cell wall synthesis in *A. nidulans* (Guest et al., 2004). The RhoA^{G14V} mutant with ectopic copies of the constitutively active *rhoA*^{G14V} allele showed the same susceptibility to NFAP and to the cell wall stressing agent CFW as its isogenic control GR5 strain (Table 1 and Table 3), but proved to be less susceptible to caffeine (Table 3). These results demonstrate that RhoA is not directly involved in NFAP toxicity, and NFAP may target the downstream effectors of RhoA, e.g. protein kinase C (Pkc). Pkc plays an important role in suppression of apoptosis through activation of the mitogen-activated protein kinase (Mpk) cascade, and also in polarity establishment independently of the Mpk cascade during germination in *A. nidulans* (Katayama et al., 2012). To investigate a possible role of Pkc/Mpk signalling in NFAP antifungal activity, we tested a conditional *A. nidulans alcA*-PkcA mutant for NFAP susceptibility. In this mutant, the expression of PkcA was repressed by glucose, but induced by glycerol (Ronen et al., 2007). There was no significant difference ($P > 0.05$) between the growth of this strain at different NFAP concentrations under PkcA repressing conditions in contrast to the isogenic recipient *A. nidulans* R153 ($P < 0.0001$) (Table 1). In the presence of glycerol the *alcA*-PkcA mutant showed the same susceptibility to NFAP as the R153 strain, and significant differences between the growth percentages at different NFAP concentrations were observed ($P < 0.0001$) (Table 1). These results suggest that NFAP has no influence on the PkcA signalling, which affects

Table 3. Growth percentages of *Aspergillus nidulans* strains in presence of caffeine and CFW in *in vitro* broth microdilution test after 48 h of incubation at 30°C or 37°C (depending on the investigated strain)

Compounds/ strain	Caffeine		CFW		Type
	10 mmol/L	20 mmol/L	10 μg/mL	20 μg/mL	
RhoA^{G14V}	78% ± 1.2%***	60% ± 8.4%*	92% ± 7.1%^{ns}	55% ± 7.7%^{ns}	Mutant
Δ*mpkA*	17% ± 3.9%***	15% ± 2.6%***	76% ± 2.5%***	49% ± 4.0%**	Mutant
GR5	59% ± 1.1%	48% ± 2.7%	97% ± 2.1%	60% ± 3.3%	Isogenic recipient of RhoAG^{14V} and Δ*mpkA*
alcA-PkcA	29% ± 0.3%***	21% ± 4.4%***	44% ± 5.9%***	22% ± 5.2%***	Mutant
R153	87% ± 2.9%	79% ± 14.4%	95% ± 3.0%	74% ± 2.6%	Isogenic recipient of *alcA*-Pkca

The untreated control is taken as 100% of growth. The mean growth percentages and their standard deviations from three replicates (n = 3) are indicated in the cells. Significant differences (P-values) were determined based on the comparison with the growth percentages of parental strain. ***: $P < 0.0001$; **: $P < 0.005$; *: $P < 0.05$, ^{ns}: no significant differences. CFW: calcofluor white.

polarity establishment and induces apoptosis suppression (Katayama et al., 2012). Further evidence was gained by using a MpkA deletion mutant. MpkA is involved in the polarized growth of *A. nidulans* and the cell wall integrity (CWI) signalling pathway (Bussink and Osmani, 1999; Fujioka et al., 2007). Cell wall stressing agents (such as caffeine and CFW) induce the CWI pathway by increasing the Pkc and Mpk phosphorylation in *A. nidulans* (Fujioka et al., 2007). Indeed, we could show that the *alcA*-PkcA and Δ*mpkA* strains were hypersensitive to caffeine and CFW compared to their isogenic control strains R153 and GR5 (Table 3). The Δ*mpkA* strain, however, exhibited resistance to NFAP, while the recipient GR5 strain was susceptible (Table 1). This observation suggests that NFAP does not interfere directly with the activation of MpkA, but may affect a CWI pathway-independent MpkA-activated unknown target playing a role in the antifungal effect of NFAP.

NFAP does not activate cell wall integrity pathway

Based on our results we hypothesized that NFAP does not activate the CWI pathway by activation of downstream targets. To prove this assumption we investigated the sensitivity of *A. nidulans* FGSC A4 strain towards 100 μg/mL CFW, and 100 μg/mL CFW in combination with 200 μg/mL NFAP. CFW induces the CWI pathway by activation of the Rho/Pkc/Mpk signalling cascade in *A. nidulans* (Fujioka et al., 2007). In the presence of CFW and NFAP alone, this strain showed 9% ± 0.3% and 52% ± 3.2% growth, respectively; while significant increase in the growth was detected in combination of the two compounds (61.3% ± 1.3%, $P < 0.0001$). This ameliorated growth effect indicates that CFW and NFAP antagonize each other, and NFAP does not activate the CWI pathway; however, further experiments (e.g. investigation of CWI gene expression) are needed for verification of this assumption.

DISCUSSION

From the 1990s several extracellular cysteine-rich proteins with remarkable antifungal activity against moulds have been isolated and characterized from filamentous Ascomycetes. The antifungal mode of action is extensively studied only in the case of three representatives, the *Aspergillus giganteus* antifungal protein (AFP) (Meyer, 2008), the *A. giganteus* A3274 antifungal protein (AFP_{NN5353} a very similar protein to AFP) (Binder et al., 2011) and the *Penicillium chrysogenum* antifungal protein (PAF) (Hegedus et al., 2011). Results presented in this study allow us to compare the detailed antifungal mechanism of NFAP with these proteins. By studying an NFAP-producing *A. nidulans* strain we previously proved that NFAP interferes with the organization of the cell wall, destroys the chitin filaments as it was observed also in the case of AFP, and triggers apoptotic-necrotic events through ROS accumulation as described for PAF (Galgóczy et al., 2013). However, it was not clarified whether these changes are the primary, short-time effect of the NFAP or if they are the consequences of its secondary, cell-killing effect.

NFAP caused metabolic inactivation and induced apoptotic cell death in *A. nidulans*, but no plasma membrane disruption could be observed within one hour of exposure. This observation is comparable with our previous finding when majority of the NFAP-producing *A. nidulans* germlings showed apoptotic phenotype after 8 h of cultivation suggesting the significance of apoptosis-inducing effect for antifungal activity (Galgóczy et al., 2013). Previously, it was proved that PAF evokes metabolic inactivity in *Aspergillus niger* (Kaiserer et al., 2003), and a PCD-like phenotype in *A. nidulans* (Leiter et al., 2005). Plasma membrane disruption by AFP was detected in *A. niger* (Theis et al., 2003).

Similar to our investigations with NFAP, disordered actin patch distribution, reduced chitin content and delocalized chitin deposition at the hyphal tips were also observed previously when *A. nidulans* was treated with PAF (Binder et al., 2010). During the polarized growth of *A. nidulans* exocytic vesicles (containing cell wall biosynthetic enzymes and wall precursors) and endocytic recycling vesicles are delivered by myosin molecules on actin cables to and from the hyphal tip. Therefore actin functions in the spatially coupled exo- and endocytosis at the tip and in holding the tip apparatus together (Taheri-Talesh et al., 2008). Since the correct localization of actin filaments is essential for polarized growth, the random distribution of actin as a consequence of NFAP treatment results in disturbed apical growth which leads to the previously well-described hyperbranching, swollen hyphal tips and destructed cell wall of the sensitive fungus in the presence of NFAP (Kovács et al., 2011; Galgóczy et al., 2013) (Fig. S3).

The indirect immunofluorescence staining experiments revealed that NFAP is internalized by passive transport across the disrupted sites of the plasma membrane and cell wall, not by an endocytotic uptake. The membrane disruption at hyphal fractures and bubbles where NFAP is accumulated after 16 h of treatment were confirmed by PI staining. These sites showed intensive red fluorescence indicating the loss of membrane integrity after more than one hour of incubation with NFAP (Fig. 2B). In contrast to NFAP, PAF and AFP_{NN5353} are internalized by endocytosis and distributed in the cytoplasm of *A. nidulans* (Oberparleiter et al., 2003; Binder et al., 2011), while AFP accumulates only in the outer layer, the cell wall and/or the plasma membrane of *A. niger* (Theis et al., 2003, 2005).

Based on the above mentioned observations NFAP may get into the cell within 30 min in a concentration below the detectable level which might be enough to interfere with the actin localization and chitin distribution and to cause apoptosis. It could be also possible that these effects are secondary consequences of the NFAP action: NFAP may bind to a surface receptor which activates a signalling pathway that leads to apoptosis and defective polar growth. The susceptibility tests with *A. nidulans* signalling mutants indeed point towards signalling pathways affected by NFAP.

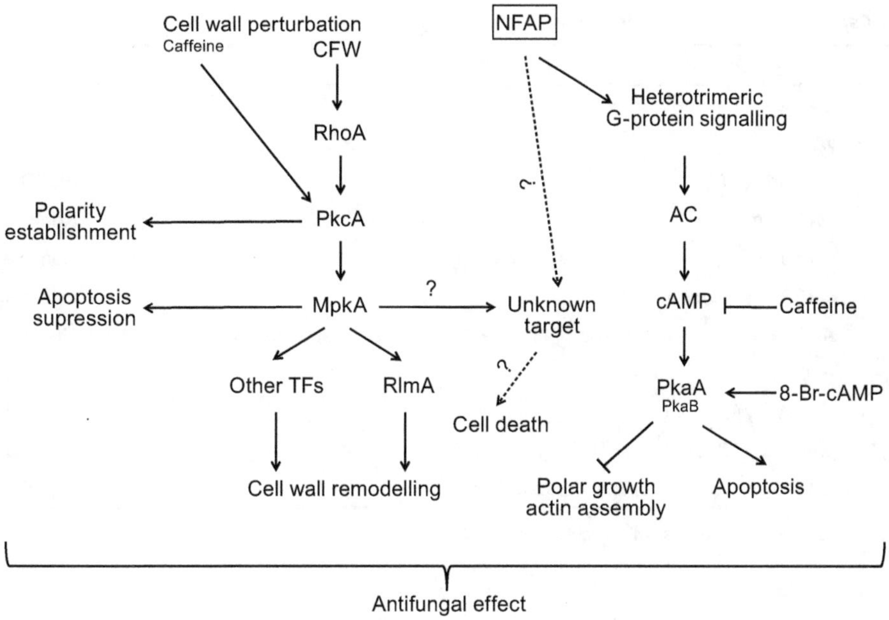

Figure 3. Tentative model for the antifungal mechanism of *Neosartorya fischeri* NFAP in *Aspergillus nidulans* modified from Binder et al. (2010; 2011) for *Penicillium chrysogenum* PAF and *Aspergillus giganteus* A3274 AFP$_{NN5353}$, respectively. 8-Br-cAMP: 8-bromoadenosine 3',5'-cyclic monophosphate, AC: adenylate cyclase, cAMP: cyclic adenosine monophosphate, CFW: calcofluor white, Mpk: mitogen activated protein kinase, Pka: protein kinase A, Pkc: protein kinase C, RhoA: small GTP binding protein, RlmA: transcription factor, TFs: transcription factors.

The antifungal mechanism of NFAP proved to be independent from the RhoA and PkcA, members of the CWI pathway. The small GTPase RhoA is an essential protein involved in the polar growth (Guest et al., 2004). PkcA plays a role in the polarity establishment independently of MpkA (Katayama et al., 2012) and in the suppression of apoptosis via MpkA (Katayama et al., 2012) (Fig. 3). Similarly, the toxicity of the *P. chrysogenum* antifungal protein PAF was also shown to be independent from RhoA, instead inhibition of RhoA-GAP targets was supposed (Binder et al., 2010). In contrast to NFAP, PAF possibly inactivates the Pkc signalling (Binder et al., 2010). Binder et al. (2011) suggested that the toxicity of AFP$_{NN5353}$ is transmitted by RhoA-GAP targets and not by RhoA itself. The role of RhoA-GAP targets in the NFAP toxicity is awaiting further investigations. In our study NFAP did not induce the CWI pathway in *A. nidulans* similar to PAF (Binder et al., 2010). In contrast, AFP and AFP$_{NN5353}$ activates the CWI pathway by increasing the MpkA/RlmA-activated α-glucan synthase *agsA* expression, a main enzyme in the cell wall remodelling of *A. niger* (Hagen et al., 2007; Binder et al., 2011) (Fig. 3).

Our results indicate that NFAP might bind to a G-protein coupled receptor similarly to other antifungal proteins and drugs (Van Dijck, 2009; Hegedüs and Marx, 2013) that results in the induction of the cAMP/Pka cascade, which inhibits the polar growth (Bencina et al., 2005) and modulates PCD (Semighini et al., 2006) (Fig. 3). Similar to NFAP, PAF also activates the cAMP/Pka signalling cascade via the heterotrimeric G-protein signal transduction pathway (Binder et al., 2010). The identification of a G-protein coupled receptor of NFAP would further accelerate the elucidation of the mechanism of its antifungal activity.

Presumably, activation of a cell wall integrity pathway-independent, but MpkA-dependent unknown factor, which induces PCD, may also play a role in the antifungal mechanism of NFAP (Fig. 3). This latter assumption is supported by the report of the existence of an MpkA-mediated cell death pathway in *A. nidulans* by Colabardini et al. (2010). This group observed an increased tolerance of the *A. nidulans* Δ*mpkA* strain to farnesol-induced cell death compared to Δ*mpkC* and Δ*hogA* strains. Proving the existence of this MpkA-activated apoptosis factor is waiting for further investigations. In contrast to NFAP, PAF fails to activate the Mpk (Binder et al., 2010), and AFP$_{NN5353}$ evokes growth inhibition via activation of Pkc/Mpk signalling pathway in *A. nidulans* (Binder et al., 2011).

The features of the cysteine-rich antifungal proteins render them exceptionally suitable compounds as commercial preservatives, bio-pesticides and drugs against moulds and offer an alternative, safely applicable solution for the recent antifungal challenges (Marx et al., 2008; Meyer, 2008). Our present results significantly contribute to the understanding

Table 4. Investigated *Aspergillus nidulans* strains and their genotype

Strain	Relevant genotype	Reference or source
FGSC A4	Glasgow wild type	FGSC
FGSC 116	*yA2*	FGSC
FGSC 1035	*yA2 fadAG203R*	FGSC
R153	*wA2; pyroA4*	Ronen et al. (2007)
alcA-PkcA	*wA2; pyroA4; pyrG89::pyr4alcA(p)::pkcAΔp*	Ronen et al. (2007)
GR5	*pyrG89; wA3; pyroA4*	Guest et al. (2004)
$RhoA^{G14V}$	*A773 + pGG2* (*$rhoA^{G14}$*) and pRG3AMA1 (co-transformation plasmid)	Guest et al. (2004)
ΔmpkA	*ΔmpkA*	Bussink and Osmani (1999)
RKIS1	*papaA1; yA2*	Shimizu and Keller (2001)
ΔpkaA	*papaA1; yA2; ΔpkaA::argB; ΔargB::trpC; trpC801; veA1*	Shimizu and Keller (2001)
Actin-GFP	*wA3; pyroA4; actin_GFP::pyrG (pyr G 89)*	Taheri-Talesh et al. (2008)

FGSC: Fungal Genetics Stock Center, Kansas, MO, USA.

of the antifungal mechanism not only of NFAP, but also of other cysteine-rich antifungal proteins from Ascomycetes in general.

MATERIALS AND METHODS

Strains and media

A. nidulans strains used in the present study are listed in Table 4. All signalling mutants, their isogenic recipients, and the actin-GFP strain were grown and maintained in minimal medium (MM, Ronen et al., 2007) with respective supplementation (Bussink and Osmani, 1999; Shimizu and Keller, 2001; Guest et al., 2004, Ronen et al., 2007; Taheri-Talesh et al., 2008). The Glasgow wild type *A. nidulans* FGSC 4A was grown and maintained in complete medium (CM, Kuranda et al., 2006).

NFAP production and purification

Heterologous expression and purification of NFAP was carried out in *P. pastoris* KM71H as described previously (Virágh et al., 2014).

Microscopic investigation

Microscopic investigation of the NFAP-specific antifungal effect was carried out on *A. nidulans* FGSC A4 and *A. nidulans* actin-GFP conidia (10^5 conidia/mL) which were immobilised on cover slips, and incubated in CM at 37°C (or at 30°C in case of *A. nidulans* actin-GFP strain) for 16 h. For the investigation of the short-time antifungal effect, cover slips were treated with CM supplemented with sublethal concentration of NFAP (25 µg/mL) for 30 and 60 min at 37°C (or at 30°C in case of *A. nidulans* actin-GFP strain). CM without NFAP was used as control. To reveal the long-time antifungal effect cover slips were incubated in the presence of NFAP for 16 h at 37°C (or at 30°C in case of *A. nidulans* actin-GFP strain).

Germinated conidia and/or hyphae were examined and photographed with a light microscope with fluorescence lamp (LR 66238C, Carl Zeiss, Axiolab) equipped with a digital microscope camera (AxioCam ERc 5s, Carl Zeiss, Axiolab). All experiments were repeated three times.

Viability staining and detection of apoptotic/necrotic events

To compare the metabolic activity of NFAP-treated and untreated *A. nidulans* FGSC A4 hyphae and to reveal the apoptotic/necrotic events, FUN1 viability staining (Invitrogen-Life Technologies, Eugene, OR, USA) and Annexin V-FITC Apoptosis Detection Kit (Sigma-Aldrich, St. Louis, MO, USA) were adapted to cover slips, respectively. After incubation, the hyphae on cover slips were washed with 10 mmol/L HEPES pH = 7.5 (Sigma-Aldrich, St. Louis, MO, USA), then incubated for 30 min at 37°C in 10 mmol/L HEPES containing 2% (*w*/*v*) glucose (pH = 7.5), and stained with 5 µmol/L FUN1 stain diluted in 10 mmol/L HEPES pH = 7.5 in the dark at room temperature for 30 min, and finally, washed again with 10 mmol/L HEPES supplemented with 2% (*w*/*v*) glucose (pH = 7.5). In case of the Annexin V-FITC Apoptosis Detection Kit the hyphae on cover slips were washed with CM and stained with PI and Annexin V-FITC diluted in binding buffer in the dark at room temperature for 10 min, and then washed again with CM.

Detection of membrane disruption

PI staining was used to observe NFAP-specific membrane disruption on *A. nidulans* FGSC A4. Cover slips treated with 70% (*v*/*v*) ethanol for 30 min at 4°C were used as positive staining controls. After washing with CM cover slips were stained with 5 µg/mL PI for 10 min at room temperature in the dark, and then washed again with CM.

Analysis of actin distribution and chitin content

To investigate the impact of NFAP on the actin distribution a derivate of the *A. nidulans* GR5 strain was used, which carries an additional copy of actin tagged with green fluorescent protein on its N-terminus (Taheri-Talesh et al., 2008). Chitin content of the *A. nidulans* FGSC A4 in the presence of NFAP was visualized with CFW staining (Sigma-Aldrich, St. Louis, MO, USA). Cover slips were washed with

CM, then stained with 10 μmol/L CFW at room temperature in the dark for 10 min, and finally, washed again with CM.

Generation and testing of anti-NFAP polyclonal sera

Two-months-old female WISTAR rats were immunized with 100 μg NFAP in complete, then twice in incomplete Freund's adjuvant. The animal experiments we performed according to institutional and national ethical guidelines, in possession of ethical clearances. Individual sera from three rats were screened by Western blotting, and the serum with the strongest specific signal was used in further experiments. For the screening of the sera, purified NFAP was separated by SDS-PAGE (XCell SureLock Mini-Cell, Invitrogen-Life Technologies, Eugene, OR, USA) and transferred onto PVDF membranes (Immobilon-P, IPVH00010, Millipore, Billerica, MA, USA). Membrane was incubated overnight at 4°C with NFAP-immunized rat serum (1:100), washed three times, and then incubated with horseradish peroxidase conjugated anti-rat secondary antibody (1:1000; R&D, Minneapolis, MN, USA). Immunoreactive signals were developed using Supersignal West Pico Chemiluminscent Substrates (Thermo Scientific, MR, USA) and detected with LI-COR ODYSSEY® Fc (Dual-mode imaging system) imager followed by analysis with Odyssey v1.2 software.

Immunofluorescence staining

Localization of NFAP in *A. nidulans* FGSC A4 was investigated using an indirect immunofluorescence staining method described by Fischer and Timberlake (1995). Cover slips with germlings were transferred to a solution containing 50 mmol/L PIPES (pH = 6.7), 25 mmol/L EGTA, 5 mmol/L $MgSO_4$, 5% (*v*/*v*) dimethyl sulfoxide, 8% (*v*/*v*) formaldehyde. After this fixation step, cover slips were washed four times with PEM buffer (50 mmol/L PIPES pH = 6.7, 25 mmol/L EGTA, 5 mmol/L $MgSO_4$). For wall digestion coverslips were transferred to PEM buffer containing Glucanex (500 mg/mL; Sigma-Aldrich, St. Louis, MO, USA) and incubated at room temperature for 2 h, then washed four times in PEM buffer (pH = 6.7) and transferred into an extraction solution containing 100 mmol/L PIPES (pH = 6.7), 20 mmol/L EGTA, 0.1% (*v*/*v*) Triton X-100. After this treatment the cover slips were incubated in the presence of rat anti-NFAP serum diluted 1:600 in Tris-buffered saline containing bovine serum albumin (TBS/B, 20 mmol/L Tris-HCl pH = 8.0, 20 mmol/L NaCl, 0.1% (*v*/*v*) Tween-20, 3% (*v*/*v*) bovine serum albumin (Sigma-Aldrich, St. Louis, MO, USA)) for 60 min. Immunocomplexes were detected with FITC-conjugated goat-anti-rat-IgG (Sigma-Aldrich, St. Louis, MO, USA) diluted 1:40 in TBS/B. For studying the uptake mechanism of NFAP, immunofluorescence staining was performed after treatment of the samples with NFAP and 5 μg/mL latB (inhibitor of actin polymerization; Sigma-Aldrich, St. Louis, MO, USA). As a positive control, indirect immunofluorescence staining was repeated with monoclonal mouse anti-actin antibody (Sigma-Aldrich, St. Louis, MO, USA) and polyclonal rabbit anti-mouse immunoglobulins/TRITC (Agilent Technologies, Dako Denmark A/S, Glostrup, Denmark).

Growth inhibition assay

The antifungal effect of NFAP (25–200 μg/mL), 8-Br-cAMP (5 mmol/L), CFW (10, 20, and 100 μg/mL), and caffeine (10 and 20 mmol/L) (all from Sigma-Aldrich, St. Louis, MO, USA) was investigated in a 96-well microtiter plate bioassay on *A. nidulans* strains in CM or MM containing 10^5 conidia/mL and supplemented with the required ingredients for the growth of certain strains (Bussink and Osmani, 1999; Shimizu and Keller, 2001; Guest et al., 2004, Ronen et al., 2007). In the combination experiments 100 μL aliquots of a serial NFAP dilution (50–200 μg/mL) containing 2×10^5 conidia/mL were mixed with 100 μL 100 μg/mL CFW, 5 mmol/L 8-Br-cAMP, or 20 mmol/L caffeine at final concentrations in the wells. Plates were incubated at the optimal temperature of the tested strain for 48 h without shaking and then the absorbances (OD_{620}) were measured after gently shaking for two seconds with a microtiter plate reader in well scanning mode (SPECTROstar Nano, BMG Labtech, Ortenberg, Germany) (Tables S1, S2, and S3). Fresh medium was used for background calibration. For calculation of the growth rates, the absorbances of the untreated control cultures were set to be 100% growth. All susceptibility tests were repeated three times.

Statistical analysis

All statistical analyses were performed using GraphPad Prism version 5.01 for Windows (GraphPad Software, San Diego, CA, USA). The significant differences between sets of data were determined by One-way analysis of variance with Bonferroni's multiple comparison posttest according to the data.

ACKNOWLEDGEMENTS

The research of V.Cs. and L.G. was supported by the European Union and the State of Hungary, co-financed by the European Social Fund in the framework of TÁMOP 4.2.4.A/2-11-1-2012-0001 'National Excellence Program'. The relating research groups were also supported by the Hungarian Scientific Research Fund (OTKA; grant reference number PD 83355) providing infrastructure and research equipment. The authors would like to thank Sándor Kocsubé for the presented pictures.

ABBREVIATIONS

8-Br-cAMP, 8-bromoadenosine 3',5'-cyclic monophosphate; AFP, *Aspergillus giganteus* antifungal protein; AFP_{NN5353}, *Aspergillus giganteus* A3274 antifungal protein; CFW, calcofluor white; CM, complete medium; CWI, cell wall integrity; latB, latrunculin B; MM, minimal medium; Mpk, mitogen activated protein kinase; NFAP, *Neosartorya fischeri* antifungal protein; PAF, *Penicillium chrysogenum* antifungal protein; PCD, programmed cell death; PI, propidium iodide; Pka, protein kinase A; Pkc, protein kinase C.

COMPLIANCE WITH ETHICS GUIDELINES

Máté Virágh, Annamária Marton, Csaba Vizler, Liliána Tóth, Csaba Vágvölgyi, Florentine Marx, László Galgóczy declare that they have no conflict of interest. All institutional and national guidelines for the care and use of laboratory animals were followed.

REFERENCES

Bencina M, Legisa M, Read ND (2005) Cross-talk between cAMP and calcium signalling in *Aspergillus niger*. Mol Microbiol 56:268–281

Binder U, Oberparleiter C, Meyer V, Marx F (2010) The antifungal protein PAF interferes with PKC/MPK and cAMP/PKA signalling of *Aspergillus nidulans*. Mol Microbiol 75:294–307

Binder U, Bencina M, Eigentler A, Meyer V, Marx F (2011) The *Aspergillus giganteus* antifungal protein AFP_{NN5353} activates the cell wall integrity pathway and perturbs calcium homeostasis. BMC Microbiol 11:209

Bussink HJ, Osmani SA (1999) A mitogen-activated protein kinase (MPKA) is involved in polarized growth in the filamentous fungus, *Aspergillus nidulans*. FEMS Microbiol Lett 173:117–125

Colabardini AC, De Castro PA, De Gouvêa PF, Savoldi M, Malavazi I, Goldman MH, Goldman GH (2010) Involvement of the *Aspergillus nidulans* protein kinase C with farnesol tolerance is related to the unfolded protein response. Mol Microbiol 78:1259–1279

Duncan VMS, O'Neil DA (2013) Commercialization of antifungal peptides. Fungal Biol. Rev. 26:156–165

Fischer R, Timberlake WE (1995) *Aspergillus nidulans apsA* (anucleate primary sterigmata) encodes a coiled-coil protein required for nuclear positioning and completion of asexual development. J Cell Biol 128:485–498

Fujioka T, Mizutani O, Furukawa K, Sato N, Yoshimi A, Yamagata Y, Nakajima T, Abe K (2007) MpkA-dependent and -independent cell wall integrity signaling in *Aspergillus nidulans*. Eukaryot Cell 6:1497–1510

Galgóczy L, Kovács L, Karácsony Z, Virágh M, Hamari Zs, Vágvölgyi Cs (2013) Investigation of the antimicrobial effect of *Neosartorya fischeri* antifungal protein (NFAP) after heterologous expression in *Aspergillus nidulans*. Microbiol-SGM 159:411–419

Gorovits R, Yarden O (2003) Environmental suppression of *Neurospora* crassa cot-1 hyperbranching: a link between COT1 kinase and stress sensing. Eukaryot Cell 2:699–707

Guest GM, Lin X, Momany M (2004) *Aspergillus nidulans* RhoA is involved in polar growth, branching, and cell wall synthesis. Fungal Genet Biol 41:13–22

Hagen S, Marx F, Ram AF, Meyer V (2007) The antifungal protein AFP from *Aspergillus giganteus* inhibits chitin synthesis in sensitive fungi. Appl Environ Microbiol 73:2128–2134

Hegedüs N, Marx F (2013) Antifungal proteins: More than antimicrobials? Fungal Biol. Rev. 26:132–145

Hegedus N, Leiter E, Kovács B, Tomori V, Kwon NJ, Emri T, Marx F, Batta G, Csernoch L et al (2011) The small molecular mass antifungal protein of *Penicillium chrysogenum*–a mechanism of action oriented review. J Basic Microbiol 51:561–571

Kaiserer L, Oberparleiter C, Weiler-Görz R, Burgstaller W, Leiter E, Marx F (2003) Characterization of the *Penicillium chrysogenum* antifungal protein PAF. Arch Microbiol 180:204–210

Katayama T, Uchida H, Ohta A, Horiuchi H (2012) Involvement of protein kinase C in the suppression of apoptosis and in polarity establishment in *Aspergillus nidulans* under conditions of heat stress. PLoS One 7:e50503

Kovács L, Virágh M, Takó M, Papp T, Vágvölgyi Cs, Galgóczy L (2011) Isolation and characterization of *Neosartorya fischeri* antifungal protein (NFAP). Peptides 32:1724–1731

Kuranda K, Leberre V, Sokol S, Palamarczyk G, François J (2006) Investigating the caffeine effects in the yeast *Saccharomyces cerevisiae* brings new insights into the connection between TOR, PKC and Ras/cAMP signalling pathways. Mol Microbiol 61:1147–1166

Leiter E, Szappanos H, Oberparleiter C, Kaiserer L, Csernoch L, Pusztahelyi T, Emri T, Pócsi I, Salvenmoser W, Marx F (2005) Antifungal protein PAF severely affects the integrity of the plasma membrane of *Aspergillus nidulans* and induces an apoptosis-like phenotype. Antimicrob Agents Chemother 49:2445–2453

Marx F, Binder U, Leiter E, Pócsi I (2008) The *Penicillium chrysogenum* antifungal protein PAF, a promising tool for the development of new antifungal therapies and fungal cell biology studies. Cell Mol Life Sci 65:445–454

Meyer V (2008) A small protein that fights fungi: AFP as a new promising antifungal agent of biotechnological value. Appl Microbiol Biotechnol 78:17–28

Montesinos E (2007) Antimicrobial peptides and plant disease control. FEMS Microbiol Lett 270:1–11

Oberparleiter C, Kaiserer L, Haas H, Ladurner P, Andratsch M, Marx F (2003) Active internalization of the *Penicillium chrysogenum* antifungal protein PAF in sensitive aspergilli. Antimicrob Agents Chemother 47:3598–3601

Ouedraogo JP, Hagen S, Spielvogel A, Engelhardt S, Meyer V (2011) Survival strategies of yeast and filamentous fungi against the antifungal protein AFP. J Biol Chem 286:13859–13868

Ronen R, Sharon H, Levdansky E, Romano J, Shadkchan Y, Osherov N (2007) The *Aspergillus nidulans pkcA* gene is involved in polarized growth, morphogenesis and maintenance of cell wall integrity. Curr Genet 51:321–329

Semighini CP, Hornby JM, Dumitru R, Nickerson KW, Harris SD (2006) Farnesol-induced apoptosis in *Aspergillus nidulans* reveals a possible mechanism for antagonistic interactions between fungi. Mol Microbiol 59:753–764

Shimizu K, Keller NP (2001) Genetic involvement of a cAMP-dependent protein kinase in a G protein signaling pathway regulating morphological and chemical transitions in *Aspergillus nidulans*. Genetics 157:591–600

Taheri-Talesh N, Horio T, Araujo-Bazán L, Dou X, Espeso EA, Peñalva MA, Osmani SA, Oakley BR (2008) The tip growth apparatus of *Aspergillus nidulans*. Mol Biol Cell 19:1439–1449

Theis T, Wedde M, Meyer V, Stahl U (2003) The antifungal protein from *Aspergillus giganteus* causes membrane permeabilization. Antimicrob Agents Chemother 47:588–593

Theis T, Marx F, Salvenmoser W, Stahl U, Meyer V (2005) New insights into the target site and mode of action of the antifungal protein of *Aspergillus giganteus*. Res Microbiol 156:47–56

Van Dijck P (2009) Nutrient sensing G protein-coupled receptors: interesting targets for antifungals? Med Mycol 47:671–680

Virágh M, Vörös D, Kele Z, Kovács L, Fizil Á, Lakatos G, Maróti G, Batta G, Vágvölgyi Cs, Galgóczy L (2014) Production of a defensin-like antifungal protein NFAP from *Neosartorya fischeri* in *Pichia pastoris* and its antifungal activity against filamentous fungal isolates from human infections. Protein Expr Purif 94:79–84

Yu JH, Wieser J, Adams TH (1996) The *Aspergillus* FlbA RGS domain protein antagonizes G protein signaling to block proliferation and allow development. EMBO J 15:5184–5190

6

Structural characterization of coatomer in its cytosolic state

Shengliu Wang[1,4], Yujia Zhai[1], Xiaoyun Pang[1], Tongxin Niu[1,4], Yue-He Ding[2], Meng-Qiu Dong[2], Victor W. Hsu[5], Zhe Sun[1✉], Fei Sun[1,3,4✉]

[1] National Key Laboratory of Biomacromolecules, Institute of Biophysics, Chinese Academy of Sciences, Beijing 100101, China
[2] National Institute of Biological Sciences, Beijing, Beijing 102206, China
[3] Center for Biological Imaging, Institute of Biophysics, Chinese Academy of Sciences, Beijing 100101, China
[4] University of Chinese Academy of Sciences, Beijing 100049, China
[5] Department of Medicine, Harvard Medical School, Brigham and Women's Hospital, Boston, MA 02115, USA
✉ Correspondence: zsun@moon.ibp.ac.cn (Z. Sun), feisun@ibp.ac.cn (F. Sun)

ABSTRACT

Studies on coat protein I (COPI) have contributed to a basic understanding of how coat proteins generate vesicles to initiate intracellular transport. The core component of the COPI complex is coatomer, which is a multimeric complex that needs to be recruited from the cytosol to membrane in order to function in membrane bending and cargo sorting. Previous structural studies on the clathrin adaptors have found that membrane recruitment induces a large conformational change in promoting their role in cargo sorting. Here, pursuing negative-stain electron microscopy coupled with single-particle analyses, and also performing CXMS (chemical cross-linking coupled with mass spectrometry) for validation, we have reconstructed the structure of coatomer in its soluble form. When compared to the previously elucidated structure of coatomer in its membrane-bound form we do not observe a large conformational change. Thus, the result uncovers a key difference between how COPI versus clathrin coats are regulated by membrane recruitment.

KEYWORDS coatomer, COPI, human, single-particle electron microscopy, membrane trafficking

Shengliu Wang and Yujia Zhai have made equal contributions to this work.

INTRODUCTION

Coat proteins play a key role in intracellular transport by initiating the formation of transport vesicles. They achieve this role by coupling two major functions, bending membrane that generates transport carriers from organellar membrane, and binding to cargoes for the proper packaging of newly formed vesicles. The ADP-ribosylation factor (ARF) family of small GTPases has been shown to regulate these function of coat proteins by dictating their recruitment from the cytosol to membrane (Bonifacino and Glick, 2004).

Studies on coat protein I (COPI) have contributed to a basic understanding of how coat proteins achieve its roles in vesicle formation and cargo sorting (Faini et al., 2013; Hsu et al., 2009; Jackson, 2014; Pucadyil and Schmid, 2009). The core component of the COPI complex is coatomer, which exists as a multimeric complex containing seven subunits: α-, β-, β′-, γ-, δ-, ε-, and ζ-COP (Hara-Kuge et al., 1994; Malhotra et al., 1989; Waters et al., 1991). Large efforts have been made to elucidate the structure of coatomer over the years, including several crystal structures of fragments of the seven subunits (Jackson et al., 2012; Lee and Goldberg, 2010; Ma and Goldberg, 2013; Suckling et al., 2015; Yu et al., 2012), and electron microscopy (EM)-based reconstruction of the entire structure of coatomer (Dodonova et al., 2015; Yip and Walz, 2011).

Coatomer can be further divided into two subcomplexes, with the B-subcomplex containing α, β′, and ε subunits and the F-subcomplex containing β, δ, γ, and ζ subunits. The F-subcomplex shows significant structural similarity to the clathrin adaptors and interacts with two molecules of activated (GTP-bound) form of ARF1 (ADP-ribosylation factor 1) via the N-termini of β- and γ-COP (Faini et al., 2013; Serafini et al., 1991b; Yu et al., 2012). In the B-subcomplex, α- and β′-COP are considered structural homologs. Both adopt the domain structure of β-propeller-β-propeller-α-solenoid, and they form a heterodimer via the α-solenoid region (Lee and Goldberg, 2010).

In recent years, structural studies on the clathrin adaptors have shed important additional insights into how membrane recruitment regulates the function of coat proteins. Clathrin AP2 has been found to undergo a major conformational change upon its recruitment from the cytosol to membrane (Jackson et al., 2010). This change is instigated by AP2 binding to a key lipid, phosphatidylinositol (4,5) bis-phosphate (PIP2), at the plasma membrane. The soluble form of AP2 exists in a "closed" conformation, which prevents cargo binding by AP2 (Collins et al., 2002). Binding to PIP2 on membrane induces an "open" conformation of AP2 to allow cargo binding (Jackson et al., 2010). Structural study on clathrin AP1 also suggests that it undergoes a major conformational change upon membrane recruitment (Ren et al., 2013). In the soluble state, AP1 adopts a similarly "closed" conformation to prevent cargo binding (Heldwein et al., 2004). Upon binding to ARF1 on membrane, AP1 shifts into an "open" conformation to allow cargo binding (Ren et al., 2013).

As basic mechanisms of intracellular transport are generally conserved (Bonifacino and Glick, 2004), coatomer is predicted to undergo a similar large conformational change during membrane recruitment. However, recent structural reconstructions of membrane-bound coatomer have begun to identify key differences between how COPI and the clathrin complexes are organized on membrane for vesicle formation (Dodonova et al., 2015; Faini et al., 2012). As such, whether the large conformational changes that have been observed for the membrane recruitment of clathrin adaptors would also apply to the COPI complex becomes an open question. In this study, we address this key mechanistic issue by elucidating the structure of coatomer in its soluble form. By comparing this structure with the previously solved structure of coatomer in its membrane-bound form (Dodonova et al., 2015), we have uncovered yet another key difference between coatomer and the clathrin adaptors.

RESULTS

Expression and purification of recombinant human coatomer

Coatomer exists as a single complex of approximately 560 kDa (Waters et al., 1991). The subunits of coatomer and their major domains are shown in Fig. 1A. To facilitate the isolation of highly purified coatomer for structural analysis, recent studies have generated recombinant mouse coatomer using a MultiBac expression system (Dodonova et al., 2015; Faini et al., 2012; Sahlmuller et al., 2011). In the present study, we developed a novel protein complex expression system, ViperTEC (to be published elsewhere), to express recombinant human coatomer in insect cells. Specifically, cDNAs of α-, β′-, and ε-COP were inserted into one expression cassette containing a p6.9 promoter and a SV40 terminator, while cDNAs for the other four subunits (β-, γ-, δ-, and ζ-COP) were inserted into another expression cassette containing a p10 promoter and a SV40 terminator. Based on previous structural studies (Jackson et al., 2012; Lee and Goldberg, 2010; Ma and Goldberg, 2013; Suckling et al., 2015; Yu et al., 2012), tags were introduced at the 5′ end of each subunit to minimize potential effects on the subsequent structure and function of coatomer. As will be seen below, this has facilitated our ability to track the expression of individual subunits and their assembly state during purification. Moreover, we did not detect significant effects on coatomer function.

To purify the recombinant human coatomer expressed in insect cells, we performed Strep tag II affinity chromatography followed by Mono Q anion exchange chromatography. Multiple observations suggested that this approach resulted in intact coatomer being purified. First, Coomassie-stained SDS-PAGE gel showed the presence of all seven subunits of coatomer (Fig. 1B), which was further verified by Western blotting (Fig. S1). Second, gel filtration showed a single peak during elution (Fig. 1C). Third, native PAGE reveals a monodisperse band, indicating that the recombinant coatomer was isolated with high purity and homogeneity (Fig. 1D). Fourth, analysis using liquid chromatography mass spectrometry (LC-MS) of the single band observed in native page resulted in all seven subunits of coatomer being identified (Table 1). Our yield of purified coatomer was approximately 1.5 mg per liter of cell culture.

Our previous experiences in purifying native coatomer from tissue suggested that the linkage between the μ-homology domain and the N-terminal domain (NTD) of δ-COP is flexible, as the μ-homology domain was frequently degraded (unpublished data). In the purification of recombinant coatomer, we also observed the degradation of δ-COP, which appeared as an additional band in Coomassie-stained gel (Fig. 1B). Western blotting (Fig. S1) and mass spectrometry analysis (Table S1) identified this additional band as the N-terminal portion of δ-COP. Collectively, these observations suggested that a portion of δ-COP is likely to be highly exposed on the surface of coatomer, as this portion shows consistent partial degradation during the purification of coatomer.

In vitro functional assays of the recombinant human coatomer

We next assessed whether the purified recombinant human coatomer is functional. For this goal, we generated recombinant human N-myristoylated ARF1 (Fig. S2A). Moreover, we used liposomal membrane rather than Golgi membrane,

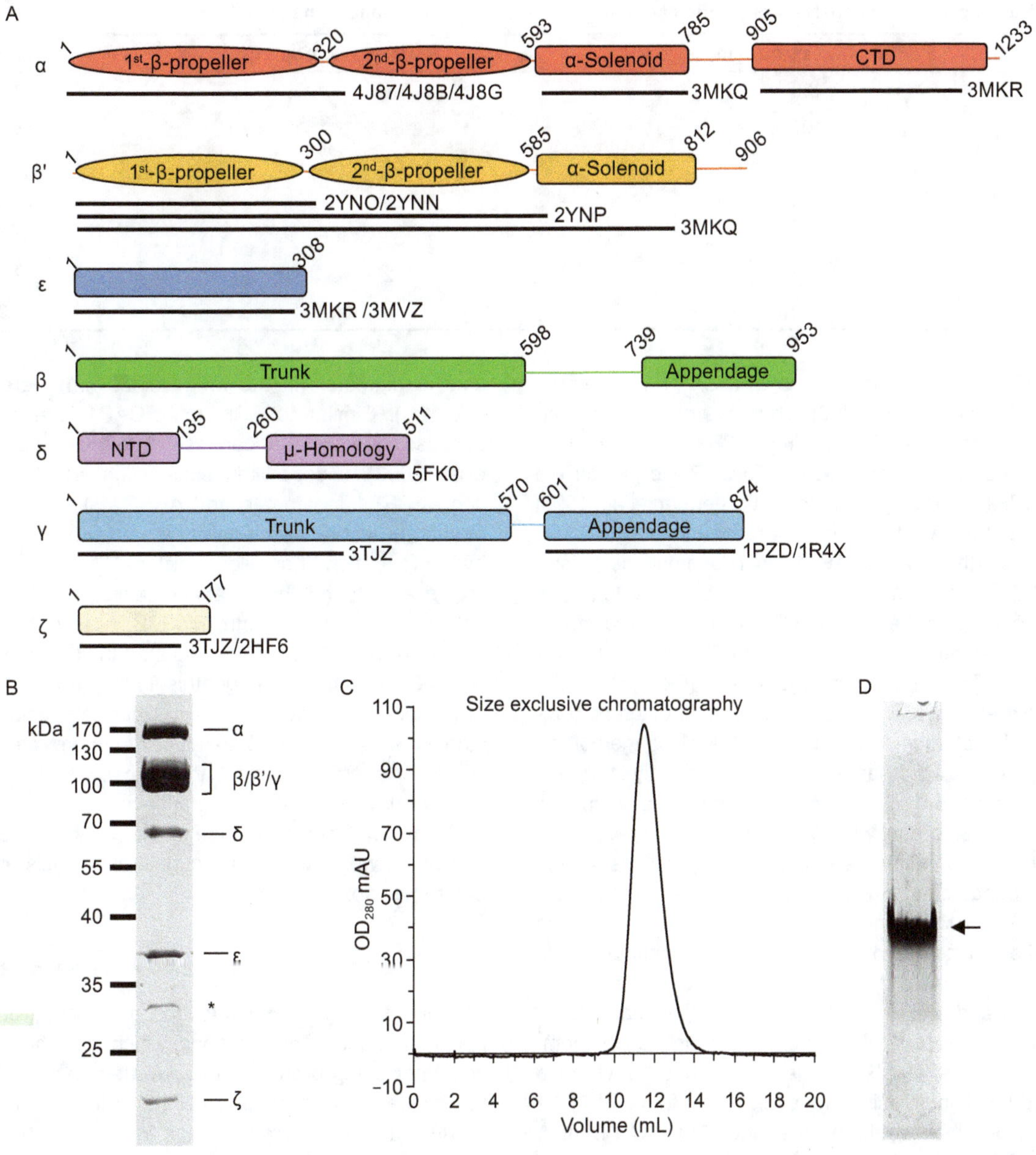

Figure 1. Preparation of purified recombinant human coatomer. (A) Schematic diagram showing the domain organization of each subunits of human coatomer. The available atomic structures with their PDB IDs and sequence coverage are also labeled. (B) Coomassie-blue stained SDS-PAGE of recombinant coatomer with all the subunits labeled. Asterisk indicates the N-terminal fragment of δ-COP, which is confirmed by mass spectrometry and Western blot. (C) Gel filtration of purified coatomer. The retention volume of the single peak on the column of Superose 6 (GE healthcare) is 11.4 mL. (D) Native PAGE of the purified coatomer. Arrow indicates the intact coatomer.

as native membrane contains a myriad of other proteins that can confound the ability to assess directly the roles of ARF1 and coatomer in COPI vesicle formation. When recombinant forms of ARF1 and coatomer were incubated with liposomes that contain a Golgi-like composition of pure lipids, we confirmed that ARF1 enhanced the recruitment of coatomer onto membrane (Fig. 2A). We also confirmed that the GTP-bound form of ARF1 was more efficient in binding to liposomes and in recruiting coatomer to this membrane (Fig. 2A). Thus, the recombinant forms of ARF1 and coatomer that we had generated are functional with respect to membrane recruitment.

We next considered that the GTPase-activating protein (GAP) responsible for catalyzing the deactivation of ARF1

Table 1. Mass spectrometry analysis of purified coatomer extracted from the band of native PAGE

Subunit	Molecular mass (kDa)	Unique peptide (n)	Coverage (%)
α-COP	139	108	77.5
β-COP	107	84	68.5
β′-COP	102	64	74.2
γ-COP	98	69	74.6
δ-COP	57	33	54.2
ε-COP	34	22	83.4
ζ-COP	20	6	37.9

during COPI vesicle formation is ARFGAP1 (Cukierman et al., 1995; Yang et al., 2002). Thus, we also purified the recombinant form of ARFGAP1 (Fig. S2B). Consistent with previous studies that have found ARFGAP1 to induce the release ARF1 from Golgi membrane (Cukierman et al., 1995; Yang et al., 2002), we found that our recombinant ARFGAP1 also reduced the level of ARF1 on liposomal membrane (Fig. 2B). This effect of ARFGAP1 could be attributed to its catalytic activity, as the use of GTPγS (a non-hydrolyzable GTP analogue) prevented the ability of ARFGAP1 to reduce the level of ARF1 on liposomal membrane (Fig. 2B).

We next examined the effect of recruiting different combinations of coatomer, ARF1, and ARFGAP1 on membrane morphogenesis. ARF1 has been found previously to induce membrane curvature by inserting its N-terminal amphipathic helix into the membrane, which is reflected by the generation of tubules that sometimes also contain constrictions, resulting in a "beads-on-string" morphology (Beck et al., 2008; Krauss et al., 2008; Lundmark et al., 2008). We observed similar effects when recombinant ARF1 was incubated with Golgi-like liposomes (Fig. 2C). As the co-incubation of activated ARF1 and coatomer has been found previously to be sufficient to produce COPI vesicles from liposomal membrane (Bremser et al., 1999; Spang et al., 1998), we also confirmed this finding using the recombinant forms of ARF1 and coatomer that we had generated (Fig. 2C). Further scrutiny of these vesicles revealed that they are coated (Fig. S3A) and have diameter in the range of 50–80 nm (Fig. 3). Western blotting confirmed the presence of ARF1 and β-COP on these vesicles (Fig. S3B).

We also noted that COPI vesicles reconstituted from Golgi membrane have been found to be relatively depleted of cholesterol (Brugger et al., 2000). Consistent with this previous finding, we found that liposomal membrane was more readily vesiculated with lower concentrations of cholesterol (Fig. S4A, and also Supplemental text). The diameters of vesicles were also affected by the level of cholesterol (Fig. S4B, and also Supplemental text).

Besides ARF1 and coatomer, ARFGAP1 has been found previously to promote COPI vesicle formation (Yang et al., 2002). Consistent with this finding, we observed vesiculation when ARFGAP1 was added in conjunction with ARF1 and coatomer to liposomes (Fig. 2C). In this case, vesicles of smaller diameter (in the range of 30 nm) were observed (Fig. 2C). Similar to ARF1, ARFGAP1 is also known to possess a domain that can insert into the membrane (Bigay et al., 2003). Thus, a likely explanation is that the combination of ARF1, coatomer, and ARFGAP1 results in greater membrane curvature being generated, which is manifested by the formation of smaller vesicles.

We also found that coatomer alone could be recruited onto liposomal membrane (Fig. 2B), with EM examination revealing that this condition resulted in liposome tubulation (Fig. 2C). To our knowledge, this finding revealed a property of coatomer that had not been widely appreciated. Future studies will be needed to elucidate more precisely how coatomer exerts this effect. In any case, as the cumulative results above represent substantial support that the recombinant form of human coatomer that we have generated is functional, we next pursued structural analysis of this coatomer in its soluble form.

Electron microscopy of coatomer in its soluble form

We initially pursued cryo-electron microscopy but failed to obtain high-quality micrographs, which could be attributed to coatomer being prone to collapse after cryo-vitrification, as the air-water interface that occurs in this procedure could produce damaging effects. Thus, we next pursued negative-stain electron microscopy coupled with computational image-processing methods.

We detected soluble coatomer as mono-dispersed particles with diameter in the range of 13–20 nm (Fig. S5A). These were selected for further image processing. Using reference-free 2D classification, we found that class averaging of most particles revealed a central density that is relatively invariant (Fig. S5B), while more peripheral densities appear variable in both location and shape (Fig. S5B). These findings were consistent with previous observations on the structure of yeast coatomer (Yip and Walz, 2011). Further scrutiny of the central density revealed that it exists as two parts (Fig. S5B and S5C). We also deduced that the variable extra densities could be attributed to flexible domains known to exist in coatomer. Besides δ-COP, these would include portions of the α^{CTD}/ε-COP heterodimer and the appendage domains of β/γ-COP.

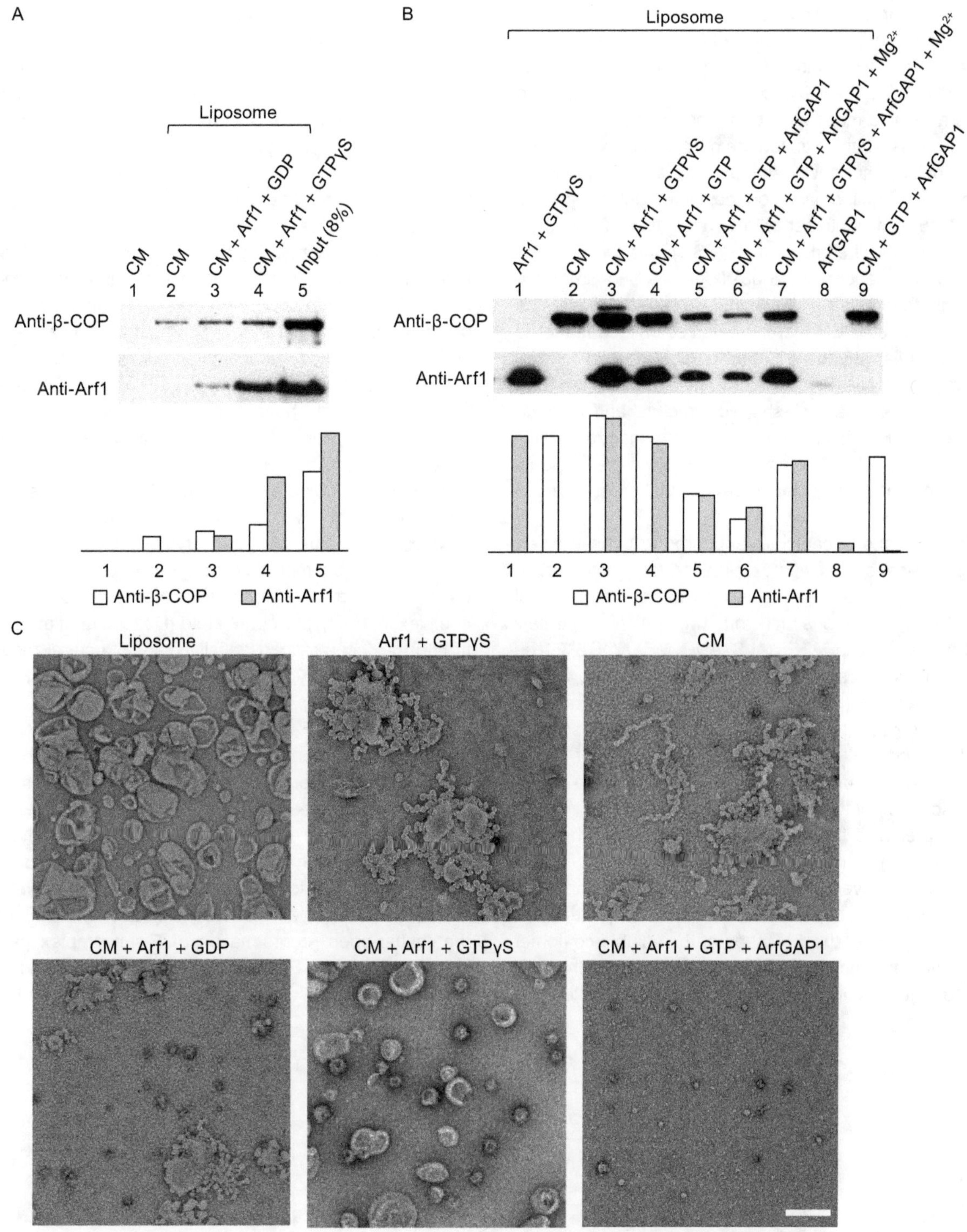

Figure 2. Membrane binding and vesiculation assays. (A and B) Membrane binding co-sedimentation assays (See MATERIALS AND METHODS). The amount of co-sedimented coatomer (β-COP) and ARF1 is also quantified in lower panels. (C) Negative-stain electron microscopy of liposomes incubated with coatomer and other factors. Scale bar, 200 nm. CM, human coatomer; ARF1, N-myristoylated human ARF1; ARFGAP1, rat ARFGAP1.

We then pursued single-particle analysis. To avoid potential modeling bias, we collected the tilt series of negatively stained coatomer and performed electron tomographic reconstruction followed by sub-tomogram averaging to generate a reliable model. We performed intensive iterative 2D and 3D classifications to sort the particle sets. We filtered out overlapping, imperfectly stained, or disrupted particles. The final reconstruction based on the filtered dataset yielded a 3D structure of coatomer with a resolution of 19 Å, as judged by Fourier shell correlation (FSC) criterion of 0.143 (Rosenthal and Henderson, 2003) (Figs. 4A and S5). Consistent with the results from the 2D class averages, the overall reconstructed EM map revealed soluble coatomer to possess two relatively independent parts. One part (part 1) adopts an arch shape (Fig. 4A, pink), while another part (part 2) sits adjacent to the arch-shaped structure and appears more globular (Fig. 4A, yellow). The arch-shaped density (part 1) was invariant during iterative 3D classifications, while the globular density (part 2) varied in different cycles of 3D classifications.

As the two parts of coatomer appeared to behave differently and independently, we next pursued local optimization (Shan et al., 2016) and performed "focus" refinement (Bai et al., 2015) to reconstruct the two parts separately (Fig. S6A; see also "MATERIALS AND METHODS"). The "focus" refinement resulted in a similar density map for the arch-shaped portion (part 1), and a different density map for the globular portion (part 2) (Fig. S6A). The final resolution was 18 Å and 20 Å, respectively, with FSC of 0.143 (Rosenthal and Henderson, 2003) (Fig. S6B). There was a flexible oscillation between the two parts, and the alignment procedure preferred the more stable part1 during the conventional refinement (Figs. 4C and S5A).

The recently solved structure of membrane-bound coatomer (Dodonova et al., 2015) provided an opportunity to analyze and interpret the EM maps of the soluble coatomer that we have generated. In order to detect potential structural difference between the two states, we initially fitted the complete structural model of the membrane-bound coatomer (PDB ID: 5A1U) as a rigid body onto the integrated density map before the "focus" refinement using UCSF *Chimera* (Pettersen et al., 2004). A good match was observed between the α/β′-COP heterodimer and the arch-shaped density. Thus, we assigned the arch-shaped density to correspond to the B-subcomplex of coatomer (Fig. 4B). In further support of this assignment, we noted that the α/β′-COP heterodimer has been suggested to be rigid, showing little difference when comparing the structure in solution (Lee and Goldberg, 2010) versus that on membrane (Dodonova et al., 2015). Thus, when taken collectively, we also assigned the other part of coatomer that appears as a globular density to correspond to the F-subcomplex. However, the fit in this case was relatively weak (Fig. 4B). This could be attributed to conformational changes of F-subcomplex between two states, or the potential flexible oscillation between the two subcomplexes of coatomer.

With the improved density map after "focus" classification and refinement, we also compared our reconstructed F-subcomplex with the three crystal structures of clathrin AP1 (the F-subcomplex homologue) in different conformations as well as the structure of membrane-bound F-subcomplex in the "hyper-open" state (Dodonova et al., 2015) to fit into the focus-refined density map. Structural models of AP1 included its "hyper-unlocked" state (PDB ID: 4P6Z) (Jia et al., 2014), "unlocked" state (PDB ID: 4HMY) (Ren et al., 2013) and "locked" state (PDB ID: 1W63) (Heldwein et al., 2004). As the C-terminal μ-homology domain of δ-COP is prone to degradation, and also considering that this domain was omitted in the previous reconstruction of membrane-bound coatomer, we omitted this domain during the fitting. The appendage domains of β- and γ-COP were also omitted, as they are known to be linked flexibly to the trunk domains. After incorporating all these considerations, we found that the "hyper-open" conformation of F-subcomplex gave the best fit (Fig. 4D). Besides the densities occupied by the F-subcomplex, the unassigned density located underneath

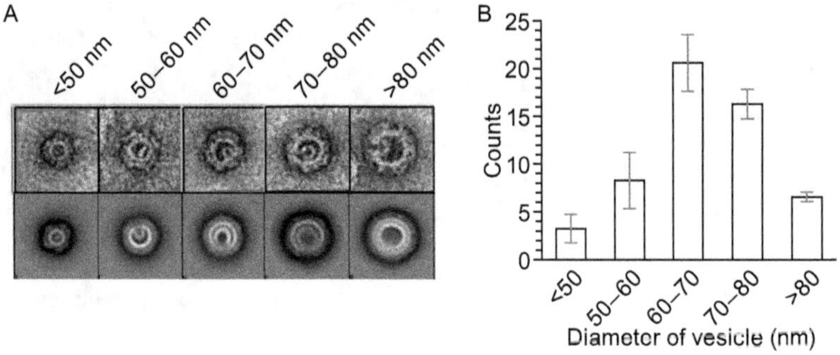

Figure 3. Diameter distribution of reconstituted COPI-coated vesicles. (A) The raw images of vesicles (first row) and the corresponding 2D class averages of vesicles with the same diameter (second row). The diameters are indicated on the top. (B) A statistical histogram of the vesicle number vs. different diameters. All error bars represent standard deviation (s.d.) from three independent vesicle reconstitution experiments. For each experiment, the number (N) of micrographs is 90.

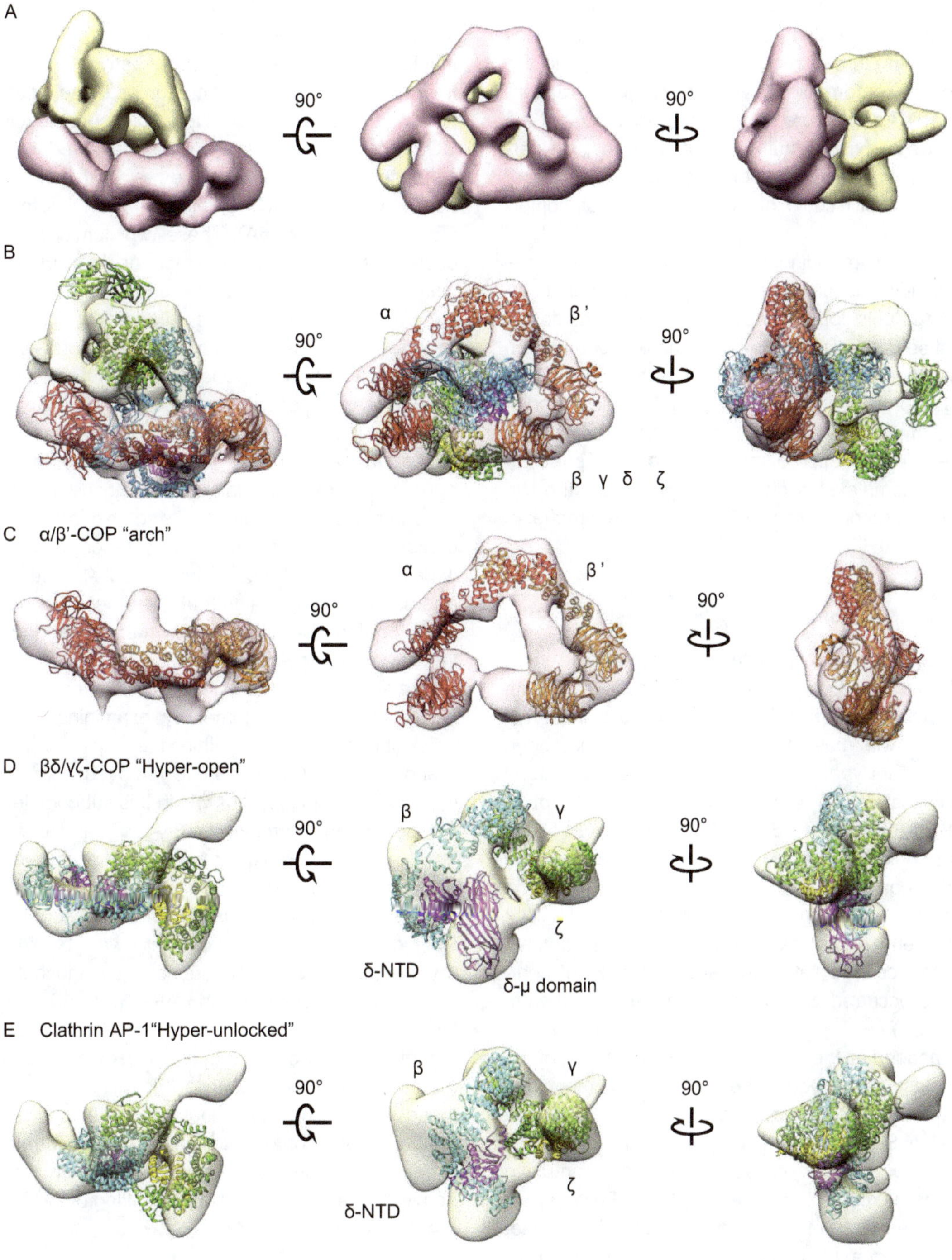

Figure 4. 3D reconstruction of the recombinant human coatomer in its soluble form. (A) Three views of surface-rendered density map of coatomer obtained by single-particle reconstruction. The map is segmented using UCSF Chimera, grouped, and colored with pink for the arch-shape portion and yellow for the more globular portion. (B) Fitting the membrane-bound model of coatomer (PDB ID: 5A1U) into the EM map. (C) Fitting the B-subcomplex αβ′-COP heterodimer into the "focus" refined arch-shape density map. (D) Fitting the F-subcomplex βδ/γζ-COP into the "focus" refined globular density map. (E) Fitting the homologue model of F-subcomplex in "hyper-unlocked" form into the "focus" refined globular density map. α-, β′-COP are colored in red and orange. β-, δ-, γ-, and ζ-COP and their homologues are colored in cyan, purple, green, and yellow respectively.

the "arch" on the β/γ-COP side could be assigned to the C-terminal μ-homology domain of δ-COP with good fit (Fig. 4D). As for the three structural models of AP1, only the "hyper-unlocked" conformation was found to give a reasonable fit (Fig. 4E). Furthermore, to test the possibility that the F-subcomplex can exist in multiple conformations, we performed 3D classification. However, after several cycles of classifications, we still could not detect any other conformation (Fig. S7).

Thus, the collective findings led us to conclude that the F-subcomplex exhibits an open conformation when coatomer is in its soluble form. Moreover, this conformation is similar to that seen previously for the membrane-bound form of coatomer (Dodonova et al., 2015). We further considered that the negative-stain EM approach provides a low-resolution structure of soluble coatomer. Thus, when also considering that flexible oscillation exists between the F-subcomplex and B-subcomplex, we did not attempt to join the two parts of structural reconstruction into one complete model of coatomer.

CXMS analysis of coatomer in its cytosolic state

We next sought to further validate the structural reconstruction of coatomer we had generated above by taking another independent approach. Chemical cross-linking of proteins coupled with mass spectrometry (CXMS) has been developed in recent years as a way of validating structural models (Leitner et al., 2014; Yang et al., 2012). We found that there were 82 pairs of DSS or BS^3 cross-linked lysine residues that could be detected with high confidence in the soluble form of coatomer, of which 50 were within the same domain and the remaining 32 occurred as cross-links between different domains. Among these inter-domain cross-links, 14 occurred within the subcomplexes of coatomer, while 18 occurred as links between the two subcomplexes (Table S2).

We next mapped all the CXMS pairs into the model of the previously elucidated membrane-bound coatomer and measured the Cα–Cα distances between the cross-linked residues (Table S2). The generally accepted Cα–Cα distance between two cross-linked lysine residues, when using DSS or BS^3 as a cross-linker, is below ~35 Å (Politis et al., 2014). Since the model of the membrane-bound coatomer was generated from a 13-Å cryo-EM map (Dodonova et al., 2015), we relaxed the distance restriction of Cα–Cα to 45 Å. This resulted in all the intra-domain cross-links having Cα–Cα distances below 30 Å except for one pair that was below 35 Å (Table S2 and Fig. 5A). These findings suggested that the conformation of individual domains changes very little from the soluble to the membrane-bound form of coatomer.

As for the inter-domain cross-links, we found that 11 had distances greater than the cut-off value of 45 Å. Ten of these links were located between α^{NTD}-COP and the α^{CTD}/ε-COP heterodimer (Table S2 and Fig. 5A, orange). Furthermore, the cross-link distances between α^{CTD}-COP and ε-COP were below the cut-off value (Table S2). Thus, these results suggested that the heterodimer of α^{CTD}/ε-COP is structurally stable but changes its location between the soluble and membrane-bound forms of coatomer. In the soluble form, the heterodimer of α^{CTD}/ε-COP interacts more closely with α^{NTD}-COP. The 11th cross-link that was observed to have greater distance than the cut off value resides between the appendage domain of β-COP and the N-terminal domain of δ-COP (Table S2 and Fig. 5A). These two domains were far from each other in the membrane-bound form of coatomer, but were sufficiently close in the soluble form of coatomer to become cross-linked, thus suggesting a conformational change. Since the appendage domain of β-COP interacts with the N-terminal β-propeller domains of α-COP and contributes to the interaction between the two subcomplexes of coatomer in the membrane-bound state, we speculated that the conformational changes predicted by the crosslinking approach corresponds to the flexible oscillation between two subcomplexes of coatomer in its soluble form. Aside from the cross-links described above, all other inter-domain links were below the cut-off value of 45 Å. Thus, the CXMS approach also suggested that coatomer does not undergo a large conformational change when comparing its soluble versus membrane-bound form.

We also sought to validate our fitting results above on the F-subcomplex of coatomer by examining all 16 pairs of cross-linked residues within the main structure of the F-subcomplex. To do this, we removed from consideration the known flexible regions within this subcomplex, which are the appendage domains of β/γ-COP and the μ-homology domain of δ-COP. The three structural models of AP1 in different conformations were used as reference, as well as the membrane-bound form of coatomer. We found that all the Cα–Cα distances of the cross-links were below 35 Å for the "hyper-open" coatomer F-subcomplex and below 40 Å for the "hyper-unlocked" clathrin AP-1 (Fig. 5B and Table S3). However for the "unlocked" and the "locked" clathrin AP-1 models, there were some cross-links with larger distance than the cut-off value of 45 Å (Fig. 5B and Table S3). Thus, these results suggested that the F-subcomplex exhibits a relatively open conformation in the soluble form of coatomer. Notably, this conclusion is also consistent with the result above derived from the rigid-body fitting of the negative-stain EM map (Fig. 4D and 4E).

DISCUSSION

Using negative-stain EM and single-particle analysis, we have reconstructed the 3D structure of recombinant human coatomer in its soluble form (Fig. 4). This structure has been further validated by an independent approach using CXMS analysis. We then compared our reconstructed structure of soluble coatomer with a previously reconstructed structure of membrane-bound coatomer to reach a key conclusion, which is that soluble coatomer adopts largely a similar overall conformation as the membrane-bound form.

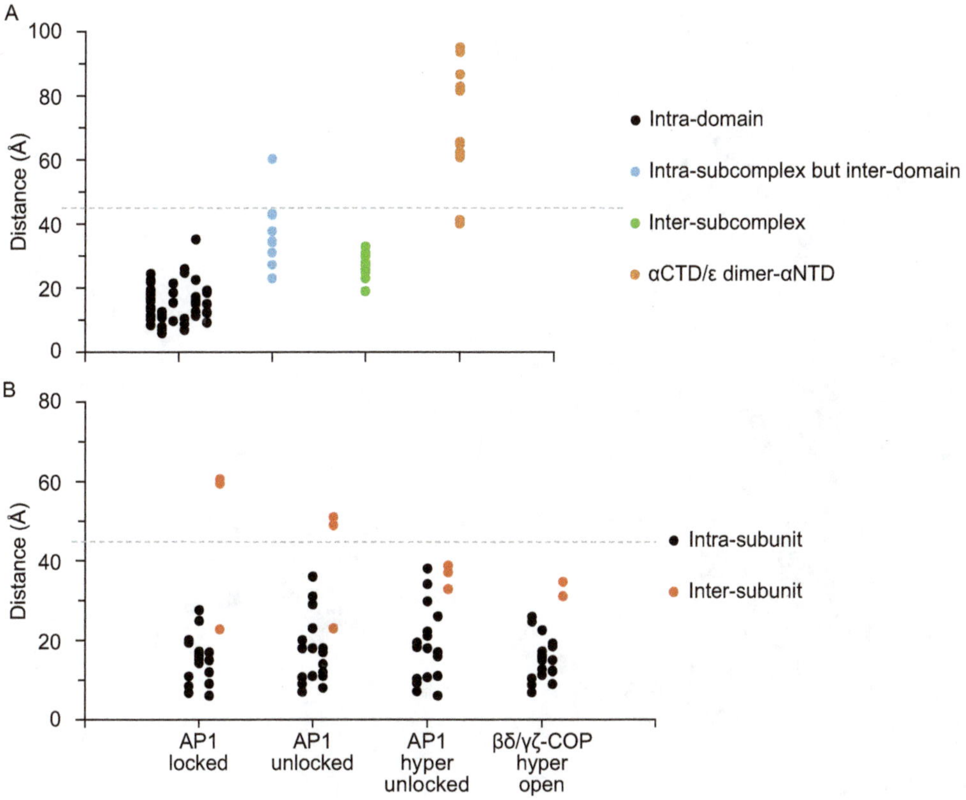

Figure 5. Mapping CXMS cross-linked residue pairs onto structural models. (A) The distances between Cα–Cα atoms of cross-linked residues pairs were measured based on the membrane-bound coatomer model (PDB ID: 5A1U), classified and plotted. (B) The distances between Cα–Cα atoms of cross-linked residues pairs within the adaptor F-subcomplex were measured based on four models as indicated, classified into intra-subunit (black) and inter-subunit (red) ones and plotted. The cut-off distance of 45 Å is represented by the dashed line.

Coatomer is comprised of two subcomplexes. We find that the B-subcomplex forms an arch-like structure, which does not vary significantly when comparing the soluble versus the membrane-bound form of coatomer (Fig. 4C). The F-subcomplex also appears largely similar when comparing the two states of coatomer. Moreover, when compared with the previously elucidated structures of clathrin AP1, the F-subcomplex adopts a conformation that is similar to the open conformation of AP1. Thus, in contrast to the case of clathrin adaptors, for which membrane recruitment induces a large conformational change from "closed" to "open" conformation, coatomer remains in the "open" conformation regardless of whether it resides in the cytosol or on membrane (Fig. 4D).

We further note that flexibility in coatomer has been noted previously (Langer et al., 2008; Reinhard et al., 1999; Yip and Walz, 2011). Our current results reveal that flexibility can come from the two subcomplexes of coatomer having mobile interactions. We also acknowledge that flexibility can come domains within individual subunits of coatomer. These would include δ-COP, as well as portions of the α^{CTD}/ε-COP heterodimer and the appendage domain of β/γ-COP. However, due to the limited resolution of the negative-stain EM approach, a more detailed assessment of these flexibilities will require future studies that can achieve a higher-resolution structure of coatomer.

In any case, our current results suggest a speculative model for how coatomer is recruited onto membrane to participate in COPI vesicle formation (Fig. 6). Considering the flexible oscillation between the two subcomplexes of the coatomer, we propose that the interactions between membrane-anchored ARF1 and the F-subcomplex, together with the interaction between the scaffolding B-subcomplex and membrane, help to immobilize the interaction between the two subcomplexes of coatomer. This then triggers a conformational transition into the "active" form of coatomer on membrane. The subsequent self-assembly of coatomer, together with regions of ARF1 and coatomer interacting with the underlying membrane, induces curvature on this membrane, which eventually results in vesicle formation.

A new insight on coatomer has also come from our functional studies. Although early studies on COPI vesicle

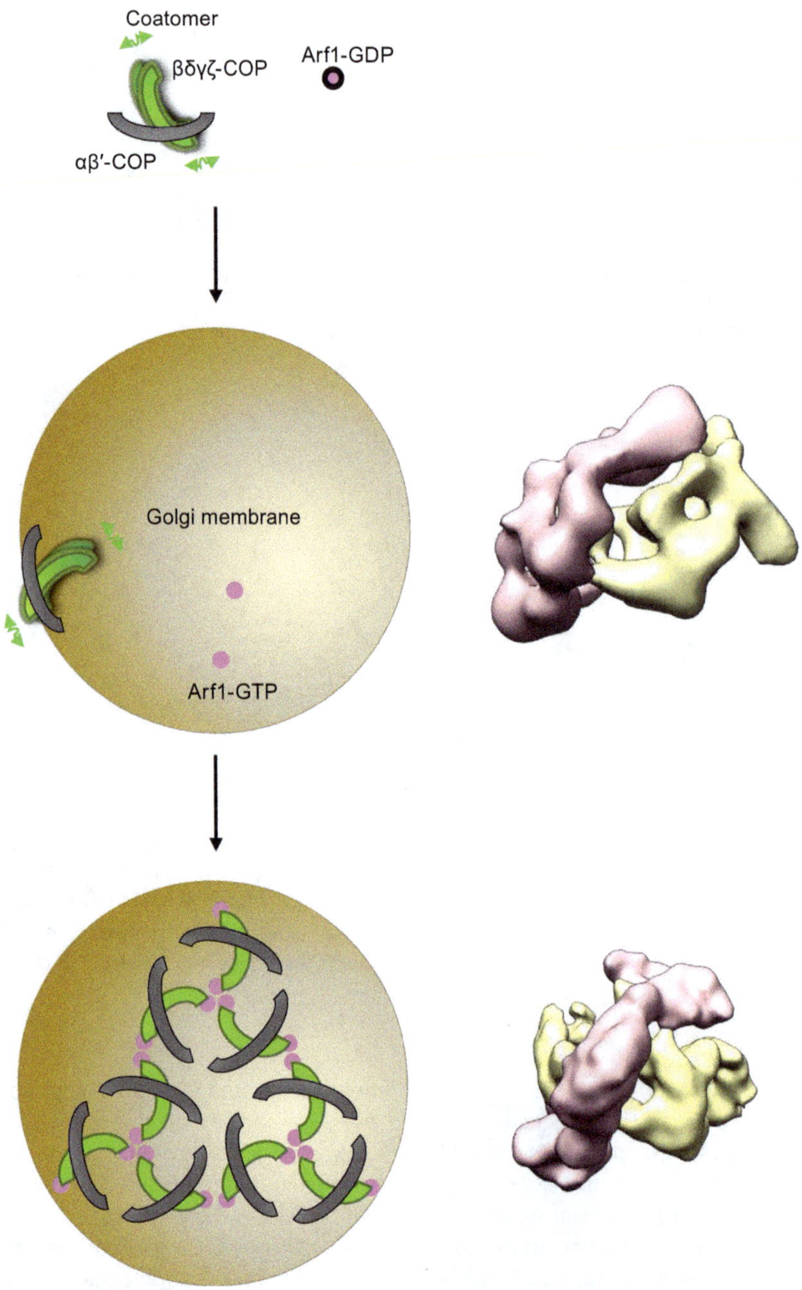

Figure 6. A working model for the membrane recruitment of coatomers in promoting COPI vesicle formation. The two subcomplexes (B-subcomplex in gray and F-subcomplex in green) of coatomer can oscillate in its cytosolic state. In the absence of ARF1-GTP (magenta), coatomer binds to membrane (brown) via the B-subcomplex, remodeling the membrane into tubules or "beads-on-string" morphology, for which the precise mechanism remains to be elucidated. Oscillation of the membrane-bound coatomer prevents its ability to drive vesicle formation. In the presence of membrane-anchored ARF1-GTP, the direct interaction between F-subcomplex and ARF1-GTP reduces the oscillation between the two subcomplexes, thereby driving the membrane-bound coatomer to a more stabilized (less oscillation) form. This stabilized coatomer is then able to assemble into oligomers, which together with ARF1, drives COPI vesicle formation.

formation have used Golgi membrane, the use of simpler liposomal membranes in recent years has provided a more direct way of assessing how key factors, such as coatomer and ARF1, act to achieve vesicle formation (Bremser et al., 1999; Spang et al., 1998). In this regard, whereas ARF1 had been known to recruit coatomer from the cytosol to Golgi membrane in regulating the function of coatomer (Donaldson et al., 1992; Serafini et al., 1991a), the use of the simpler liposomal membrane has uncovered that ARF1 also possess a direct ability to induce membrane curvature (Beck et al., 2008; Krauss et al., 2008; Lundmark et al., 2008). In the current study, we have used of the simpler liposomal membrane to uncover that coatomer also possesses a direct ability to induce membrane curvature. To our knowledge, this behavior of coatomer has not been noted previously.

Studies on different coat complexes have revealed that single factors typically convert liposomes into tubules or "beads-on-string" morphology (Beck et al., 2008; Farsad et al., 2001; Lee et al., 2005a; Peter et al., 2004), while multiple factors are needed to convert liposomes into vesicles (Bremser et al., 1999; Matsuoka et al., 1998; Spang et al., 1998; Takei et al., 1998). We have observed similar effects, as the incubation of Golgi-like liposomes with either coatomer or ARF1 induces "beads-on-string" morphology, while their incubation together induces a more complete fission of liposomal membrane to generate COPI vesicles. These observations can be explained by the fission process requiring considerable force/strain being imposed on membrane. Thus, multiple factors acting in concert provide a way to generate such formidable force/strain.

At the mechanistic level, how protein factors induce membrane curvature is explained currently by two general mechanisms: i) introducing a wedge defect into one leaflet of the membrane bilayer that results in strain to promote curvature or ii) exerting a scaffolding mechanism that involves a curved protein structure binding to the underlying membrane through electrostatic interactions (McMahon and Gallop, 2005). The first mechanism explains how the ARF-like small GTPases induce membrane curvature, as they possess an N-terminal amphipathic helix that can insert into one leaflet of the membrane bilayer to induce membrane curvature (Beck et al., 2008; Bielli et al., 2005; Hariri et al., 2014; Lee et al., 2005a). How coatomer induces membrane curvature is less clear. Based on the recent structural reconstruction of coatomer on membrane, four sites in coatomer are predicted to contact the underlying membrane (Dodonova et al., 2015). Two of them are located in F-subcomplex and interact with the underlying membrane indirectly via ARF1. The other two are located in the B-subcomplex, and can potentially interact directly with membrane. However, because a high-resolution structure for these potential direct binding sites is currently unavailable, the precise mechanism by which coatomer induces membrane curvature remains to be determined in the future.

It is also notable that, besides ARF1, coatomer has been found to interact with multiple other proteins upon recruitment to membrane. These include cargo proteins (Cosson and Letourneur, 1994; Fiedler et al., 1996) and the catalytic regulators of ARF1, its GEF (Deng et al., 2009), and GAP (Goldberg, 1999; Lee et al., 2005b). Thus, an important future goal will be to pursue structural studies of these interactions, as they have the prospect of uncovering different ways by which the function of coatomer is regulated.

MATERIALS AND METHODS

Antibodies

For Western blotting, the rabbit polyclonal antibody specific for C-terminus of β-COP (H-300) was purchased from Santa Cruz Biotechnology. The anti-ARF1 antibody was raised against human ARF1(ΔN17) in rabbits.

Molecular cloning, protein expression, and purification

The following cDNAs encoding for subunits of coatomer were synthesized commercially (Genewitz): α-COP isoform 1 (accession number NM_001098398.1), β-COP (accession NM_001144061.1), β′-COP (accession number NM_004766.2), γ-COP (accession number NM_016128.3), δ-COP isoform 1 (accession number NM_001655.4), ε-COP isofom 1 (accession number NM_007263.3), and ζ-COP isoform 1 (accession number NM_016057.2). The large plasmid containing all seven genes separated into two independent expression cassettes was developed by using ViperTEC method (to be published elsewhere) and transformed into DH10Bac competent *E. coli* for bacmid generation. The presence of all seven subunits of coatomer in the bacmid was verified by PCR before transfecting into Sf9 insect cells. The generation of baculovirus was performed by standard procedures. All Sf9 cells were cultured in serum free media (either Sf900-II, Life Technologies or ESF921, Expression Systems).

For protein expression, we infected 600 mL spinner flask cultures of Sf9 cell at a density of 2×10^6 cells/mL with 5 mL of P2 virus. After 72 h, cells were harvested by centrifugation and suspended in cold lysis buffer (50 mmol/L Tris-Cl, pH 8.0, 100 mmol/L KCl, 10% glycerol, 1 mmol/L DTT) supplemented with protease inhibitors (*Cocktail*, complete-EDTA free, Roche Applied Science). Cells were lysed by ultra-sonication. After centrifugation, the cleared lysate was incubated with 1.5 mL pre-equilibrated Strep-Tactin Superflow beads (IBA). After washing with 50 mL lysis buffer, coatomer complex was eluted with 5 mL elution buffer (lysis buffer supplied with 5 mmol/L desthiobiotin). The elution was then loaded onto a 1 mL MonoQ anion exchange column (GE healthcare) pre-equilibrated with buffer A (20 mmol/L Bicine, pH 8.0, 100 mmol/L KCl, 10% glycerol, 1 mmol/L DTT) and eluted using a gradient from 100 mmol/L to 1 mol/L KCl with buffer B (20 mmol/L Bicine, pH 8.0, 1 mol/L KCl, 10% glycerol, 1 mmol/L DTT). Pooled fractions containing coatomer (approximately 0.35–0.4 mol/L KCl) were concentrated and then diluted in order for the final KCl concentration to be approximately 100 mmol/L. The two-step chromatography purification yielded above 90% pure coatomer at a protein concentration of ~10 mg/mL.

Recombinant human N-myristoylated Arf1 was purified from *E. coli* as previously described (Franco et al., 1996). Rat ArfGAP1 with a 10× His tag was expressed using BacMagic system and purified by affinity chromatography using Ni-NTA resin (GE Healthcare) and eluted by McAc300 buffer (50 mmol/L Tris-Cl, pH 8.0, 300 mmol/L KCl, 300 mmol/L imidazole).

LC-MS protein identification

The gel band of interest was manually excised from native PAGE gel and then digested following standard procedures. The digested solution was then transferred to a sample vial for LC-MS analysis.

LC-MS analysis was performed using a Thermo Finnigan LTQ linear ion trap mass spectrometer in line with a Thermo Finnigan Surveyor MS Pump Plus HPLC system. Peptides generated above were loaded onto a trap column (300SB-C18, 5 × 0.3 mm, 5 μm particle) (Agilent Technologies, Santa Clara CA) which was connected through a zero dead volume union to the self-packed analytical column (C18, 100 μm i.d × 100 mm, 3 μm particle) (SunChrom, Germany). The peptides were then eluted over a gradient (0%–45% B in 55 min, 45%–100% B in 10 min, where B = 80% Acetonitrile, 0.1% formic acid) at a flow rate of 500 nL/min and introduced online into the linear ion trap mass spectrometer (ThermoFisher Corporation, San Jose, CA) using nano electrospray ionization (ESI). Data dependent scanning was incorporated to select the 5 most abundant ions (one microscan per spectra; precursor isolation width 1.0 m/z, 35% collision energy, 30 ms ion activation, exclusion duration: 90 s; repeat count: 1) from a full-scan mass spectrum for fragmentation by collision induced dissociation (CID). MS data were analyzed using SEQUEST against NCBI human protein database; and results were filtered, sorted, and displayed using the Bioworks 3.2. Peptides with +1, +2, or +3 charge states were accepted if they were fully enzymatic and had a cross correlation (Xcorr) of 1.90, >2.5, and >3.0, respectively. The following residue modifications were allowed in the search: carbamidomethylation on cysteine and oxidation on methionine. Sequest was searched with a peptide tolerance of 3 Amu and a fragment ion tolerance of 1.0 Amu.

Preparations of liposomes

Pure lipids were purchased from Avanti Polar lipids. Lipids were dissolved and mixed in either chloroform or methanol with composition similar to Golgi membrane (van Meer, 1998; van Meer et al., 2008): 45 mol% DOPC, 19 mol% DOPE, 5 mol% DOPS, 8 mol% PI, 7 mol% SM, 16 mol% cholesterol. After evaporation, the dry lipid film was suspended and hydrated in Assay buffer (25 mmol/L HEPES, pH 7.2, 25 mmol/L KCl, 2.5 mmol/L $Mg(OAc)_2$). After vortexing and repeated freezing/thawing, liposomes were further extruded through polycarbonate filter with 0.2 μm pore.

Membrane binding and vesiculation assays

Incubation was carried out in 50 μL AS buffer (25 mmol/L Hepes, pH 7.2, 25 mmol/L KCl, 0.2 mol/L sucrose, 2.5 mmol/L $Mg(OAc)_2$) for 30 min at 37°C with myr-Arf1 (4.8 μmol/L), coatomer (182 nmol/L), GDP (1 mmol/L)/GTPγS (25 μmol/L), and liposome (0.2 mg/mL). EDTA (5 mmol/L) was added firstly for guanine nucleotides exchange of Arf1, and then $Mg(OAc)_2$ (10 mmol/L) was supplied followed by the addition of other proteins into the incubation reactions. For studies of ArfGAP1, these incubation was carried out for 25 min at 37°C followed by additional incubation with ArfGAP1 (1 μmol/L) for 5 min at 37°C. After incubation, 10 μL of the reaction mix was reserved for negative staining and EM observation, and the remaining 40 μL reaction mix was ultracentrifuged (250,000 ×*g*, 30 min, 4°C) through 500 μL AS buffer to pellet membranes. After resuspending with AS buffer, the membranes and their bound proteins were analyzed by SDS-PAGE and Western blotting.

For purification of COPI-coated vesicles, the incubation volume was scaled up to 250 μL with liposome at a concentration of 1.5 mg/mL. After incubation, the reconstituted COPI vesicles were isolated from membranes and soluble proteins by subsequent centrifugation through two sucrose cushions in AS buffer (37.5%, 50 μL; 45%, 10 μL), essentially as previously described (Beck et al., 2009). After centrifugation for 50 min at 100,000×*g*, the fraction at the interface of 37.5%/45% sucrose cushion was collected and equally divided into two aliquots for either Western blotting or negative stain EM analysis.

Negative stain electron microscopy

For soluble coatomer, holy carbon grids coated with thin (5 nm) continuous carbon films, which was manufactured in-house, were used. Samples were applied to the glow-discharged grids and adsorbed for 1 min at RT. After washing with buffer, the grids were stained with 2% (*w*/*v*) uranyl acetate for 1 min at RT. For liposomes and vesicles, the incubation reaction solutions were directly applied onto glow-discharged EM grids (LifeTrust) coated with continuous carbon film and stained with 2% (*w*/*v*) uranyl acetate for 1 min at RT.

Images were collected on a 200 kV FEI F200C Talos field emission transmission electron microscopy equipped with a 4 k × 4 K Ceta charge-coupled device camera (FEI, Netherlands). Single particle data were collected automatically with SerialEM software package (Mastronarde, 2005) at the magnification of 57,000 with a pixel size of 1.81 Å.

Electron tomography and sub-tomogram averaging

The same grid from the single particle data collection was applied for tilt series tomography data collection. Data were collected using SerialEM software package (Mastronarde, 2005) at the magnification of 57,000 with a pixel size of 1.81 Å. The angular range was from −55° to +55° with an increment of 5° and the total electron dose was 80–100 e/Å^2. Processing was performed on binned data with a pixel size of 3.6 Å. Gctf (Zhang, 2016) was used to estimate the defocus, which ranges from 5.0 μm to 8.0 μm. The tilt series were aligned using MarkerAuto (Han et al., 2015) before reconstruction using IMOD (Kremer et al., 1996). 632 sub-volumes with a box size of 84 pixels were picked using IMOD (Kremer et al., 1996) and extracted using RELION1.4 (Bharat et al., 2015). Subsequently, sub-volume alignment and refinement were performed using RELION1.4 without CTF model. The initial model used in RELION1.4 refinement was a random rotational average of all the extracted sub-volumes (Bharat et al., 2015). The reconstruction after refinement was used as an initial model for the further single particle data processing.

Single particle analysis

Micrographs were screened out due to severe astigmatism, drift, inappropriate defocus, contamination, particle aggregation, and imperfect staining. Particles were first manually picked from 20 micrographs and class averaged using RELION1.3 (Scheres, 2012). Next, the class average results were used for template-based particle autopicking by Gautomatch (by Kai Zhang, MRC). To limit model bias, we applied a 40 Å low-pass filter to these templates. An initial dataset of 353,777 particles was submitted to multiple cycles of reference-free 2D classification in RELION1.3 to exclude bad particles, such as contamination or overlapping particles. The 109,342 selected particles were used for 3D reconstruction with the initial model that was obtained from sub-tomogram average and filtered to 60 Å. After several cycles of 3D classification and refinement, a 2D classification step using the alignment parameters from the 3D refinement was performed to check the alignment accuracy and to further exclude particles liable to misalignment. The resulting 60,007 qualified particles were used for final 3D refinement and the subsequent "focus" refinement (Bai et al., 2015). For "focus" refinement, we segmented the final reconstruction map using the "Segmentation" tool in UCSF Chimera (Pettersen et al., 2004). The segmented densities were grouped into two sets, which were assigned as the two subcomplexes according to the rigid-body fitting of the membrane-bound coatomer structure. Then, the grouped densities were further extracted to a new volume data, and used for subtraction of the corresponding signals from the raw particles images using "relion_project" commands. The resulting new particle stacks were used as input in the subsequent 3D auto refine and classification processes.

CXMS

CXMS analysis was performed essentially as described previously (Ding et al., 2016). Briefly, in each 20 μL cross-linking reaction, about 12 μg of the purified COPI complex was incubated with BS^3, DSS, or Sulfo-GMBS at a molar ratio of 1:8 or 1:4 (protein : cross-linker) for one hour at room temperature. The reactions were quenched with 20 mmol/L NH_4HCO_3 for BS^3 and DSS reactions, or with 20 mmol/L DTT, 20 mmol/L NH_4HCO_3 for Sulfo-GMBS cross-linking. Proteins were precipitated with ice-cold acetone, resuspended in 8 mol/L urea, 100 mmol/L Tris, pH 8.5. After Trypsin digestion, the LC-MS/MS analysis was performed on an Easy-nLC 1000 UPLC (Thermo Fisher Scientific) coupled to a Q Exactive-Orbitrap mass spectrometer (Thermo Fisher Scientific). Peptides were loaded onto a pro column (75 μm ID, 8 cm long, packed with ODS-AQ 12 nm–10 μm beads from YMC Co., Ltd.) and separated on an analytical column (75 μm ID, 11 cm long, packed with Luna C18 1.8 μm 100 Å resin from Welch Materials) using an acetonitrile gradient from 0%–21% in 83 min at a flow rate of 200 nL/min. The top 15 most intense precursor ions from each full scan (resolution = 70,000) were isolated for HCD MS2 (resolution = 17,500; normalized collisional energy = 27) with a dynamic exclusion time of 20 s. Precursors with 1+, 2+, 7+ or above, or unassigned charge states were excluded. Cross-linked peptides were identified using pLink (Yang et al., 2012) with a 5% FDR cutoff at the spectral level and then an *E*-value cutoff at 0.001.

The Cα–Cα distance between two cross-linked residues were measured in PyMOL (The PyMOL Molecular Graphics System, Version 1.74, Schrodinger, LLC.). For the three structural models of clathrin AP-1 homologues, we first built models with human coatomer sequence, taking the crystal structure models (PDB ID: 4P6Z/4HMY/1W63) as reference using *Modeller* software (Šali et al., 1995), and then the Cα–Cα distance between cross-linked residues were measured in PyMOL.

ACKNOWLEDGMENTS

We would like to thank Dr. Yan Zhang (F.S. lab) for help with image processing. We also thank Dr. Xiaojun Huang, Dr. Gang Ji, and Shuoguo Li from Center for Biological Imaging, Institute of Biophysics, CAS, for their help with EM data collection, and Dr. Xiang Ding from Laboratory of Proteomics, Institute of Biophysics, CAS, for her help with the mass spectrometry analysis. This work was supported by grants from the Strategic Priority Research Program of Chinese Academy of Sciences (XDB08030202) to F.S., the National Basic Research Program (973 Program) (No. 2014CB910700 to F.S.,2014CB849800 to M.-Q.D.), the National Natural Science Foundation of China (Grant No. 21375010 to M.-Q.D.), and the National Institutes of Health USA (GM058651 to V.W.H.). All the EM work as well as related image processing were performed at Center for Biological Imaging (CBI, http://cbi.ibp.ac.cn), Institute of Biophysics, Chinese Academy of Sciences.

ABBREVIATIONS

ARF, ADP-ribosylation factor; COPI, coat protein I; CXMS, chemical cross-linking coupled with mass spectrometry; EM, electron microscopy; GAP, GTPase-activating protein; LC-MS, liquid chromatography mass spectrometry; NTD, N-terminal domain; CTD, C-terminal domain; PIP2, phosphatidylinositol (4,5) bis-phosphate.

COMPLIANCE WITH ETHICS GUIDELINES

Shengliu Wang, Yujia Zhai, Xiaoyun Pang, Tongxin Niu, Yue-He Ding, Meng-Qiu Dong, Victor W. Hsu, Zhe Sun, and Fei Sun declare that they have no conflict of interest.

All procedures followed were in accordance with the ethical standards of the responsible committee on human experimentation (institutional and national) and with the Helsinki Declaration of 1975, as revised in 2000 (5). Informed consent was obtained from all patients for being included in the study.

AUTHOR CONTRIBUTIONS

FS and ZS supervised the project. YZ performed molecular cloning and protein expression. SW performed protein expression and purification and all the EM works. SW, XP, and ZS performed functional studies. TN performed tomography works. YD and MD performed cross-linking and mass spectrometry. SW, ZS, and FS analyzed the data and wrote the manuscript with help from VWH.

REFERENCES

Bai XC, Rajendra E, Yang GH, Shi YG, Scheres SHW (2015) Sampling the conformational space of the catalytic subunit of human gamma-secretase. eLife 4

Beck R, Sun Z, Adolf F, Rutz C, Bassler J, Wild K, Sinning I, Hurt E, Brugger B, Bethune J et al (2008) Membrane curvature induced by Arf1-GTP is essential for vesicle formation. Proc Natl Acad Sci USA 105:11731–11736

Beck R, Adolf F, Weimer C, Bruegger B, Wieland FT (2009) ArfGAP1 Activity and COPI Vesicle Biogenesis. Traffic 10:307–315

Bharat TA, Russo CJ, Lowe J, Passmore LA, Scheres SH (2015) Advances in single-particle electron cryomicroscopy structure determination applied to sub-tomogram averaging. Structure 23:1743–1753

Bielli A, Haney CJ, Gabreski G, Watkins SC, Bannykh SI, Aridor M (2005) Regulation of Sar1 NH_2 terminus by GTP binding and hydrolysis promotes membrane deformation to control COPII vesicle fission. J Cell Biol 171:919–924

Bigay J, Gounon P, Robineau S, Antonny B (2003) Lipid packing sensed by ArfGAP1 couples COPI coat disassembly to membrane bilayer curvature. Nature 426:563–566

Bonifacino JS, Glick BS (2004) The mechanisms of vesicle budding and fusion. Cell 116:153–166

Bremser M, Nickel W, Schweikert M, Ravazzola M, Amherdt M, Hughes CA, Sollner TH, Rothman JE, Wieland FT (1999) Coupling of coat assembly and vesicle budding to packaging of putative cargo receptors. Cell 96:495–506

Brugger B, Sandhoff R, Wegehingel S, Gorgas K, Malsam J, Helms JB, Lehmann WD, Nickel W, Wieland FT (2000) Evidence for segregation of sphingomyelin and cholesterol during formation of COPI-coated vesicles. J cell biol 151:507–518

Collins BM, McCoy AJ, Kent HM, Evans PR, Owen DJ (2002) Molecular architecture and functional model of the endocytic AP2 complex. Cell 109:523–535

Cosson P, Letourneur F (1994) Coatomer interaction with di-lysine endoplasmic reticulum retention motifs. Science 263:1629–1631

Cukierman E, Huber I, Rotman M, Cassel D (1995) The ARF1-GTPase-activating protein: Zinc finger motif and Golgi complex localization. Science 270:1999–2002

Deng Y, Golinelli-Cohen MP, Smirnova E, Jackson CL (2009) A COPI coat subunit interacts directly with an early-Golgi localized Arf exchange factor. EMBO Rep 10:58–64

Ding Y, Fan S, Li S, Feng B, Gao N, Ye K, He S-M, Dong M-Q (2016) Increasing the depth of mass spectrometry-based structural analysis of protein complexes through the use of multiple cross-linkers. Anal Chem 88(8):4461–4469

Dodonova SO, Diestelkoetter-Bachert P, von Appen A, Hagen WJ, Beck R, Beck M, Wieland F, Briggs JA (2015) VESICULAR TRANSPORT. A structure of the COPI coat and the role of coat proteins in membrane vesicle assembly. Science 349:195–198

Donaldson JG, Cassel D, Kahn RA, Klausner RD (1992) ADP-ribosylation factor, a small GTP-binding protein, is required for binding of the coatomer protein beta-COP to Golgi membranes. Proc Natl Acad Sci USA 89:6408–6412

Faini M, Prinz S, Beck R, Schorb M, Riches JD, Bacia K, Brugger B, Wieland FT, Briggs JA (2012) The structures of COPI-coated vesicles reveal alternate coatomer conformations and interactions. Science 336:1451–1454

Faini M, Beck R, Wieland FT, Briggs JA (2013) Vesicle coats: structure, function, and general principles of assembly. Trends Cell Biol 23:279–288

Farsad K, Ringstad N, Takei K, Floyd SR, Rose K, De Camilli P (2001) Generation of high curvature membranes mediated by direct endophilin bilayer interactions. J Cell Biol 155:193–200

Fiedler K, Veit M, Stamnes MA, Rothman JE (1996) Bimodal interaction of coatomer with the p24 family of putative cargo receptors. Science 273:1396–1399

Franco M, Chardin P, Chabre M, Paris S (1996) Myristoylation-facilitated binding of the G protein ARF1(GDP) to membrane phospholipids is required for its activation by a soluble nucleotide exchange factor. J Biol Chem 271:1573–1578

Goldberg J (1999) Structural and functional analysis of the ARF1-ARFGAP complex reveals a role for coatomer in GTP hydrolysis. Cell 96:893–902

Han R, Wang L, Liu Z, Sun F, Zhang F (2015) A novel fully automatic scheme for fiducial marker-based alignment in electron tomography. J Struct Biol 192:403–417

Hara-Kuge S, Kuge O, Orci L, Amherdt M, Ravazzola M, Wieland FT, Rothman JE (1994) En bloc incorporation of coatomer subunits during the assembly of COP-coated vesicles [published erratum appears in J Cell Biol 1994 Jul; 126(2):589]. J Cell Biol 124:883–892

Hariri H, Bhattacharya N, Johnson K, Noble AJ, Stagg SM (2014) Insights into the mechanisms of membrane curvature and vesicle scission by the small GTPase Sar1 in the early secretory pathway. J Mol Biol 426:3811–3826

Heldwein EE, Macia E, Wang J, Yin HL, Kirchhausen T, Harrison SC (2004) Crystal structure of the clathrin adaptor protein 1 core. Proc Natl Acad Sci USA 101:14108–14113

Hsu VW, Lee SY, Yang JS (2009) The evolving understanding of COPI vesicle formation. Nat Rev Mol Cell Biol 10:360–364

Jackson LP (2014) Structure and mechanism of COPI vesicle biogenesis. Curr Opin Cell Biol 29:67–73

Jackson LP, Kelly BT, McCoy AJ, Gaffry T, James LC, Collins BM, Honing S, Evans PR, Owen DJ (2010) A large-scale conformational change couples membrane recruitment to cargo binding in the AP2 clathrin adaptor complex. Cell 141:1220–1229

Jackson LP, Lewis M, Kent HM, Edeling MA, Evans PR, Duden R, Owen DJ (2012) Molecular basis for recognition of dilysine trafficking motifs by COPI. Dev Cell 23:1255–1262

Jia, X., Weber, E., Tokarev, A., Lewinski, M., Rizk, M., Suarez, M., Guatelli, J., and Xiong, Y. (2014). Structural basis of HIV-1 Vpu-mediated BST2 antagonism via hijacking of the clathrin adaptor protein complex 1. eLife *3*, e02362

Krauss M, Jia JY, Roux A, Beck R, Wieland FT, De Camilli P, Haucke V (2008) Arf1-GTP-induced tubule formation suggests a function of Arf family proteins in curvature acquisition at sites of vesicle budding. J Biol Chem 283:27717–27723

Kremer JR, Mastronarde DN, McIntosh JR (1996) Computer visualization of three-dimensional image data using IMOD. J Struct Biol 116:71–76

Langer JD, Roth CM, Bethune J, Stoops EH, Brugger B, Herten DP, Wieland FT (2008) A conformational change in the alpha-subunit of coatomer induced by ligand binding to gamma-COP revealed by single-pair FRET. Traffic 9:597–607

Lee C, Goldberg J (2010) Structure of coatomer cage proteins and the relationship among COPI, COPII, and clathrin vesicle coats. Cell 142:123–132

Lee MC, Orci L, Hamamoto S, Futai E, Ravazzola M, Schekman R (2005a) Sar1p N-terminal helix initiates membrane curvature and completes the fission of a COPII vesicle. Cell 122:605–617

Lee SY, Yang JS, Hong W, Premont RT, Hsu VW (2005b) ARFGAP1 plays a central role in coupling COPI cargo sorting with vesicle formation. J Cell Biol *168*, 281-290

Leitner A, Walzthoeni T, Aebersold R (2014) Lysine-specific chemical cross-linking of protein complexes and identification of cross-linking sites using LC-MS/MS and the xQuest/xProphet software pipeline. Nat Protoc 9:120–137

Lundmark R, Doherty GJ, Vallis Y, Peter BJ, McMahon HT (2008) Arf family GTP loading is activated by, and generates, positive membrane curvature. Biochem J 414:189–194

Ma W, Goldberg J (2013) Rules for the recognition of dilysine retrieval motifs by coatomer. EMBO J 32:926–937

Malhotra V, Serafini T, Orci L, Shepherd JC, Rothman JE (1989) Purification of a novel class of coated vesicles mediating biosynthetic protein transport through the Golgi stack. Cell 58:329–336

Mastronarde DN (2005) Automated electron microscope tomography using robust prediction of specimen movements. J Struct Biol 152:36–51

Matsuoka K, Orci L, Amherdt M, Bednarek SY, Hamamoto S, Schekman R, Yeung T (1998) COPII-coated vesicle formation reconstituted with purified coat proteins and chemically defined liposomes. Cell 93:263–275

McMahon HT, Gallop JL (2005) Membrane curvature and mechanisms of dynamic cell membrane remodelling. Nature 438:590–596

Peter BJ, Kent HM, Mills IG, Vallis Y, Butler PJ, Evans PR, McMahon HT (2004) BAR domains as sensors of membrane curvature: the amphiphysin BAR structure. Science 303:495–499

Pettersen EF, Goddard TD, Huang CC, Couch GS, Greenblatt DM, Meng EC, Ferrin TE (2004) UCSF chimera - A visualization system for exploratory research and analysis. J Comput Chem 25:1605–1612

Politis A, Stengel F, Hall Z, Hernandez H, Leitner A, Walzthoeni T, Robinson CV, Aebersold R (2014) A mass spectrometry-based hybrid method for structural modeling of protein complexes. Nat Methods 11:403–406

Pucadyil TJ, Schmid SL (2009) Conserved functions of membrane active GTPases in coated vesicle formation. Science 325:1217–1220

Reinhard C, Harter C, Bremser M, Brugger B, Sohn K, Helms JB, Wieland F (1999) Receptor-induced polymerization of coatomer. Proc Natl Acad Sci USA 96:1224–1228

Ren X, Farias GG, Canagarajah BJ, Bonifacino JS, Hurley JH (2013) Structural basis for recruitment and activation of the AP-1 clathrin adaptor complex by Arf1. Cell [illegible]

Rosenthal PB, Henderson R (2003) Optimal determination of particle orientation, absolute hand, and contrast loss in single-particle electron cryomicroscopy. J Mol Biol 333:721–745

Sahlmuller MC, Strating JR, Beck R, Eckert P, Popoff V, Haag M, Hellwig A, Berger I, Brugger B, Wieland FT (2011) Recombinant heptameric coatomer complexes: novel tools to study isoform-specific functions. Traffic 12:682–692

Šali A, Potterton L, Yuan F, van Vlijmen H, Karplus M (1995) Evaluation of comparative protein modeling by MODELLER. Proteins: Structure. Funct Bioinform 23:318–326

Scheres SHW (2012) RELION: Implementation of a Bayesian approach to cryo-EM structure determination. J Struct Biol 180:519–530

Serafini T, Orci L, Amherdt M, Brunner M, Kahn RA, Rothman JE (1991a) ADP-ribosylation factor is a subunit of the coat of Golgi-derived COP-coated vesicles: a novel role for a GTP-binding protein. Cell 67:239–253

Serafini T, Stenbeck G, Brecht A, Lottspeich F, Orci L, Rothman JE, Wieland FT (1991b) A coat subunit of Golgi-derived non-clathrin-coated vesicles with homology to the clathrin-coated vesicle coat protein beta-adaptin. Nature 349:215–220

Shan H, Wang Z, Zhang F, Xiong Y, Yin CC, Sun F (2016) A local-optimization refinement algorithm in single particle analysis for macromolecular complex with multiple rigid modules. Protein Cell 7:46–62

Spang A, Matsuoka K, Hamamoto S, Schekman R, Orci L (1998) Coatomer, Arf1p, and nucleotide are required to bud coat protein complex I-coated vesicles from large synthetic liposomes. Proc Natl Acad Sci USA 95:11199–11204

Suckling RJ, Poon PP, Travis SM, Majoul IV, Hughson FM, Evans PR, Duden R, Owen DJ (2015) Structural basis for the binding of tryptophan-based motifs by delta-COP. Proc Natl Acad Sci USA 112:14242–14247

Takei K, Haucke V, Slepnev V, Farsad K, Salazar M, Chen H, De Camilli P (1998) Generation of coated intermediates of clathrin-mediated endocytosis on protein-free liposomes. Cell 94:131–141

van Meer G (1998) Lipids of the Golgi membrane. Trends Cell Biol 8:29–33

van Meer G, Voelker DR, Feigenson GW (2008) Membrane lipids: where they are and how they behave. Nat Rev Mol Cell Biol 9:112–124

Waters MG, Serafini T, Rothman JE (1991) 'Coatomer': a cytosolic protein complex containing subunits of non-clathrin-coated Golgi transport vesicles. Nature 349:248–251

Yang JS, Lee SY, Gao M, Bourgoin S, Randazzo PA, Premont RT, Hsu VW (2002) ARFGAP1 promotes the formation of COPI vesicles, suggesting function as a component of the coat. J Cell Biol 159:69–78

Yang B, Wu YJ, Zhu M, Fan SB, Lin J, Zhang K, Li S, Chi H, Li YX, Chen HF et al (2012) Identification of cross-linked peptides from complex samples. Nat Methods 9:904–906

Yip CK, Walz T (2011) Molecular structure and flexibility of the yeast coatomer as revealed by electron microscopy. J Mol Biol 408:825–831

Yu X, Breitman M, Goldberg J (2012) A structure-based mechanism for Arf1-dependent recruitment of coatomer to membranes. Cell 148:530–542

Zhang K (2016) Gctf: Real-time CTF determination and correction. J Struct Biol 193:1–12

Identification of natural compounds targeting Annexin A2 with an anti-cancer effect

Yu-Shi Wang, He Li, Yang Li, Hongyan Zhu, Ying-Hua Jin✉

Key Laboratory for Molecular Enzymology and Engineering of the Ministry of Education, College of Life Science, Jilin University, Changchun 130012, China
✉ Correspondence: yhjin@jlu.edu.cn (Y.-H. Jin)

ABSTRACT

Annexin A2, a multifunctional tumor associated protein, promotes nuclear factor-kappa B (NF-κB) activation by interacting with NF-κB p50 subunit and facilitating its nuclear translocation. Here we demonstrated that two ginsenosides Rg5 (G-Rg5) and Rk1 (G-Rk1), with similar structure, directly bound to Annexin A2 by molecular docking and cellular thermal shift assay. Both Rg5 and Rk1 inhibited the interaction between Annexin A2 and NF-κB p50 subunit, their translocation to nuclear and NF-κB activation. Inhibition of NF-κB by these two ginsenosides decreased the expression of inhibitor of apoptosis proteins (IAPs), leading to caspase activation and apoptosis. Over expression of K302A Annexin A2, a mutant version of Annexin A2, which fails to interact with G-Rg5 and G-Rk1, effectively reduced the NF-κB inhibitory effect and apoptosis induced by G-Rg5 and G-Rk1. In addition, the knockdown of Annexin A2 largely enhanced NF-κB activation and apoptosis induced by the two molecules, indicating that the effects of G-Rg5 and G-Rk1 on NF-κB were mainly mediated by Annexin A2. Taken together, this study for the first time demonstrated that G-Rg5 and G-Rk1 inhibit tumor cell growth by targeting Annexin A2 and NF-κB pathway, and G-Rg5 and G-Rk1 might be promising natural compounds for targeted cancer therapy.

KEYWORDS Annexin A2, G-Rg5, G-Rk1, HCC, NF-κB

INTRODUCTION

Annexin A2, a Ca^{2+}-dependent lipid binding protein, widely distributed in nuclear, cytoplasm, endosomes and extracellular space and participates in various cellular process including signal transduction, endocytosis and exocytosis, proliferation, differentiation and apoptosis in mammalian cells (Klionsky et al., 2012; Moreau et al., 2011; Gerke et al., 2005; Moss and Morgan, 2004). A lot of evidence suggests that Annexin A2 may be a promising therapeutic target for cancer treatment (Chen et al., 2015; Staquicini et al., 2017; Kesavan et al., 2010; Wang et al., 2012). First, an accumulating observations have shown that Annexin A2 overexpressed in various types of cancer cells such as hepatocellular carcinoma (Zhang et al., 2012; Longerich et al., 2011; Sun et al., 2013), breast cancer (Deng et al., 2013; Chuthapisith et al., 2009; Shetty et al., 2012), lung cancer (Yao et al., 2009), gastric carcinoma (Zhang et al., 2012), pancreatic cancer (Takano et al., 2008), and colorectal cancer (Yang et al., 2013; Duncan et al., 2008). Second, the abnormal up-regulation of Annexin A2 enhances cancer development with higher aggression and poorer prognosis (Ohno et al., 2009; Ma et al., 2014; Yang et al., 2014). Third, forced inhibition of Annexin A2 effectively reduces tumor progression both *in vitro* and *in vivo* studies. Furthermore, signal transduction study has shown that Annexin A2 either promotes cancer cell metastasis by forming heterotetramer with S100A10 and binding with cell membrane (Réty et al., 1999), or largely enhances pro-survival capability and chemoresistance by activating transcription factors STAT-3 and nuclear factor-kappa B (NF-κB) (Tong et al., 2015; Jung et al., 2015).

NF-κB is a transcription factor activating the expression of genes involved in inflammation, immune response, angiogenesis, cell proliferation and apoptosis (Perkins, 2012). Constitutive activation of NF-κB is a hallmark of several types of cancers, which not only promotes tumorigenesis and cancer development but also enhances drug resistance (Dolcet et al., 2005; Hanahan and Weinberg, 2000). Recent reports have shown that inhibition of NF-κB significantly inhibited cancer growth by directly inducing cancer cell apoptosis (Woo et al., 2016; Alexandersavino et al., 2016; Kwon et al., 2016; Yin et al., 2016; Hayden et al., 2008). Interestingly, the multifunctional tumor associated protein Annexin A2 promotes NF-κB activation via directly binding to NF-κB p50 subunit with its N-terminal sequences, and inhibition of Annexin A2 provides a new regulatory tool on NF-κB activity (Jung et al., 2015).

Ginsenosides are the major bio-active component of ginseng (Attele et al., 1999) and ginsenoside Rg5 (G-Rg5) and Rk1 (G-Rk1) are the two main ginsenosides identified in heat processed ginseng (Toh et al., 2010). In the present study, we identified G-Rg5 and G-Rk1 that inhibited NF-κB activation, down-regulated the expression of IAPs and induced apoptosis by specifically binding to Annexin A2.

RESULTS

G-Rg5 and G-Rk1 interacted with Annexin A2

Molecular docking was first performed as a primary screening for natural compounds targeting Annexin A2. G-Rg5 bound to Annexin A2 with Glu296 and Lys302 responsible to this interaction (Fig. 1A). Cellular thermal shift assay (CTSA) with HepG2 cells demonstrated that G-Rg5 and G-Rk1 largely improved the thermal stability of Annexin A2, indicating the binding of G-Rg5 and G-Rk1 with Annexin. Exogenous wild-type Annexin A2 (Annexin A2-WT) showed a similar thermal stability shift under treatment of G-Rg5 and G-Rk1, while G-Rg5 and G-Rk1 didn't change the thermal stability of exogenous K302A mutant of Annexin A2 (Annexin A2-K302A), indicating Lys302 might be responsible for the interaction between Annexin A2 and ginsenosides (Fig. 1B), as predicted by molecular docking. To rule out the effects from other cellular content and signaling towards the thermal stability of Annexin A2, we carried out the thermal shift assay with purified Annexin A2 and the two ginsenosides. Both G-Rg5 and G-Rk1 obviously enhanced the thermal stability of Annexin A2-WT but not Annexin A2-K302A in a dose-dependent manner (Fig. 1C). These data indicated that both G-Rg5 and G-Rk1 directly bound to Annexin A2, and Lys302 is a critical residue for this interaction.

G-Rg5 and G-Rk1 inhibited the interaction and nuclear co-localization of Annexin A2 and NF-κB p50 subunit

In order to tell whether the G-Rg5 and G-Rk1 binding to Annexin A2 modulated the action of Annexin A2 in cells, we determined the interaction and nuclear co-localization of

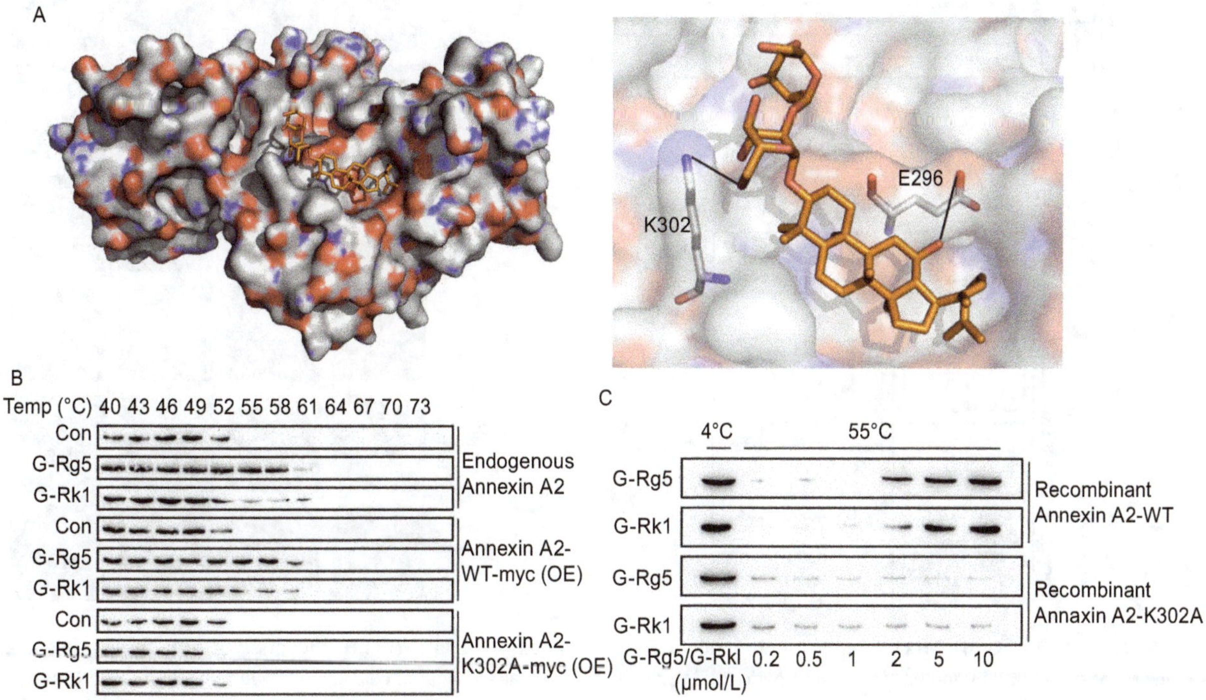

Figure 1. G-Rg5 and G-Rk1 bound to Annexin A2. (A) Results of molecular docking shows the interaction between Annexin A2 and G-Rg5. (B) Cellular thermal shift assay presented the thermal stability of Annexin A2 under treatment with G-Rg5 (12 μmol/L) and G-Rk1 (12 μmol/L). (C) Thermal shift *in vitro* showed the thermal stability of purified Annexin A2 (0.2 μmol/L) under treatment with G-Rg5 and G-Rk1.

Annexin A2 and NF-κB p50 subunit under G-Rg5 and G-Rk1 treatment. Immuno-precipitation was performed using HepG2 cells treated with 6 μmol/L G-Rg5 or 6 μmol/L G-Rk1, with or without NF-κB activator PMA (100 ng/mL) or etoposide (25 μg/mL). The interaction between Annexin A2 and NF-κB p50 was significantly attenuated under G-Rg5 and G-Rk1 treatment, both in resting cells and NF-κB activator treated cells (Fig. 2A). To rule out the possibility that the interaction was regulated by other eukaryotic cellular content, another immunoprecipitation was carried out with prokaryotic cell-expressed Annexin A2 and p50, and the interaction was also inhibited largely by G-Rg5 and G-Rk1 (Fig. 2B). Next, immunofluorescence was employed to examine the sub-cellular distribution of Annexin A2 and p50 under ginsenoside treatment. The nuclear co-localization of Annexin A2 and p50 was inhibited in both resting state and NF-κB activator-stimulating cells under G-Rg5 and G-Rk1 treatment (Fig. 2C). Taken together, G-Rg5 and G-Rk1 inhibited the interaction between Annexin A2 and p50, and their nuclear co-localization.

G-Rg5 and G-Rk1 inhibited NF-κB activation and down-regulated downstream anti-apoptosis gene expression

The activation of NF-κB was examined by dual luciferase reporter assay with HepG2 cells treated with G-Rg5 (6 μmol/L), G-Rk1 (6 μmol/L), etoposide (25 μg/mL) and PMA (100 ng/mL). G-Rg5 and G-Rk1 inhibited NF-κB

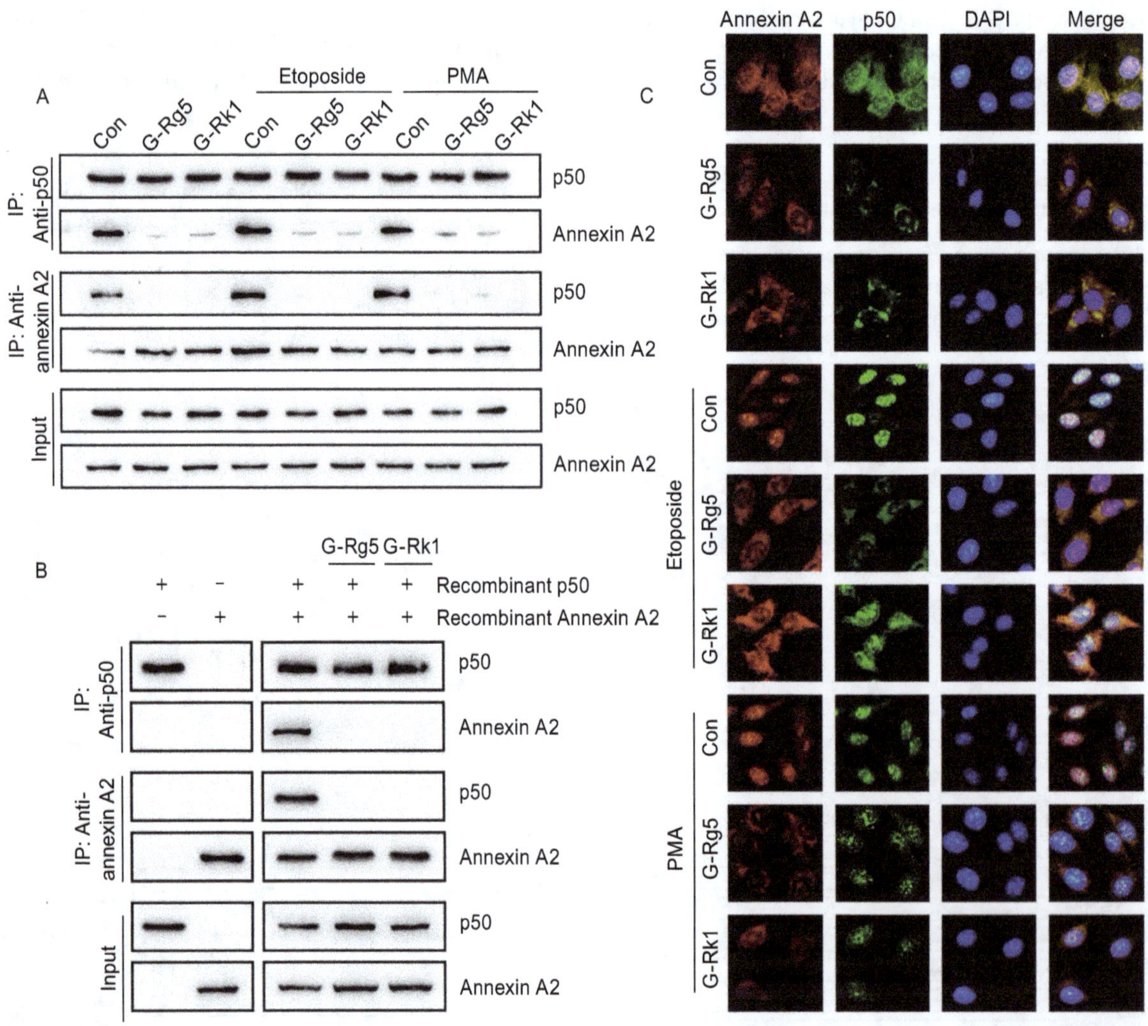

Figure 2. G-Rg5 and G-Rk1 inhibited interaction between Annexin A2 and NF-κB p50 subunit and their nuclear co-localization. (A) Immunoprecipitation was performed with whole-cell lysate of HepG2 cells under treatment with G-Rg5 (6 μmol/L), G-Rk1 (6 μmol/L), etoposide (25 μg/mL) and PMA (100 ng/mL), and the interaction was analyzed by an immunoblot. (B) Immunoprecipitation was performed with prokaryotic cells-expressed Annexin A2 and NF-κB p50 subunit under treatment with G-Rg5 (6 μmol/L) and G-Rk1 (6 μmol/L), and the interaction was analyzed by an immunoblot. (C) The subcellular distribution of Annexin A2 and NF-κB p50 subunit was examined by immunofluorescence under treatment with G-Rg5 (6 μmol/L), G-Rk1 (6 μmol/L), etoposide (25 μg/mL) and PMA (100 ng/mL), and DAPI showed the nuclear region.

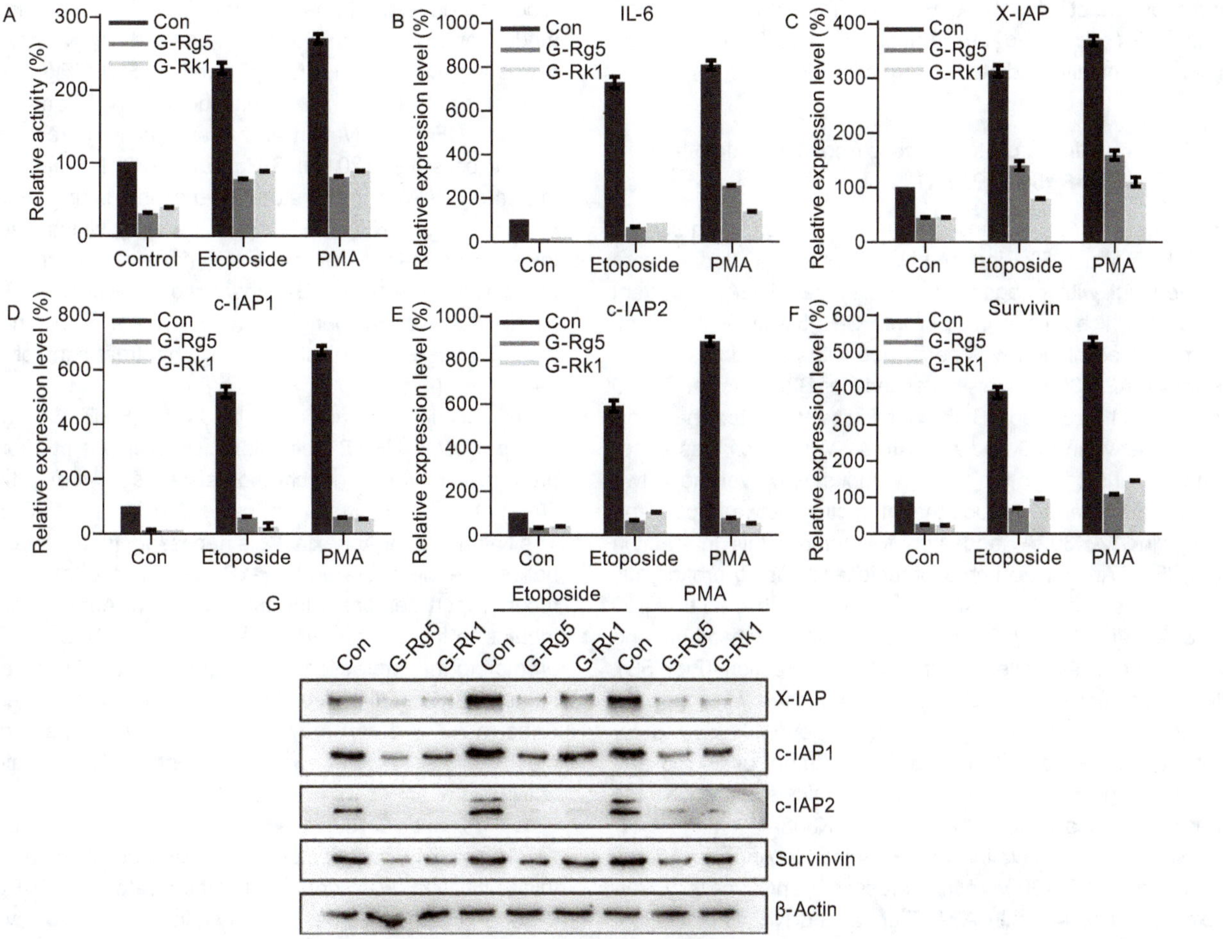

Figure 3. G-Rg5 and G-Rk1 inhibited NF-κB activation and down-regulated downstream anti-apoptosis genes. (A) NF-κB activity was examined by luciferase reporter assay under treatment with G-Rg5 (6 μmol/L), G-Rk1 (6 μmol/L), etoposide (25 μg/mL) and PMA (100 ng/mL). (B–F) Relative gene expression levels of IL-6 (B), X-IAP (C), c-IAP1 (D), c-IAP2 (E) and survivin (F) were examined by qRT-PCR under treatment with G-Rg5 (6 μmol/L), G-Rk1 (6 μmol/L), etoposide (25 μg/mL) and PMA (100 ng/mL). (G) Protein levels of X-IAP, c-IAP1, c-IAP2 and survivin were examined by immunoblot under treatment with G-Rg5 (6 μmol/L), G-Rk1 (6 μmol/L), etoposide (25 μg/mL) and PMA (100 ng/mL), and β-actin was shown as a loading control. All data are shown as the mean ± SD and the experimental points show the average of at least triplicates. All experiments were repeated at least 3 times.

activation in resting state and under treatment with etoposide and PMA, activator enhancing NF-κB activity (Fig. 3A). The similar inhibitory effect was observed to downstream genes of NF-κB like IL-6 (Fig. 3B), a classic NF-κB activating gene, and IAPs genes including X-IAP, c-IAP1, c-IAP2 and survivin (Fig. 3C–F). The protein levels of X-IAP, c-IAP1, c-IAP2 and survivin were also down-regulated by G-Rg5 and G-Rk1, in both resting and NF-κB activator-stimulated cells (Fig. 3G).

Annexin A2 knock-down enhanced anti-cancer activity of G-Rg5 and G-Rk1 in HepG2 cells

To examine whether Rg5 inhibits NF-κB activation primarily by Annexin A2, knock-down of Annexin A2 was employed with shRNA vectors. Both Annexin A2 knock-down and G-Rg5 inhibited NF-κB activation, and the inhibitory effect of G-Rg5 was largely enhanced by Annexin A2 knock-down at the moderate concentration (6 μg/mL), and at the high dose of G-Rg5 (6 μg/mL), Annexin A2 knock-down showed little effect (Fig. 4A), which indicates G-Rg5 regulating NF-κB mainly through Annexin A2. Caspase 3 and 9 was activated under G-Rg5 treatment, which was also enhanced by Annexin A2 knock-down (Fig. 4B and 4C). MTT assay showed the anti-survival activity of G-Rg5, which was also enhanced by Annexin A2 knock-down (Fig. 4D). G-Rk1 behaved in a similar manner to G-Rg5 on NF-κB inhibition (Fig. 4E), caspase activation (Fig. 4F and 4G) and anti-proliferation activity (Fig. 4H). A plate clone formation assay showed that G-Rg5 and G-Rk1 inhibited clone formation of HepG2 cells and Annexin A2 knock-down intensified this inhibitory effect (Fig. 4I), as expected. Taken together, G-Rg5 and G-Rk1 inhibited proliferation of HepG2 cells, which might be mainly mediated by Annexin A2.

K302A Annexin A2 protected cells from cytotoxic effect induced by G-Rg5 and G-Rk1

To further provide evidence for the major role of Annexin A2 in G-Rg5 and G-Rk1-induced NF-κB inhibition, studies were carried out with exogenous Annexin A2-K302A, a mutant version, which fails to interact with G-Rg5 and G-Rk1. First, immunoprecipitation was performed with Annexin A2-WT or Annexin A2-K302A over-expressed HepG2 cells with G-Rg5 and G-Rk1 treatment. Both chemicals attenuated the interaction between p50 and Annexin A2-WT, which was either endogenous or myc-tagged exogenous. However these two chemicals failed to reduce the interaction between p50 and Annexin A2-K302A, appeared to promote this interaction (Fig. 5A). An *in-vitro* immunoprecipitation using prokaryotic cell expressed p50 and Annexin A2-K302A showed that p50 and Annexin A2-K302A did interact with each other, and either chemicals failed to affect this interaction (Fig. 5B). Immunocytochemistry clearly showed that wild type of Annexin A2 (either endogenous or exogenous) failed to co-localize to nuclear with p50 under treatment of G-Rg5 and G-Rk1 (Fig. 5C), while the two chemicals promoted the nuclear co-localization of Annexin A2-K302A.

Next, NF-κB activation was examined in Annexin A2-WT or Annexin A2-K302A overexpressed HepG2 cells. Over-expression of Annexin A2-WT enhanced NF-κB activation, and G-Rg5 inhibited NF-κB activity in a dose-dependent manner both in mock transfected and in Annexin A2-WT expressed HepG2 cells. Interestingly, G-Rg5 didn't inhibit, even enhance NF-κB activity in Annexin A2-K302A overexpressed HepG2 cells (Fig. 6A). Activation of caspase 9 and 3 was attenuated when Annexin A2-WT was overexpressed, and only slight activation was detected with an over-expression of Annexin-K302A (Fig. 6B and 6C). Similarly, Annexin A2-K302A abolished the inhibition of NF-κB induced by G-Rk1 (Fig. 6D). G-Rk1 induced weaker caspase activation in Annexin A2-WT over-expressing cells and a little caspase activation was observed in Annexin A2-K302A over-expressing cells compared with mock transfected cells (Fig. 6E and 6F). Anti-proliferation activity was examined by MTT assay and plate clone formation assay with Annexin A2-WT or Annexin A2-K302A over-expressing HepG2 cells. Both Annexin A2-WT and Annexin A2-K302A over-expression rendered cells to resist the cytotoxic effect of G-Rg5 and G-Rk1, and the activity of Annexin A2-K302A showed higher than that of Annexin A2-WT (Fig. 6G and 6H).

DISCUSSION

Annexin A2, a multi-functional cancer associated protein, promotes cancer progress in a list of mechanisms. In this report we developed an assay for screening small-molecule modulator of Annexin A2 and identified two Annexin A2 regulating compounds G-Rg5 and G-Rk1 (Fig. 1A). A recent report has shown that both G-Rg5 and G-Rk1 inhibited NF-κB activation and related immuno-response in HepG2 cells (Lee, 2015). G-Rg5 and G-Rk1 directly bound to Annexin A2 and disturbed the interaction between Annexin A2 and NF-κB p50 (Figs. 1 and 2), which leaded to the cytosol-retention of these two proteins and resulted in the inactivation of NF-κB transcriptional activity. IAPs, the NF-κB downstream genes were down-regulated, and caspase-3 and -9 were activated by the treatment of either compound (Fig. 3).

As described, Annexin A2 binds to NF-κB p50 subunit, and promotes NF-κB activation, which might promote the pro-survival effect and chemoresistance in tumor tissues (Tong et al., 2015; Staquicini et al., 2017; Das et al., 2010). The N-terminal of Annexin A2 is necessary for the interaction between Annexin A2 and NF-κB p50 (Jung et al., 2015). A recent report demonstrates that Lys302 of Annexin A2 contributes to the conformational stability of its N-terminal, and promoting the interactions with other proteins driven by this sequence (Ecsédi et al., 2017). The interaction between Annexin A2 and p50 may possibly follow this pattern, and modifications towards Lys302 will probably regulate the activation of NF-κB.

G-Rg5 and G-Rk1 triggered multiple cellular events, which possibly participated in NF-κB activation, however, this activation was dominantly inactivated by G-Rg5 or G-Rk1 binding to Annexin A2-WT in Annexin A2 over-expressed cancer cells. As G-Rg5 and G-Rk1 failed to interact with Annexin A2-K302A, G-Rg5 or G-Rk1 could not inactivate the other pathway triggered NF-κB activation (Figs. 5, 6A and 6B), presenting little anti-cancer activity in Annexin A2-K302A over-expressed HepG2 cells (Fig. 6). We further examined the effect of G-Rg5 and Rk1 on NF-κB activity in various cancer cells, including SW480, PC3, HeLa and Huh7, as well as noncancerous HEK-293T cells and normal CCC-HEL1 cells. G-Rg5 or G-Rk1 significantly inhibited NF-κB activity in Annexin A2 over-expressed cells like SW480, PC3, HeLa as did in HepG2 cells, but not in Huh7, a liver cancer cell line with low expression of Annexin A2. Interestingly neither compound inhibited the NF-κB activity in noncancerous HEK-293T cell and normal CCC-HEL1 cells (Fig. S1). These data suggested that G-Rg5 or G-Rk1 may serve as a specific anticancer reagent for the treatment of Annexin A2 over-expressed cancers, and this possibility should be examined in future study.

In conclusion, we have identified two specific small-molecule inhibitors of Annexin A2, G-Rg5 and G-Rk1, which inhibited NF-κB activity and promoted apoptosis. These data supported the notion that the inhibition of Annexin A2 may be a useful strategy to prevent and treat cancers and G-Rg5 and G-Rk1 can serve as a leading compound for targeted cancer treatment.

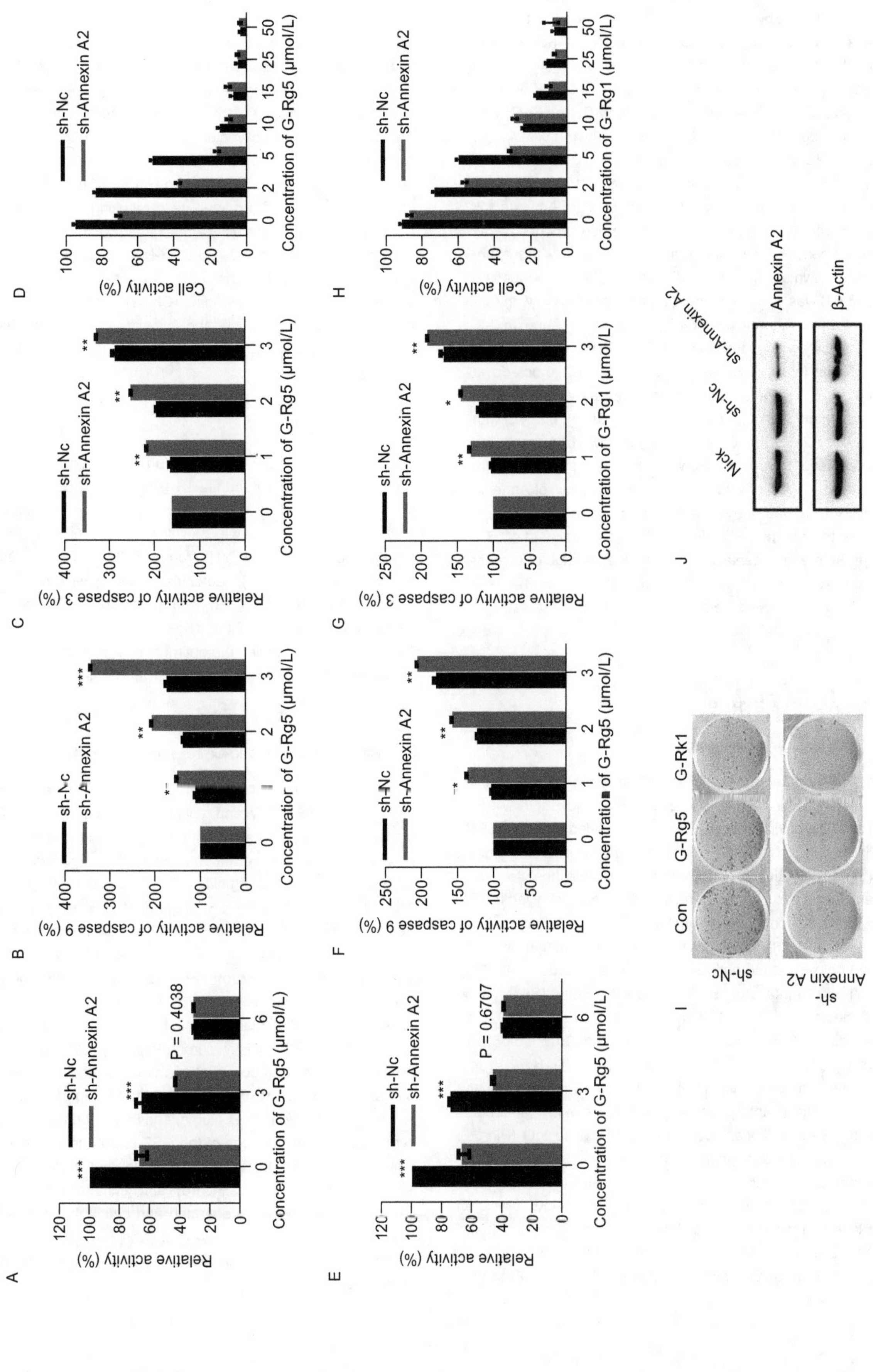
A
B
C
D
E
F
G
H
I
J
sh-Nc
sh-Annexin A2
Relative activity (%)
Relative activity of caspase 9 (%)
Relative activity of caspase 3 (%)
Cell activity (%)
P = 0.4038
P = 0.6707
Concentration of G-Rg5 (μmol/L)
Concentration of G-Rg1 (μmol/L)
Con
G-Rg5
G-Rk1
Nick
Annexin A2
β-Actin

◄ **Figure 4. Knock-down of Annexin A2 enhanced anti-proliferation effect of G-Rg5 and G-Rk1.** (A) NF-κB activity was examined by luciferase reporter assay under treatment of G-Rg5 with (sh-Annexin A2) or without (sh-NC) knock-down of Annexin A2. (B and C) Activity of caspase 9 (B) and 3 (C) was examined under treatment of G-Rg5 with (sh-Annexin A2) or without (sh-NC) knock-down of Annexin A2. (D) Cell viability was examined by MTT for 48 h under treatment of G-Rg5 with (sh-Annexin A2) or without (sh-NC) knock-down of Annexin A2. (E) NF-κB activity was examined by luciferase reporter assay under treatment of G-Rk1 with (sh-Annexin A2) or without (sh-NC) knock-down of Annexin A2. (F and G) Activity of caspase 9 (F) and 3 (G) was examined under treatment of G-Rk1 with (sh-Annexin A2) or without (sh-NC) knock-down of Annexin A2. (H) Cell viability was examined by MTT for 48 h under treatment of G-Rk1 with (sh-Annexin A2) or without (sh-NC) knock-down of Annexin A2. (I) Plate clone formation assay was examined under treatment of G-Rg5 (6 μmol/L) and G-Rk1 (6 μmol/L) with (sh-Annexin A2) or without (sh-NC) knock-down of Annexin A2. (J) Protein level of Annexin A2 was examined by immunoblot with (sh-Annexin A2) or without (sh-NC) knock-down of Annexin A2, and β-actin was shown as a loading control. All data are shown as the mean ± SD. $^{*}P < 0.05$, $^{**}P < 0.01$ and $^{***}P < 0.001$ and the experimental points show the average of at least triplicates. All experiments were repeated at least 3 times. Statistical analyses were performed using Student's *t*-test.

MATERIALS AND METHODS

Cell lines, reagents and plasmids

HepG2 cells were obtained from the American Type Culture Collection (ATCC, Rockville, MA, USA). Dulbecco's Modified Eagle Medium (DMEM) and fetal bovine serum (FBS) were obtained from Gibco BRL (Grand Island, NE, USA). Chemical reagents were listed below: G-Rg5 (Sigma), G-Rk1 (Sigma), phorbol myristate acetate (PMA, Sigma) and etoposide (Sigma). Ginsenosides were dissolved in 75% alcohol with a final concentration of 12 mmol/L. PMA (10 μg/mL) and etoposide (25 mg/mL) were dissolved in DMSO. Primary antibodies were listed below: mouse anti-Annexin A2 (Santa Cruz, sc-47696), mouse anti-p50 (Santa Cruz, sc-8414), rabbit anti-p50 (Santa Cruz, sc-7178), mouse anti-C-myc (Santa Cruz, sc-49), rabbit anti-C-myc (Santa Cruz, sc-789), mouse anti-X-IAP (Santa Cruz, sc-55551), rabbit anti-c-IAP1 (Santa Cruz, sc-7943), rabbit anti-c-IAP2 (Santa Cruz, sc-7944), mouse anti-survivin (Santa Cruz, sc-17779) and mouse anti-β-actin (Santa Cruz, sc-47778). Secondary antibodies were listed below: HRP-conjugated goat anti-mouse IgG (Pierce), HRP-conjugated goat anti-rabbit IgG (Pierce), Cy™3 affinipure donkey anti-mouse IgG (Jackson ImmunoResearch Inc., PA, USA), and Alexa Fluor® 488 affinipure donkey anti-Rabbit IgG (Jackson ImmunoResearch Inc., PA, USA). Genes of human Annexin A2 and human NF-κB p50 subunit were amplified by PCR, and cloned into pEXS-CG (GST-tag at C-terminal, provided by Professor Fei Sun's group, at the Institute of Biophysics of Chinese Academy of Sciences). Prokaryotic vectors were gained that allowed expression of non-tagged Annexin A2 (pEXS-Annexin A2-WT), non-tagged NF-κB p50 subunit (pEXS-p50), and C-terminal GST-tagged Annexin A2 (pEXS-CG-Annexin A2-WT). A single point mutation was engineered within Annexin A2-expressing vector, with resulting vectors called pEXS-Annexin A2-K302A and pEXS-CG-Annexin A2-K302A. The sequences of both Annexin A2-WT and Annexin A2-K302A were cloned into pcs4 (C-myc-tag at C-terminal), with resulting vectors called pcs4-Annexin A2-WT-myc and pcs4-Annexin A2-K302A-myc. Short hairpin RNA (shRNA) vector, pGPU6-GFP-Neo-Annexin A2 (1496), was obtained to knock down Annexin A2 expression in RNA interference assays (GenePharma, Jiangsu, China). The luciferase reporter assays were performed with the following plasmids: pNFκB-TA-luc (Beyotime, Shanghai, China) and pRL-CMV (Promega, WI, USA).

Molecular docking

We downloaded the three-dimensional structure of G-Rg5 (PubChem CID: 44416768) and G-Rk1 (PubChem CID: 11499198) from the NCBI Pubchem Compound database (http://www.ncbi.nlm.nih.gov/pccompound), and we downloaded the crystal structure of Annexin A2 (PDB ID: 2HYU) from the RCSB Protein Data Bank (http://www.rcsb.org/pdb). Molecular docking was performed with AutoDock tools (version 4.2.6) with the default setting, based on the Lamarckian Genetic Algorithm (Scripps Research Institute, La Jolla, CA, USA). We processed the optimum structure of the complex with the Discovery Studio 4.0 Visualizer (BIOVIA, CA, USA).

Purification of prokaryotic-expressed Annexin A2

The *E. coli* expression strain, BL21 (DE3), was transformed with pEXS-CG-Annexin A2, and cultured in Luria-Bertani (LB) liquid medium with 50 μg/mL ampicillin at 37°C until the density reached an OD_{600} of 1.5. Cells were cooled to 16°C and cultured for another 12 h at 16°C with 1.0 mmol/L IPTG for protein expression. Then, cells were harvested by centrifuging at 6,000 rpm (JA10 rotor, Beckman) for 12 min. The cell pellet was resuspended in pre-cooled lysis buffer (PBS containing 1 mg/mL lysozyme, 1 mmol/L DTT and 1 mmol/L PMSF), placed on ice, and ultra-sonicated for cell lysis. The lysed cells were separated with centrifugation at 12,000 rpm (JA25.50 rotor, Beckman) for 40 min, and the supernatant was loaded onto a GST-affinity chromatography column. The column was washed with pre-cooled PBS containing 1 mol/L NaCl, 1 mmol/L DTT and 1 mmol/L PMSF. Then, we added 10 μg of human HRV 3C protease (TAKARA) to the column, and incubated the column at 4°C overnight to allow cleavage of the GST-tag at the C-terminus of the protein. The eluted fraction was then loaded onto a Superdex75 16/600 column (GE Healthcare) and eluted at a flow rate of 1.0 mL/min. Fractions were pooled and concentrated to 10 mg/mL with a 10-kDa cut-off centrifuge filter (Millipore).

The Annexin A2-K302A mutant was purified with the same procedure.

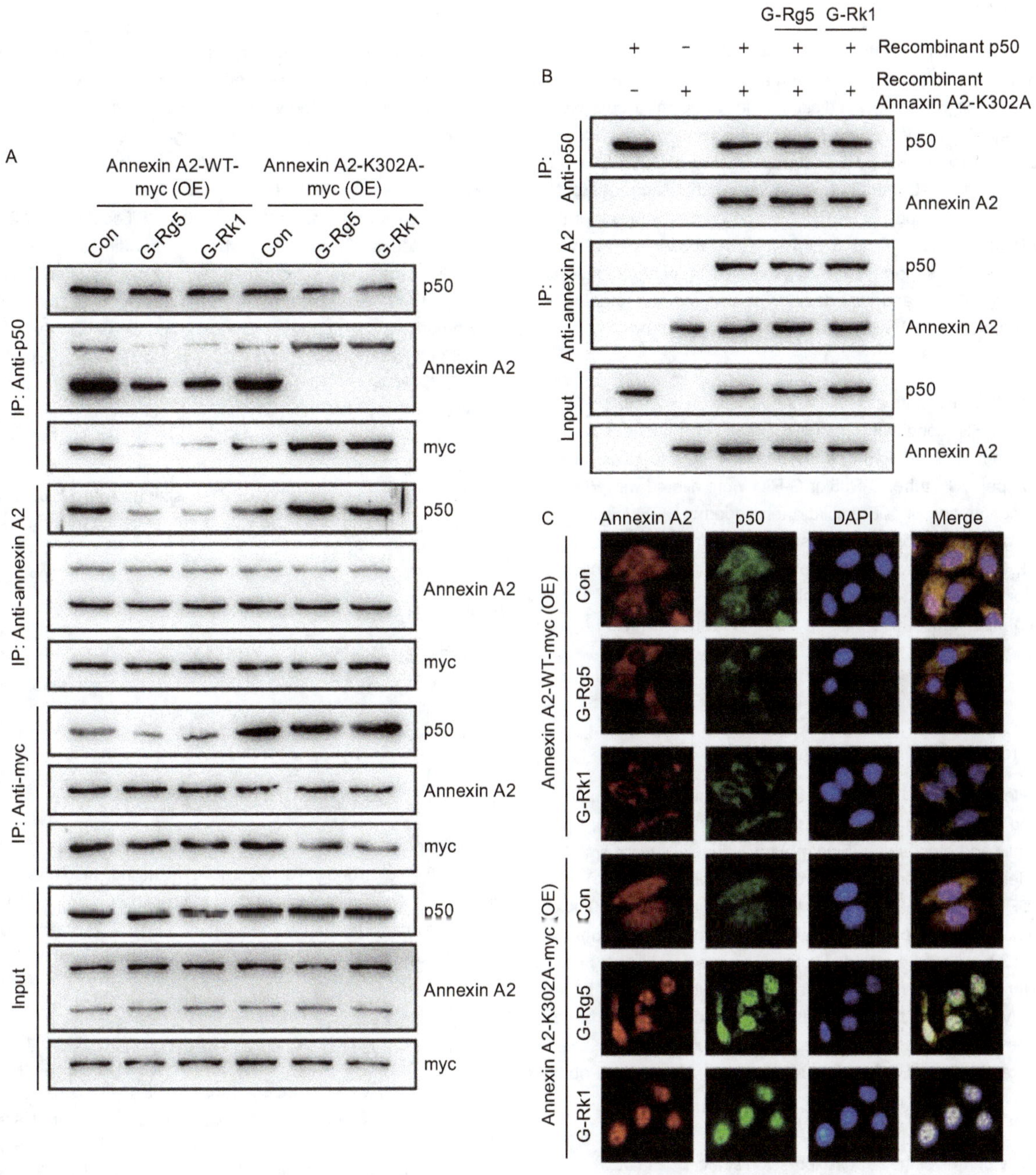

Figure 5. Annexin A2-K302A blocked the inhibitory effect of G-Rg5 and G-Rk1 on the interaction between Annexin A2 and NF-κB p50 subunit and their nuclear co-localization. (A) HepG2 cells were transfected with Annexin A2-WT-myc and Annexin A2-K302A-myc, and immunoprecipitation was performed with whole-cell lysate under treatment with G-Rg5 (6 μmol/L), G-Rk1 (6 μmol/L) and examined by an immunoblot. (B) Immunoprecipitation was performed with prokaryotic cells-expressed Annexin A2-K302A and NF-κB p50 subunit under treatment with G-Rg5 (6 μmol/L) and G-Rk1 (6 μmol/L), and the interaction was analyzed by an immunoblot. (C) HepG2 cells were transfected with Annexin A2-WT-myc and Annexin A2-K302A-myc, and the subcellular distribution of Annexin A2 and NF-κB p50 subunit was examined by immunofluorescence under treatment with G-Rg5 (6 μmol/L) and G-Rk1 (6 μmol/L), and DAPI showed the nuclear region.

Thermal shift assay

In vivo thermal shift assay (cellular thermal shift assay, CTSA): HepG2 cells (3×10^7) were seeded into a 100-mm culture plate with 12 μmol/L of G-Rg5 or G-Rk1 and cultured for 1 h. Control cells were incubated with the same volume. Cells were cultivated and counted, followed by resuspending in PBS (containing 1 mmol/L PMSF) to a final density of 2×10^7/mL. Then cells were subpackaged into 12 PCR tubes, with 100 μL per PCR tube, and heated with a thermal gradient from 40°C to 73°C, for 3 min. After freeze-thawed twice with liquid nitrogen, the supernatant was separated by centrifugation at 20,000 ×*g* for 20 min and collected. 20 μL of the supernatant was loaded onto an SDS-PAGE gel, followed by an immunoblot.

In vitro thermal shift assay: Purified Annexin A2 protein was diluted with PBS (containing 1 mmol/L PMSF) to a final concentration of 0.2 mmol/L (~7 μg/mL) and subpackaged into PCR tubes, with 90 μL per PCR tube. G-Rg5 or G-Rk1 were diluted with PBS to a final concentration of 100 μmol/L, then added to PCR tubes with Annexin A2 protein, forming a dose gradient from 0.2 μmol/L to 10 μmol/L. PCR tubes were made up to 100 μL with PBS and heated at 55°C for 3 min, with the control tube on ice for 3 min. Then 10 μL from each tube was loaded onto an SDS-PAGE gel, followed by an immunoblot.

Immunoprecipitation

In vivo: HepG2 cells (1.6×10^6) were plated into 100-mm culture plates and treated with G-Rg5 (6 μmol/L), G-Rk1 (6 μmol/L), PMA (100 ng/mL) and etoposide (25 μg/mL) for 12 h. Cells were lysed in lysis buffer (Pierce) containing protease inhibitors (Roche) and 1 mmol/L PMSF. 2 mg of total protein lysates were mixed with 10 μL of anti-Annexin A2, anti-p50 antibody or anti-C-myc, and incubated at 4°C for 3 h on a tube rotator. Protein A/G beads (Millipore) were washed three times with lysis buffer, and then incubated at 4°C for 4 h with the lysate-antibody complexes. The protein-agarose beads complexes were washed three times with lysis buffer. Samples were then separated with an SDS-PAGE and analyzed with an immunoblot.

In vitro: Cultures of the *E. coli* expression strain, BL21 (DE3), were transformed with pEXS-CG, pEXS-Annexin A2 (label-free) and pEXS-p50. Cells were cultivated in LB liquid medium containing 50 μg/mL ampicillin at 37°C, until the cell density reached an OD_{600} of 1.5. Then, cells were cultivated with 1 mmol/L IPTG for another 12 h. Cells were lysed with ultra-sonication in lysis buffer (100 mmol/L PBS, 125 mmol/L NaCl, 1 mg/mL lysozyme, and 1 mmol/L PMSF). 1 mg of the protein lysate containing Annexin A2 and another 1 mg of the protein lysate containing p50 were incubated at 4°C overnight with 12 μmol/L G-Rg5 or G-Rk1. As a control, 1 mg of protein lysate that contained neither Annexin A2 nor p50 was incubated in parallel. The remaining steps were performed as described above for the *in vivo* assay.

Immunofluorescence

Glass cover slips were placed into the wells of a 24-well plate, and 5×10^4 HepG2 cells were seeded into each well. After a 16-h incubation, cells were treated with G-Rg5 (6 μmol/L), G-Rk1 (6 μmol/L), PMA (100 ng/mL) and etoposide (25 μg/mL) for 2 h. Next, cells were fixed with pre-cool methanol at 4°C for 5 min. Then, cells were washed twice with PBST (100 mmol/L PBS with 0.5% Tween-20). The fixed cells were permeabilized with PBST containing 0.2% Triton X-100 at 4°C for 20 min, then washed three times with PBST. Permeabilized cells were incubated with blocking buffer (PBST containing 3% donkey serum) at room temperature for 1 h, then incubated with a primary antibody (mouse anti-Annexin A2 and rabbit anti-NF-κB, each diluted 1:200 in PBST with 5% BSA) for 3 h at 4°C. Cells were washed three times with PBST and incubated with a secondary antibody (Cy™3 affinipure donkey anti-mouse IgG and Alexa Fluor® 488 affinipure donkey anti-Rabbit IgG, each diluted 1:200 in PBST with 5% BSA) for 2 h at room temperature. After the cells were washed twice with PBST, stained with PBST and 0.1% DAPI (Sigma), and analyzed with a fluorescence microscope.

Cell viability assay

HepG2 cells were plated (5×10^3 per well) onto 96-well plates and treated with G-Rg5 or G-Rk1 at the indicated concentrations for 48 h. Cell viability was determined with the MTT assay.

Plate clone formation assay

HepG2 were plated (1×10^3 per well) onto 6-well plates and treated with G-Rg5 (6 μmol/L) and G-Rk1 (6 μmol/L) at the indicated concentrations for 1 week. Then cells were washed twice with PBS and fixed with precooled methanol for 10 min, followed by crystal violet staining for 5 min. Then cells were washed twice with PBS and photographed.

Quantitative real-time PCR (qRT-PCR)

Whole-cell RNA was isolated with TRIzol (Invitrogen). Then, 5-μg aliquots of whole-cell RNA were used for cDNA synthesis with the EasyScript Reverse Transcriptase kit (Transgen). QRT-PCR was performed on Applied Biosystem 7500 Real-time PCR system (Applied Biosystem. Inc) and TransStart Tuo Green qPCR SuperMix (Transgen). The amplification was performed with a three-step program, 1 cycle at 94°C for 30 s, followed by 45 cycles of 94°C for 5 s, 50°C for 15 s, and 72°C for 10 s, with signal collecting steps after annealing and extending for 34 s.

Statistical analysis

Data are presented as the mean ± S.D. Statistical significance was calculated with the Student's *t*-test.

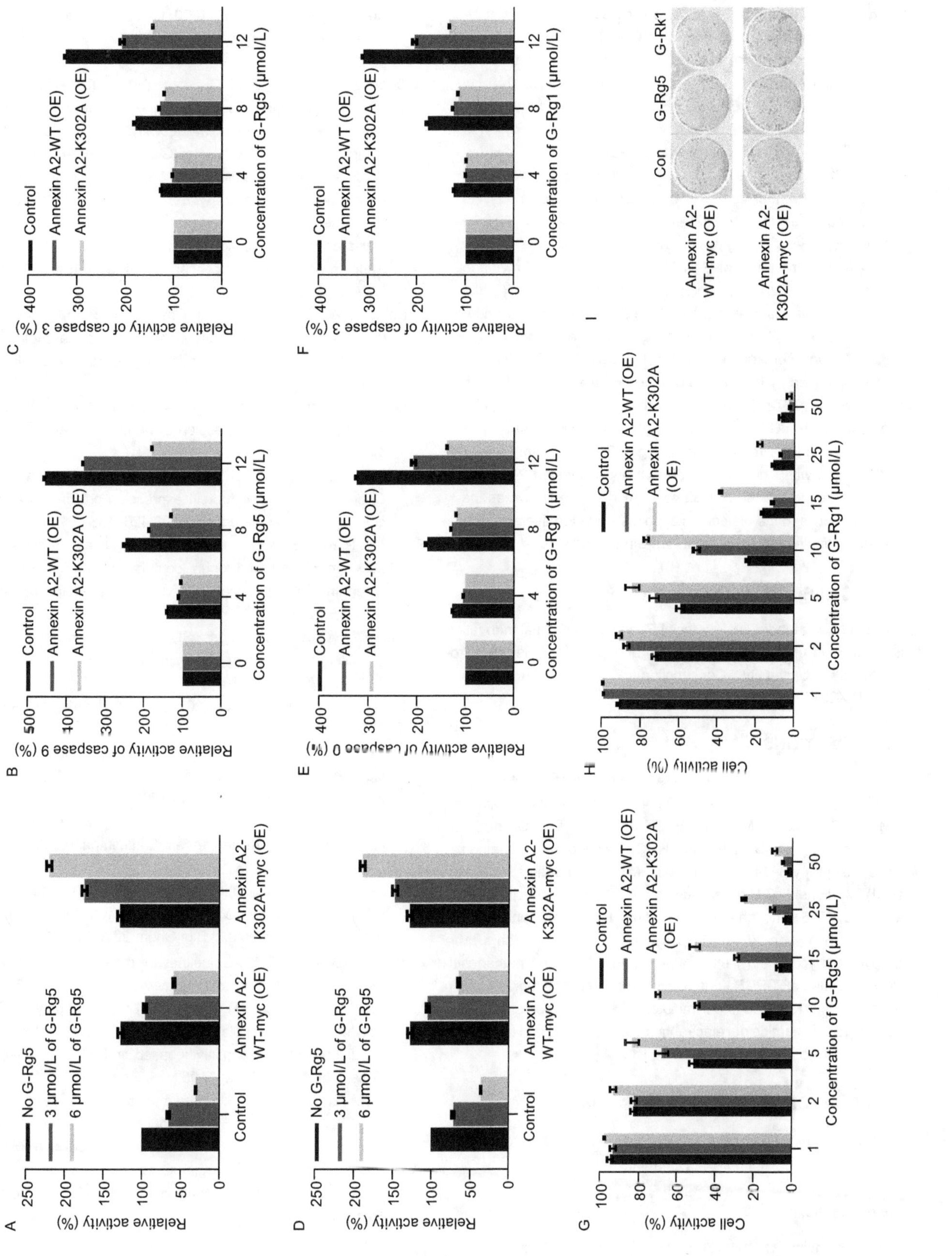
A
Relative activity (%)
No G-Rg5
3 μmol/L of G-Rg5
6 μmol/L of G-Rg5
Control
Annexin A2-WT-myc (OE)
Annexin A2-K302A-myc (OE)
B
Relative activity of caspase 9 (%)
Control
Annexin A2-WT (OE)
Annexin A2-K302A (OE)
Concentration of G-Rg5 (μmol/L)
C
Relative activity of caspase 3 (%)
Control
Annexin A2-WT (OE)
Annexin A2-K302A (OE)
Concentration of G-Rg5 (μmol/L)
D
Relative activity (%)
No G-Rg5
3 μmol/L of G-Rg5
6 μmol/L of G-Rg5
Control
Annexin A2-WT-myc (OE)
Annexin A2-K302A-myc (OE)
E
Control
Annexin A2-WT (OE)
Annexin A2-K302A (OE)
Concentration of G-Rg1 (μmol/L)
F
Relative activity of caspase 3 (%)
Control
Annexin A2-WT (OE)
Annexin A2-K302A (OE)
Concentration of G-Rg1 (μmol/L)
G
Cell activity (%)
Control
Annexin A2-WT (OE)
Annexin A2-K302A (OE)
Concentration of G-Rg5 (μmol/L)
H
Control
Annexin A2-WT (OE)
Annexin A2-K302A (OE)
Concentration of G-Rg1 (μmol/L)
I
Con
G-Rg5
G-Rk1
Annexin A2-WT-myc (OE)
Annexin A2-K302A-myc (OE)

◀ **Figure 6. Annexin A2-K302A protected cells from anti-proliferation effect of G-Rg5 and G-Rk1.** (A) NF-κB activity was examined by luciferase reporter assay under treatment of G-Rg5 with Annexin A2-WT-myc and Annexin A2-K302A-myc overexpressed in HepG2 cells. (B and C) Activity of caspase 9 (B) and 3 (C) was examined under treatment of G-Rg5 with Annexin A2-WT-myc and Annexin A2-K302A-myc overexpressed in HepG2 cells. (D) Cell viability was examined by MTT for 48 h under treatment of G-Rg5 with Annexin A2-WT-myc and Annexin A2-K302A-myc overexpressed in HepG2 cells. (E) NF-κB activity was examined by luciferase reporter assay under treatment of G-Rk1 with Annexin A2-WT-myc and Annexin A2-K302A-myc overexpressed in HepG2 cells. (F and G) Activity of caspase 9 (F) and 3 (G) was examined under treatment of G-Rk1 with Annexin A2-WT-myc and Annexin A2-K302A-myc overexpressed in HepG2 cells. (H) Cell viability was examined by MTT for 48 h under treatment of G-Rk1 with Annexin A2-WT-myc and Annexin A2-K302A-myc overexpressed in HepG2 cells. (I) Plate clone formation assay was examined under treatment of G-Rg5 (6 μmol/L) and G-Rk1 (6 μmol/L) with Annexin A2-WT-myc and Annexin A2-K302A-myc overexpressed in HepG2 cells. All data are shown as the mean ± SD and the experimental points show the average of at least triplicates. All experiments were repeated at least 3 times.

ACKNOWLEDGMENTS

This study was funded by Provincial Project for Industrial Innovation Special Fund of Jilin Province (2017) and Special Project for Province and University Construction Plan of Jilin Province (SXGJXX2017-13).

ABBREVIATIONS

BSA, bovine serum albumin; cDNA, complementary DNA; CTSA, cellular thermal shift assay; DAPI, 4',6-diamidino-2-phenylindole; DMEM, Dulbecco's Modified Eagle Medium; DMSO, dimethyl sulfoxide; DTT, DL-Dithiothreitol; HCC, hepatocellular carcinoma; IPTG, isopropy-β-D-thiogalactoside; MTT, 3-(4,5-dimethyl-2-thiazolyl)-2,5-diphenyl-2-H-tetrazolium bromide; NF-κB, nuclear factor kappa B; PBS, phosphate buffer saline; PBST, phosphate buffer saline with Tween-20; PCR, polymerase chain reaction; PMA, propylene glycol methyl ether acetate; PMSF, phenylmethanesulfonyl fluoride; qRT-PCR, quantitative real-time polymerase chain reaction; rpm, revolutions per minute; SDS-PAGE, sodium dodecyl sulfate polyacrylamide gel electrophoresis; STAT3, signal transducer and activator of transcription 3

COMPLIANCE WITH ETHICS GUIDELINES

Yu-Shi Wang, He Li, Yang Li, Hongyan Zhu and Ying-Hua Jin declare they have no conflict of interest.

REFERENCES

Alexandersavino CV, Hayden MS, Richardson C et al (2016) Doxycycline is an NF-κB inhibitor that induces apoptotic cell death in malignant T-cells. Oncotarget 7(46):75954–75967

Attele AS, Wu JA, Yuan CS (1999) Ginseng pharmacology: multiple constituents and multiple actions. Biochem Pharmacol 58:1685–1693

Chen CY, Lin YS, Chen CL et al (2015) Targeting annexin A2 reduces tumorigenesis and therapeutic resistance of nasopharyngeal carcinoma. Oncotarget 6(29):26946–26959

Chuthapisith S, Bean BE, Cowley G et al (2009) Annexins in human breast cancer: Possible predictors of pathological response to neoadjuvant chemotherapy. Eur J Cancer 45:1274–1281

Das S, Shetty P, Valapala M et al (2010) Signal transducer and activator of transcription 6 (STAT6) is a novel interactor of Annexin A2 in prostate cancer cells. Biochemistry 49(10):2216–2226

Deng S, Wang J, Hou L, Li J, Chen G, Jing B, Zhang X, Yang Z (2013) Annexin A1, A2, A4 and A5 play important roles in breast cancer, pancreatic cancer and laryngeal carcinoma, alone and/or synergistically. Oncol Lett 5:107–112

Dolcet X, Llobet D, Pallares J, Matias-Guiu X (2005) NF-κB in development and progression of human cancer. Virchows Arch 446:475–482

Duncan R, Carpenter B, Main LC, Telfer C, Murray GI (2008) Characterisation and protein expression profiling of annexins in colorectal cancer. Br J Cancer 98:426–433

Ecsédi P, Kiss B, Gógl G et al (2017) Regulation of the equilibrium between closed and open conformations of Annexin A2 by N-terminal phosphorylation and S100A4-binding. Structure 25 (8):1195.e5–1207.e5

Gerke V, Creutz CE, Moss SE (2005) Annexins: linking Ca^{2+} signalling to membrane dynamics. Nat Rev Mol Cell Biol 6:449–461 [PubMed: 15928709]

Hanahan D, Weinberg RA (2000) The hallmarks of cancer. Cell 100:57–70

Hayden MS, Ghosh S et al (2008) Shared principles in NF-κB signaling. Cell 132(3):344–362

Jung H, Kim JS, Kim WK et al (2015) Intracellular annexin A2 regulates NF-κB signaling by binding to the p50 subunit: implications for gemcitabine resistance in pancreatic cancer. Cell Death Dis 6(1):e1606

Kesavan K, Ratliff J, Johnson EW et al (2010) Annexin A2 is a molecular target for TM601, a peptide with tumor-targeting and anti-angiogenic effects. J Biol Chem 285(7):4366

Klionsky DJ et al (2012) Guidelines for the use and interpretation of assays for monitoring autophagy. Autophagy 8:445–544

Kwon M, Jang H, Kim EH et al (2016) Efficacy of poly (ADP-ribose) polymerase inhibitor olaparib against head and neck cancer cells: Predictions of drug sensitivity based on PAR-p53-NF-κB interactions. Cell Cycle 15(22):3105

Lee SM (2015) Anti-inflammatory effects of ginsenosides Rg5, Rz1, and Rk1: inhibition of TNF-α-induced NF-κB, COX-2, and iNOS transcriptional expression. Phytother Res 28(12):1893–1896

Longerich T, Haller MT, Mogler C et al (2011) Annexin A2 as a differential diagnostic marker of hepatocellular tumors. Pathol Res Pract 207(1):8–14

Ma RL, Shen LY, Chen KN (2014) Coexpression of ANXA2, SOD2 and HOXA13 predicts poor prognosis of esophageal squamous cell carcinoma. Oncol Rep 31:2157–2164

Moreau K, Ravikumar B, Renna M, Puri C, Rubinsztein DC (2011) Autophagosome precursor maturation requires homotypic fusion. Cell 146:303–317

Moss SE, Morgan RO (2004) The annexins. Genome Biol 5:219. [PubMed: 15059252]

Ohno Y, Izumi M, Kawamura T, Nishimura T, Mukai K, Tachibana M (2009) Annexin II represents metastatic potential in clearcell renal cell carcinoma. Br J Cancer 101:287–294

Perkins ND (2012) The diverse and complex roles of NF-κB subunits in cancer. Nat Rev Cancer 12:121–132

Réty S, Sopkova J, Renouard M, Osterloh D, Gerke V, Tabaries S, Russo-Marie F, Lewit-Bentley A (1999) The crystal structure of a complex of p11 with the annexin II N-terminal peptide. Nat Struct Biol 6:89–95

Shetty PK, Thamake SI, Biswas S, Johansson SL, Vishwanatha JK (2012) Reciprocal regulation of annexin A2 and EGFR with Her-2 in Her-2 negative and herceptin-resistant breast cancer. PLoS One 7:e44299

Staquicini DI, Rangel R, Guzmanrojas L et al (2017) Intracellular targeting of annexin A2 inhibits tumor cell adhesion, migration, and in vivo grafting. Sci Rep 7(1):4243

Sun Y, Gao G, Cai J et al (2013) Annexin A2 is a discriminative serological candidate in early hepatocellular carcinoma. Carcinogenesis 34(3):595–604

Takano S, Togawa A, Yoshitomi H et al (2008) Annexin II overexpression predicts rapid recurrence after surgery in pancreatic cancer patients undergoing gemcitabine-adjuvant chemotherapy. Ann Surg Oncol 15:3157–3168

Toh DF, New LS, Koh HL, Chan EC (2010) Ultra-high performance liquid chromatography/time-of-flight mass spectrometry (UHPLC/TOFMS) for time-dependent profiling of raw and steamed *Panax notoginseng*. J Pharm Biomed Anal 52:43–50

Tong W, Jie Y, Jie Z et al (2015) Anxa2 binds to STAT3 and promotes epithelial to mesenchymal transition in breast cancer cells. Oncotarget 6(31):30975–30992

Wang CY, Chen CL, Tseng YL et al (2012) Annexin A2 silencing induces G2 arrest of non-small cell lung cancer cells through p53-dependent and -independent mechanisms. J Biol Chem 287 (39):32512–32524

Woo SM, Min KJ, Seo BR, Kwon TK (2016) Ym155 sensitizes trail-induced apoptosis through cathepsin s-dependent down-regulation of mcl-1 and n NF-κB down-regulation of c-flip expression in human renal carcinoma caki cells. Oncotarget 7(38):61520–61532

Yang T, Peng H, Wang J, Yang J, Nice EC, Xie K, Huang C (2013) Prognostic and diagnostic significance of annexin A2 in colorectal cancer. Colorectal Dis 15:e373–e381

Yang SF, Hsu HL, Chao TK, Hsiao CJ, Lin YF, Cheng CW (2014) Annexin A2 in renal cell carcinoma: Expression, function, and prognostic significance. Urol Oncol 33:22.e11–22.e21

Yao H, Zhang Z, Xiao Z et al (2009) Identification of metastasis associated proteins in human lung squamous carcinoma using two-dimensional difference gel electrophoresis and laser capture microdissection. Lung Cancer 65:41–48

Yin DL, Liang YJ, Zheng TS et al (2016) EF24 inhibits tumor growth and metastasis via suppressing NF-kappaB dependent pathways in human cholangiocarcinoma. Sci Rep 6:32167

Zhang HJ, Yao DF, Yao M, Huang H, Wu W, Yan MJ, Yan XD, Chen J (2012a) Expression characteristics and diagnostic value of annexin A2 in hepatocellular carcinoma. World J Gastroenterol 18:5897–5904

Zhang Q, Ye Z, Yang Q, He X, Wang H, Zhao Z (2012b) Upregulated expression of annexin II is a prognostic marker for patients with gastric cancer. World J Surg Oncol 10:103

8

IgG Fc engineering to modulate antibody effector functions

Xinhua Wang, Mary Mathieu, Randall J. Brezski

Genentech, Antibody Engineering, South San Francisco, CA 94080, USA
Correspondence: brezski.randall@gene.com (R. J. Brezski)

ABSTRACT

Therapeutic monoclonal antibodies are among the most effective biotherapeutics to date. An important aspect of antibodies is their ability to bind antigen while at the same time recruit immune effector functions. The majority of approved recombinant monoclonal antibody therapies are of the human IgG1 subclass, which can engage both humoral and cellular components of the immune system. The wealth of information generated about antibodies has afforded investigators the ability to molecularly engineer antibodies to modulate effector functions. Here, we review various antibody engineering efforts intended to improve efficacy and safety relative to the human IgG isotype. Further, we will discuss proposed mechanisms by which engineering approaches led to modified interactions with immune components and provide examples of clinical studies using next generation antibodies.

KEYWORDS antibody-dependent cell-mediated cytotoxicity, antibody-dependent cellular phagocytosis, complement-dependent cytotoxicity, Fc engineering, Fc gamma receptor, monoclonal antibody, neonatal Fc receptor

INTRODUCTION

Antibodies are a critical component of the immune response, and the advent of the monoclonal antibody (mAb) technology has afforded the biotechnology and pharmaceutical industries the ability to develop antibodies as drugs. The success of mAb-based therapies is based on their safety, selectivity, diversity, solubility, tolerability, stability, and long circulating half-life. This success is further exemplified by the fact that there are currently over 50 investigational antibodies undergoing evaluation in late-stage clinical trials (Reichert, 2017) and approved antibodies generate an annual market of over $60 billion (Ecker et al., 2015).

To date, all of the currently approved antibodies are of the IgG isotype, which is further divided into four subclasses in humans (i.e., IgG1, IgG2, IgG3, and IgG4) (Brezski and Georgiou, 2016; Lefranc, 2001). The structure of an IgG antibody is comprised of two antigen-binding Fab arms linked to a single Fc domain via the hinge region. This structural arrangement allows antibodies to link bound antigen with humoral and cellular components of the immune system. Engagement of the humoral immune response is governed, in large part, by interactions with C1q and the initiation of a series of proteolytic events known as the complement cascade (Meyer et al., 2014). The cellular immune response occurs mostly due to the interactions between the antibody and Fc gamma receptors (FcγRs). There are 5 activating FcγRs: the high affinity FcγRI (HGNC:3613) that can bind to monovalent antibody, and the lower affinity FcγRIIa (HGNC:3616) and IIc (HGNC:15626), and FcγRIIIa (HGNC:3619) and IIIb (HGNC:3620) that require avidity-based interactions. There is one inhibitory receptor: FcγRIIb (HGNC:3618). Intracellular signaling through the activating receptors is modulated through the phosphorylation of immunoreceptor tyrosine-based activation motifs (ITAMs), which leads to effector functions such as antibody-dependent cell-mediated cytotoxicity (ADCC), antibody-dependent cellular phagocytosis (ADCP), and inflammation via the induction of cytokine secretion. In contrast, intracellular signaling through the inhibitory FcγRIIb is modulated through the phosphorylation of immunoreceptor tyrosine-based inhibitory motifs (ITIMs), which recruit phosphatases that counter-balance activating signaling pathways (Nimmerjahn and Ravetch, 2008). Antibody interactions with FcγRs and C1q are dependent on the hinge and proximal CH2 amino acid sequence as well as glycosylation

---

Xinhua Wang and Mary Mathieu have contributed equally to this work.

Table 1. Examples of modifications to modulate antibody effector function. Unless otherwise noted, the mutations are on the IgG1 subclass.

Engineering and intended function	Mutation	Reference
Enhance ADCC		
Increased FcγRIIIa binding	F243L/R292P/Y300L/V305I/P396L	(Stavenhagen et al., 2007)
Increased FcγRIIIa binding	S239D/I332E	(Lazar et al., 2006)
Increased FcγRIIIa binding, Decreased FcγRIIb binding	S239D/I332E/A330L	(Lazar et al., 2006)
Increased FcγRIIIa binding	S298A/E333A/K334A	(Shields et al., 2001)
	In one heavy chain: L234Y/L235Q/G236W/S239M/H268D/D270E/S298A In the opposing heavy chain: D270E/K326D/A330M/K334E	(Mimoto et al., 2013a)
Enhance ADCP		
Increased FcγRIIa binding, Increased FcγRIIIa binding	G236A/S239D/I332E	(Richards et al., 2008)
Enhance CDC		
Increased C1q binding	K326W/E333S	(Idusogie et al., 2001)
Increased C1q binding	S267E/H268F/S324T	(Moore et al., 2010)
Increased C1q binding	IgG1/IgG3 cross subclass	(Natsume et al., 2008)
Hexamerization	E345R/E430G/S440Y	(Diebolder et al., 2014)
Reduce effector function		
Aglycosylated	N297A or N297Q or N297G	(Bolt et al., 1993; Leabman et al., 2013; Tao and Morrison, 1989; Walker et al., 1989)
Reduced FcγR and C1q binding	L235E	(Alegre et al., 1992)
Reduced FcγR and C1q binding	IgG1: L234A/L235A; IgG4:F234A/L235A	(Xu et al., 2000)
Reduced FcγR and C1q binding	IgG2/IgG4 cross isotype	(Rother et al., 2007)
Reduced FcγR and C1q binding	IgG2: H268Q/V309L/A330S/P331S	(An et al., 2009)
Reduced FcγR and C1q binding	IgG2: V234A/G237A/P238S/H268A/V309L/A330S/P331S	(Vafa et al., 2014)
Increase half-life		
Increased FcRn binding at pH 6.0	M252Y/S254T/T256E	(Dall'Acqua et al., 2002)
Increased FcRn binding at pH 6.0	M428L/N434S	(Zalevsky et al., 2010)
Increased coengagement		
Increased FcγRIIb binding	S267E/L328F	(Chu et al., 2008)
Increased FcγRIIa binding, Decreased FcγRIIIa binding	N325S/L328F	(Shang et al., 2014)

at the conserved amino acid N297 (EU numbering (Edelman et al., 1969)) in the CH2 region (Fig. 1).

Although many antibody-based therapies have had clinical and commercial success, antibody-based therapeutics are often only effective in a subset of patients. Therefore, considerable effort has been put into fine-tuning the properties of antibodies in attempts to achieve efficacy in larger patient populations. Among the many efforts, one area of focus has been modulating the ability of antibodies to interact with the humoral and cellular components of the immune system through a process known as Fc engineering (Carter, 2006; Presta, 2008; Strohl, 2011). The purpose of this review is to detail Fc engineering efforts that modulate effector function, and where appropriate, provide examples of clinical

Table 2. An alignment of the hinge, CH2, and CH3 region for human IgG1 according to IMGT unique numbering (http://www.imgt.org/IMGTScientificChart/Numbering/Hu_IGHGnber.html) and EU numbering is shown.

Core Hinge																					
EU numbering	226	227	228	229																	
IMGT unique	11	12	13	14																	
Human IgG1	C	P	P	C																	
Lower Hinge																					
EU numbering	230	231	232	233	234	235	236	237	238												
IMGT unique	15	1.6	1.5	1.4	1.3	1.2	1.1	1	2												
Human IgG1	P	A	P	E	L	L	G	G	P												
CH3 Region																					
EU numbering	239	240	241	242	243	244	245	246	247	248	249	250	251	252	253	254	255	256	257	258	259
IMGT unique	3	4	5	6	7	8	9	10	11	12	13	14	15	15.1	15.2	16	17	18	19	20	21
Human IgG1	S	V	F	L	F	P	P	K	P	K	D	T	L	M	I	S	R	T	P	E	V
CH2 Region																					
EU numbering	260	261	262	263	264	265	266	267	268	269	270	271	272	273	274	275	276	277	278	279	280
IMGT unique	22	23	24	25	26	27	28	29	30	31	34	35	36	37	38	39	40	41	42	43	44
Human IgG1	T	C	V	V	V	D	V	S	H	E	D	P	E	V	K	F	N	W	Y	V	D
CH2 Region																					
EU numbering	281	282	283	284	285	286	287	288	289	290	291	292	293	294	295	296	297	298	299	300	301
IMGT unique	45	45.1	45.2	45.3	45.4	77	78	79	80	81	82	83	84	84.1	84.2	84.3	84.4	85.4	85.3	85.2	85.1
Human IgG1	G	V	E	V	H	N	A	K	T	K	P	R	E	E	Q	Y	N	S	T	Y	R
CH2 Region																					
EU numbering	302	303	304	305	306	307	308	309	310	311	312	313	314	315	316	317	318	319	320	321	322
IMGT unique	85	86	87	88	89	90	91	92	93	94	95	96	97	98	99	100	101	102	103	104	105
Human IgG1	V	V	S	V	L	T	V	L	H	Q	D	W	L	N	G	K	E	Y	K	C	K
CH2 Region																					
EU numbering	323	324	325	326	327	328	329	330	331	332	333	334	335	336	337	338	339	340			
IMGT unique	106	107	108	109	110	113	114	115	116	117	118	119	120	121	122	123	124	125			
Human IgG1	V	S	N	K	A	L	P	A	P	I	E	K	T	I	S	K	A	K			
CH3 Region																					
EU numbering	341	342	343	344	345	346	347	348	349	350	351	352	353	354	355	356	357	358	359	360	361
IMGT unique	1.4	1.3	1.2	1.1	1	2	3	4	5	6	7	8	9	10	11	12	13	14	15	16	17
Human IgG1	G	Q	P	R	E	P	Q	V	Y	T	L	P	P	S	R	D	E	L	T	K	N
CH3 Region																					
EU numbering	362	362	364	365	366	367	368	369	370	371	372	373	374	375	376	377	378	379	380	381	382
IMGT unique	18	19	20	21	22	23	24	25	26	27	28	29	30	35	36	37	38	39	40	41	42
Human IgG1	Q	V	S	L	T	C	L	V	K	G	F	Y	P	S	D	I	A	V	E	W	E
CH3 Region																					
EU numbering	383	384	385	386	387	388	389	390	391	392	393	394	395	396	397	398	399	400	401	402	403
IMGT unique	43	44	45	45.1	45.2	45.3	45.4	77	78	79	80	81	82	83	84	84.1	84.2	84.3	84.4	85.4	85.3
Human IgG1	S	N	G	Q	P	E	N	N	Y	K	T	T	P	P	V	L	D	S	D	G	S
CH3 Region																					
EU numbering	404	405	406	407	408	409	410	411	412	413	414	415	416	417	418	419	420	421	422	423	424
IMGT unique	85.2	85.1	85	86	87	88	89	90	91	92	93	94	95	96	97	98	99	100	101	102	103
Human IgG1	F	F	L	Y	S	K	L	T	V	D	K	S	R	W	Q	Q	G	N	V	F	S
CH3 Region																					
EU numbering	425	426	427	428	429	430	431	432	433	434	435	436	437	438	439	440	441	442	443	444	445
IMGT unique	104	105	106	107	108	109	110	112	113	114	115	116	117	118	119	120	121	122	123	124	125
Human IgG1	C	S	V	M	H	E	A	L	H	N	H	Y	T	Q	K	S	L	S	L	S	P
CH3 Region																					
EU numbering	446	447																			
IMGT unique	129	130																			
Human IgG1	G	K																			

Note: The core hinge region is defined by the N-terminal and C-terminal cysteines that link the two heavy chains together (Liu and May, 2012). The lower hinge region ends with the final amino acid prior to the start of the first CH2 β-sheet domain (as shown in PDBID: 3AVE, 1HZH, 5KD3, and 4HAG). The red, green, and blue coloring scheme is the same that was used in Fig. 1.

studies where Fc engineered antibodies have been employed.

ENHANCED COMPLEMENT-BASED EFFECTOR FUNCTION

Several approved antibodies have demonstrated potent *in vitro* CDC activity such as the anti-CD20 mAbs rituximab and ofatumumab (Teeling et al., 2006), and there are numerous ways that investigators have utilized Fc engineering to enhance complement-based effector function (Table 1). Idusogie et al. demonstrated that K326W/E333S enhanced C1q binding and CDC activity relative to an IgG1, leading the authors to suggest that these two residues play a structural role in interactions between C1q and IgG (Idusogie et al., 2001). Moore et al. demonstrated that the Fc mutations S267E/H268F/S324T enhanced C1q binding 47-fold and CDC activity 6.9-fold over IgG1 (Moore et al., 2010). The authors suggested that the S267E mutant altered the charge of the Fc, potentially increasing ionic interactions with the C1q subunit B. Among the four human IgG subclasses, IgG3 has the most potent CDC activity (Tao et al., 1993). However, there are no approved IgG3 therapeutics due in part to the relatively short half-life of IgG3 as well as bioprocessing issues associated with IgG3's long hinge region (Brezski and Georgiou, 2016). To overcome the liabilities associated with IgG3, Natsume et al. generated a cross IgG subclass variant combining the CH1 and hinge regions of IgG1 with the CH2 and CH3 regions of IgG3 (Natsume et al., 2008), which resulted in up to 50-fold enhanced CDC activity relative to IgG1 depending on the cell line used to assess CDC activity. This enhanced activity may be attributable to the differences between IgG1 and IgG3 in the CH2 region (i.e., K274Q/N276K/Y300F), which are structurally proximal to the residues D270/K322/P329/P331 deemed critical for C1q interactions (Idusogie et al., 2000; Thommesen et al., 2000). Diebolder et al. employed an approach where they engineered the Fc region with the mutations E345R/E430G/S440Y, which resulted in preformed IgG hexamers (Diebolder et al., 2014) and enhanced CDC activity. Structural data indicated that the hexamer oriented K322 in a position where it could favorably accommodate interactions with the hexameric C1q headpiece. Together, these studies demonstrated that several different mutational approaches can enhance C1q binding and the resultant CDC activity.

ENGINEERING FOR FcγR-BASED EFFECTOR FUNCTION

Early studies defining key amino acid binding sites on IgGs for FcγRs were performed by mutational analyses, and it was determined that the lower hinge and proximal CH2 regions were critical (Burton et al., 1988; Canfield and Morrison, 1991; Chappel et al., 1991; Duncan et al., 1988; Hulett et al., 1994; Jefferis et al., 1990; Lund et al., 1991; Partridge et al., 1986; Shields et al., 2001; Tamm et al., 1996; Wines et al., 2000; Woof et al., 1986). Additionally, multiple co-crystal structures of human IgG1 Fc with the low affinity FcγRs have allowed high resolution mapping of the binding interfaces. A recurrent finding among the low affinity FcγRs was that P329 packs between two conserved tryptophan residues found in all of the FcγRs (Mimoto et al., 2013b; Radaev et al., 2001; Ramsland et al., 2011; Sondermann et al., 2000)—an interaction termed the "proline sandwich". More recently, there have been several reports of human IgG1 Fc co-crystals with the high affinity FcγRI (Kiyoshi et al., 2015; Lu et al., 2015; Oganesyan et al., 2015). A common feature among these structures was that a hydrophobic pocket on the surface of FcγRI well-accommodated L235 on the Fc. Together, these mutational studies and crystal structures have resulted in a wealth of knowledge about how IgGs interact with their respective FcγRs.

The efficacy of several anti-cancer mAbs is thought to rely, in part, on recruitment of FcγR-based effector functions. This includes activation of NK cells via FcγRIIIa and resultant ADCC and inflammatory cytokine release, macrophage-mediated ADCP through interactions with multiple FcγRs, and recruitment and activation of other immune cells (e.g., neutrophils). For FcγRIIIa, the key receptor for NK cell-mediated ADCC, there are two polymorphic variants: V158 with higher affinity for IgG1, and F158 that has lower affinity for IgG1. The potential significance of these polymorphisms was reported in several clinical trials where cancer patients with the high-affinity V158 polymorphism showed better outcomes from cetuximab (Bibeau et al., 2009), trastuzumab (Gavin et al., 2017), and rituximab (Cartron et al., 2002) therapy compared to patients with the low-affinity F158 polymorphism, although there were also clinical trials where there was no apparent benefit from expressing the high affinity polymorphism. Due to the potential benefits of augmenting binding to FcγRs and the resultant enhanced innate immune cell function, investigators have utilized multiple approaches to engineer mAbs including glyco-engineering and mutating amino acids within the Fc region.

Glyco-engineering

IgGs contain a conserved glycosylation site at amino acid N297 in the CH2 domain. The core structure of the glycan is comprised of *N*-acetylglucosamine (GlcNAc) and mannose, where additional modifications can include bisecting GlcNAc, fucose, galactose, and sialic acid. One of the first reports linking glyco-engineering with enhanced Fc effector function demonstrated that IgG1 antibodies produced in a Chinese hamster ovary (CHO) cell line expressing b (1,4)-N-acetylglucosaminyltransferase III to express bisecting GlcNAc augmented ADCC activity relative to IgG1 (Umana et al., 1999). A study by Shields et al. demonstrated that IgG1s deficient in fucose had an up to 50-fold increase in FcγRIIIa binding relative to IgG1 as well as enhanced ADCC (Shields et al., 2002). Shinkawa et al. later demonstrated that fucose deficient antibodies had improved ADCC function compared to antibodies containing bisecting GlcNAc (Shinkawa et al., 2003). The putative mechanism for enhanced interactions between afucosylated antibodies and FcγRIIIa was shown in a crystallographic study (Ferrara et al., 2011). Amino acid N162 in FcγRIIIa contains a glycan, and the absence of fucose allows greater carbohydrate-carbohydrate interactions with the Fc, which increases the overall binding strength. At present, there are two glyco-engineered antibodies that have been approved, the anti-CD20 mAb, obinutuzumab, and the anti-CCR4 mAb, mogamulizumab (Beck and Reichert, 2012). These two antibodies demonstrate that glyco-engineering for enhanced effector function can translate into clinically approved therapeutics.

Amino acid mutations

There are several ways in which investigators have mutated the amino acid sequence of the IgG Fc region to modulate FcγR-based effector function. Some examples include point mutations, design algorithms, yeast display, and asymmetric engineering. The results from each of these different technologies have yielded numerous mutations that modify Fc-FcγR interactions and the resultant effector functions (Table 1).

Shields et al. performed single alanine point mutations of all solvent-exposed amino acids in CH2 and CH3 domains of human IgG1. They further demonstrated that combining several identified mutations (e.g., S298A/E333A/K334A) resulted in enhanced ADCC relative to IgG1 (Shields et al., 2001). Lazar et al. engineered a series of Fc variants with optimized Fcγ receptor affinity using computational design algorithms and high-throughput screening. The authors identified S239D/I332E and S239D/I332E/A330L as two variants with enhanced ADCC activity (Lazar et al., 2006). A crystal structure of an Fc fragment with the mutations S239D/A330L/I332E was solved and modeling studies suggested that additional hydrogen bonds, hydrophobic

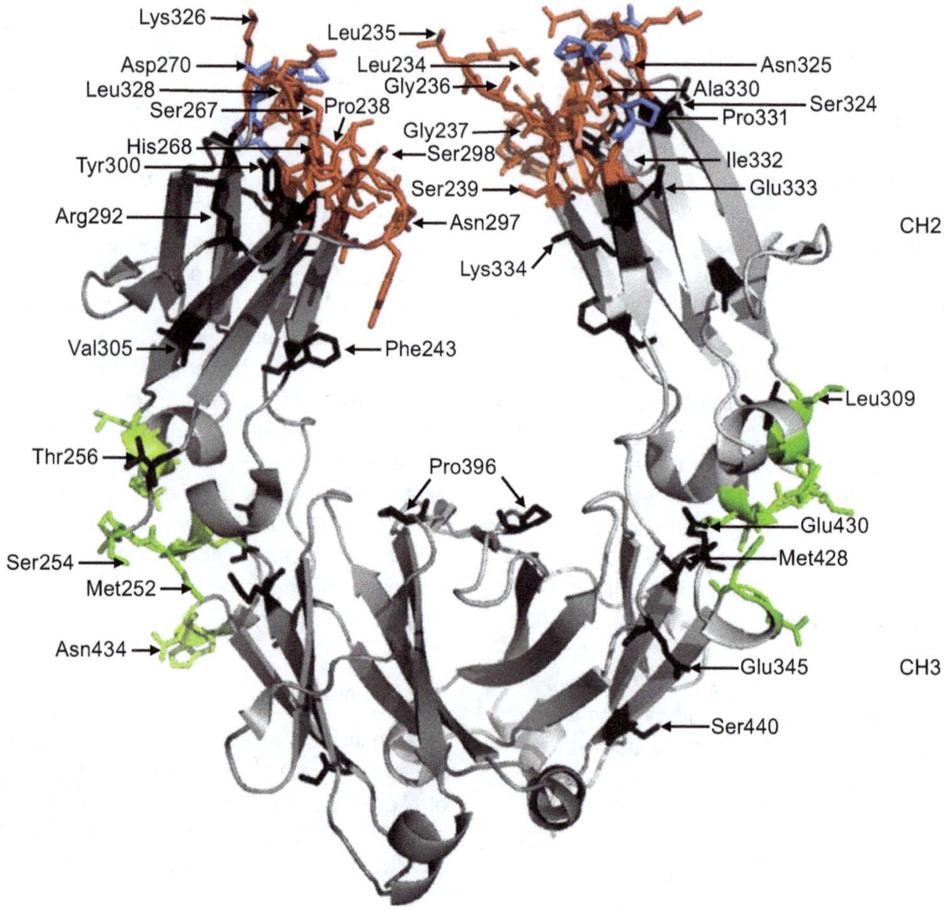

Figure 1. The Fc domain of human IgG1 is shown (PDBID:3AVE). The positions of amino acids described in Table 1 are shown. Key sequence motifs for FcRn interactions are shown in green (L251 S254; L309-Q311; N434-H435) (Oganesyan et al., 2014). Amino acids 5 angstroms proximal to the FcγR:Fc interface for FcγRI (PDBID:4W4O) (Kiyoshi et al., 2015), FcγRIIa (PDBID:3RY6) (Ramsland et al., 2011), FcγRIIb (PDBID:3WJJ) (Mimoto et al., 2013b), and FcγRIIIb (PDBID:1E4K) (Sondermann et al., 2000) as determined with PyMOL are shown in red (P232-S239; D265-D270; Y296-T299; N325-I332). The amino acids that are critical for interactions with C1q are shown in blue (D270, K322, P329, P331) (Idusogie et al., 2000). Amino acids depicted in black were not identified as critical for interactions with FγRs, FcRn, or C1q. Figure 1 was generated with PyMOL.

contacts, and/or electrostatic interactions resulted in enhanced binding to FcγRIIIa (Oganesyan et al., 2008). Richards demonstrated that the addition of G236A to the S239D/I332E mutations resulted in up to 70-fold improved binding to FcγRIIa, a 13-fold improvement in the FcγRIIa/FcγRIIb binding ratio (activating/inhibitory ratio), and enhanced phagocytosis of antibody-coated target cells by macrophages (Richards et al., 2008). Stavenhagen et al. utilized yeast surface display to identify the variant F243L/R292P/Y300L/V305I/P396L, which showed >100 fold increased ADCC activity (Stavenhagen et al., 2007). Margetuximab, an anti-HER2 mAb with these 5 mutations completed a phase I clinical trial for patients with HER2-overexpressing carcinomas (Bang et al., 2017). Mimoto et al. designed antibody variants with an asymmetrically engineered Fc domain by introducing different amino acid changes in each Fc domain. They screened ~1000 variants and demonstrated that L234Y/L235Q/G236W/S239M/H268D/D270E/S298A changes in one Fc domain and D270E/K326D/A330M/K334E changes in the other increased affinity for FcγRIIIa F158 by more than 2000-fold and FcγRIIIa V158 by more than 1000-fold (Mimoto et al., 2013a). Together, these investigations demonstrate that various technologies can be employed to obtain Fc variants with enhanced FcγR-dependent effector function.

FC ENGINEERING FOR REDUCED EFFECTOR FUNCTION

For cases where mAbs are intended to engage cell surface receptors and prevent receptor-ligand interactions (i.e., antagonists), it may be desirable to reduce or eliminate effector function for example to prevent target cell death or unwanted cytokine secretion. Other examples where reduced effector function may be warranted include preventing antibody-drug conjugates from interacting with FcγRs leading to off-target cytotoxicity (Uppal et al., 2015). The need for reducing or eliminating effector function was recognized with the first approved mAb, the murine IgG2a anti-CD3ε mAb, OKT3, muromonab, which was intended to prevent T cell activation in tissue transplant patients receiving a donor kidney, lung, or heart (Chatenoud and Bluestone, 2007). Many patients receiving muromonab had adverse events including the induction of pro-inflammatory cytokines (i.e., cytokine storm), which was attributed in part to muromonab's interactions with FcγRs (Alegre et al., 1992). In order to reduce this unintended effector function, Bluestone and colleagues generated the human IgG4 variant L235E (Alegre et al., 1992) or F234A/L235A (Xu et al., 2000), and the human IgG1 variant L234A/L235A (Xu et al., 2000), all of which reduced inflammatory cytokine release.

Another early approach intended to reduce effector function was to mutate the glycosylation site at N297 with mutations such as N297A, N297Q, and N297G (Bolt et al., 1993; Leabman et al., 2013; Tao and Morrison, 1989; Walker et al., 1989). This aglycosylation approach has proven successful in abrogating Fc interactions with the low affinity FcγRs and effector functions such as CDC and ADCC. However, it was recognized early on that under avidity-based binding conditions, afucosylated IgGs have been shown to retain effector function when monocytes were employed as effector cells (Bolt et al., 1993). More recently, it was demonstrated that aglycosylated IgG1 mAbs had macrophage-mediated ADCP similar to IgG1 in the absence of competing serum via interactions with FcγRI (Lo et al., 2017; Nesspor et al., 2012). However, in the presence of 10% serum, the aglycosylated IgG1 had undetectable ADCP activity, suggesting that aglycosylated IgGs would have reduced effector function in circulation, which has serum IgG levels ranging from 10 to 20 mg/mL (Nesspor et al., 2012).

Among the four IgG subclasses, each has a different ability to elicit immune effector functions. For instance, IgG1 and IgG3 have been recognized to recruit complement more effectively than IgG2 and IgG4 (Tao et al., 1993). Additionally, IgG2 and IgG4 have very limited ability to elicit ADCC (Brezski et al., 2014). Therefore, several investigators have employed a cross-subclass approach to reduce effector function. The approved anti-C5 therapeutic, eculizumab, has the IgG2 amino acids 118 to 260 and the IgG4 amino acids 261 to 447 and has been shown to have limited or undetectable effector function (Rother et al., 2007). As such, eculizumab represents the first proof-of-concept that reduced effector function mAbs can be safely utilized in an approved mAb-based therapeutic. In a further refinement of the cross-subclass approach, An et al. generated an IgG2 variant with point mutations from IgG4 (i.e., H268Q/V309L/A330S/P331S). This variant had reduced effector function as well as a circulating half-life comparable to that of IgG1 in rhesus monkeys (An et al., 2009). In a similar approach, a variant was reported that contained the IgG2 to IgG4 cross-subclass mutations V309L/A330S/P331S combined with the non-germline mutations V234A/G237A/P238S/H268A, which resulted in undetectable CDC, ADCC, and ADCP (Vafa et al., 2014). The crystal structure of the mutated Fc was solved and modeling studies suggested that conformational changes at P329 would not be favorable for formation of the "proline sandwich" and changes at D270 could negatively impact C1q interactions. Together, these studies demonstrate the numerous approaches that can be taken to reduce or eliminate effector function in hope of improving the safety of mAb therapeutics.

FC ENGINEERING FOR COENGAGEMENT OF ANTIGEN AND FcγRS

There are some cases where a potential therapeutic mAb's target antigen is found on cells that also express FcγRs. Chu et al. engineered a variant to capitalize on such a situation by augmenting binding to FcγRIIb, the primary Fc receptor expressed on B cells. The variant S267E/L328F has greater than 400-fold increased binding to FcγRIIb compared to IgG1 (Chu et al., 2008). Szili et al. showed that coengagement of CD19 and FcγRIIb with a mAb containing the S267E/L328F variant resulted in FcγRIIb ITIM phosphorylation, recruitment of SH2 domain-containing inositol polyphosphate 5-phosphatase (SHIP), and inhibition of downstream signaling pathways in B cells (e.g., Akt) (Szili et al., 2014). In contrast to utilizing coengagement of antigen with the inhibitory FcγRIIb, there are situations where antigen and FcγR coengagement can dramatically increase mAb potency through avidity-based interactions. Loyau et al. showed that a full length IgG1 anti-TLR4 mAb was 1000-fold more potent than the $F(ab')_2$ fragment version (Loyau et al., 2014). This was attributable to the Fab arms engaging TLR4 while the Fc simultaneously engaged FcγRI on the same cell. Coengagement of TLR4 and FcγRIIa resulted in increased potency (i.e., half maximal inhibitory concentration), but to a lesser effect (~25-fold). The antibody was engineered with two Fc mutations, N325S/L328F, which enhanced binding to FcγRIIa (Shang et al., 2014). The Fc engineered anti-TLR4 mAb has completed phase I clinical trials (Monnet et al., 2017).

One consideration for engineering mAbs for coengagement of antigen with FcγRs is the induction of FcγR signaling pathways. For the S267E/L328F variant, engagement of FcγRIIb ITIM signaling was intended (Chu et al., 2008). However, for other receptors such as FcγRIIa, not only is there the potential to engage ITAM signaling, there is also the possibility to engage inhibitory ITAM (ITAMi) signaling—an event where instead of recruiting kinases, the ITAM recruits phosphatases (Blank et al., 2009). Shang et al. tested if the anti-TLR4 mAb engaged ITAMi signaling and showed that there was no activation of Src homology 2 domain containing protein tyrosine phosphatase (SHP-1) (Shang et al., 2014). Novel cases where coengagement is employed should assess downstream ITIM, ITAM, and ITAMi signaling pathways to determine whether or not the antibodies engage unintended signaling events.

FCRN ENGINEERING TO ALTER HALF-LIFE

The neonatal Fc receptor (FcRn) is a MH1Like heterodimer consisting of a major histocompatibility MH1Like alpha chain non-covalently associated with beta-2-microglobulin (β2m) (Burmeister et al., 1994a) that plays a central role in the cellular trafficking and serum half-life of IgGs (Rath et al., 2013). Its name originates from historical observations leading to the search for a receptor mediating the transfer of humoral immunity from mother to neonatal rats (Brambell, 1966). Research in the early 1960s supported a mechanism of IgG transport specifically involving the Fc region (Brambell, 1966; Brambell et al., 1964; Fahey and Robinson, 1963). It was postulated that a single receptor might control both the transport of IgG during early life and the protection of IgG from catabolism later in life (Brambell et al., 1964). These studies led to the current understanding that the long circulating half-life of antibodies (7–21 days depending on the IgG subclass (Morell et al., 1970)) is attributed, in large part, to their ability to bind FcRn through the Fc. The role of FcRn in IgG homeostasis was further supported by genetic studies where β2m knockout mice had lower levels of circulating IgG (Ghetie et al., 1996; Israel et al., 1996; Junghans and Anderson, 1996).

FcRn is primarily expressed within endosomes where it is capable of binding to IgG internalized through pinocytosis. Conserved histidine residues in the CH2-CH3 domains of IgG become protonated at acidic endosomal pH (6.0–6.5), driving pH-dependent binding to the α-chain of FcRn. Subsequent recycling and release of IgG to the bloodstream at physiological pH (7.4) salvages the antibody from lysosomal degradation (Roopenian and Akilesh, 2007). Mutational and x-ray crystallographic analyses provided insights into the amino acids involved in the Fc:FcRn interaction (Burmeister et al., 1994b; Kim et al., 1999; Martin et al., 2001; Medesan et al., 1997). More recently, the first co-crystal structure of an engineered human Fc with human FcRn provided a molecular understanding of Fc:FcRn interactions (Oganesyan et al., 2014). Key contact positions on the Fc region with FcRn include L251, M252Y (one of the Fc mutations), I253, L309, H310, L314, Q311, and N434 (Fig. 1 and Table 2). The authors further implicated strong hydrogen bonding through H310 as critical for pH dependent binding. Indeed, mutation of this amino acid to any other amino acid (excluding cysteine) resulted in undetectable binding to FcRn at pH 6.0. These characterizations of Fc:FcRn interactions combined with understanding pH dependent Fc:FcRn interactions ignited interest in modulating the pharmacokinetic (PK) properties of therapeutic antibodies through FcRn-mediated recycling mechanisms. Mutations at various positions proximal to the CH2-CH3 interface have been engineered with the intent that prolonged serum half-life may benefit patients by lowering the therapeutic dose and/or frequency of administration and reducing the cost of care (Table 1).

Screening of a phage display Fc variant library identified mutations (e.g., M252Y/S254T/T256E—termed YTE) with enhanced binding to human FcRn at pH 6.0 (Dall'Acqua et al., 2002). A later study demonstrated that the YTE variant had 10-fold increased binding to human and cynomolgus FcRn at pH 6.0, which translated into a 4-fold increase in half-life in a cynomolgus PK study (Dall'Acqua et al., 2006). Crystal structure analysis indicated that the T256E mutation provided 2 novel salt bridges between the Fc and β2m subunit of FcRn (Oganesyan et al., 2014). In another report, Zalesky et al. showed that the mutations M428L/N434S resulted in an 11-fold increase in affinity for human FcRn at pH 6.0 (Zalevsky et al., 2010). When human FcRn transgenic tumor-bearing mice were treated with either an anti-EGFR antibody or an anti-VEGF antibody containing the M428L/N434S mutations, an increased reduction in tumor burden was observed compared with IgG1 treated animals

(Zalevsky et al., 2010). This was the first *in vivo* demonstration of improved anti-tumor activity resulting from an FcRn-dependent increase in half-life. It was postulated that the N434S mutation allowed additional hydrogen bonds with FcRn (Oganesyan et al., 2014), resulting in the increased binding. Overall, these cases demonstrated that increasing the affinity of IgG for FcRn at pH 6.0 resulted in extended half-life and efficacy in pre-clinical studies.

Motavizumab-YTE, a humanized antibody targeting respiratory syncytial virus, is the first antibody engineered for FcRn mediated half-life extension to be tested in human subjects (Robbie et al., 2013). The results from this phase I clinical trial in healthy adults demonstrated a 2- to 4-fold increase in half-life relative to motavizumab IgG1 depending on the dose, which provided proof-of-concept that FcRn low pH enhancing modifications translated into increased half-life in humans.

CONCLUSIONS

As there are currently 70 phase III and 575 phase I/II antibody-based drugs in clinical trials (Strohl, 2017), competition for targets is increasing, and the need to differentiate antibodies is becoming imperative. The successful approval of the effector function modulated mAbs eculizumab, obinutuzumab, and mogamulizumab has demonstrated that altering effector function can be a clinically viable way to differentiate mAbs from standard human IgG subclasses. With an increasing number of Fc engineered mAbs progressing through clinical trials, it will be interesting to see if modulating antibody effector function leads to the next generation of approved mAb-based therapeutics.

ACKNOWLEDGEMENTS

The authors would like to thank Yee Seir (Sheil) Kee for critical comments.

ABBREVIATIONS

ADCC, antibody-dependent cell-mediated cytotoxicity; ADCP, antibody-dependent cellular phagocytosis; β2m, beta-2-microglobulin; CDC, complement-dependent cytotoxicity; CHO, Chinese hamster ovary; EGFR, epidermal growth factor; FcγR, Fc gamma receptor; FcRn, neonatal Fc receptor; ITAM, immunoreceptor tyrosine-based activation motif; ITAMi, inhibitory immunoreceptor tyrosine-based activation motif; ITIM, immunoreceptor tyrosine-based inhibitory motifs; mAb, monoclonal antibody; MHC, major histocompatibility; SHIP, SH2 domain-containing inositol polyphosphate 5-phosphatase; TLR, toll-like receptor; VEGF, vascular endothelial growth factor

COMPLIANCE WITH ETHICS GUIDELINES

Xinhua Wang is an employee of Genentech. Mary Mathieu is an employee of Genentech. Randall J. Brezski is an employee of Genentech.

REFERENCES

Alegre ML, Collins AM, Pulito VL, Brosius RA, Olson WC, Zivin RA, Knowles R, Thistlethwaite JR, Jolliffe LK, Bluestone JA (1992) Effect of a single amino acid mutation on the activating and immunosuppressive properties of a “humanized” OKT3 monoclonal antibody. J Immunol 148:3461–3468

An Z, Forrest G, Moore R, Cukan M, Haytko P, Huang L, Vitelli S, Zhao JZ, Lu P, Hua J et al (2009) IgG2m4, an engineered antibody isotype with reduced Fc function. MAbs 1:572–579

Bang YJ, Giaccone G, Im SA, Oh DY, Bauer TM, Nordstrom JL, Li H, Chichili GR, Moore PA, Hong S et al (2017) First-in-human phase 1 study of margetuximab (MGAH22), an Fc-modified chimeric monoclonal antibody, in patients with HER2-positive advanced solid tumors. Ann Oncol 28:855–861

Beck A, Reichert JM (2012) Marketing approval of mogamulizumab: a triumph for glyco-engineering. MAbs 4:419–425

Bibeau F, Lopez-Crapez E, Di Fiore F, Thezenas S, Ychou M, Blanchard F, Lamy A, Penault-Llorca F, Frebourg T, Michel P et al (2009) Impact of Fc{gamma}RIIa-Fc{gamma}RIIIa polymorphisms and KRAS mutations on the clinical outcome of patients with metastatic colorectal cancer treated with cetuximab plus irinotecan. J Clin Oncol 27:1122–1129

Blank U, Launay P, Benhamou M, Monteiro RC (2009) Inhibitory ITAMs as novel regulators of immunity. Immunol Rev 232:59–71

Bolt S, Routledge E, Lloyd I, Chatenoud L, Pope H, Gorman SD, Clark M, Waldmann H (1993) The generation of a humanized, non-mitogenic CD3 monoclonal antibody which retains in vitro immunosuppressive properties. Eur J Immunol 23:403–411

Brambell FW (1966) The transmission of immunity from mother to young and the catabolism of immunoglobulins. Lancet 2:1087–1093

Brambell FW, Hemmings WA, Morris IG (1964) A theoretical model of gamma globulin catabolism. Nature 203:1352–1354

Brezski RJ, Georgiou G (2016) Immunoglobulin isotype knowledge and application to Fc engineering. Curr Opin Immunol 40:62–69

Brezski RJ, Kinder M, Grugan KD, Soring KL, Carton J, Greenplate AR, Petley T, Capaldi D, Brosnan K, Emmell E et al (2014) A monoclonal antibody against hinge-cleaved IgG restores effector function to proteolytically-inactivated IgGs in vitro and in vivo. MAbs 6:1265–1273

Burmeister WP, Gastinel LN, Simister NE, Blum ML, Bjorkman PJ (1994a) Crystal structure at 2.2 A resolution of the MHC-related neonatal Fc receptor. Nature 372:336–343

Burmeister WP, Huber AH, Bjorkman PJ (1994b) Crystal structure of the complex of rat neonatal Fc receptor with Fc. Nature 372:379–383

Burton DR, Jefferis R, Partridge LJ, Woof JM (1988) Molecular recognition of antibody (IgG) by cellular Fc receptor (FcRI). Mol Immunol 25:1175–1181

Canfield SM, Morrison SL (1991) The binding affinity of human IgG for its high affinity Fc receptor is determined by multiple amino acids in the CH2 domain and is modulated by the hinge region. J Exp Med 173:1483–1491

Carter PJ (2006) Potent antibody therapeutics by design. Nat Rev Immunol 6:343–357

Cartron G, Dacheux L, Salles G, Solal-Celigny P, Bardos P, Colombat P, Watier H (2002) Therapeutic activity of humanized anti-CD20 monoclonal antibody and polymorphism in IgG Fc receptor FcgammaRIIIa gene. Blood 99:754–758

Chappel MS, Isenman DE, Everett M, Xu YY, Dorrington KJ, Klein MH (1991) Identification of the Fc gamma receptor class I binding site in human IgG through the use of recombinant IgG1/IgG2 hybrid and point-mutated antibodies. Proc Natl Acad Sci USA 88:9036–9040

Chatenoud L, Bluestone JA (2007) CD3-specific antibodies: a portal to the treatment of autoimmunity. Nat Rev Immunol 7:622–632

Chu SY, Vostiar I, Karki S, Moore GL, Lazar GA, Pong E, Joyce PF, Szymkowski DE, Desjarlais JR (2008) Inhibition of B cell receptor-mediated activation of primary human B cells by coengagement of CD19 and FcgammaRIIb with Fc-engineered antibodies. Mol Immunol 45:3926–3933

Dall'Acqua WF, Woods RM, Ward ES, Palaszynski SR, Patel NK, Brewah YA, Wu H, Kiener PA, Langermann S (2002) Increasing the affinity of a human IgG1 for the neonatal Fc receptor: biological consequences. J Immunol 169:5171–5180

Dall'Acqua WF, Kiener PA, Wu H (2006) Properties of human IgG1s engineered for enhanced binding to the neonatal Fc receptor (FcRn). J Biol Chem 281:23514–23524

Diebolder CA, Beurskens FJ, de Jong RN, Koning RI, Strumane K, Lindorfer MA, Voorhorst M, Ugurlar D, Rosati S, Heck AJ et al (2014) Complement is activated by IgG hexamers assembled at the cell surface. Science 343:1260–1263

Duncan AR, Woof JM, Partridge LJ, Burton DR, Winter G (1988) Localization of the binding site for the human high-affinity Fc receptor on IgG. Nature 332:563–564

Ecker DM, Jones SD, Levine HL (2015) The therapeutic monoclonal antibody market. MAbs 7:9–14

Edelman GM, Cunningham BA, Gall WE, Gottlieb PD, Rutishauser U, Waxdal MJ (1969) The covalent structure of an entire gammaG immunoglobulin molecule. Proc Natl Acad Sci USA 63:78–85

Fahey JL, Robinson AG (1963) Factors controlling serum gamma-globulin concentration. J Exp Med 118:845–868

Ferrara C, Grau S, Jager C, Sondermann P, Brunker P, Waldhauer I, Hennig M, Ruf A, Rufer AC, Stihle M et al (2011) Unique carbohydrate-carbohydrate interactions are required for high affinity binding between FcgammaRIII and antibodies lacking core fucose. Proc Natl Acad Sci USA 108:12669–12674

Gavin PG, Song N, Kim SR, Lipchik C, Johnson NL, Bandos H, Finnigan M, Rastogi P, Fehrenbacher L, Mamounas EP et al (2017) Association of polymorphisms in FCGR2A and FCGR3A with degree of trastuzumab benefit in the adjuvant treatment of ERBB2/HER2-positive breast cancer: analysis of the NSABP B-31 trial. JAMA Oncol 3:335–341

Ghetie V, Hubbard JG, Kim JK, Tsen MF, Lee Y, Ward ES (1996) Abnormally short serum half-lives of IgG in beta 2-microglobulin-deficient mice. Eur J Immunol 26:690–696

Hulett MD, Witort E, Brinkworth RI, McKenzie IF, Hogarth PM (1994) Identification of the IgG binding site of the human low affinity receptor for IgG Fc gamma RII. Enhancement and ablation of binding by site-directed mutagenesis. J Biol Chem 269:15287–15293

Idusogie EE, Presta LG, Gazzano-Santoro H, Totpal K, Wong PY, Ultsch M, Meng YG, Mulkerrin MG (2000) Mapping of the C1q binding site on rituxan, a chimeric antibody with a human IgG1 Fc. J Immunol 164:4178–4184

Idusogie EE, Wong PY, Presta LG, Gazzano-Santoro H, Totpal K, Ultsch M, Mulkerrin MG (2001) Engineered antibodies with increased activity to recruit complement. J Immunol 166:2571–2575

Israel EJ, Wilsker DF, Hayes KC, Schoenfeld D, Simister NE (1996) Increased clearance of IgG in mice that lack beta 2-microglobulin: possible protective role of FcRn. Immunology 89:573–578

Jefferis R, Lund J, Pound J (1990) Molecular definition of interaction sites on human IgG for Fc receptors (huFc gamma R). Mol Immunol 27:1237–1240

Junghans RP, Anderson CL (1996) The protection receptor for IgG catabolism is the beta2-microglobulin-containing neonatal intestinal transport receptor. Proc Natl Acad Sci USA 93:5512–5516

Kim JK, Firan M, Radu CG, Kim CH, Ghetie V, Ward ES (1999) Mapping the site on human IgG for binding of the MHC class I-related receptor, FcRn. Eur J Immunol 29:2819–2825

Kiyoshi M, Caaveiro JM, Kawai T, Tashiro S, Ide T, Asaoka Y, Hatayama K, Tsumoto K (2015) Structural basis for binding of human IgG1 to its high-affinity human receptor FcgammaRI. Nat Commun 6:6866

Lazar GA, Dang W, Karki S, Vafa O, Peng JS, Hyun L, Chan C, Chung HS, Eivazi A, Yoder SC et al (2006) Engineered antibody Fc variants with enhanced effector function. Proc Natl Acad Sci USA 103:4005–4010

Leabman MK, Meng YG, Kelley RF, DeForge LE, Cowan KJ, Iyer S (2013) Effects of altered FcgammaR binding on antibody pharmacokinetics in cynomolgus monkeys. MAbs 5:896–903

Lefran MPL (2001) The immunoglobulin factsbook. Elsevier, Amsterdam

Liu H, May K (2012) Disulfide bond structures of IgG molecules: structural variations, chemical modifications and possible impacts to stability and biological function. MAbs 4:17–23

Lo M, Kim HS, Tong RK, Bainbridge TW, Vernes JM, Zhang Y, Lin YL, Chung S, Dennis MS, Zuchero YJ et al (2017) Effector-attenuating substitutions that maintain antibody stability and reduce toxicity in mice. J Biol Chem 292:3900–3908

Loyau J, Malinge P, Daubeuf B, Shang L, Elson G, Kosco-Vilbois M, Fischer N, Rousseau F (2014) Maximizing the potency of an anti-TLR4 monoclonal antibody by exploiting proximity to Fcgamma receptors. MAbs 6:1621–1630

Lu J, Chu J, Zou Z, Hamacher NB, Rixon MW, Sun PD (2015) Structure of FcgammaRI in complex with Fc reveals the importance of glycan recognition for high-affinity IgG binding. Proc Natl Acad Sci USA 112:833–838

Lund J, Winter G, Jones PT, Pound JD, Tanaka T, Walker MR, Artymiuk PJ, Arata Y, Burton DR, Jefferis R et al (1991) Human Fc gamma RI and Fc gamma RII interact with distinct but overlapping sites on human IgG. J Immunol 147:2657–2662

Martin WL, West AP Jr, Gan L, Bjorkman PJ (2001) Crystal structure at 2.8 A of an FcRn/heterodimeric Fc complex: mechanism of pH-dependent binding. Mol Cell 7:867–877

Medesan C, Matesoi D, Radu C, Ghetie V, Ward ES (1997) Delineation of the amino acid residues involved in transcytosis and catabolism of mouse IgG1. J Immunol 158:2211–2217

Meyer S, Leusen JH, Boross P (2014) Regulation of complement and modulation of its activity in monoclonal antibody therapy of cancer. MAbs 6:1133–1144

Mimoto F, Igawa T, Kuramochi T, Katada H, Kadono S, Kamikawa T, Shida-Kawazoe M, Hattori K (2013a) Novel asymmetrically engineered antibody Fc variant with superior FcgammaR binding affinity and specificity compared with afucosylated Fc variant. MAbs 5:229–236

Mimoto F, Katada H, Kadono S, Igawa T, Kuramochi T, Muraoka M, Wada Y, Haraya K, Miyazaki T, Hattori K (2013b) Engineered antibody Fc variant with selectively enhanced FcgammaRIIb binding over both FcgammaRIIa(R131) and FcgammaRIIa (H131). Protein Eng Des Sel 26:589–598

Monnet E, Lapeyre G, Poelgeest EV, Jacqmin P, Graaf K, Reijers J, Moerland M, Burggraaf J, Min C (2017) Evidence of NI-0101 pharmacological activity, an anti-TLR4 antibody, in a randomized phase I dose escalation study in healthy volunteers receiving LPS. Clin Pharmacol Ther 101:200–208

Moore GL, Chen H, Karki S, Lazar GA (2010) Engineered Fc variant antibodies with enhanced ability to recruit complement and mediate effector functions. MAbs 2:181–189

Morell A, Terry WD, Waldmann TA (1970) Metabolic properties of IgG subclasses in man. J Clin Invest 49:673–680

Natsume A, In M, Takamura H, Nakagawa T, Shimizu Y, Kitajima K, Wakitani M, Ohta S, Satoh M, Shitara K et al (2008) Engineered antibodies of IgG1/IgG3 mixed isotype with enhanced cytotoxic activities. Cancer Res 68:3863–3872

Nesspor TC, Raju TS, Chin CN, Vafa O, Brezski RJ (2012) Avidity confers FcgammaR binding and immune effector function to aglycosylated immunoglobulin G1. J Mol Recognit 25:147–154

Nimmerjahn F, Ravetch JV (2008) Fcgamma receptors as regulators of immune responses. Nat Rev Immunol 8:34–47

Oganesyan V, Damschroder MM, Leach W, Wu H, Dall'Acqua WF (2008) Structural characterization of a mutated, ADCC-enhanced human Fc fragment. Mol Immunol 45:1872–1882

Oganesyan V, Damschroder MM, Cook KE, Li Q, Gao C, Wu H, Dall'Acqua WF (2014) Structural insights into neonatal Fc receptor-based recycling mechanisms. J Biol Chem 289:7812–7824

Oganesyan V, Mazor Y, Yang C, Cook KE, Woods RM, Ferguson A, Bowen MA, Martin T, Zhu J, Wu H et al (2015) Structural insights into the interaction of human IgG1 with FcgammaRI: no direct role of glycans in binding. Acta Crystallogr D 71:2354–2361

Partridge LJ, Woof JM, Jefferis R, Burton DR (1986) The use of anti-IgG monoclonal antibodies in mapping the monocyte receptor site on IgG. Mol Immunol 23:1365–1372

Presta LG (2008) Molecular engineering and design of therapeutic antibodies. Curr Opin Immunol 20:460–470

Radaev S, Motyka S, Fridman WH, Sautes-Fridman C, Sun PD (2001) The structure of a human type III Fcgamma receptor in complex with Fc. J Biol Chem 276:16469–16477

Ramsland PA, Farrugia W, Bradford TM, Sardjono CT, Esparon S, Trist HM, Powell MS, Tan PS, Cendron AC, Wines BD et al (2011) Structural basis for Fc gammaRIIa recognition of human IgG and formation of inflammatory signaling complexes. J Immunol 187:3208–3217

Rath T, Kuo TT, Baker K, Qiao SW, Kobayashi K, Yoshida M, Roopenian D, Fiebiger E, Lencer WI, Blumberg RS (2013) The immunologic functions of the neonatal Fc receptor for IgG. J Clin Immunol 33(Suppl 1):S9–17

Reichert JM (2017) Antibodies to watch in 2017. MAbs 9:167–181

Richards JO, Karki S, Lazar GA, Chen H, Dang W, Desjarlais JR (2008) Optimization of antibody binding to FcgammaRIIa enhances macrophage phagocytosis of tumor cells. Mol Cancer Ther 7:2517–2527

Robbie GJ, Criste R, Dall'acqua WF, Jensen K, Patel NK, Losonsky GA, Griffin MP (2013) A novel investigational Fc-modified humanized monoclonal antibody, motavizumab-YTE, has an extended half-life in healthy adults. Antimicrob Agents Chemother 57:6147–6153

Roopenian DC, Akilesh S (2007) FcRn: the neonatal Fc receptor comes of age. Nat Rev Immunol 7:715–725

Rother RP, Rollins SA, Mojcik CF, Brodsky RA, Bell L (2007) Discovery and development of the complement inhibitor eculizumab for the treatment of paroxysmal nocturnal hemoglobinuria. Nat Biotechnol 25:1256–1264

Shang L, Daubeuf B, Triantafilou M, Olden R, Depis F, Raby AC, Herren S, Dos Santos A, Malinge P, Dunn-Siegrist I et al (2014) Selective antibody intervention of Toll-like receptor 4 activation through Fc gamma receptor tethering. J Biol Chem 289:15309–15318

Shields RL, Namenuk AK, Hong K, Meng YG, Rae J, Briggs J, Xie D, Lai J, Stadlen A, Li B et al (2001) High resolution mapping of the binding site on human IgG1 for Fc gamma RI, Fc gamma RII, Fc gamma RIII, and FcRn and design of IgG1 variants with improved binding to the Fc gamma R. J Biol Chem 276:6591–6604

Shields RL, Lai J, Keck R, O'Connell LY, Hong K, Meng YG, Weikert SH, Presta LG (2002) Lack of fucose on human IgG1 N-linked oligosaccharide improves binding to human Fcgamma RIII and antibody-dependent cellular toxicity. J Biol Chem 277:26733–26740

Shinkawa T, Nakamura K, Yamane N, Shoji-Hosaka E, Kanda Y, Sakurada M, Uchida K, Anazawa H, Satoh M, Yamasaki M et al (2003) The absence of fucose but not the presence of galactose or bisecting N-acetylglucosamine of human IgG1 complex-type oligosaccharides shows the critical role of enhancing antibody-dependent cellular cytotoxicity. J Biol Chem 278:3466–3473

Sondermann P, Huber R, Oosthuizen V, Jacob U (2000) The 3.2-A crystal structure of the human IgG1 Fc fragment-Fc gammaRIII complex. Nature 406:267–273

Stavenhagen JB, Gorlatov S, Tuaillon N, Rankin CT, Li H, Burke S, Huang L, Vijh S, Johnson S, Bonvini E et al (2007) Fc optimization of therapeutic antibodies enhances their ability to kill tumor cells in vitro and controls tumor expansion in vivo via low-affinity activating Fcgamma receptors. Cancer Res 67:8882–8890

Strohl WR (2011) Isotype selection and Fc engineering: design and construction of fit-for-purpose therapeutic antibodies. Antibody drug discovery Imperial College Press, London

Strohl WR (2017) Current progress in innovative engineered antibodies. Protein Cell. doi:10.1007/s13238-017-0457-8

Szili D, Cserhalmi M, Banko Z, Nagy G, Szymkowski DE, Sarmay G (2014) Suppression of innate and adaptive B cell activation pathways by antibody coengagement of FcgammaRIIb and CD19. MAbs 6:991–999

Tamm A, Kister A, Nolte KU, Gessner JE, Schmidt RE (1996) The IgG binding site of human FcgammaRIIIB receptor involves CC' and FG loops of the membrane-proximal domain. J Biol Chem 271:3659–3666

Tao MH, Morrison SL (1989) Studies of aglycosylated chimeric mouse-human IgG. Role of carbohydrate in the structure and effector functions mediated by the human IgG constant region. J Immunol 143:2595–2601

Tao MH, Smith RI, Morrison SL (1993) Structural features of human immunoglobulin G that determine isotype-specific differences in complement activation. J Exp Med 178:661–667

Teeling JL, Mackus WJ, Wiegman LJ, van den Brakel JH, Beers SA, French RR, van Meerten T, Ebeling S, Vink T, Slootstra JW et al (2006) The biological activity of human CD20 monoclonal antibodies is linked to unique epitopes on CD20. J Immunol 177:362–371

Thommesen JE, Michaelsen TE, Loset GA, Sandlie I, Brekke OH (2000) Lysine 322 in the human IgG3 C(H)2 domain is crucial for antibody dependent complement activation. Mol Immunol 37:995–1004

Umana P, Jean-Mairet J, Moudry R, Amstutz H, Bailey JE (1999) Engineered glycoforms of an antineuroblastoma IgG1 with optimized antibody-dependent cellular cytotoxic activity. Nat Biotechnol 17:176–180

Uppal H, Doudement E, Mahapatra K, Darbonne WC, Bumbaca D, Shen BQ, Du X, Saad O, Bowles K, Olsen S et al (2015) Potential mechanisms for thrombocytopenia development with trastuzumab emtansine (T-DM1). Clin Cancer Res 21:123–133

Vafa O, Gilliland GL, Brezski RJ, Strake B, Wilkinson T, Lacy ER, Scallon B, Teplyakov A, Malia TJ, Strohl WR (2014) An engineered Fc variant of an IgG eliminates all immune effector functions via structural perturbations. Methods 65:114–126

Walker MR, Lund J, Thompson KM, Jefferis R (1989) Aglycosylation of human IgG1 and IgG3 monoclonal antibodies can eliminate recognition by human cells expressing Fc gamma RI and/or Fc gamma RII receptors. Biochem J 259:347–353

Wines BD, Powell MS, Parren PW, Barnes N, Hogarth PM (2000) The IgG Fc contains distinct Fc receptor (FcR) binding sites: the leukocyte receptors Fc gamma RI and Fc gamma RIIa bind to a region in the Fc distinct from that recognized by neonatal FcR and protein A. J Immunol 164:5313–5318

Woof JM, Partridge LJ, Jefferis R, Burton DR (1986) Localisation of the monocyte-binding region on human immunoglobulin G. Mol Immunol 23:319–330

Xu D, Alegre ML, Varga SS, Rothermel AL, Collins AM, Pulito VL, Hanna LS, Dolan KP, Parren PW, Bluestone JA et al (2000) In vitro characterization of five humanized OKT3 effector function variant antibodies. Cell Immunol 200:16–26

Zalevsky J, Chamberlain AK, Horton HM, Karki S, Leung IW, Sproule TJ, Lazar GA, Roopenian DC, Desjarlais JR (2010) Enhanced antibody half-life improves in vivo activity. Nat Biotechnol 28:157–159

9

Transmembrane domain dependent inhibitory function of FcγRIIB

Junyi Wang[1], Zongyu Li[1], Liling Xu[2✉], Hengwen Yang[3✉], Wanli Liu[1✉]

[1] MOE Key Laboratory of Protein Sciences, Collaborative Innovation Center for Diagnosis and Treatment of Infectious Diseases, School of Life Sciences, Tsinghua University, Beijing 100084, China
[2] Ragon Institute of Massachusetts General Hospital, Massachusetts Institute of Technology, Harvard University, 400 Technology Square, Cambridge, MA 02139, USA
[3] The First Affiliate Hospital, Biomedical Translational Research Institute, Guangdong Province Key Laboratory of Molecular Immunology and Antibody Engineering, Jinan University, Guangzhou 510632, China
✉ Correspondence: xuliling87@gmail.com (L. Xu), benyang97@gmail.com (H. Yang), liulab@tsinghua.edu.cn (W. Liu)

ABSTRACT

FcγRIIB, the only inhibitory IgG Fc receptor, functions to suppress the hyper-activation of immune cells. Numerous studies have illustrated its inhibitory function through the ITIM motif in the cytoplasmic tail of FcγRIIB. However, later studies revealed that in addition to the ITIM, the transmembrane (TM) domain of FcγRIIB is also indispensable for its inhibitory function. Indeed, recent epidemiological studies revealed that a non-synonymous single nucleotide polymorphism (rs1050501) within the TM domain of FcγRIIB, responsible for the I232T substitution, is associated with the susceptibility to systemic lupus erythematosus (SLE). In this review, we will summarize these epidemiological and functional studies of FcγRIIB-I232T in the past few years, and will further discuss the mechanisms accounting for the functional loss of FcγRIIB-I232T. Our review will help the reader gain a deeper understanding of the importance of the TM domain in mediating the inhibitory function of FcγRIIB and may provide insights to a new therapeutic target for the associated diseases.

KEYWORDS B cell, FcγRIIB, transmembrane domain, systemic lupus erythematosus, autoimmune disease

INTRODUCTION

Receptors for the Fc portion of IgG molecules (FcγRs) play important roles in regulating the activation of immune cells for the purpose of balancing immunoprotection and immunopathology. The human immune system contains six types of canonical FcγRs, including the high affinity receptor FcγRI, and low affinity receptors FcγRIIA, FcγRIIB, FcγRIIC, FcγRIIIA, as well as FcγRIIIB, among which FcγRIIB is the only one possessing an inhibitory function (Amigorena et al., 1989; Nimmerjahn and Ravetch, 2011; Pincetic et al., 2014b; Smith and Clatworthy, 2010b). Activating FcγRs transduces a signal through the immunoreceptor tyrosine-based activation motif (ITAM) in its intracellular domain, or the associated signaling subunit called FcR common γ-chain (Blank et al., 2009; Daeron, 1997). In marked contrast, FcγRIIB possesses an immunoreceptor tyrosine-based inhibition motif (ITIM) in its intracellular domain (Bolland and Ravetch, 2000). When activated by antigen-antibody immune complex, the tyrosine within the ITIM could be phosphorylated by the Src-family kinase Lyn, leading to recruitment of SH2-containing SHIP and/or SHP to transduce inhibitory signal cascades (Nimmerjahn and Ravetch, 2008). FcγRIIB is ubiquitously expressed on most types of immune cells including dendritic cells (DCs), monocytes, macrophages, mast cells, neutrophils, basophils and even memory CD8$^+$ T cells (Daeron et al., 1993; Ravetch and Kinet, 1991; Starbeck-Miller et al., 2014). In addition, FcγRIIB was found to be expressed in microglia, participating in the phagocytosis in central nervous system (CNS) infection (Peress, 1993). In B cells, FcγRIIB serves as the exclusive FcγR expressed on the plasma membrane, which potently suppresses B cell activation, antigen internalization and presentation to T cells after binding to the antigen-antibody complex. All of these studies reinforce the universal and crucial role of FcγRIIB in down-regulating the activation and function of all these immune cells (Lehmann et al., 2012). Thus, malfunction of FcγRIIB is usually detrimental for the immune system (Niederer et al., 2010a; Pincetic et al., 2014a; Smith and Clatworthy, 2010a).

Indeed, single nucleotide polymorphisms (SNPs) of the human FcγRIIB gene significantly influence the susceptibility to autoimmune diseases (Kyogoku et al., 2002b; Niederer et al., 2010a; Smith and Clatworthy, 2010a). Recent epidemiological studies revealed a non-synonymous single nucleotide polymorphism (rs1050501) within the transmembrane (TM) domain of FcγRIIB, causing the I232T substitution, is associated with the susceptibility to systemic lupus erythematosus (SLE) (Kyogoku et al., 2002a). In this review, we will summarize these epidemiological and functional studies of FcγRIIB-I232T in the past few years and will further discuss the mechanisms accounting for the functional loss of FcγRIIB-I232T. We will also discuss the evolutionary conservation of the TM domain among different FcγRs in different species. Our review will help the reader gain a deeper understanding of the importance of the TM domain in mediating the inhibitory function of FcγRIIB and may provide insights to a new therapeutic target for the associated diseases.

EPIDEMIOLOGY STUDIES OF FCΓRIIB-I232T WITH AUTOIMMUNE DISEASES

Human *FCGR2B* gene possesses a number of SNPs, including several SNPs in the promoter region and eight missense SNPs. Amongst the nonsynonymous SNPs, the T-to-C transition in exon 5 (rs1050501), which leads to a replacement of isoleucine at position 232 by threonine (FcγRIIB-I232T variant), occurs at a notable frequency and is associated with autoimmune disease. In 2002, Kyogoku first demonstrated the correlation of FcγRIIB-I232T with the susceptibility to SLE in Japanese (Kyogoku et al., 2002a), while the epidemiology studies of FcγRIIB-I232T in the context of autoimmune diseases sprung up in the years to follow. Later studies further confirmed that FcγRIIB-I232T was also associated with the susceptibility to SLE in Thai (Siriboonrit et al., 2003), Chinese (Chu et al., 2004) and also Caucasian (Willcocks et al., 2010) populations. According to Chu's work, FcγRIIB-I232T is highly associated with nephritis in SLE patients (Chu et al., 2004). Interestingly, it is also found that homozygosity for this polymorphism is associated with protection against severe malaria, where selection pressure may contribute to the higher frequency of FcγRIIB-T232 and, hence, SLE in Africans and Southeast Asians (Willcocks et al., 2010).

Aside from SLE, rheumatoid arthritis (RA) is another common systemic autoimmune disease. Although epidemiology studies in literature did not find a significant correlation between FcγRIIB-I232T homozygotes with RA (Chen et al., 2008; Kyogoku et al., 2002c; Radstake et al., 2006), it is striking to observe that RA patients carrying homozygous FcγRIIB-T232 would develop severe radiologic joint damage during the first six-year course of the disease (Radstake et al., 2006). Besides, the FcγRIIB-I232T is associated with susceptibility to RA in Chinese population according to our unpublished statistical data.

Though epidemiological studies revealed the association of FcγRIIB-I232T with SLE and RA, the correlation studies with other autoimmune diseases are still scarce. It has already been proved that there is a lower expression level of FcγRIIB in multiple sclerosis (MS) patients (Tackenberg et al., 2009). And among idiopathic thrombocytopenic purpura (ITP) patients, a higher frequency of hetero FcγRIIB-I232/I232T genotype is observed in chronic disease patients compared to that in acute disease patients (Bruin et al., 2004). All these results indicate that FcγRIIB may play a sophisticated role in different autoimmune diseases.

TRANSMEMBRANE DOMAIN-DEPENDENT INHIBITORY FUNCTION OF FCΓRIIB

Although numerous studies have reported the critical role of ITIM as a prerequisite for the inhibitory function of FcγRIIB, recent reports also revealed that the unique property of the TM domain is also essential for its role of inhibition. In this section, we will discuss the four possible mechanisms through which the TM-domain is required for the inhibitory function of FcγRIIB: (1) impeding the conformational changes of BCR, which is crucial for the initiation of B cell receptor (BCR) downstream signaling; (2) co-localizing with lipid raft to activate the inhibitory signaling; (3) blocking the synaptic co-localization of BCR and CD19 microclusters; (4) conferring fast lateral mobility.

The initiation of BCR signaling is reported to be accompanied by a "closed to open" form transition of the cytoplasmic domains of the BCR (Sohn et al., 2008b; Tolar et al., 2005) (Fig. 1A). According to model 1, blocking the conformational changes in the cytoplasmic domains of the BCRs might be one of the approaches for FcγRIIB to play its inhibitory role. Indeed, FcγRIIB was proved to block the transition to active signaling conformation when colligated with BCR, however FcγRIIB-I232T failed to inhibit such conformational changes of BCR (Liu et al., 2010), even though FcγRIIB-I232T maintains an intact extracellular domain for ligand recognition and an intact intracellular ITIM (Fig. 1B).

It was indicated by fluorescence resonance energy transfer (FRET) assay that the interaction between BCR and lipid raft transiently take place at the periphery of the B cell immunological synapse where BCR-antigen clusters are newly formed (Sohn et al., 2008b). In contrast, the interaction between FcγRIIB and lipid raft is much more stable and the co-localization of FcγRIIB with raft lipid is found to be independent of the activity of Src family kinase (Liu et al., 2010). In model 2, when co-localizing with BCR in the lipid raft, FcγRIIB inhibits the early events of BCR activation, such as BCR recruitment and oligomerization in the lipid raft, in addition to BCR downstream signaling. In 2002, Kono et al. indicated that spatial association between the FcγRIIB TM

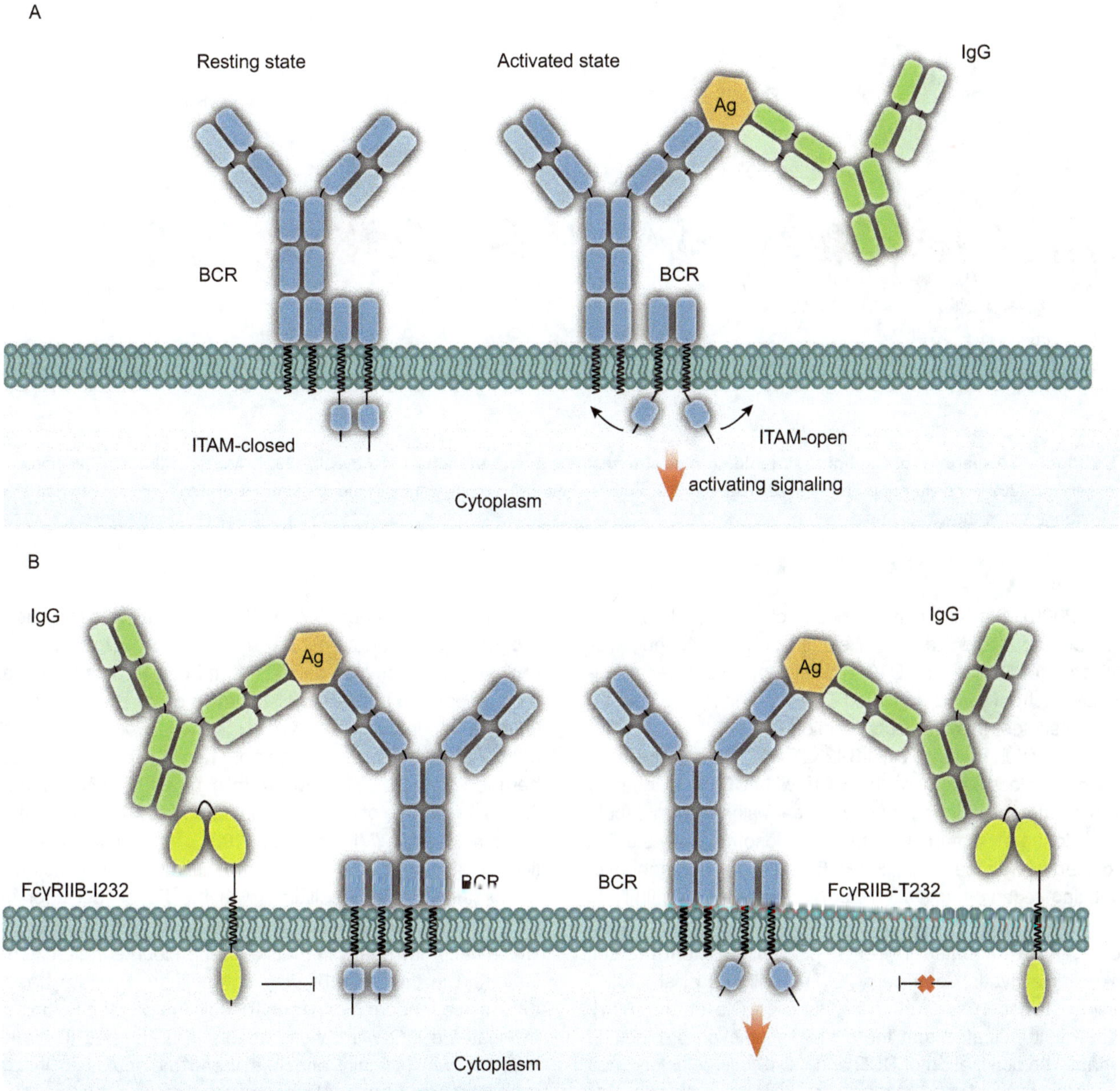

Figure 1. A conformational change blocking model. (A) Shown are the "close to open" conformational changes within the cytoplasmic tail of BCR during the cross-membrane signal transduction. (B) FcγRIIB-I232 blocks such conformational changes in the cytoplasmic domains of BCR, whereas FcγRIIB-T232 fails to do so.

domain and lipid raft is essential for its inhibitory signaling (Kono et al., 2002). However, FcγRIIB-I232T exhibited defective membrane proximal signaling compared to FcγRIIB-WT (Floto et al., 2005). The subsequent study by Western blot and live cell imaging with fluorescently labeled cholera toxin B, a lipid raft marker, discovered more FcγRIIB co-localized with lipid raft than FcγRIIB-I232T, confirming the necessity of the residence to sphingolipid rafts (Floto et al., 2005; Kono et al., 2005). As the signal pathway dependent on the interaction between FcγRIIB-I232T and protein kinases that reside preferentially in lipid-raft-rich domains might be interfered, thus the inhibitory function is impaired (Fig. 2).

Model 3 relies on the finding that the TM domain of FcγRIIB is responsible for blocking the synaptic colocalization of BCR and CD19 microclusters. CD19 is the activating co-receptor of BCR, which promotes BCR signaling when co-localizing with BCR in the lipid raft during B cell activation. In 2000, Fong et al. noticed that dephosphorylation of CD19 was independent of the cytoplasmic tail of FcγRIIB (Fong

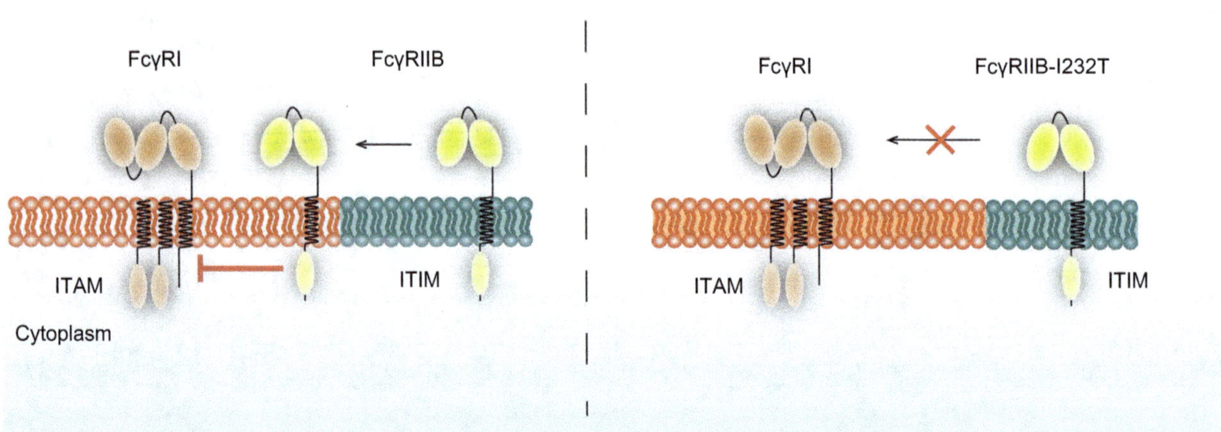

Figure 2. A diagram of the lipid raft-residing model explaining the dysfunction of FcγRIIB-I232T. Model of the FcγRIIB and FcγRIIB-I232T showing that FcγRIIB-I232T fails to distribute to lipid raft, which results in the impairment of the inhibitory function. Orange color shows lipid raft.

et al., 2000). Our study showed that both of FcγRIIB with or without cytoplasmic tail could significantly impair the colocalization of BCR and CD19 microclusters, however a chimeric construct with a truncated TM from fruit fly N-Cadherin completely lost this inhibitory function (Xu et al., 2014). The loss-of-function mutant FcγRIIB-I232T also failed to block the synaptic colocalization of the BCR with CD19, leading to dysregulated recruitment of downstream signaling molecule pPI3K to the membrane proximal signalosome. Early studies provided evidence that FcγRIIB-I232T is incapable of inhibiting receptor activation due to the reduced affinity to sphingolipid rafts, whereas strikingly, the chimeric construct containing a truncated TM of a lipid rafts resident protein, linker for activation of T cells (LAT) (Tanimura et al., 2003) could not rescue such inhibitory function (Xu et al., 2014). This result indicated that the inhibitory function of blocking synaptic colocalization of BCR and CD19 microclusters may depend on its unique TM sequence instead of its affinity to lipid rafts. Therefore, a possible functional model is that FcγRIIB prevents the colocalization of BCR and CD19 to attenuate the boosting of BCR signaling from CD19 (Fig. 3).

Recent studies also revealed the polymorphism FcγRIIB-I232T affects the lateral mobility of FcγRIIB, and its enrichment to the immune complexes (ICs) significantly decreases after stimulation. However, if FcγRIIB-I232T was given sufficient time to diffuse and interact with ICs or its TM domain was substituted with one of a fast floating CD86 molecule, its inhibitory function can be reinstated (Xu et al., 2016). Hence, the lateral mobility could be one of the key factors for FcγRIIB to downregulate the BCR signaling. This mechanistic explanation of how a single amino acid change in the TM domain of FcγRIIB results in its dysfunction, which can be described as a "catch-me-if-you-can" model (model 4, Fig. 4). This model suggests that the inhibitory function of FcγRIIB is dependent on its lateral mobility, which is a property mainly determined by the TM domain of the receptor. It is generally believed that FcγRIIB acts mainly as a co-receptor to regulate the activation of the main activating immune receptors like FcγRI and BCR, which show high affinity to their ligand at nanomole range (Batista and Neuberger, 2000; Lu et al., 2015). In contrast, FcγRIIB and its ligand, Fc portion of human IgG antibody, show low-affinity at micromole range (Mimura et al., 2001). In this case, it is very likely that BCR binding to antigens is much more instantaneous than FcγRIIB colliding with the Fc of IgG upon IC stimulation. BCR is a relatively slow diffusion molecule, which becomes nearly immobile after encountering antigens. Defeated in the initial binding competition, FcγRIIB has to diffuse in a random manner until it collides with the Fc portion of clustered IgG with avidity effects in preformed IC microclusters, which finally enriches these molecules to initiate their inhibitory function. In order to collide with the Fc region of IgG in the immune-complex that has been caught by the BCRs, the faster that FcγRIIB diffuses, the better of a chance it would have at catching the Fc region and induce inhibition to the activation signaling from BCR in a timely manner. The model is also supported by the study using the PALM-based super-resolution imaging and Monte Carlo simulation analysis, which suggested that the lack of sufficient lateral mobility of FcγRIIB-I232T hinders its successful collision and enrichment in the IC microclusters and the subsequent assembly of larger-sized FcγRIIB nanoclusters in response to ICs (Xu et al., 2016).

UNIQUE SEQUENCE OF FCΓRIIB TM DOMAIN

The TM domain sequences of human FcγRII members are highly conserved among different subclasses but differing

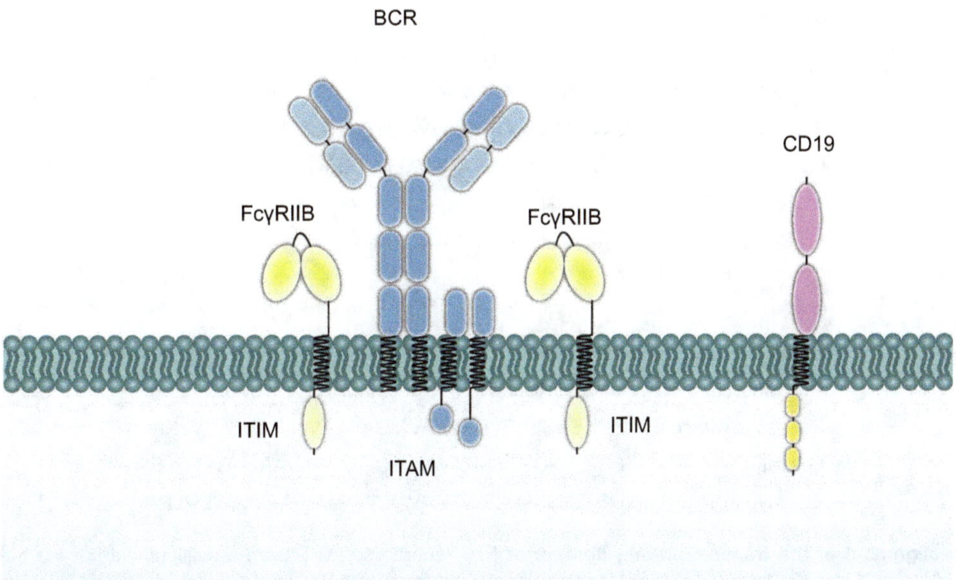

Figure 3. A CD19-blocking model that reveals the inhibitory function of FcγRIIB. The FcγRIIB can inhibit B cell activation by preventing CD19, which is an activating co-receptor, from colocalizing with BCR microclusters. The impaired function of FcγRIIB-I232T can also be explained as the failure in blocking the colocalization.

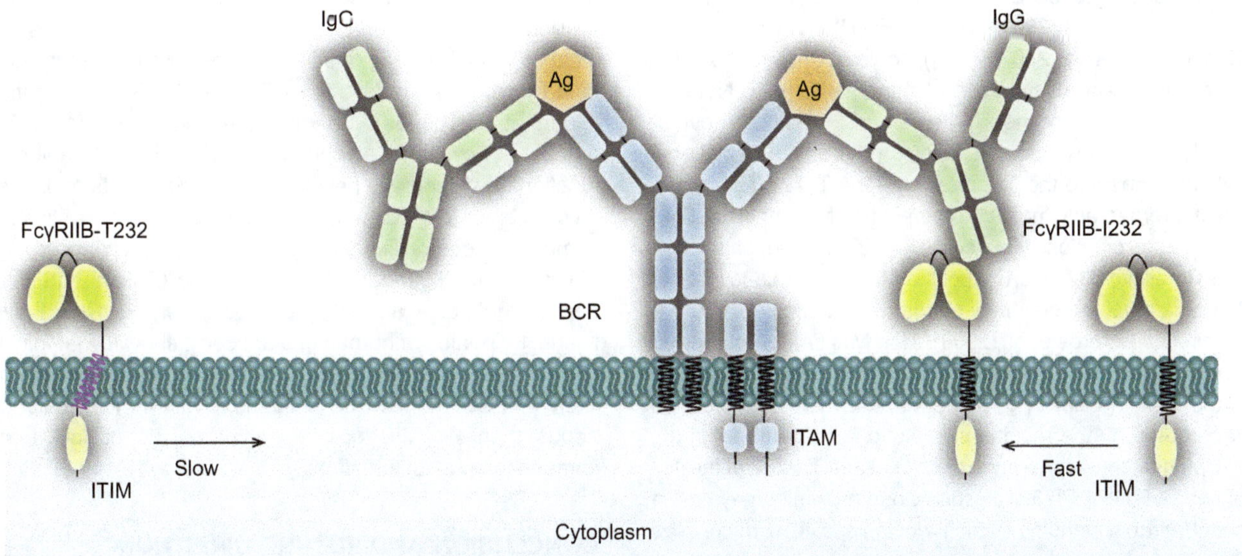

Figure 4. The "catch-me-if-you-can" model. The model indicates that the fast lateral mobility of FcγRIIB is crucial for its inhibitory function. Due to the high affinity between BCR and antigen, the BCR is more likely to catch the antigen before FcγRIIB catch the IgG upon immunocomplex stimulation. Then the low affinity FcγRIIB-WT has to diffuse fast enough to catch the IgG and stop the BCR signaling in time. In marked contrast, FcγRIIB-I232T fails to "catch" the IC due to its slow diffusion and thus loses its inhibitory function.

A

FcγR-I	-----------------------LAV GI------------------------------
FcγR-IIA	SSPMGVIVAVVIATAVAAIVAAVVALIYCRKKRISA
FcγR-IIB	SSPMGIIVAVVTGIAVAAIVAAVVALIYCRKKRISA
FcγR-IIC	SSPMGIIVAVVTGIAVAAIVAAVVALIYCRKKRISA
FcγR-IIIA	---------------------GLAVSTI------------------------------
FcγR-IIIB	---------------------GLAVSTI------------------------------

B

Homo sapiens	AP-SS- SPMG I IVAVVTGIAVAAI VAAVVAL IYCRK
Mus musculus	GPKSSRSLPVLTIVAAVTGIAVAAI VI I LVSLVYLKK
Rattus norvegicus	EPKSSSSLPVLTIVAAVAGIAVAAI VI I LVSLVYLKK
Bos taurus	ES-SSSGPSSMTAVAIGTCFAAVAI VAAI I TWFRLRR
Sus scrofa	GS-SLNNLLVTI VVAVVAWIVAMAIAAAIAAWFRLRR

Figure 5. An alignment of the transmembrane domain of Fcγ receptors. (A) Shown are alignments of the transmembrane domain sequences of low affinity Fcγ receptors in NCBI. The conserved regions are highlighted in red color. (B) Shown are alignments of the transmembrane domain sequences from different species that are available in NCBI. The conserved regions are highlighted in red color.

from that of FcγRI and FcγRIII (Fig. 5A). Amongst the FcγRII family, FcγRIIC, which has the same extracellular and TM domains as FcγRIIB and the same cytoplasmic tail as FcγRIIA, is believed to generate from the gene recombination between FcγRIIB and FcγRIIA (Warmerdam et al., 1993). The SNP Q57stop in FcγRIIC, up-regulating the expression of this activating FcγRIIC, leads to enhanced immune response and autoimmune disease in both human patients and transgenic mice (Li et al., 2013). Different, but analogous to FcγRIIB, which possesses Thr230 and Ile232, FcγRIIA has a Thr at site 232 and an Ile at site 230 (Fig. 5A), however, thus far no study has reported any dysfunction of this TM domain. Notably, protein sequences of the TM domain of FcγRIIB are evolutionarily conserved amongst species (Fig. 5B). All functional FcγRIIBs conserve the Thr residue in the TM domain, however, it is remarkable that none of them has two Thr distributed in such close positions like FcγRIIB-I232T. The substitution of Ile to Thr at site 232 inserts a polar amino acid residue into the TM domain, which definitely bends the α-helices of the TM domain (Ballesteros et al., 2000). This was further supported by the molecular dynamics simulation results which suggested that the bending of the FcγRIIB TM helix can be exacerbated by the residue substitution of I232T (Xu et al., 2016). Thus, the appearance of Thr230 and Thr232 at the TM domain of FcγRIIB-I232T in such close proximity may cause an unbearable bending, which would lead to a decreased lateral mobility.

Indeed, when we swapped the TM domain of FcγRIIB-I232T into different type-I TM molecules, such as TfR, MHC-1 and CD86 (Xu et al., 2014), we found only TM-CD86 could rescue the inhibitory function like wild type FcγRIIB (Xu et al., 2016). Among them, TfR was shown to be a non-lipid-raft associated protein (Harder et al., 1998) and MHC-1 had a slow lateral mobility (Georgiou et al., 2002), while CD86 had the shortest predicted TM domain and fastest lateral mobility (Xu et al., 2016). As expected, the chimeric constructs FcγRIIB-TM-TfR showed a loss-of-function phenomenon (Xu et al., 2014). This result combined with the result of FcγRIIB-TM-LAT suggests that a sole non-lipid raft residing or constitutive lipid raft residing TM cannot rescue the inhibitory function of FcγRIIB. There could be two explanations for this result according to the four models above: First, there was an early report by Sohn et al. showing that FcγRIIB is not located in lipid raft regions in quiescent B cells, but translocated to the lipid rafts upon BCR and FcγRIIB co-ligation (Sohn et al., 2008a). Second, the sequence of the TM domain of FcγRIIB is also important in exerting its inhibitory function. It is generally believed that FcγRIIB acts mainly as a co-receptor to regulate the activation of the main activating immune receptors like FcγRI and BCR, which show high affinity to their ligand at nanomole range (Batista and Neuberger, 2000; Lu et al., 2015). Comparatively, FcγRIIB and its ligand, Fc portion of human IgG antibody, show low-affinity at micromole range (Mimura et al., 2001). In this case, it is very likely that BCR binding to antigens is much more instantaneous than FcγRIIB colliding with the Fc of IgG upon immunocomplex stimulation.

CONCLUSION AND FUTURE DIRECTION

The association between FcγRIIB-I232T and autoimmune diseases has been identified for over 14 years. Lots of studies have revealed the correlation, but the mechanism by which I232T substitution in the TM domain causes the loss-of-function phenotype is not entirely known yet. A better understanding of this SNP shall give insight to the subtle regulation of the TM domain of type-I transmembrane receptor to their functions and may provide a new approach and target for autoimmune disease diagnosis and therapies.

A recent study revealed that the inhibitory function of FcγRIIB is dependent on its TM domain, rather than the ITIM motif present in the cytoplasmic domain, to block the synaptic colocalization of BCR and CD19 microclusters (Xu et al., 2014). We propose that the TM domain provides an extra layer of inhibitory function for FcγRIIB to secure its unique position as the only inhibitory IgG FcR to downregulate immune responses (Hippen et al., 1997). Although the exact molecular mechanism remains unclear, it is revealed that the proper TM domain of FcγRIIB can impede BCR's conformational changes, activate the inhibitory signaling distributed to lipid rafts, inhibit the recruitment of CD19 to B cell IS, which depends on its fast lateral mobility. The importance of the FcγRIIB TM domain is also supported by a rich amount of epidemiological studies showing a strong positive correlation of the FcγRIIB-I232T mutant with SLE disease (Chen et al., 2006; Chu et al., 2004; Clatworthy et al., 2007; Kyogoku et al., 2002a; Kyogoku et al., 2004; Niederer et al., 2010a; Niederer et al., 2010b; Pan et al., 2006; Siriboonrit et al., 2003; Willcocks et al., 2010).

In conclusion, the inhibitory role of the FcγRIIB TM domain in BCR signaling may functionally explain the pathological mechanism of SLE and other autoimmune diseases.

ACKNOWLEDGEMENTS

This work is supported by funds from the National Basic Research Program (973 Program) (No. 2014CB542500) and the National Natural Science Foundation China (Grant Nos. 81422020 and 01621002).

ABBREVIATIONS

BCR, B cell receptor; CNS, central nervous system; DC, dendritic cell; FcγR, receptor for Fc portion of IgG molecule; FRET, fluorescence resonance energy transfer; IC, immune complex; ITAM, immunoreceptor tyrosine-based activation motif; ITIM, immunoreceptor tyrosine-based inhibition motif; ITP, idiopathic thrombocytopenic purpura; LAT, linker for activation of T cells; RA, rheumatoid arthritis; SLE, systemic lupus erythematosus; SNP, single nucleotide polymorphism; TM, transmembrane.

COMPLIANCE WITH ETHICS GUIDELINES

Junyi Wang, Zongyu Li, Liling Xu, Hengwen Yang and Wanli Liu declare that they have no conflict of interest. This article does not contain any studies with human or animal subjects performed by the any of the authors.

REFERENCES

Amigorena S, Bonnerot C, Choquet D, Fridman WH, Teillaud JL (1989) Fc-Gamma-Rii expression in resting and activated lymphocytes-B. Eur J Immunol 19:1379–1385

Ballesteros JA, Deupi X, Olivella M, Haaksma EEJ, Pardo L (2000) Serine and threonine residues bend alpha-helices in the chi(1) = g(−) conformation. Biophys J 79:2754–2760

Batista FD, Neuberger MS (2000) B cells extract and present immobilized antigen: implications for affinity discrimination. EMBO J 19:513–520

Blank U, Launay P, Benhamou M, Monteiro RC (2009) Inhibitory ITAMs as novel regulators of immunity. Immunol Rev 232:59–71

Bolland S, Ravetch JV (2000) Spontaneous autoimmune disease in Fc(gamma)RIIB-deficient mice results from strain-specific epistasis. Immunity 13:277–285

Bruin M, Bierings M, Uiterwaal C, Revesz T, Bode L, Wiesman ME, Kuijpers T, Tamminga R, de Haas M (2004) Platelet count, previous infection and FCGR2B genotype predict development of chronic disease in newly diagnosed idiopathic thrombocytopenia in childhood: results of a prospective study. Br J Haematol 127:561–567

Chen JY, Wang CM, Wu JM, Ho HH, Luo SF (2006) Association of rheumatoid factor production with FcgammaRIIIa polymorphism in Taiwanese rheumatoid arthritis. Clin Exp Immunol 144:10–16

Chen JY, Wang CM, Ma CC, Hsu LA, Ho HH, Wu YJJ, Kuo SN, Wu J (2008) A transmembrane polymorphism in Fc gamma RIIb (FCGR2B) is associated with the production of anti-cyclic citrullinated peptide autoantibodies in Taiwanese RA. Genes Immun 9:680–688

Chu ZT, Tsuchiya N, Kyogoku C, Ohashi J, Qian YP, Xu SB, Mao CZ, Chu JY, Tokunaga K (2004) Association of Fcgamma receptor IIb polymorphism with susceptibility to systemic lupus erythematosus in Chinese: a common susceptibility gene in the Asian populations. Tissue Antigens 63:21–27

Clatworthy MR, Willcocks L, Urban B, Langhorne J, Williams TN, Peshu N, Watkins NA, Floto RA, Smith KG (2007) Systemic lupus erythematosus-associated defects in the inhibitory receptor FcgammaRIIb reduce susceptibility to malaria. Proc Natl Acad Sci USA 104:7169–7174

Daeron M (1997) Fc receptor biology. Annu Rev Immunol 15:203–234

Daeron M, Malbec O, Latour S, Bonnerot C, Segal DM, Fridman WH (1993) Distinct intracytoplasmic sequences are required for endocytosis and phagocytosis via murine Fc gamma RII in mast cells. Int Immunol 5:1393–1401

Floto RA, Clatworthy MR, Heilbronn KR, Rosner DR, MacAry PA, Rankin A, Lehner PJ, Ouwehand WH, Allen JM, Watkins NA, Smith KGC (2005) Loss of function of a lupus-associated Fc gamma RIIb polymorphism through exclusion from lipid rafts. Nat Med 11:1056–1058

Fong DC, Brauweiler A, Minskoff SA, Bruhns P, Tamir I, Mellman I, Daeron M, Cambier JC (2000) Mutational analysis reveals multiple distinct sites within Fc gamma receptor IIB that function in inhibitory signaling. J Immunol 165:4453–4462

Georgiou G, Bahra SS, Mackie AR, Wolfe CA, O'Shea P, Ladha S, Fernandez N, Cherry RJ (2002) Measurement of the lateral diffusion of human MHC class I molecules on HeLa cells by fluorescence recovery after photobleaching using a phycoerythrin probe. Biophys J 82:1828–1834

Harder T, Scheiffele P, Verkade P, Simons K (1998) Lipid domain structure of the plasma membrane revealed by patching of membrane components. J Cell Biol 141:929–942

Hippen KL, Buhl AM, Dambrosio D, Nakamura K, Persin C, Cambier JC (1997) Fc gamma RIIB1 inhibition of BCR-mediated phosphoinositide hydrolysis and Ca2+ mobilization is integrated by CD19 dephosphorylation. Immunity 7:49–58

Kono H, Suzuki T, Yamamoto K, Okada M, Yamamoto T, Honda Z (2002) Spatial raft coalescence represents an initial step in Fc gamma R signaling. J Immunol 169:193–203

Kono H, Kyogoku C, Suzuki T, Tsuchiya N, Honda H, Yamamoto K, Tokunaga K, Honda Z (2005) FcgammaRIIB Ile232Thr transmembrane polymorphism associated with human systemic lupus erythematosus decreases affinity to lipid rafts and attenuates inhibitory effects on B cell receptor signaling. Hum Mol Genet 14:2881–2892

Kyogoku C, Dijstelbloem HM, Tsuchiya N, Hatta Y, Kato H, Yamaguchi A, Fukazawa T, Jansen MD, Hashimoto H, van de Winkel JG et al (2002a) Fcgamma receptor gene polymorphisms in Japanese patients with systemic lupus erythematosus: contribution of FCGR2B to genetic susceptibility. Arthritis Rheum 46:1242–1254

Kyogoku C, Dijstelbloem HM, Tsuchiya N, Hatta Y, Kato H, Yamaguchi A, Fukazawa T, Jansen MD, Hashimoto H, van de Winkel JGJ et al (2002b) Fc gamma receptor gene polymorphisms in Japanese patients with systemic lupus erythematosus - Contribution of FCGR2B to genetic susceptibility. Arthritis Rheum 46:1242–1254

Kyogoku C, Tsuchiya N, Matsuta K, Tokunaga K (2002c) Studies on the association of Fc gamma receptor IIA, IIB, IIIA and IIIB polymorphisms with rheumatoid arthritis in the Japanese: evidence for a genetic interaction between HLA-DRB1 and FCGR3A. Genes Immun 3:488–493

Kyogoku C, Tsuchiya N, Wu H, Tsao BP, Tokunaga K (2004) Association of Fcgamma receptor IIA, but not IIB and IIIA, polymorphisms with systemic lupus erythematosus: a family-based association study in Caucasians. Arthritis Rheum 50:671–673

Lehmann B, Schwab I, Bohm S, Lux A, Biburger M, Nimmerjahn F (2012) FcgammaRIIB: a modulator of cell activation and humoral tolerance. Expert Rev Clin Immunol 8:243–254

Li XR, Wu JM, Ptacek T, Redden DT, Brown EE, Alarcon GS, Ramsey-Goldman R, Petri MA, Reveille JD, Kaslow RA, et al. (2013) Allelic-dependent expression of an activating Fc receptor on B cells enhances humoral immune responses. Sci Transl Med 5

Liu WL, Sohn HW, Tolar P, Meckel T, Pierce SK (2010) Antigen-induced oligomerization of the B cell receptor is an early target of Fc gamma RIIB inhibition. J Immunol 184:1977–1989

Lu J, Chu J, Zou Z, Hamacher NB, Rixon MW, Sun PD (2015) Structure of FcgammaRI in complex with Fc reveals the importance of glycan recognition for high-affinity IgG binding. Proc Natl Acad Sci USA 112:833–838

Mimura Y, Sondermann P, Ghirlando R, Lund J, Young SP, Goodall M, Jefferis R (2001) Role of oligosaccharide residues of IgG1-Fc in Fc gamma RIIb binding. J Biol Chem 276:45539–45547

Niederer HA, Clatworthy MR, Willcocks LC, Smith KG (2010a) FcgammaRIIB, FcgammaRIIIB, and systemic lupus erythematosus. Ann N Y Acad Sci 1183:69–88

Niederer HA, Willcocks LC, Rayner TF, Yang W, Lau YL, Williams TN, Scott JA, Urban BC, Peshu N, Dunstan SJ et al (2010b) Copy number, linkage disequilibrium and disease association in the FCGR locus. Hum Mol Genet 19:3282–3294

Nimmerjahn F, Ravetch JV (2008) Fc gamma receptors as regulators of immune responses. Nat Rev Immunol 8:34–47

Nimmerjahn F, Ravetch JV (2011) FcγRs in health and disease. Curr Top Microbiol Immunol 350:105–125

Pan F, Zhang K, Li X, Xu J, Hao J, Ye D (2006) Association of Fcgamma receptor IIB gene polymorphism with genetic susceptibility to systemic lupus erythematosus in Chinese populations–a family-based association study. J Dermatol Sci 43:35–41

Peress NS (1993) Identification of Fc gamma RI, II and III on normal human brain ramified microglia and on microglia in senile plaques in Alzheimer's disease. J Neuroimmunol 48:71–79

Pincetic A, Bournazos S, DiLillo DJ, Maamary J, Wang TT, Dahan R, Fiebiger BM, Ravetch JV (2014a) Type I and type II Fc receptors regulate innate and adaptive immunity. Nat Immunol 15:707–716

Pincetic A, Bournazos S, Dilillo DJ, Maamary J, Wang TT, Dahan R, Fiebiger BM, Ravetch JV (2014b) Type I and type II Fc receptors regulate innate and adaptive immunity. Nat Immunol 15:707

Radstake TRDJ, Franke B, Wenink MH, Nabbe KCAA, Coenen MJH, Welsing P, Bonvini E, Koenig S, van den Berg WB, Barrera P, van Riel PLCM (2006) The functional variant of the inhibitory Fc gamma receptor IIb (CD32B) is associated with the rate of radiologic joint damage and dendritic cell function in rheumatoid arthritis. Arthritis Rheum 54:3828–3837

Ravetch JV, Kinet JP (1991) Fc receptors. Annu Rev Immunol 9:457–492

Siriboonrit U, Tsuchiya N, Sirikong M, Kyogoku C, Bejrachandra S, Suthipinittharm P, Luangtrakool K, Srinak D, Thongpradit R, Fujiwara K et al (2003) Association of Fcgamma receptor IIb and IIIb polymorphisms with susceptibility to systemic lupus erythematosus in Thais. Tissue Antigens 61:374–383

Smith KG, Clatworthy MR (2010a) FcgammaRIIB in autoimmunity and infection: evolutionary and therapeutic implications. Nat Rev Immunol 10:328–343

Smith KGC, Clatworthy MR (2010b) Fc[gamma]RIIB in autoimmunity and infection: evolutionary and therapeutic implications. Nat Rev Immunol 10:328–343

Sohn HW, Pierce SK, Tzeng SJ (2008a) Live cell imaging reveals that the inhibitory Fc gamma RIIB destabilizes B cell receptor membrane-lipid blocks immune synapse formation. J Immunol 180:793–799

Sohn HW, Tolar P, Pierce SK (2008b) Membrane heterogeneities in the formation of B cell receptor-Lyn kinase microclusters and the immune synapse. J Cell Biol 182:367–379

Starbeck-Miller GR, Badovinac VP, Barber DL, Harty JT (2014) Cutting edge: expression of FcgammaRIIB tempers memory CD8 T cell function in vivo. J Immunol 192:35–39

Tackenberg B, Jelcic I, Baerenwaldt A, Oertel WH, Sommer N, Nimmerjahn F, Lunemann JD (2009) Impaired inhibitory Fc gamma receptor IIB expression on B cells in chronic inflammatory demyelinating polyneuropathy. Proc Natl Acad Sci USA 106:4788 4792

Tanimura N, Nagafuku M, Minaki Y, Umeda Y, Hayashi F, Sakakura J, Kato A, Liddicoat DR, Ogata M, Hamaoka T, Kosugi A (2003) Dynamic changes in the mobility of LAT in aggregated lipid rafts upon T cell activation. J Cell Biol 160:125–135

Tolar P, Sohn HW, Pierce SK (2005) The initiation of antigen-induced B cell antigen receptor signaling viewed in living cells by fluorescence resonance energy transfer. Nat Immunol 6:1168–1176

Warmerdam PAM, Nabben NMJM, Vandegraaf SAR, Vandewinkel JGJ, Capel PJA (1993) The human low affinity immunoglobulin-G Fc receptor-Iic gene is a result of an unequal crossover event. J Biol Chem 268:7346–7349

Willcocks LC, Carr EJ, Niederer HA, Rayner TF, Williams TN, Yang WL, Scott JAG, Urban BC, Peshu N, Vyse TJ et al (2010) A defunctioning polymorphism in FCGR2B is associated with protection against malaria but susceptibility to systemic lupus erythematosus. Proc Natl Acad Sci USA 107:7881–7885

Xu L, Li G, Wang J, Fan Y, Wan Z, Zhang S, Shaheen S, Li J, Wang L, Yue C et al (2014) Through an ITIM-independent mechanism the FcgammaRIIB blocks B cell activation by disrupting the colocalized microclustering of the B cell receptor and CD19. J Immunol 192:5179–5191

Xu L, Xia M, Guo J, Sun X, Li H, Xu C, Gu X, Zhang H, Yi J, Fang Y et al (2016) Impairment on the lateral mobility induced by structural changes underlies the functional deficiency of the lupus-associated polymorphism FcgammaRIIB-T232. J Exp Med 213:2707–2727

10

Ligand binding and conformational changes of SUR1 subunit in pancreatic ATP-sensitive potassium channels

Jing-Xiang Wu[1,2], **Dian Ding**[1,2,3], **Mengmeng Wang**[1,2,3], **Yunlu Kang**[1,2], **Xin Zeng**[1,2,3], **Lei Chen**[1,2]✉

[1] State Key Laboratory of Membrane Biology, Institute of Molecular Medicine, Peking University, Beijing Key Laboratory of Cardiometabolic Molecular Medicine, Beijing 100871, China
[2] Peking-Tsinghua Center for Life Sciences, Peking University, Beijing 100871, China
[3] Academy for Advanced Interdisciplinary Studies, Peking University, Beijing 100871, China
✉ Correspondence: chenlei2016@pku.edu.cn (L. Chen)

ABSTRACT

ATP-sensitive potassium channels (K_{ATP}) are energy sensors on the plasma membrane. By sensing the intracellular ADP/ATP ratio of β-cells, pancreatic K_{ATP} channels control insulin release and regulate metabolism at the whole body level. They are implicated in many metabolic disorders and diseases and are therefore important drug targets. Here, we present three structures of pancreatic K_{ATP} channels solved by cryo-electron microscopy (cryo-EM), at resolutions ranging from 4.1 to 4.5 Å. These structures depict the binding site of the antidiabetic drug glibenclamide, indicate how Kir6.2 (inward-rectifying potassium channel 6.2) N-terminus participates in the coupling between the peripheral SUR1 (sulfonylurea receptor 1) subunit and the central Kir6.2 channel, reveal the binding mode of activating nucleotides, and suggest the mechanism of how Mg-ADP binding on nucleotide binding domains (NBDs) drives a conformational change of the SUR1 subunit.

KEYWORDS K_{ATP}, SUR, ABC transporter, glibenclamide, sulfonylurea, diabetes

INTRODUCTION

ATP-sensitive potassium channels (K_{ATP}) are ion channels that selectively allow potassium ions to permeate the cell. Their channel activities are tightly regulated by endogenous nucleotide metabolites. Specifically, they are inhibited by ATP and activated by Mg-ADP. By sensing the intracellular ADP/ATP ratio, K_{ATP} channels tune the potassium ion efflux across the plasma membrane and adjust the membrane potential. Therefore, K_{ATP} channels convert the cellular metabolic status into electrical signals, a unique output that has broad physiological effects. K_{ATP} channels are widely distributed in many tissues, including the pancreas, brain, heart, and smooth muscle (Hibino et al., 2010), and they play important roles in many physiological processes, such as hormone secretion and vasodilatation (Hibino et al., 2010). Genetic mutation of genes that encode K_{ATP} channel subunits can lead to several metabolic diseases and neuronal diseases (Ashcroft et al., 2017; Hibino et al., 2010). Therefore, K_{ATP} channels are important drug targets. Clinically relevant sulfonylureas drugs inhibit pancreatic K_{ATP} channels and serve as insulin secretagogues for the treatment of type II diabetes (Bryan et al., 2005), while K_{ATP} activators, such as potassium channel openers (KCOs), are used for hypoglycemia and show promise for myoprotection (Flagg et al., 2010; Hibino et al., 2010).

Previous studies have established that the functional K_{ATP} channel is a hetero-octamer composed of four inward-rectifying potassium channel 6 (Kir6) subunits and four sulfonylurea receptor (SUR) regulatory subunits (Clement et al., 1997; Shyng and Nichols, 1997; Ueda et al., 1997) . Kir6 subunits are encoded by either *KCNJ8* (Kir6.1) or *KCNJ11* (Kir6.2). Kir6 subunits harbor sites for inhibitory ATP binding. The activities of Kir6 can be enhanced by PIP_2, which is a signaling lipid present in the inner leaflets of the plasma membrane. SUR subunits are composed of the N-terminal transmembrane domain 0-loop 0 (TMD0-L0) and ATP-binding cassettes (ABC) transporter-like modules. They are encoded by either *ABCC8* (SUR1) or *ABCC9* (SUR2). SUR subunits bind stimulatory Mg-ADP, KCOs, and inhibitory sulfonylureas and regulate Kir6 channel activity. K_{ATP} channels composed of different isoform combinations are distributed in different tissues and have distinct pharmacological profiles, which are primarily determined by the identity of associated SUR subunits.

The ABC transporter has two nucleotide binding domains (NBDs). Each NBD harbors two adjacent halves of a catalytic site for ATP hydrolysis. Therefore, one NBD must pair with the other NBD to form two functional ATP hydrolysis sites when they are in proximity (Locher, 2016). SUR proteins belong to subfamily C of the ABC family transporters (ABCC). The other two family members of ABCC are multidrug resistance-associated proteins (MRPs) and the cystic fibrosis transmembrane conductance regulator (CFTR). MRP is a transporter with known substrates, and the transporter activity of MRP depends on the hydrolysis of Mg-ATP, whereas the CFTR is an Mg-ATP-gated chloride channel. The hallmark of the ABCC family is that the conserved glutamate residue on NBD1 Walker B motif of canonical ABC transporters is replaced by an aspartate (D855 in SUR1). Therefore, the catalytic activity of one ATP hydrolysis site (degenerate site or nucleotide binding site 1, NBS1) is lost (Aittoniemi et al., 2009), and only one functional ATP hydrolysis site (consensus site or NBS2) remains. The activities of most ABC transporters depend on the hydrolysis of Mg-ATP. However, Mg-ADP is the primary nucleotide species that activates SUR1 to enhance K_{ATP} channel activity (Aittoniemi et al., 2009).

Because of the physiological importance of pancreatic K_{ATP} channels, they have served as a model to study the biochemical and physiological properties of K_{ATP} channels for decades. K_{ATP} channels in pancreatic islets are composed of Kir6.2 channel subunits and SUR1 regulatory subunits. Recently, several structures of pancreatic K_{ATP} channel are solved by us and other groups, including the GBM bound inhibited state (abbreviated, GBM state, EMD-6689) (Li et al., 2017), the ATP + GBM bound inhibited state (abbreviated, ATP + GBM state, EMD-7073 and EMD-8470) (Martin et al., 2017a; Martin et al., 2017b) and the Mg-ATP and Mg-ADP bound state (abbreviated, Mg-ATP&ADP state, EMD-7338 and EMD-7339) (Lee et al., 2017). These structures reveal the architecture of K_{ATP} channels, the details of how different subunits are assembled together, the binding site of sulfonylureas on SUR1, ATP on Kir6.2, and Mg-ADP/ATP on SUR1. However, the mechanisms of the sulfonylureas inhibition and Mg-ADP activation are still lacking.

In this paper, we present three structures of the pancreatic K_{ATP} channel solved using an SUR1-Kir6.2 fusion construct. The first structure is the K_{ATP} channel in complex with ATPγS and GBM, and it is solved to a resolution of 4.3 Å (abbreviated, ATP + GBM state). The second structure is the K_{ATP} channel in complex with ATPγS alone, which is solved to 4.4 Å resolution (abbreviated, ATP state). The third structure is the K_{ATP} channel in complex with Mg-ADP, VO_4^{3-}, PIP_2 and NN414 (Carr et al., 2003), a ligand combination that favors the NBD-dimerized state of SUR1, which is solved to 4.3 Å (abbreviated, Mg-ADP state). This new structural information, together with previously solved structures, provides mechanistic insights into how the pancreatic K_{ATP} channel works.

RESULTS AND DISCUSSION

A K_{ATP} channel fusion construct for cryo-EM structure determination

To overcome the compositional heterogeneity of the K_{ATP} channel when heterologously co-expressing SUR1 with Kir6.2 (Li et al., 2017), we created a SUR1-Kir6.2 fusion construct, in which the intracellular C-terminus of SUR1 is covalently linked to the N-terminus of Kir6.2 by a linker (Fig. S1A). Fusion construct with a 6 amino acids (6AA) linker was previously used to elucidate the association and stoichiometry of K_{ATP} channels (Clement et al., 1997) and was recently used to solve the structure of K_{ATP} channel in complex with Mg-ATP and Mg-ADP (Lee et al., 2017). To avoid potential artifacts on the structure of the K_{ATP} channel because of limited linker length, we used a flexible linker with 39 amino acids (39AA), which should be sufficient for a 130 Å linear distance when fully extended. Similar to the wild-type channel, the fusion construct can be inhibited by ATP and activated by Mg-ADP in an inside-out patch clamp, which is the characteristic property of K_{ATP} channel (Fig. S1B). However, inhibition by GBM is markedly decreased (Fig. S1B). Because it has been reported that the individually expressed SUR1 subunit is sufficient for high affinity GBM binding (Aguilar-Bryan et al., 1995) and our construct has an intact SUR1 subunit, we reasoned that the reduced GBM inhibition is due to insufficient coupling between SUR1 and Kir6.2 of our fusion construct, as discussed later. To be noted, it is reported that 1 μmol/L GBM can fully inhibit the 6AA linker construct in whole cell $^{86}Rb^{+}$ efflux assays (Shyng and Nichols, 1997), this is probably because the GBM can bind SUR1 to counteract the activation effect of intracellular Mg-ADP or Mg-ATP, rather than directly inhibits the 6AA fusion channel. It is also reported that 300 μmol/L tolbutamide can attenuate the currents of 6AA construct in inside-out patch clamp recording (Lee et al., 2017). This is possibly due to the low affinity block of sulfonylurea on Kir6.2 channel (Reimann 1999). Whether the 6AA construct can be directly inhibited by the GBM bound on SUR1 remains to be thoroughly determined.

Structures of the K_{ATP} channel in the ATP + GBM and ATP states

We aimed to obtain structures of K_{ATP} with and without GBM at higher resolution than our previous GBM state (Li et al., 2017), which can be used to locate the GBM density by

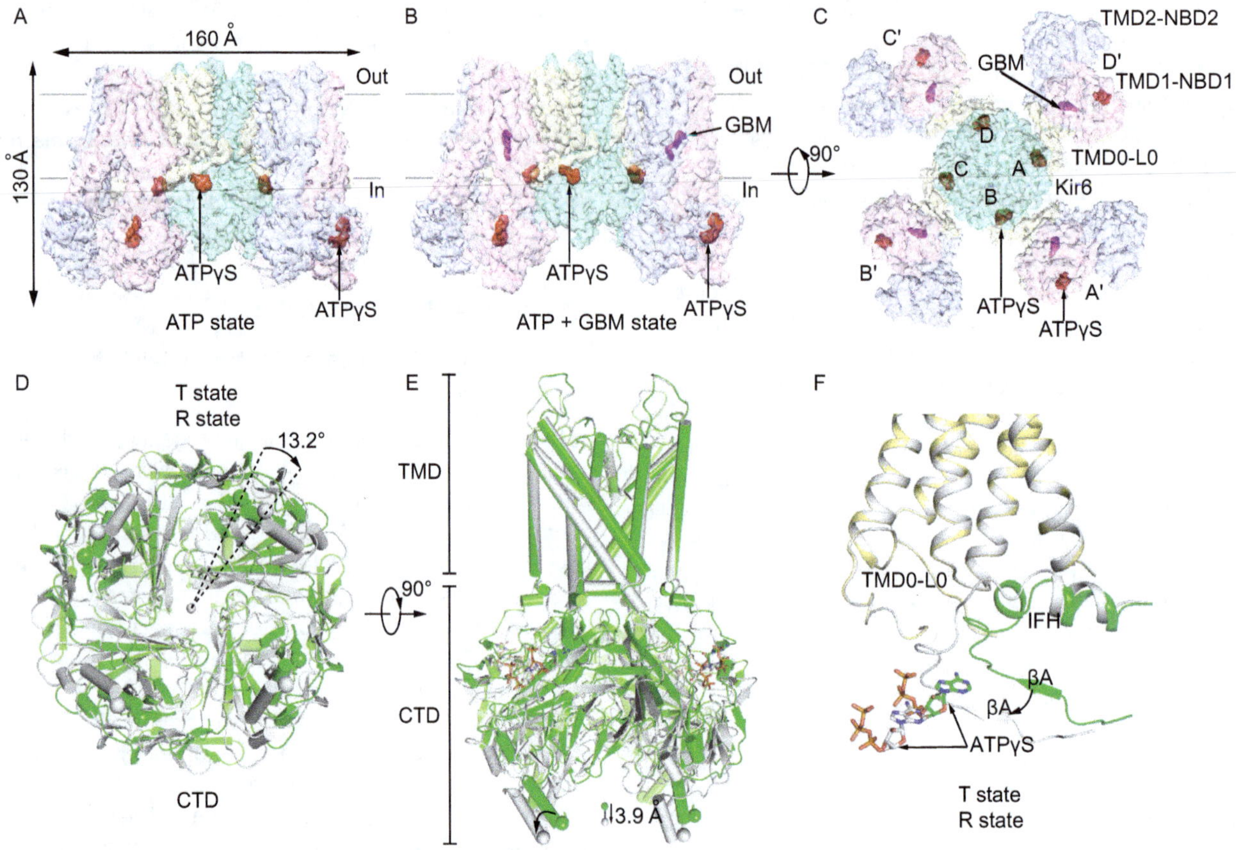

Figure 1. Structures of the pancreatic K_{ATP} channel in complex with ATPγS or ATPγS and glibenclamide (GBM). (A) Side view of the K_{ATP} channel in complex with ATPγS (ATP state). Approximate extent of the phospholipid bilayer is shown as thick gray lines. Kir6.2, SUR1 TMD0-L0 (transmembrane domain 0-loop 0) fragment, TMD1-NBD1 (nucleotide binding domain1), TMD2-NBD2, ATPγS and GBM are shown in green, yellow, magenta, blue, red and purple, respectively. (B) Side view of the K_{ATP} channel in complex with ATPγS and GBM (ATP + GBM state), colored the same as in (A). (C) Bottom view of the K_{ATP} channel in complex with ATPγS and GBM from the intracellular side. (D) Bottom view of the aligned Kir6.2 structures between "T state" (green) and "R state" (gray) of the ATP + GBM state. Rotation angle between CTDs (cytoplasmic domain) was measured using Cα positions of L356 of Kir6.2, shown as spheres. (E) Side view of the aligned Kir6.2 structures between "T state" (green) and "R state" (gray) of the ATP + GBM state. Vertical movement of CTDs is measured between the centers of mass of L356 Cαs. (F) Positional difference of Kir6.2 βA and IFH (interfacial helix) between "T state" and "R state" structures of the ATP + GBM state.

direct comparison. Our previous study demonstrated that the K_{ATP} channel displays a large degree of conformational heterogeneity when in complex with the high affinity inhibitor GBM alone (Li et al., 2017). The additional ATP molecule, an endogenous inhibitor of K_{ATP}, can further stabilize the Kir6.2 subunit to reduce conformational heterogeneity (Martin et al., 2017b). Therefore, we applied ATPγS, a slowly hydrolyzable and functional ATP analog, to stabilize the structures of the K_{ATP} channel with and without GBM (Schwanstecher et al., 1994a).

The fusion channel protein was purified and then subjected to cryo-EM single particle analysis (Fig. S1C and S1D). The structure of K_{ATP} in complex with ATPγS and GBM was solved at an average resolution of 4.3 Å, slightly higher than the structure in complex with ATPγS alone at a resolution of 4.5 Å (Figs. 1A–C, S2, S3A, and Tables S1 and S2). The overall K_{ATP} channel structure and SUR1 conformation in the ATP + GBM state is almost identical to that in the ATP state based on our density maps (Fig. S3B). Therefore, we focus the discussion on the structure of the ATP + GBM state with a higher resolution, unless stated otherwise.

Focused three-dimensional (3D) classification on the Kir6.2 cytosolic globular domain (CTD) isolated two major classes with distinct conformations (Fig. S2 and Table S1). The CTDs of one class have relative rotation to the other, as observed previously in the structure of the ATP + GBM state (Martin et al., 2017b). Each class was individually refined to reach resolutions of 4.3 Å or 4.6 Å. In contrast, the transmembrane domain (TMD) of the two classes shares the same structure and focused refinement with all particles reached an overall resolution of 4.1 Å (Fig. S2C and

Table S1). The maps demonstrate side chain densities for most residues and enable further improvement of our previous model based on medium resolution maps. The overall structures of both classes are similar to the previous symmetric Class 1.1 structure of the GBM state (EMD-6689) (Li et al., 2017) or that of the ATP + GBM state (EMD-8470) (Martin et al., 2017b), with the central Kir6.2 in a closed state and the peripheral SUR1 in an inward-facing inactive state (Fig. 1B and 1C). By comparing the structures of the two classes, we found that the CTDs of Kir6.2 (R32 to L66 and K170 to L356) move in a rigid body fashion relative to the transmembrane domains, resulting in a 3.9-Å downward movement and 13.2° clockwise rotation (Fig. 1D and 1E). In contrast, the structure of transmembrane domain of Kir6.2 and SUR1 is largely unchanged (Fig. 1D and 1E). Therefore, the interface between Kir6.2 CTD and SUR1 TMD0-L0 fragment is altered and the conformations of the linkers between the two rigid bodies also change. Here, we refer to class 1 structure as the "T state" (the tense state) and class 2 as the "R state" (the relaxed state). Specifically, in the "R state" structure, the linker between βA and the interfacial helix (IFH) of Kir6.2 moves away from the TMD0-L0 of SUR1 (Fig. 1F). The differences between the two conformations described herein are not due to differences in ATPγS occupancy on CTDs, because we observed similarly strong ATPγS densities in the maps of both classes (Fig. S4A). Instead, they possibly reflect the endogenous mobility of the CTDs in the Kir6.2-inhibited state, because we observed similar conformational heterogeneity of the K_{ATP} channel in the ATP state and the Mg-ADP state as described later. The co-existence of "T state" and "R state" might be functionally important since the rotation of CTD relates to channel gating in another Kir channel member GIRK2 (Whorton and MacKinnon, 2011, 2013) .

GBM binding site on SUR1

Previous studies demonstrated that mutations of Y230A and W232A on the lasso motif reduce GBM affinity (Vila-Carriles et al., 2007), indicating the lasso motif might be in proximity to or on the binding site of GBM. Therefore, we compared the density maps around lasso motif between the ATP state and the ATP + GBM state, and found that the extra density near the lasso motif in the previous medium resolution map is present in both states. With improved resolution and map quality, we found that this extra density was contributed by both the M5-Lh1 loop and an unknown ligand, probably a lipid head group that constitutively binds K_{ATP}, because it is also present in the Mg-ADP state described later (Fig. S4B). Instead, a previously unmodeled density inside the SUR1 central cavity in the ATP + GBM state map has a distinct elongated shape that matches the GBM molecule in an extended conformation (Figs. 2A and S4C–E). In contrast, this density is not present in the ATP state, where GBM was not supplied into the sample (Fig. 2B). This unambiguously demonstrates that the extra density inside SUR1 central cavity in the map of the ATP + GBM state represents the GBM molecule (Figs. 2A and S4C–E). The extended conformation of GBM bound to SUR1 is dramatically different from the compact conformation of GBM bound to human aldo-keto reductase 1C3 subfamily (AKR1Cs) (Zhao et al., 2015) (Fig. S4C). Our results confirmed the GBM binding site reported by another group recently (PDB: 6BAA, EMD-7073) (Martin et al., 2017a). Accompanied by mutagenesis results, they suggested indirect allosteric effects of Y230A or W232A. These mutations on the lasso motif might change the structure of adjacent transmembrane helices that are essential for GBM binding (Martin et al., 2017a). Therefore, mutagenesis results might be affected by allostery, and comparison between the maps of K_{ATP} solved from samples

Figure 2. Binding sites of ATPγS and GBM on SUR1. ▶ (A) Occupied GBM binding site in the 4.4 Å ATP + GBM state map after focus refinement of SUR1 ABC transporter module. The density of GBM is shown as green mesh with GBM molecule shown in stick representation. The putative KNtp (Kir6.2 N-terminus) binding site is outlined by green dash lines. (B) Apo GBM binding site in the 4.4 Å ATP state map. Maps in (A) and (B) are low-pass filtered to 4.5 Å for comparison. (C–F) EM densities of putative KNtp. Extra continuous density close to GBM molecule is colored in green and GBM density is shown in purple. For better visualization of low resolution features, maps were sharpened with lower B factor than determined by post-processing process. Maps of the GBM state (two half maps for EMD-6689 entry) (C), ATP + GBM state (EMD-7073) (D and E), and ATP + GBM state with SUR1-Kir6.2 fusion construct (F) were sharpened with B factors −200, −50, and +200, and contoured at 0.4, 1.4 and 0.015 levels, respectively. To illustrate the position of KNtp relative to the rest of the Kir6.2 channel, the side view of a single Kir6.2 and SUR1 subunit is shown in (E). (G) Sequence alignment of KNtp of *Mus musculus* Kir6.2 (mmKir6.2), *Mus musculus* Kir6.1 (mmKir6.1), *Homo sapiens* Kir6.2 (hsKir6.2), and *Danio rerio* Kir6.2 (drKir6.2). Because KNtp is unmodeled due to limited map quality, it is shown as dashed lines above. Nonconserved residues are highlighted in red. Position of the inserted dipeptide (Gly-Ser) is indicated. (H) Comparison of GBM inhibitory effects among Kir6.2 constructs with different numbers of Gly-Ser (GS) dipeptides inserted or N-terminal deletion, when coexpressed with SUR1. The inhibitory effect of GBM on SUR1-Kir6.2 fusion construct is also shown. Currents after GBM treatment were normalized to the currents before GBM treatment. Each bar represents mean ± standard error, n = 3. (I) EM densities at the degenerate site of SUR1 in the ATP + GBM state. Densities of ATPγS and surrounding region are shown as green and blue meshes, respectively. Figure is created from map after focused refinement of SUR1 ABC transporter module.

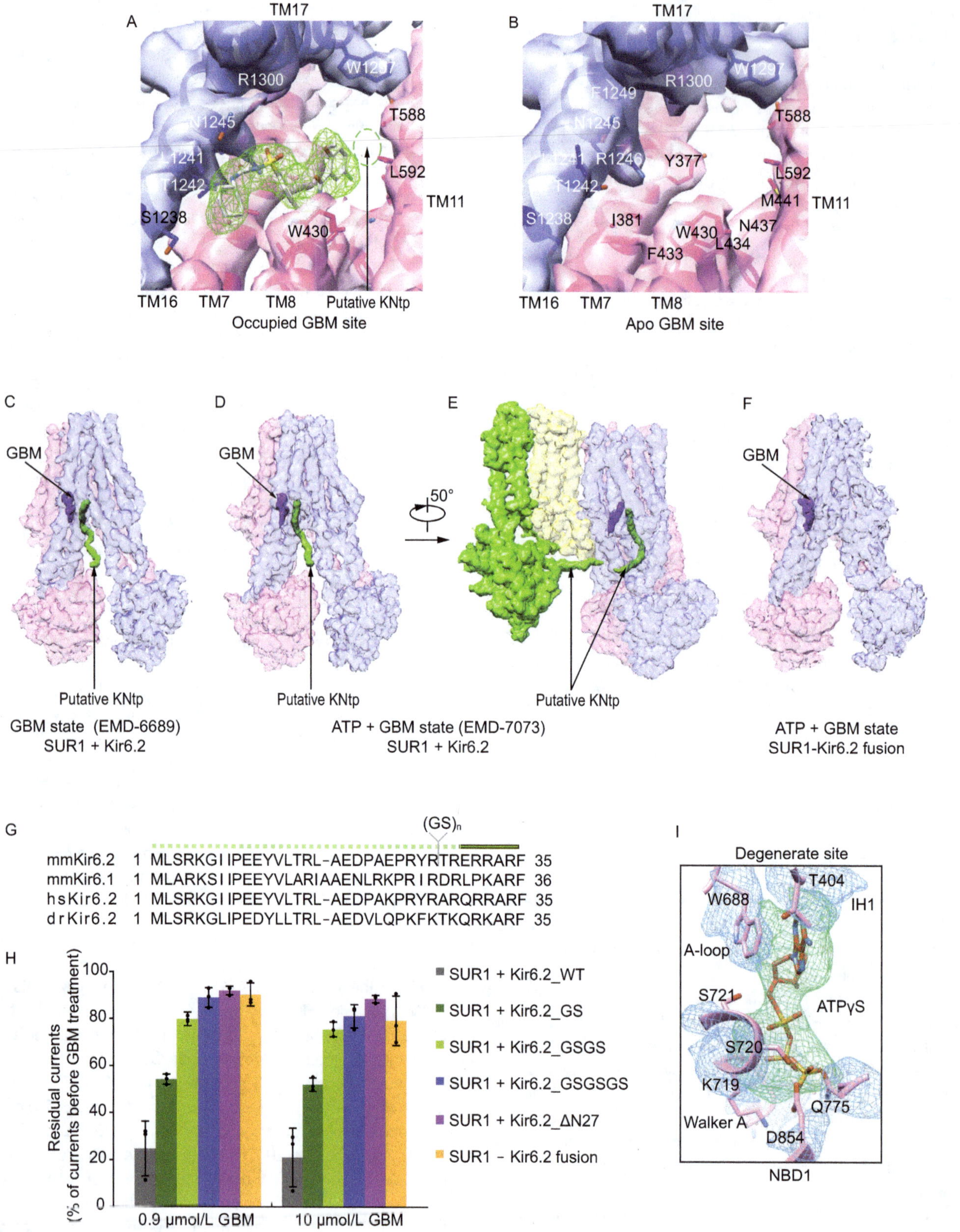
A
TM17
R1300
W1297
T588
N1245
L1241
T1242
L592
TM11
S1238
W430
TM16
TM7
TM8
Putative KNtp
Occupied GBM site
B
TM17
F1249
R1300
W1297
T588
N1245
L1241
R1246
Y377
L592
T1242
M441
TM11
S1238
I381
W430
N437
L434
F433
TM16
TM7
TM8
Apo GBM site
C
D
E
F
GBM
GBM
50°
GBM
Putative KNtp
Putative KNtp
Putative KNtp
GBM state (EMD-6689)
SUR1 + Kir6.2
ATP + GBM state (EMD-7073)
SUR1 + Kir6.2
ATP + GBM state
SUR1-Kir6.2 fusion
G
(GS)n
mmKir6.2 1 MLSRKGIIPEEYVLTRL-AEDPAEPRYRTRERRARF 35
mmKir6.1 1 MLARKSIIPEEYVLARIAAENLRKPRIRDRLPKARF 36
hsKir6.2 1 MLSRKGIIPEEYVLTRL-AEDPAKPRYRARQRRARF 35
drKir6.2 1 MLSRKGLIPEDYLLTRL-AEDVLQPKFKTKQRKARF 35
H
Residual currents
(% of currents before GBM treatment)
100
80
60
40
20
0
0.9 μmol/L GBM
10 μmol/L GBM
SUR1 + Kir6.2_WT
SUR1 + Kir6.2_GS
SUR1 + Kir6.2_GSGS
SUR1 + Kir6.2_GSGSGS
SUR1 + Kir6.2_ΔN27
SUR1 – Kir6.2 fusion
I
Degenerate site
T404
W688
IH1
A-loop
S721
ATPγS
S720
K719
Q775
Walker A
D854
NBD1

supplied with and without GBM is essential to definitively assign the non-protein density for GBM.

Location of Kir6.2 N-terminus

The fact that we observed strong GBM density inside SUR1 further reinforced our above assumption that SUR1-Kir6.2 fusion protein can bind GBM but the reduced GBM inhibition is because of insufficient coupling between SUR1 and Kir6.2 due to the covalent linker (Fig. S1B). When docking the improved model of SUR1 and GBM into our previous Class 1.1 map of the GBM state (EMD-6689) (Li et al., 2017), we found an additional positive density in the SUR1 central cavity that originates from the proximity of the GBM molecule and extends out of the SUR1 ABC transporter module in the direction of Kir6.2 CTD (Fig. 2C). Our previous GBM state structure (Li et al., 2017) and the ATP + GBM state structure by another group (Martin et al., 2017b) were solved using K_{ATP} protein generated by the coexpression of SUR1 with Kir6.2 as separated polypeptides; therefore, the Kir6.2 protein had a native N-terminus in these studies. In the map of the ATP + GBM state (EMD-7073), we observed a similar density (Fig. 2D and 2E). Unfortunately, because of limited local map quality, we cannot infer the identity of this density purely from these maps. In contrast, we did not observe any density at the same position in our map of the ATP + GBM state that was solved using the SUR1-Kir6.2 fusion construct, in which the C-terminus of SUR1 is covalently connected to the N-terminus of Kir6.2 by a linker (Fig. 2F) and the linker together with Kir6.2 N-terminus is completely disordered in our cryo-EM map. In our previous models (PDB: 5WUA) (Li et al., 2017), which were built according to the cryo-EM map, only one amino acids of the SUR1 C-terminus were unmodeled, and they are unlikely to contribute to such an elongated unaccounted density. Therefore, this positive density, which is close to the GBM binding site, might represent the N-terminus peptide of Kir6.2 with 30 amino acids omitted in the model. This observation agrees with previous biochemical data showing that Kir6.2 can be labeled by ^{125}I-azidoglibenclamide and deletion of KNtp reduced not only ^{125}I-azidoglibenclamide labeling (Babenko and Bryan, 2002; Vila-Carriles et al., 2007) but also the affinity for GBM and repaglinide (Kuhner et al., 2012). In addition, GBM enhanced the cross-linking of SUR1 with Kir6.2 with the photo cross-linkable amino acid incorporated in the N-terminus (Devaraneni et al., 2015), indicating that the KNtp should be in proximity to SUR1. This is further supported by the electrophysiological data that high affinity inhibition of tolbutamide, GBM and repaglinide to the K_{ATP} channel were either abolished or markedly decreased when KNtp is deleted (Devaraneni et al., 2015; Kuhner et al., 2012; Reimann, 1999). Moreover, the density of the putative KNtp is closed to the 1-chloro-4-methoxybenzene group of GBM (Figs. 2A and S4E), the so called "B" site, where the azido group in ^{125}I-azidoglibenclamide sits. This is in accordance with previous findings that the ^{125}I-azidoglibenclamide but not ^{125}I-glibenclamide can label Kir6.2 subunit (Aguilar-Bryan et al., 1990; Schwanstecher et al., 1994b). Because the length of KNtp is conserved and limited (Fig. 2G), the special position of the Kir6.2 N-terminus intuitively suggests that it attenuates channel gating by acting as a chain to restrain the rotation of CTD, which is important for Kir family channel gating (Whorton and MacKinnon, 2011, 2013). To further validate this model, we inserted one to three Gly-Ser dipeptides (GS) between amino acid R27 and T28 of Kir6.2, a position before the structured Kir6.2 CTD and not conserved in the primary sequence (Fig. 2G), suggesting this position is not directly involved in SUR1 binding. We hypothesized that the KNtp of these mutants are still able to bind SUR1 but the restriction of the CTD rotation by KNtp might be reduced. Indeed, we found that the inhibitory effect of GBM progressively becomes weaker as the linker length increases (Fig. 2H). The SUR1-Kir6.2 fusion construct demonstrated little GBM inhibition (Figs. 2H and S1B), a phenotype similar to the K_{ATP} channel assembled with KNtp deleted Kir6.2, suggesting the KNtp in the fusion construct cannot bind SUR1 anymore, because of the covalent linker. Moreover, without sulphonylureas, truncation of KNtp increases channel activity and reduces ATP sensitivity (Babenko et al., 1999; Koster, 1999; Reimann, 1999) and synthetic KNtp dose-dependently attenuates ATP inhibition (Babenko and Bryan, 2002), suggesting KNtp might remain bound to SUR1 when ATP, but no sulphonylurea, is present, albeit with decreased affinity. When we superposed our SUR1 ABC transporter module model of the GBM + ATP state to that from the another group (PDB: 6BAA) (Martin et al., 2017a), we found that there is a small but noticeable change of overall conformation of SUR1 and the angle between M10 and M16 is 42.7° in our structure whereas 39.3° in their structure (Fig. S4F). Whether the observed difference of SUR1 conformation is due to KNtp binding discussed above or different data processing procedure still waits further investigation.

Mg-independent ATP binding site on SUR1 NBD1

During cryo-EM data processing, 3D classification demonstrated continuous conformational heterogeneity of SUR1 in the ATP + GBM state. Specifically, four peripheral SUR1 subunits wobble around the central Kir6.2 channel in small angles. This greatly limited the resolution of the overall cryo-EM density map and deteriorated local map quality of the peripheral SUR1 ABC transporter module. Further improvement of SUR1 map quality at the ATP + GBM state benefited from focused refinement with local search on symmetry-expanded (Zhou et al., 2015) and signal-subtracted particles of the SUR1 ABC-transporter module (Bai et al., 2015). The map of "the isolated" SUR1 ABC-transporter module reached a resolution of 4.4 Å and demonstrated markedly enhanced quality, especially for NBDs (Fig. S2C). The map also suggests a register shift of residues 1,061–1,079 on M13 in the structure by another group (PDB: 6BAA) (Martin et al., 2017a). More interestingly, we observed extra density on the degenerate site of SUR1 NBD1. The shape and size of this density matched that of

ATPγS (Fig. 2I). Moreover, this density was at the expected location of nucleotides in the nucleotide-bound NBD structure of other ABC transporters (Zhang et al., 2017). Therefore, we modeled this density as an ATPγS molecule. We observed a similar density in the map of the ATP state structure (Fig. 1A). In the sample preparation procedure, we added 2 mmol/L ethylenediaminetetraacetic acid (EDTA) to avoid the activation effect of ATPγS by chelating divalent ions (Proks et al., 2010). Therefore, we observed an Mg-independent ATP binding site in the degenerate site on the NBD1 of SUR1. It is possible that in the ATP + GBM structure reported by another group (Martin et al., 2017a), ATP molecule is also bound on the NBD1, but limited local map quality (EMD-7073) precludes any conclusion.

Architecture of the K_{ATP} channel in the Mg-ADP bound state

It is previously reported that KNtp is not necessary for channel opening and activation by Mg-ADP (Reimann, 1999). The fact that our 39AA fusion construct can be reactivated by Mg-ADP (Fig. S1B) suggests this construct can be used for structural study on the Mg-ADP activation mechanism. To trap SUR1 in an NBD-dimerized conformation, we supplemented the K_{ATP} fusion protein with Mg-ADP, the endogenous activator of the K_{ATP} channel. We also added NN414, a SUR1 specific KCO; VO_4^{3-}, an ion that mimics the post-hydrolytic form of ATP γ-phosphate; and 08:0 PI(4,5)P_2 lipid, a soluble PIP_2 analogue which is required for K_{ATP} channel activation. The structure of K_{ATP} channel in the ADP bound state was solved to an overall resolution of 4.2 Å (Fig. S5 and Table S3). We observed similar "T state" and "R state" conformational heterogeneity of the Kir6.2 CTD and SUR1 ABC transporter module as in the ATP + GBM state or the ATP state. The map quality of specific regions was dramatically improved by focused classification and refinement (Fig. S5 and Table S3).

The overall architecture of K_{ATP} in the ADP bound state is also similar to the ATP + GBM state or the ATP state (Fig. 3A and 3B). Four Kir6.2 channel subunits are located at the center, and four SUR1 subunits are in the periphery. The SUR1 ABC transporter module connects to the Kir6.2 via SUR1 TMD0-L0 fragment. This is in great contrast to the "quatrefoil form" structure of Mg-ATP and Mg-ADP (Lee et al., 2017) which has a disordered lasso motif and a large rotation of the SUR1 transporter module. The "quatrefoil form" might represent a "decoupled" state, in which the conformational change of SUR1 cannot be transferred to Kir6.2 by either the TMD0-L0 or the N-terminus of Kir6.2. Our overall structure is similar to the coupled "propeller form" (Lee et al., 2017), which has an ordered lasso motif, despite some positional shift of SUR1 subunits (Fig. S6A and S6B). The large difference between the "quatrefoil form" and the "propeller form" might arise from different sample preparation procedures, because detergents were replaced by amphipols in their experiments (Lee et al., 2017). Similarly to the ATP or ATP + GBM state, the CTD of Kir6.2 in "T state" has a clockwise rotation (~11.8°) relative to that of "R state" (Fig. 3C). The two nucleotide binding sites on SUR1 are occupied by Mg-ADP, which results in an NBD-dimerized conformation (Fig. 3A and 3B). Moreover, the ATP binding sites of Kir6.2 are also occupied by ADP molecules, and the Kir6.2 channel is in the closed conformation, similarly to the ATP + GBM state (Figs. 3A, 3B and S6C), which is consistent with the result that ADP can inhibit the K_{ATP} channel albeit at a lower affinity (Hopkins et al., 1992; Schwanstecher et al., 1994a). The ADP binding mode at the inhibitory site of Kir6.2 is similar to that of ATP (Martin et al., 2017a) or ATPγS (Fig. S6C). Because ADP lacks the γ phosphate, there is no direct interaction between ADP and the γ-phosphate-coordinating residue R50 (Martin et al., 2017b). This explains why ADP binds Kir6.2 with a lower affinity and the R50 mutation shifts the IC_{50} curve to a similar range of ADPβS on wild-type channels (Schwanstecher et al., 1994a;

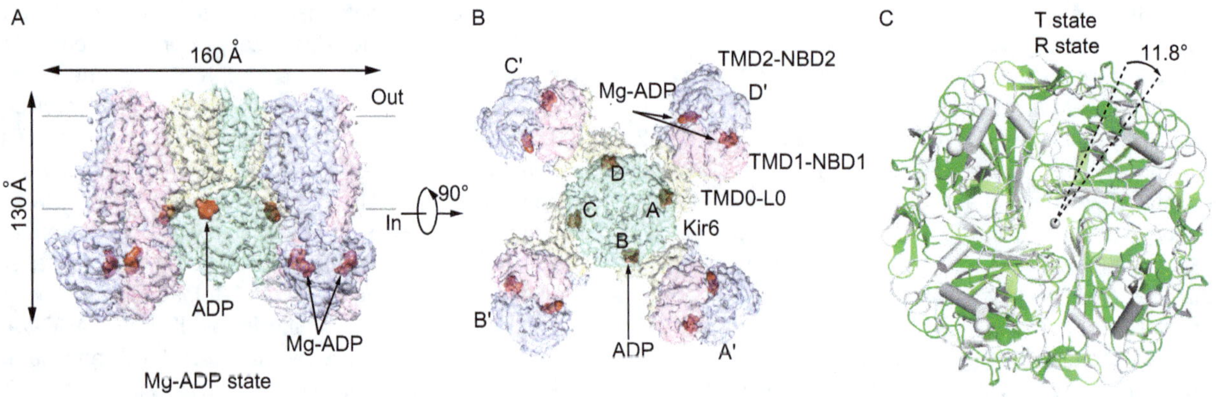

Figure 3. Structure of K_{ATP} in the Mg-ADP state. (A) Side view of the K_{ATP} structure in the Mg-ADP state, colored as in Fig. 1A. ADP and Mg-ADP densities are colored in red. (B) Bottom view of the K_{ATP} channel in the Mg-ADP state. (C) Bottom view of the aligned Kir6.2 structures between "T state" (green) and "R state" (gray) of the Mg-ADP state. Rotation angle between CTDs is measured similarly as in Fig. 1E.

Shimomura et al., 2006). We did not observe any density at the PIP_2 binding site, probably due to the low affinity of the 08:0 $PI(4,5)P_2$ to the purified protein and also the negative cooperativity of inhibitory nucleotide binding and PIP_2 binding on Kir6.2 (Baukrowitz et al., 1998; Hilgemann and Ball, 1996; Shyng and Nichols, 1998). This is in accordance with the fact that we observed a closed state Kir6.2 structure in the Mg-ADP state, because PIP_2 is necessary for channel opening (Baukrowitz et al., 1998; Shyng and Nichols, 1998).

Asymmetric NBD dimer of SUR1 in the Mg-ADP bound state

In our focused refined map of SUR1 in the Mg-ADP state, we observed the densities of Mg-ADP molecules in both the degenerate and consensus sites (Fig. 4A–C). We did not observe densities of VO_4^{3-}, correlating with the fact that VO_4^{3-} do not markedly affect SUR1 activity (Fig. S1B) (Proks, 1999; Shyng et al., 1997; Ueda et al., 1997). The Mg-ADP molecule in the degenerate site interacts with residues from both NBDs to induce a full closure of the degenerate site (Fig. 4C and 4D). In contrast, the Mg-ADP molecule in the consensus site primarily interacts with NBD2, and the consensus site is open (Fig. 4C and 4E). We use the distance between the Cα atoms of the conserved glycine on the ABC Walker A motif and the serine residue (a cysteine on SUR1 degenerate site) on the signature motif as an indicator of NBS closure. The distance between Cα of G1485 and C717 in the degenerate site is 11.5 Å, while the distance between Cα of G833 and S1383 in the consensus site is 13.0 Å (Fig. 4F). These distances are the opposites of the NBD-dimerized state structure of CFTR with Mg-ATP bound (PDB: 5W81) (Zhang et al., 2017). For CFTR, the degenerate site is partially open, and the consensus site is fully closed. The distance between G1350 and S461 in the degenerate site is 12.7 Å, while the distance between G550 and S1249 in the consensus site is 10.6 Å (Fig. 4G). The Mg-ADP state structure has the same asymmetric NBD-dimerized conformation as the Mg-ATP&ADP state structure reported by another group (PDB: 6C3P EMD-7339) (Lee et al., 2017). In particular, Mg-ADP molecule induces similar full closure of degenerate site compared to the Mg-ATP molecules (Fig. S6D).

Conformational changes of SUR1 ABC transporter module upon Mg-ADP binding

By aligning the TMD0 domain of SUR1 between the ADP and ATP + GBM states, we found a large conformational change of the SUR1 ABC transporter module (Fig. 5A). The full closure of the degenerated site induced by Mg-ADP binding leads to the asymmetric dimerization of two NBDs (Fig. 5B). These conformational changes are further conveyed to the transmembrane domain via coupling helices that interact with NBDs to drive two halves of the transmembrane domain of SUR1 ABC transporter module to move closer (Fig. 5C). Specifically, M14 and M17 move toward M8 and M11 (Fig. 5C). As a result, SUR1 central cavity shrinks, and the GBM and KNtp sites are disrupted. We added NN414, a KCO, to our sample (Fig. S7A). We observed two positive densities in SUR1. One is close to the top of SUR1, surrounded by M10-11 from TMD1, and M12 and M17 from TMD2 (Fig. S7B). The other is at the central cavity surrounded by M8 from TMD1, and M15–17 from TMD2 (Fig. S7C). Both of the densities are absence in the Mg-ATP&ADP state map, where NN414 was not added into the sample (EMD-7338, PDB: 6C3O) (Fig. S7B and S7C). By comparing the structures of transmembrane domain with (Mg-ADP state) and without NN414 (Mg-ATP&ADP state), the ligand corresponding to density 1 introduces local structural changes of M9-11 (Fig. S7D). But whether these densities represent NN414 needs further studies. Notably, we did not observe any density that might account for KNtp. Both our Mg-ADP state structure reported here and the Mg-ATP&ADP state structure reported by another group (Lee et al., 2017) are captured using the fusion construct, where the authentic KNtp is not present. It is possible that the KNtp might adopt an unknown conformation that is different from the one observed in the ATP + GBM state but compatible with our current Mg-ADP state structure.

Nucleotide binding induced NBD dimerization of SUR1

In the ATP and ATP + GBM state structures, SUR1 is in an inward-facing conformation. We observed ATP density only in the NBD1 degenerate site but not in the NBD2 consensus site, which suggests that the degenerate site has a higher affinity for Mg-free ATP, and ATP alone is not able to drive the dimerization of SUR1 NBDs. This correlates with previous studies that NBD1 of SUR1 can bind 8-azido-ATP with high affinity in an Mg-independent manner (Matsuo et al., 1999a; Matsuo et al., 1999b; Ueda et al., 1997). In addition, the subsequent binding of ATP to NBD1 can be strongly antagonized by prebound Mg-ADP (Ueda et al., 1997), indicating that the degenerate site of NBD1 can be occupied by Mg-ADP as well. Indeed, in our Mg-ADP state structure, we observed Mg-ADP molecules in both the NBD1 degenerate site and the NBD2 consensus site.

The asymmetric NBD dimer conformation observed in the Mg-ADP state structure suggests that the Mg-ADP molecule in the closed degenerate site directly mediates the formation of the NBD dimer interface, because this Mg-ADP molecule physically interacts with both NBDs. In contrast, the Mg-ADP molecule in the open consensus site interacts only with NBD2 but not with NBD1, indicating that it does not contribute directly to NBD dimer formation. By comparing the structures of individual NBD monomer in different states, we found a noticeable conformational change in NBD2 but not in NBD1 (Fig. 5D). Induced conformational changes of the NBD by nucleotide binding were previously observed in the crystal structures of isolated NBD of ABC transporter MJ1267 (Karpowich et al., 2001) and in the cryo-EM structure of CFTR (Zhang and Chen, 2016; Zhang et al., 2017)

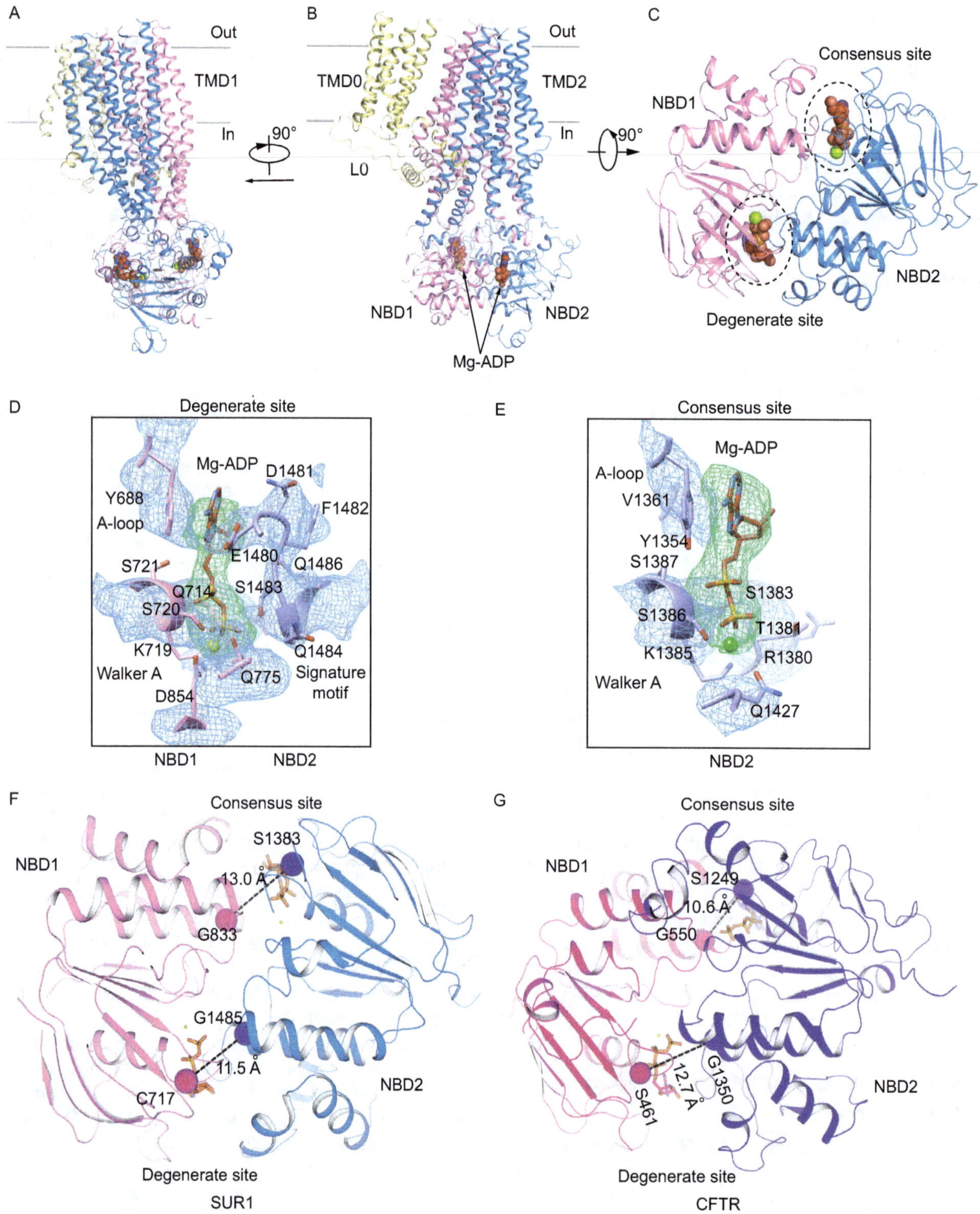

Figure 4. Mg-ADP on SUR1. (A) Side view of a single SUR1 subunit in cartoon representation. Color scheme is the same as in Fig. 1A. Mg^{2+} and ADP are shown as green and red spheres, respectively. (B) Side view of a single SUR1 subunit with a 90° rotation relative to panel (A). (C) Bottom view of a single SUR1 subunit with a 90° rotation relative to (B). For simplicity, only NBD1 and NBD2 are shown. Two Mg-ADP molecules are circled. (D and E) EM densities at the degenerate site (D) and consensus site (E), colored as in Fig. 2H. Mg^{2+} is shown as a green sphere. (F and G) Differences of NBD closures in SUR1 (F) and CFTR (PDB: 5W81) (G). Cα distances between the glycine in the Walker A motif and serine (cysteine instead at the degenerate site of SUR1) in the ABC signature motif are shown as dashed lines.

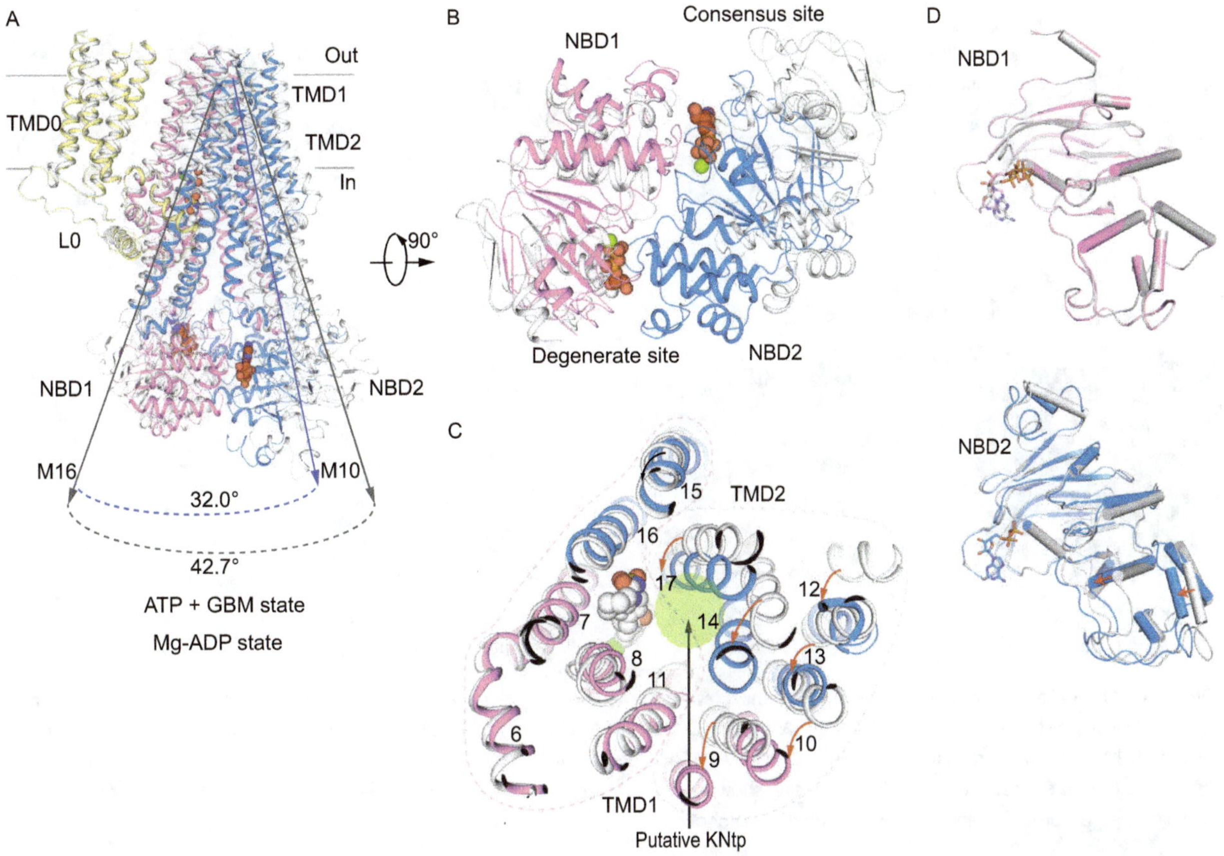

Figure 5. Conformational changes of SUR1. (A) Superposition of the Mg-ADP bound structure (colored) and the ATP + GBM bound structure (gray) by aligning the TMD0 domain of SUR1. Angles between helices M10 and M16 in the two structures are labeled. Mg-ADP, ATPγS and GBM molecules are all shown as spheres. (B) Bottom view of NBDs of the K_{ATP} channel with a 90° rotation relative to (A). (C) Bottom view of TMDs of the K_{ATP} channel with a 90° rotation relative to (A). Positional changes of transmembrane helices from the first half of TMDs (M6–10, 12–14 and 17) are indicated by red arrows. The dashed lines specify the two halves of TMDs. The putative KNtp binding site is shown as a green circle, showing the displacement of GBM-binding sites upon nucleotide binding. (D) Structural comparisons of individual NBD1 (top) or NBD2 (bottom) between the ATP + GBM state (gray) and the Mg-ADP state (colored). Conformational changes of α helices-rich subdomain in NBD2 are indicated by red arrows. NBD structures are aligned according to β sheet-rich subdomain.

and also were sampled in MD-simulations (Jones and George, 2017). The conformational changes of NBD2 observed here might partially be due to the Mg-ADP binding and full closure of the degenerate site, but we suggest that the Mg-ADP binding on the NBD2 consensus site plays a more prominent role. This is supported by the previous radioactive 8-azido-ATP photo cross-linking results. First, the NBD2 consensus site mutation K1385M greatly reduces the ability of prebound Mg-ADP to prevent further ATP binding on NBD1, when Mg-ADP and ATP were added to SUR1 sequentially (Ueda et al., 1997; Ueda et al., 1999), suggesting that Mg-ADP acts on the NBD2 consensus site to allosterically increase the affinity of Mg-ADP on the NBD1 degenerate site. Second, Mg-ADP or Mg-ATP binding on the NBD2 consensus site can stabilize prebound ATP on the NBD1 degenerate site (Ueda et al., 1999), indicating that nucleotide binding on the NBD2 consensus site can send a signal to the degenerate site to increase its affinity for nucleotides, probably via the conformational change of NBD2, as we observed here.

Model for SUR1 conformational change

On the basis of experiments and structures of K_{ATP} channels in complex with different ligand combinations solved with different constructs, we propose the following hypothetic model for the conformational change of SUR1 subunit (Fig. 6A–H). When GBM is bound, GBM and the Kir6.2 N-terminus cooperatively bind inside SUR1 to inhibit channel gating (Fig. 6A and 6E). Without GBM, the Kir6.2 N-terminus can still bind to the central cavity of SUR1 to inhibit channel gating but with decreased affinity. In the presence of ATP, ATP can occupy the NBD1 degenerate site, but the NBDs are still separated (Fig. 6B and 6F). When the ADP

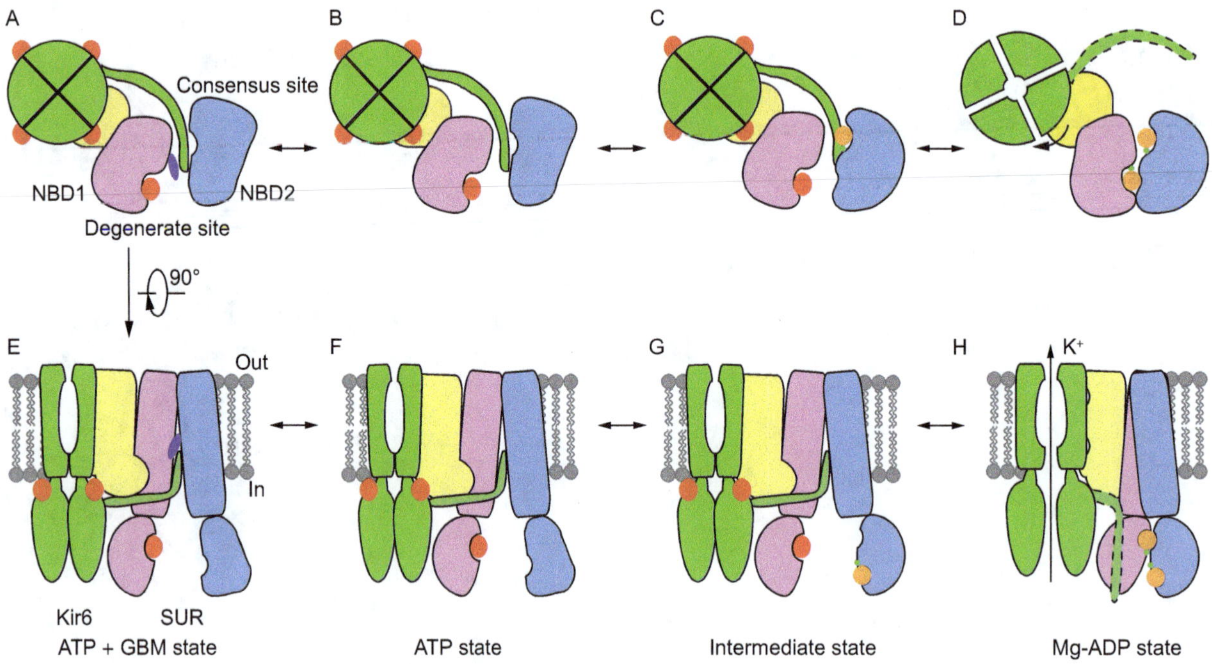

Figure 6. Model for nucleotide-induced NBD dimerization of the K_{ATP} channel. (A–D) Bottom view of SUR1 cartoon models from cytosolic side. For simplicity, only CTDs of Kir6.2 and TMD0-L0 and NBDs from one SUR1 subunit are shown and colored the same as in Fig. 1A. KNtp with an unknown conformation in Mg-ADP state is outlined by dashed lines. ATP and ADP are shown as red and orange big spheres, respectively. Mg^{2+} ions are shown as green small spheres. (E–H) Side view of SUR1 cartoon models. For simplicity, a pair of Kir6.2 and one SUR1 subunit are shown.

concentration increases, Mg-ADP first binds to the NBD2 consensus site to induce a conformational change of NBD2 and enhance the affinity of the NBD1 degenerate site for Mg-ADP or Mg-ATP (Fig. 6C and 6G). Then, prebound ATP together with Mg ion or newly bound Mg-ADP bridges two halves of the degenerate site to induce its full closure and NBD dimerization. Finally, there is a global conformational change of SUR1, which allosterically activates the Kir6.2 channel and occludes the GBM site of SUR1, and Kir6.2 N-terminus relocates accordingly (Fig. 6D and 6H). Mg-ATP binding at the NBD2 consensus site can have similar functions as Mg-ADP, albeit with lower affinity (Vedovato et al., 2015). Our model suggests that the essential elements for asymmetric NBD dimerization are the degenerate site of both NBDs and the consensus site of NBD2. Indeed, previous mutation results support this model. Mutations of K719A and D854N on the degenerate site of NBD1; G1479D, G1479R, G1485D, G1485R and Q1486H on the degenerate site of NBD2; and K1385M, D1506N, and D1506A on the consensus site of NBD2 (Gribble et al., 1997; Nichols et al., 1996; Shyng et al., 1997) can impair Mg-ADP activation, whereas mutations of G827D, G827R, and Q834H on the consensus site of NBD1 can preserve Mg-ADP activation, although with altered kinetics (Nichols et al., 1996; Shyng et al., 1997). Our model also suggests that ATP hydrolysis is not required for channel activation, which is consistent with functional studies (Choi et al., 2008; Ortiz et al., 2013).

MATERIALS AND METHODS

Cell culture

Sf9 cells (Thermo Fisher Scientific) were cultured in Sf-900 III SFM medium (Thermo Fisher Scientific) at 27 °C. FreeStyle 293-F (Thermo Fisher Scientific) suspension cells were cultured in Freestyle 293 medium (Thermo Fisher Scientific) supplemented with 1% FBS at 37 °C with 6% CO_2 and 70% humidity.

Screening of fusion constructs

Fusion constructs, SUR1-Kir6.2, were created by linking the C-terminus of SUR1 from *Mesocricetus auratus* (Aguilar-Bryan et al., 1995) to the N-terminus of Kir6.2 (Woo et al., 2013) from *Mus musculus* using flexible linkers with variable lengths. The fusion constructs were cloned into a modified C terminal GFP-tagged BacMam expression vector which contains two Strep tags after GFP as described previously (Li et al., 2017). SUR1-Kir6.2 fusion constructs were screened by transient transfection of FreeStyle 293-F cells (Thermo Fisher Scientific) cultured in SMM-293TI medium (Sino Biological Inc.) using polyethylenimine (Polyscience). Cells were harvested 60 h post-transfection and solubilized in 20 mmol/L HEPES pH8.0 + 120 mmol/L KCl + 1% digitonin (Biosynth) at 25 °C. Cell lysates were cleared by centrifuge at 40,000 rpm for 30 min and supernatants were loaded onto Superose 6 increase column (GE Healthcare) for analysis (Kawate and Gouaux, 2006). Based on expression level and SUR1-Kir6.2 tetramer assembly, SUR1-Kir6.2 fusion with a 39-residue linker (VDGSGSGSGSAAGSGSGSGSGSGAAGSGSGSGSGSGAAA) was selected and expressed in BacMam system for further structural characterization (Goehring et al., 2014).

Electrophysiology

K_{ATP} constructs were transfected into FreeStyle 293-F cells at density of 1 × 10^6 cells/mL using polyethylenimine and cells were cultured 36–48 h before recording. Patch electrodes were pulled from a horizontal microelectrode puller (P-1000, Sutter Instrument Co, USA) to a tip resistance of 3.0–5.0 MΩ. The pipette solution contained (mmol/L): 140 KCl, 1.2 $MgCl_2$, 2.6 $CaCl_2$, 10 HEPES (pH 7.4, NaOH) and the bath solution contained (mmol/L): 140 KCl, 10 EGTA, 1 $MgCl_2$, 10 HEPES (pH 7.4, NaOH). Macroscopic currents were recorded using inside-out mode at +60 mV through an Axopatch 200B amplifier (Axon Instruments, USA). Recording were started when currents were stable after patch excision. Signals were acquired at 20 kHz and lowpass filtered at 5 kHz. Data was further analyzed with pclampfit 9.0 software and filtered at 1 kHz. For the GBM inhibition assay, steady state currents after GBM treatment were normalized to the currents before GBM treatment. Run-down was not corrected during data analysis.

Protein expression and purification

K_{ATP} channels were expressed in FreeStyle 293-F cells as described previously (Li et al., 2017). For purification, membrane pellets were homogenized in 20 mmol/L HEPES pH 8.0 at 25 °C + 120 mmol/L KCl and then solubilized in 1% digitonin for 30 min at 4 °C. Unsolubilized materials were removed by centrifugation at 40,000 rpm for 50 min in Ti45 rotor (Beckman). Supernatant was loaded onto 5 mL strep-tactin superflow high capacity resin (IBA). Resin was washed by 20 mmol/L HEPES pH 8.0 at 25 °C + 120 mmol/L KCl + 0.1% digitonin and protein was eluted by 20 mmol/L HEPES pH 8.0 at 25 °C + 120 mmol/L KCl + 0.1% digitonin + 1 mmol/L EDTA + 5 mmol/L desthiobiotin. Protein eluted from strep-tactin column was concentrated by 100-kDa cut-off concentrator (Millipore) and loaded onto Superose 6 increase (GE Healthcare) running in 20 mmol/L HEPES pH 8.0 at 25 °C + 120 mmol/L KCl + 0.1% digitonin. Peak fractions that contain SUR1-Kir6.2 fusion protein were combined and concentrated to A_{280} = 8 (estimated protein concentration was 9 μmol/L K_{ATP} proteins). For cryo-EM sample preparation, the proteins were supplemented with 2 mmol/L EDTA, 200 μmol/L glibenclamide (TCI), and 2 mmol/L ATPγS (Sigma) (ATP + GBM state), or 2 mmol/L EDTA and 2 mmol/L ATPγS (Sigma) (ATP state), or 2.5 mmol/L $MgCl_2$, 1 mmol/L ADP (Sigma), 400 μmol/L 08:0 PI (4,5)P_2 (Avanti), 500 μmol/L NN414 (Sigma), and 1 mmol/L Na_3VO_4 (Mg-ADP state). Cryo-EM sample preparation was prepared as described previously (Li et al., 2017).

Cryo-EM data acquisition

Cryo-grids were screened on a Talos (Thermo Fisher Scientific) operated at a voltage of 200 kV with an eagle 4 k × 4 k CCD camera (Thermo Fisher Scientific). Good grids were transferred into a Titan Krios (Thermo Fisher Scientific) operated at 300 kV and images were collected using K2 camera (Gatan) mounted post a quantum energy filter with 20 eV slit. K2 camera was operated under super resolution mode with super resolution pixel size 0.5275 Å (ATP + GBM state and Mg-ADP state) or 0.67 Å (ATP state) at object plane. The defocus value ranged from −1.0 μm to −3.5 μm. Data acquisitions were performed automatically using Leginon software (Suloway et al., 2005) in the movie mode with a dose rate of 3.8 or 4.7 $e^-/s/Å^2$. The total exposure was 50 $e^-/Å^2$ and each 13 or 11 seconds movie was dose-fractioned into 50 frames.

Image processing

Collected movies were gain-corrected, motion-corrected, exposure-filtered and binned with MotionCor2 (Zheng et al., 2017), producing dose-weighted and summed micrographs with pixel size 1.055 Å/pixel (ATP + GBM state and Mg-ADP state) or 1.34 Å/pixel (ATP state). CTF models of dose-weighted micrographs were determined using Gctf (Zhang, 2016). Autopick was done with Gautomatch (developed by Kai Zhang, MRC-LMB) using the projections of our previous map (EMD-6689) as template. Auto-picked particles were extracted from dose-weighted micrographs by binning factor of 2 and subjected to 2D classification using cryoSPARC (Punjani et al., 2017). 3D classification was done using our former K_{ATP} map (EMD-6689) as the initial model with GPU-accelerated Relion 2.0 (Kimanius et al., 2016). Particles from selected 3D classes were re-centered and re-extracted from summed micrographs without binning (1.055 Å/pixel or 1.34 Å/pixel). Re-extracted particles were subjected to 3D auto-refinement with C4 symmetry. Additional focused classification of Kir6.2 CTD with a soft mask but without alignment was used to separate 3D classes with different CTD conformations. To improve the local map quality of SUR1 ABC transporter module (maSUR1 214-1579), alignment parameters of each SUR1 ABC transporter module particles were obtained from global refined particles by symmetry expansion using relion_particle_symmetry_expand–sym C4 (Zhou et al., 2015). Then the signals outside SUR1 ABC transporter module were subtracted from each particles using particle subtraction function in Relion 2.0 as described previously (Bai et al., 2015). The signal subtracted particles were used for further 3D refinement using local search by restricting sigma_ang to 5°, using the map of SUR1 ABC transporter module as the starting model. All of the resolution estimations were based on gold standard FSC 0.143 after correction for mask effects (Chen et al., 2013). The final map was sharpened with B factor automatically determined by Relion 2.0 (Kimanius et al., 2016).

Model building

Individual domain of previous K_{ATP} models (PDB: 5WUA and 5TWV) and CFTR in Mg-ATP bound state (PDB: 5W81) were used as the initial model, docked into cryo-EM map with best local quality of that region by chimera and manually rebuilt in Coot according to the density (Emsley et al., 2010). The model was further refined by phenix (Adams et al., 2010). Specifically, models of Kir6.2 transmembrane domain and SUR1 TMD0 were refined against the focused refined map of transmembrane domain in the ATP + GBM state. Models of SUR1 ABC transporter module were refined against the focused refined map in respective states. Models of Kir6.2 CTD were refined against maps after focused classification and refinement of Kir6.2 CTD in the ATP + GBM state. The individually refined models were then docked into the maps of the full K_{ATP} channel and refined respectively. Figures were prepared with Pymol (Schrödinger, LLC.) or Chimera (Pettersen et al., 2004).

Quantification and statistical analysis

Resolution estimations of cryo-EM density maps are based on the 0.143 Fourier Shell Correlation (FSC) criterion (Chen et al., 2013).

ACCESSION CODES

The 3D cryo-EM density maps have been deposited in the Electron Microscopy Data Bank under the accession number EMD: EMD-6831, EMD-6832, and EMD-6833 for the ATP + GBM state of K_{ATP}, EMD-6848, EMD-6849, and EMD-6850 for the ATP state of K_{ATP}, and EMD-6852, EMD-6851, and EMD-6853 for the Mg-ADP state of K_{ATP}. Coordinates of the K_{ATP} structure have been deposited in the Protein Data Bank under the accession number PDB: 5YKE, 5YKF, 5YKG, 5YW7 for the ATP + GBM state of K_{ATP}, 5YW8, 5YW9, and 5YWA for the ATP state of K_{ATP}, and 5YWC, 5YWB, and 5YWD for the Mg-ADP state of K_{ATP}.

ACKNOWLEDGMENTS

We thank all of Chen Lab members for kindly help. We also thank Joseph Bryan for sharing maSUR1 cDNA and J. Marc Simard for sharing mmKir6.2 cDNA. Cryo-EM data collection was supported by electron microscopy laboratory and Cryo-EM platform of Peking University with the assistance of Xuemei Li and Daqi Yu, the National Center for Protein Science (Shanghai) with assistance of Liangliang Kong and Zhenglin Fu, Center for Biological Imaging, Institute of Biophysics, Chinese Academy of Sciences with assistance of Zhenxi Guo. Part of structural computation was also performed on the Computing Platform of the Center for Life Science and High-performance Computing Platform of Peking University.

The work is supported by grants from the Ministry of Science and Technology of China (National Key R&D Program of China, 2016YFA0502004 to Lei Chen) and National Natural Science Foundation of China (Grant Nos. 31622021 and 31521062 to Lei Chen) and Young Thousand Talents Program of China to Lei Chen and the China Postdoctoral Science Foundation (2016M600856 and 2017T100014 to Jing-Xiang Wu). Jing-Xiang Wu is supported by the postdoctoral foundation of the Peking-Tsinghua Center for Life Sciences, Peking University.

ABBREVIATIONS

ABC, ATP-binding cassettes; cryo-EM, cryo-electron microscopy; GBM, glibenclamide; K_{ATP}, ATP-sensitive potassium channel; Kir6, inward-rectifying potassium channel 6; NBD, nucleotide binding domain; SUR, sulfonylurea receptor; TMD0-L0, transmembrane domain 0-loop 0

COMPLIANCE WITH ETHICS GUIDELINES

Jing-Xiang Wu, Dian Ding, Mengmeng Wang, Yunlu Kang, Xin Zeng and Lei Chen declare that they have no conflict of interest. This article does not contain any studies with human or animal subjects performed by the any of the authors.

AUTHOR CONTRIBUTIONS

Lei Chen initiated the project. Jiang-Xiang Wu, Dian Ding, Mengmeng Wang, Yunlu Kang, Xin Zeng and Lei Chen designed experiments. Jiang-Xiang Wu prepared the cryo-EM sample. Jiang-Xiang Wu, Mengmeng Wang, Dian Ding, Yunlu Kang and Lei Chen collected EM data. Jiang-Xiang Wu and Lei Chen performed image processing and analyzed EM data. Dian Ding performed electrophysiology experiments. Xin Zeng performed the HPLC and mass spectrum analysis. Lei Chen and Jiang-Xiang Wu built the model, prepared the figures and wrote the manuscript draft. All authors contributed to manuscript preparation.

REFERENCES

Adams PD et al (2010) PHENIX: a comprehensive Python-based system for macromolecular structure solution. Acta Crystallogr D Biol Crystallogr 66:213–221

Aguilar-Bryan L et al (1990) Photoaffinity labeling and partial purification of the beta cell sulfonylurea receptor using a novel, biologically active glyburide analog. J Biol Chem 265:8218–8224

Aguilar-Bryan L et al (1995) Cloning of the beta cell high-affinity sulfonylurea receptor: a regulator of insulin secretion. Science 268:423–426

Aittoniemi J et al (2009) Review. SUR1: a unique ATP-binding cassette protein that functions as an ion channel regulator. Philos Trans R Soc Lond B Biol Sci 364:257–267

Ashcroft FM et al (2017) Neonatal diabetes and the KATP channel: from mutation to therapy. Trends Endocrinol Metab 28:377–387

Babenko AP, Bryan J (2002) SUR-dependent modulation of KATP channels by an N-terminal KIR6.2 peptide. Defining intersubunit gating interactions. J Biol Chem 277:43997–44004

Babenko AP et al (1999) The N-terminus of KIR6.2 limits spontaneous bursting and modulates the ATP-inhibition of KATP channels. Biochem Biophys Res Commun 255:231–238

Bai XC et al (2015) Sampling the conformational space of the catalytic subunit of human gamma-secretase. Elife. https://doi.org/10.7554/eLife.11182

Baukrowitz T et al (1998) PIP2 and PIP as determinants for ATP inhibition of KATP channels. Science 282:1141–1144

Bryan J et al (2005) Insulin secretagogues, sulfonylurea receptors and K(ATP) channels. Curr Pharm Des 11:2699–2716

Carr RD et al (2003) NN414, a SUR1/Kir6.2-selective potassium channel opener, reduces blood glucose and improves glucose tolerance in the VDF Zucker rat. Diabetes 52:2513–2518

Chen S et al (2013) High-resolution noise substitution to measure overfitting and validate resolution in 3D structure determination by single particle electron cryomicroscopy. Ultramicroscopy 135:24–35

Choi KH et al (2008) Testing for violations of microscopic reversibility in ATP-sensitive potassium channel gating. J Phys Chem B 112:10314–10321

Clement JPT et al (1997) Association and stoichiometry of K(ATP) channel subunits. Neuron 18:827–838

Devaraneni PK et al (2015) Structurally distinct ligands rescue biogenesis defects of the KATP channel complex via a converging mechanism. J Biol Chem 290:7980–7991

Emsley P et al (2010) Features and development of coot. Acta Crystallogr D Biol Crystallogr 66:486–501

Flagg TP et al (2010) Muscle KATP channels: recent insights to energy sensing and myoprotection. Physiol Rev 90:799–829

Goehring A et al (2014) Screening and large-scale expression of membrane proteins in mammalian cells for structural studies. Nat Protoc 9:2574–2585

Gribble FM et al (1997) The interaction of nucleotides with the tolbutamide block of cloned ATP-sensitive K^+ channel currents expressed in Xenopus oocytes: a reinterpretation. J Physiol 504 (Pt 1):35–45

Hibino H et al (2010) Inwardly rectifying potassium channels: their structure, function, and physiological roles. Physiol Rev 90:291–366

Hilgemann DW, Ball R (1996) Regulation of cardiac Na^+, Ca^{2+} exchange and KATP potassium channels by PIP2. Science 273:956–959

Hopkins WF et al (1992) Two sites for adenine-nucleotide regulation of ATP-sensitive potassium channels in mouse pancreatic beta-cells and HIT cells. J Membr Biol 129:287–295

Jones PM, George AM (2017) How intrinsic dynamics mediates the allosteric mechanism in the ABC transporter nucleotide binding domain dimer. J Chem Theory Comput 13:1712–1722

Karpowich N et al (2001) Crystal structures of the MJ1267 ATP binding cassette reveal an induced-fit effect at the ATPase active site of an ABC transporter. Structure 9:571–586

Kawate T, Gouaux E (2006) Fluorescence-detection size-exclusion chromatography for precrystallization screening of integral membrane proteins. Structure 14:673–681

Kimanius D et al (2016) Accelerated cryo-EM structure determination with parallelisation using GPUs in RELION-2. Elife. https://doi.org/10.7554/eLife.18722

Koster JC et al (1999) ATP inhibition of KATP channels: control of nucleotide sensitivity by the N-terminal domain of the Kir6.2 subunit. J Physiol 515(Pt 1):19–30

Kuhner P et al (2012) Importance of the Kir6.2 N-terminus for the interaction of glibenclamide and repaglinide with the pancreatic K (ATP) channel. Naunyn Schmiedebergs Arch Pharmacol 385:299–311

Lee KPK et al (2017) Molecular structure of human KATP in complex with ATP and ADP. Elife. https://doi.org/10.7554/eLife.32481

Li N et al (2017) Structure of a pancreatic ATP-sensitive potassium channel. Cell 168:101–110

Locher KP (2016) Mechanistic diversity in ATP-binding cassette (ABC) transporters. Nat Struct Mol Biol 23:487–493

Martin GM et al (2017a) Anti-diabetic drug binding site in a mammalian KATP channel revealed by Cryo-EM. Elife. https://doi.org/10.7554/eLife.31054

Martin GM et al (2017b) Cryo-EM structure of the ATP-sensitive potassium channel illuminates mechanisms of assembly and gating. Elife. https://doi.org/10.7554/eLife.24149

Matsuo M et al (1999a) ATP binding properties of the nucleotide-binding folds of SUR1. J Biol Chem 274:37479–37482

Matsuo M et al (1999b) NEM modification prevents high-affinity ATP binding to the first nucleotide binding fold of the sulphonylurea receptor, SUR1. FEBS Lett 458:292–294

Nichols CG et al (1996) Adenosine diphosphate as an intracellular regulator of insulin secretion. Science 272:1785–1787

Ortiz D et al (2013) Reinterpreting the action of ATP analogs on K (ATP) channels. J Biol Chem 288:18894–18902

Pettersen EF et al (2004) UCSF Chimera—a visualization system for exploratory research and analysis. J Comput Chem 25:1605–1612

Proks P et al (1999) Involvement of the N-terminus of Kir6.2 in the inhibition of the KATP channel by ATP. J Physiol 514(Pt 1):19–25

Proks P et al (2010) Activation of the K(ATP) channel by Mg-nucleotide interaction with SUR1. J Gen Physiol 136:389–405

Punjani A et al (2017) cryoSPARC: algorithms for rapid unsupervised cryo-EM structure determination. Nat Methods 14:290–296

Reimann F et al (1999) Involvement of the n-terminus of Kir6.2 in coupling to the sulphonylurea receptor. J Physiol 518(Pt 2):325–336

Schwanstecher C et al (1994a) Interaction of tolbutamide and cytosolic nucleotides in controlling the ATP-sensitive K^+ channel in mouse beta-cells. Br J Pharmacol 111:302–310

Schwanstecher M et al (1994b) Identification of a 38-kDa high affinity sulfonylurea-binding peptide in insulin-secreting cells and cerebral cortex. J Biol Chem 269:17768–17771

Shimomura K et al (2006) Mutations at the same residue (R50) of Kir6.2 (KCNJ11) that cause neonatal diabetes produce different functional effects. Diabetes 55:1705–1712

Shyng S, Nichols CG (1997) Octameric stoichiometry of the KATP channel complex. J Gen Physiol 110:655–664

Shyng SL, Nichols CG (1998) Membrane phospholipid control of nucleotide sensitivity of KATP channels. Science 282:1138–1141

Shyng S et al (1997) Regulation of KATP channel activity by diazoxide and MgADP. Distinct functions of the two nucleotide binding folds of the sulfonylurea receptor. J Gen Physiol 110:643–654

Suloway C et al (2005) Automated molecular microscopy: the new Leginon system. J Struct Biol 151:41–60

Ueda K et al (1997) MgADP antagonism to Mg^{2+}-independent ATP binding of the sulfonylurea receptor SUR1. J Biol Chem 272:22983–22986

Ueda K et al (1999) Cooperative binding of ATP and MgADP in the sulfonylurea receptor is modulated by glibenclamide. Proc Natl Acad Sci USA 96:1268–1272

Vedovato N et al (2015) The nucleotide-binding sites of SUR1: a mechanistic model. Biophys J 109:2452–2460

Vila-Carriles WH et al (2007) Defining a binding pocket for sulfonylureas in ATP-sensitive potassium channels. FASEB J 21:18–25

Whorton MR, MacKinnon R (2011) Crystal structure of the mammalian GIRK2 K^+ channel and gating regulation by G proteins, PIP2, and sodium. Cell 147:199–208

Whorton MR, MacKinnon R (2013) X-ray structure of the mammalian GIRK2-betagamma G-protein complex. Nature 498:190–197

Woo SK et al (2013) The sulfonylurea receptor 1 (Sur1)-transient receptor potential melastatin 4 (Trpm4) channel. J Biol Chem 288:3655–3667

Zhang K (2016) Gctf: real-time CTF determination and correction. J Struct Biol 193:1–12

Zhang Z, Chen J (2016) Atomic structure of the cystic fibrosis transmembrane conductance regulator. Cell 167(1586–1597):e1589

Zhang Z et al (2017) Conformational changes of CFTR upon phosphorylation and ATP binding. Cell 170(483–491):e488

Zhao Y et al (2015) In vitro inhibition of AKR1Cs by sulphonylureas and the structural basis. Chem Biol Interact 240:310–315

Zheng SQ et al (2017) MotionCor2: anisotropic correction of beam-induced motion for improved cryo-electron microscopy. Nat Methods 14:331–332

Zhou M et al (2015) Atomic structure of the apoptosome: mechanism of cytochrome c- and dATP-mediated activation of Apaf-1. Genes Dev 29:2349–2361

11

4.4 Å Resolution Cryo-EM structure of human mTOR Complex 1

Huirong Yang[1,2,3], Jia Wang[4], Mengjie Liu[1,2,3], Xizi Chen[1,2,3], Min Huang[5], Dan Tan[6], Meng-Qiu Dong[6], Catherine C. L. Wong[5], Jiawei Wang[4✉], Yanhui Xu[1,2,3✉], Hong-Wei Wang[4✉]

1 Fudan University Shanghai Cancer Center, Institute of Biomedical Sciences, Shanghai Medical College of Fudan University, Shanghai 200032, China
2 Key Laboratory of Molecular Medicine, Ministry of Education, Department of Systems Biology for Medicine, School of Basic Medical Sciences, Shanghai Medical College of Fudan University, Shanghai 200032, China
3 State Key Laboratory of Genetic Engineering, Collaborative Innovation Center of Genetics and Development, School of Life Sciences, Fudan University, Shanghai 200433, China
4 Ministry of Education Key Laboratory of Protein Sciences, Tsinghua-Peking Joint Center for Life Sciences, Beijing Advanced Innovation Center for Structural Biology, School of Life Sciences, Tsinghua University, Beijing 100084, China
5 National Center for Protein Science, Shanghai Institute of Biochemistry and Cell Biology, Shanghai Institutes of Biological Sciences, Chinese Academy of Sciences, Shanghai 200031, China
6 National Institute of Biological Sciences, Beijing 102206, China
✉ Correspondence: jwwang@tsinghua.edu.cn (J. Wang), xuyh@fudan.edu.cn (Y. Xu), hongweiwang@tsinghua.edu.cn (H.-W. Wang)

ABSTRACT

Mechanistic target of rapamycin (mTOR) complex 1 (mTORC1) integrates signals from growth factors, cellular energy levels, stress and amino acids to control cell growth and proliferation through regulating translation, autophagy and metabolism. Here we determined the cryo-electron microscopy structure of human mTORC1 at 4.4 Å resolution. The mTORC1 comprises a dimer of heterotrimer (mTOR-Raptor-mLST8) mediated by the mTOR protein. The complex adopts a hollow rhomboid shape with 2-fold symmetry. Notably, mTORC1 shows intrinsic conformational dynamics. Within the complex, the conserved N-terminal caspase-like domain of Raptor faces toward the catalytic cavity of the kinase domain of mTOR. Raptor shows no caspase activity and therefore may bind to TOS motif for substrate recognition. Structural analysis indicates that FKBP12-Rapamycin may generate steric hindrance for substrate entry to the catalytic cavity of mTORC1. The structure provides a basis to understand the assembly of mTORC1 and a framework to characterize the regulatory mechanism of mTORC1 pathway.

KEYWORDS mTORC1, structure, cryo-electron microscopy

INTRODUCTION

Mechanistic target of rapamycin (mTOR) is a Ser/Thr kinase that belongs to the family of phosphoinositide-3-kinase-related kinases (PIKK) and is structurally and functionally conserved from yeast to mammals. mTOR exists in two distinct protein complexes: mTOR complex 1 (mTORC1) and mTOR complex 2 (mTORC2), which share two core components, the mTOR protein and the mammalian lethal with SEC13 protein 8 (mLST8, also known as GβL). mTORC1 contains a unique subunit, regulatory-associated protein of mTOR (Raptor), whereas mTORC2 is defined by rapamycin insensitive companion of mTOR (Rictor). Rapamycin inhibits mTORC1 by forming a complex with immunophilin FKBP12 (12 kDa FK506-binding protein) (Loewith et al., 2002; Sarbassov et al., 2004).

In response to multiple growth factors, energy status, and stress pathways, Tuberous Sclerosis Complex 1/2 (TSC1/2)

Huirong Yang, Jia Wang, Mengjie Liu, Xizi Chen have contributed equally to this work.

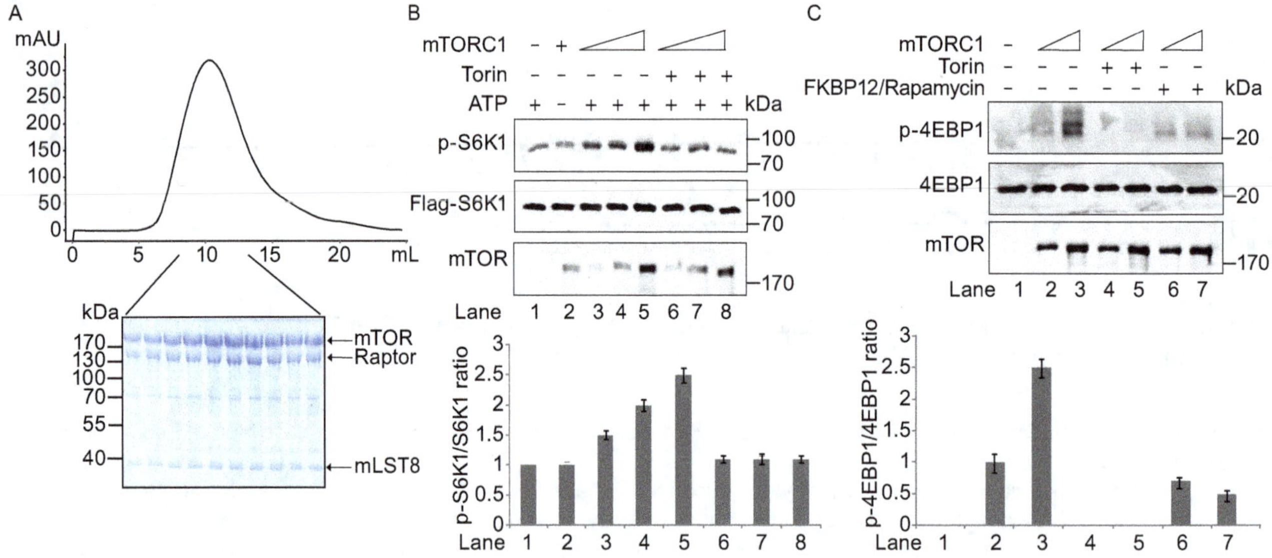

Figure 1. Size exclusion chromatogram of the human mTORC1 and kinase activity. (A) The gel-filtration was performed using a Superose 6 column (10/3004 GL, GE Healthcare). The peak fractions were subjected to SDS-PAGE and stained with Coomassie blue. (B and C) Phosphorylation of purified S6K1 (K100R) (B) and 4EBP1 (C) by mTORC1 in the presence or absence of Torin. The phosphorylation was detected by immunoblotting with antibodies targeting phospho-Thr-389 (top), Flag (middle), and mTOR (bottom) in (B), and antibodies targeting phospho-4EBP1 (top), 4EBP1 (middle), and mTOR (bottom) in (C). Below are the quantification of the immunoblots for B and C.

complex serves as a negative regulator of mTORC1 and functions as a GTPase-activating protein (GAP) to inactivate the small GTPase Ras homolog Rheb (Garami et al., 2003; Inoki et al., 2003; Tee et al., 2003), which binds to and activates mTORC1. Nutrients promote the association of Raptor and Rag GTPases, which recruits mTORC1 to the surfaces of lysosomes and late endosomes for the activation by Rheb (Kim et al., 2008). Deregulation of mTORC1 has been found in many human diseases, especially in cancers (Dazert and Hall, 2011; Inoki et al., 2005; Tee and Blenis, 2005) and mTORC1 inhibitors have been clinically used for the treatment of organ transplantation and solid tumors (Benjamin et al., 2011). mTORC1 regulates cell growth primarily by phosphorylating a large number of proteins, including the eukaryotic initiation factor 4E (eIF4E) binding protein 1 (4EBP1) and p70-S6 Kinase 1 (S6K1) (Gingras et al., 1999; Holz et al., 2005; Holz and Blenis, 2005).

mTORC1 is one of the most important regulators to control cell growth and proliferation. The cellular function and dynamic regulation of mTORC1 have been extensively studied during the past decades. In contrast, the three-dimensional structure of mTORC1 remains largely unknown due to the technical difficulties in preparing the complex to homogeneity, as well as structural determination. The low-resolution (26 Å) cryo-electron microscopy (cryo-EM) structure of mTORC1 shows a two-fold symmetric dimer of complex formation (Yip et al., 2010). The crystal structure of the mTOR protein (deletion of N-terminal 1375 residues) in complex with mLST8 at 3.2 Å resolution was reported (Yang et al., 2013). The structure shows that the kinase domain adopts a canonical protein kinase conformation and provides a model for the inhibition of mTORC1 by Rapamycin-FKBP12. Recently, the cryo-EM structures of mTORC1 at 5.9 Å resolution (Aylett et al., 2016) and Tor-Lst8 from the thermotolerant yeast *Kluyveromyces marxianus* at 6 Å (Baretic et al., 2016) were reported, respectively. However, controversial conclusions were made for the topology of mTOR according to these studies. Here we report the cryo-EM structure of mTORC1 at 4.4 Å resolution. The higher resolution structure and biochemical analyses together provide a topological interpretation of human mTOR and relatively more accurate model for understanding the assembly and function of mTORC1.

RESULTS

mTORC1 protein purification

To determine the cryo-EM structure of mTORC1, we purified the active ternary complex to homogeneity by the following procedures. The mTORC1 was transiently expressed with myc-mTOR, Flag-Raptor, Flag-mLST8 co-transfected into HEK293F cells in suspension culture. The complex was purified over an anti-Flag affinity resin, followed by ion exchange and gel filtration (Fig. 1A). The purified mTORC1 consists of mTOR, Raptor, and mLST8 in stoichiometry and exhibits kinase activity on S6K1 and 4EBP1, which can be

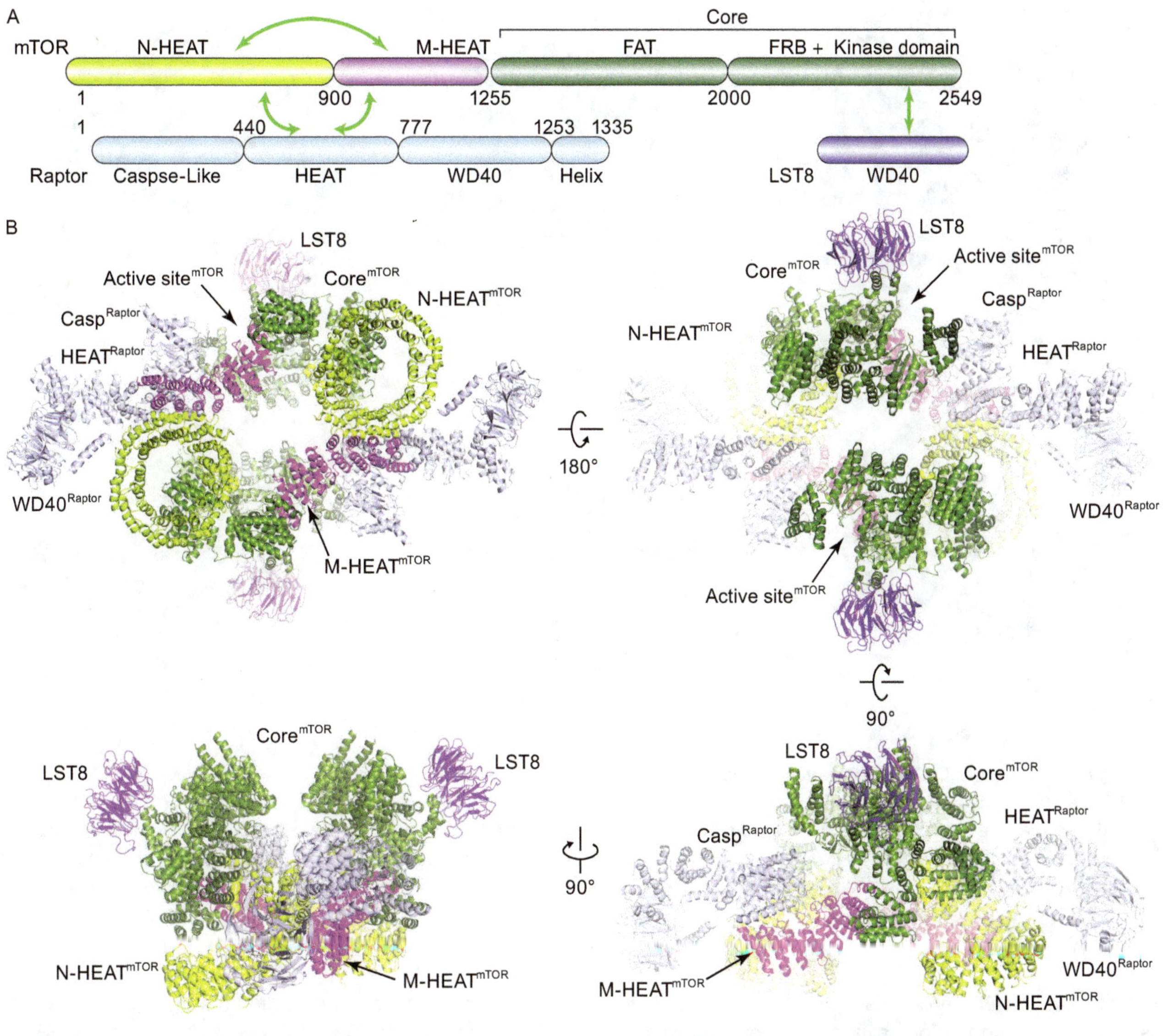

Figure 2. Overall structure of human mTORC1. (A) Colored coded domain architecture of the three essential components of human mTORC1. The same color scheme is used in all structure figures. The inter- and intramolecular interactions are indicated as arrows. B) Ribbon representation of mTORC1 structure in four different views. The proteins and domains are indicated. Casp^{Raptor} represents Caspase-like domain of Raptor.

inhibited by Torin1, a well-characterized ATP-competitive inhibitor of mTOR, and FKBP12-Rapamycin (Fig. 1B and 1C).

The mTORC1 was embedded in vitreous ice as monodispersed particles with multiple orientations, allowing us to perform single particle cryo-EM 3D reconstruction (Fig. S1A–C). Using a Titan Krios electron microscope equipped with a direct electron detector, we collected about 500,000 particle images of the mTORC1 and reconstructed the complex at an average resolution of 4.4 Å and the central portion was further refined to 4.0 Å resolution using a local mask (Methods, Supplementary information Fig. S1D–G, Table S1, and Movie 1). The Cryo-EM map showed well-defined secondary structural elements in the complex (Fig. S2), therefore allowing us to build the atomic models of the Raptor and mLST8 with unambiguous topology (Methods) except for the mTOR protein (see below).

The overall structure of mTORC1

The mTORC1 architecture reveals a hollow rhomboid shape in the front view with overall dimensions of ~280 × 210 × 130 (Å^3) (Fig. 2). The complex adopts a 2-fold symmetry with the central core primarily formed by two mTOR molecules. mLST8 is located on the distal convex along the short axis and protrudes out of the rhomboid architecture. Each Raptor

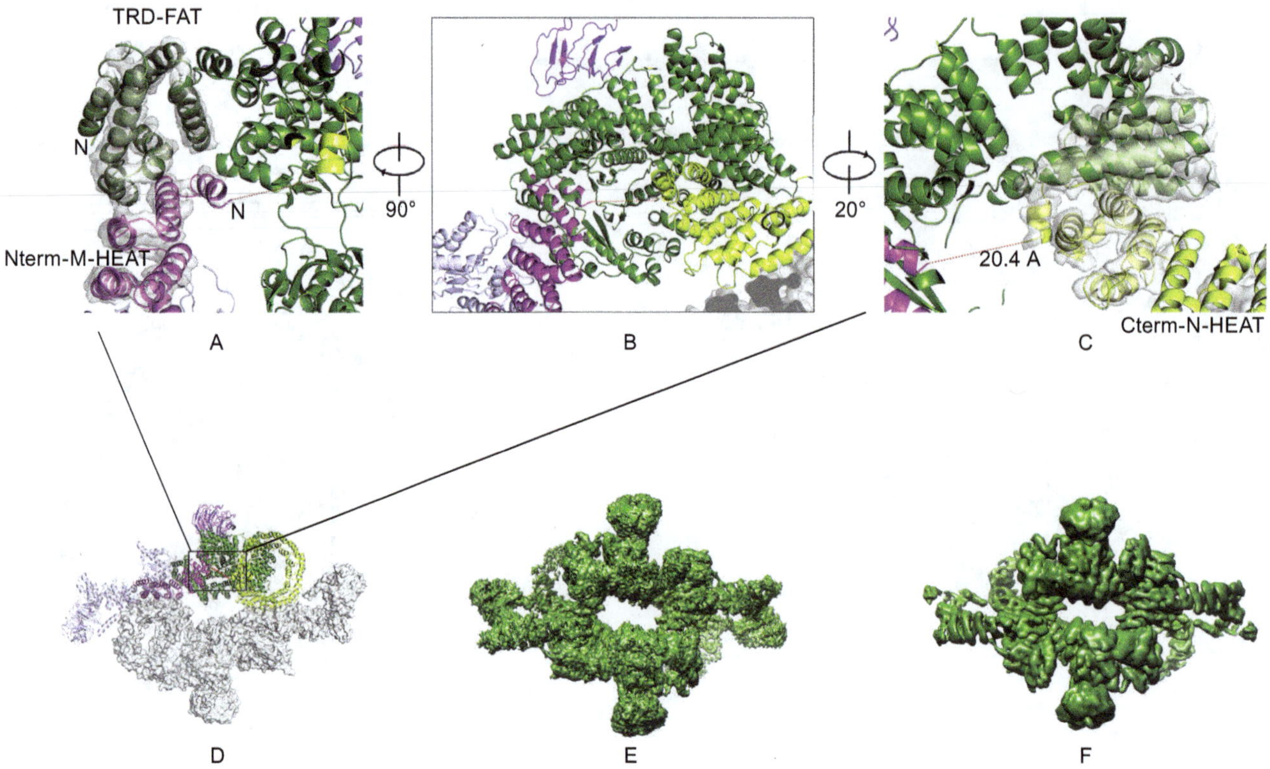

Figure 3. The cryo-EM electron potential maps for the interfaces between M-HEAT and Core and between N-HEAT and M-HEAT (D) contoured at 5 sigma level. (A–C) is the different zoom-in view. The putative linkage between N-HEAT and M-HEAT is depicted as the red dot line, which is measured to be 20.4 Å. Because of the high quality of the cryo-EM map, it is clear that the density making up the M-HEAT region does not directly join with the FAT domain. (E and F) Comparison of cryo-EM map from this study at 4.4 Å (E) and previous study at 5.9 Å (F) resolution, respectively. Both of the maps are displayed in Chimera at the same threshold after normalization.

binds to two mTOR molecules and extends towards the distal convex along the long axis of the rhomboid with the WD40 repeats domain located on the apical side. A central hole is formed within the mTOR dimer. From a side view, the N-terminal super-helical α-solenoids of mTOR and Raptor are located on one side of the rhomboid whereas the kinase domain of mTOR and the associated mLST8 are on the other side (lower panels in Fig. 2B). The kinase domain is close to mLST8 and the Caspase-like domain of Raptor and the open catalytic cavity faces towards outside of the central core region. Our cryo-EM analysis revealed a strong heterogeneous conformation of the dimeric mTORC1 with the distance between edges of two kinase domains ranging from 9 Å to 29 Å even after the mTORC1 cryo-EM sample was generated with mild glutaraldehyde gradient fixation (Fig. S3), which explains the difficulty in improving the resolution of mTORC1 reconstruction.

The C-terminal region of mTOR protein forms a compact core domain (designated Core), in which a "C"-shaped α-solenoid (FAT) wraps around the kinase domain from which the FRB domain protrudes out (Fig. 2B). The N-terminus of mTOR protein adopts a spiral (~1.3-turn) right-hand super-helix comprising of 16 HEAT repeat (designated N-HEAT). The middle region adopts extended HEAT repeat (designated M-HEAT) bridging the N-HEAT and the Core domain of mTOR. Intriguingly, no direct contact was observed between the N-HEAT and M-HEAT nearby the Core domain. Furthermore, while the M-HEAT's two ends are well defined in the EM map, no density was observed to connect the two ends to either the N-HEAT or the Core regions even at lower threshold. Therefore, the mTOR protein adopts a peculiar disconnected/branched domain architecture within the mTORC1 so that the mTOR protein's topology is difficult to be unambiguously determined solely from the 4.4 Å resolution reconstruction.

Topology of mTOR protein

We sought to use a combination of bioinformatics, crosslinking mass spectrometry and structural biology approaches to determine the topology of mTOR protein. The C-terminal regions could be clearly defined according to the

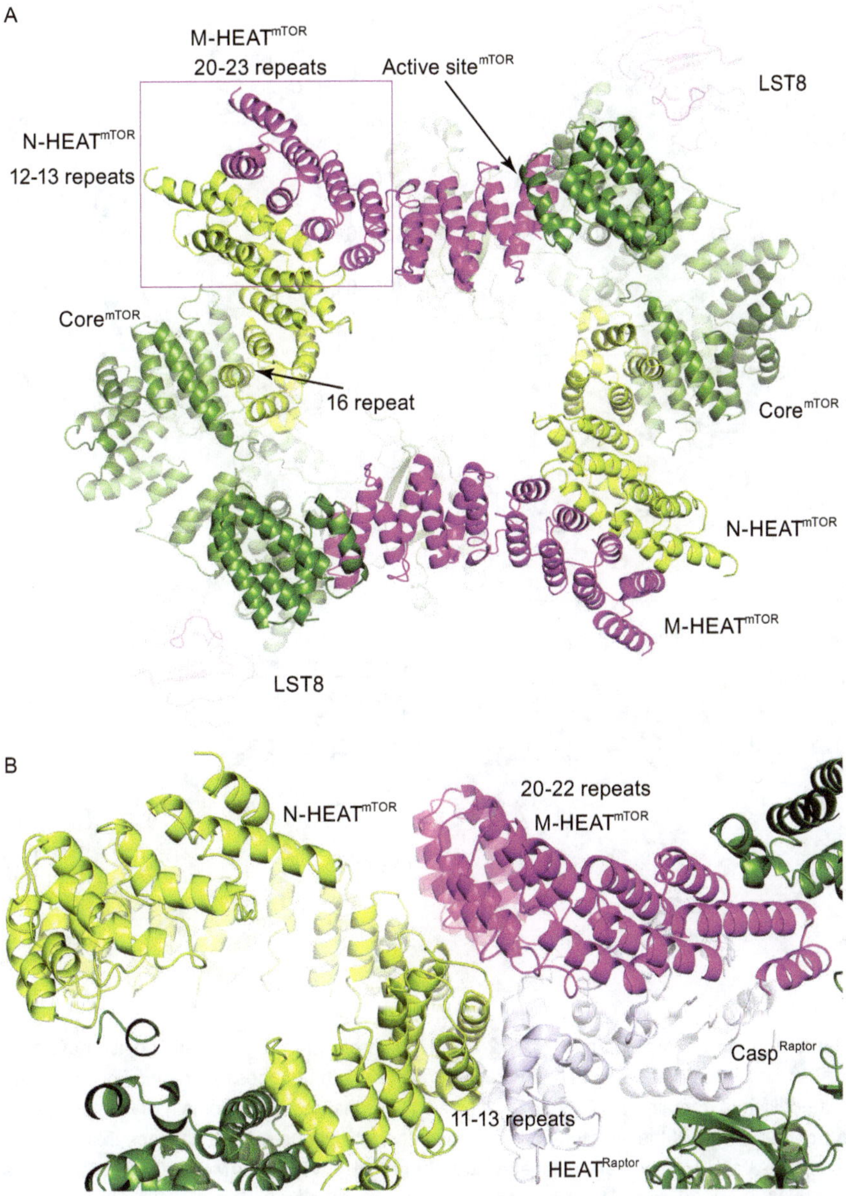

Figure 4. Intramolecular and intermolecular interactions within mTORC1. (A) Closed-up view of the dimer formation of mTORC1. Ribbon representation of the mTORC1 is shown. The mTORC1 dimer is formed through the intermolecular interaction between the M-HEAT and the Core and that between the N-HEAT and the Core. (B) Closed-up view of the interaction between Raptor and mTOR. Raptor binds to the ridge region of the M-HEAT repeats in one mTOR and the convex side of the N-HEAT region in the other mTOR.

previous crystal structure (Yang et al., 2013). However, the topology of the N-HEAT and M-HEAT in mTOR protein is hard to define. Opposite orientations of N-HEAT and M-HEAT from two mTOR molecules can generate eight possible topology models for the mTOR dimer. Taking account of the proximity likelihood among N-HEAT, M-HEAT and the Core domain, there are four possible models worth further consideration (Fig. S4). The relatively good quality and high resolution of our potential map in the central portion of the reconstruction revealed clearly that the density corresponding to the M-HEAT region does not directly join with TRD subdomain (numbering according to the crystal structure of Core^{mTOR}) (Yang et al., 2013) of FAT domain (Fig. 3, Movie 2). A short segment of helix from M-HEAT extends in the opposite direction away from the N-terminus of TRD, with a straightaway distance of 35 Å between the two termini. Interestingly, one terminus of the N-HEAT domain also extrudes out a short helix from its HEAT repeat motif, which faces toward the M-HEAT domain (Fig. 3, Movie 2). The distance between the above two short helix

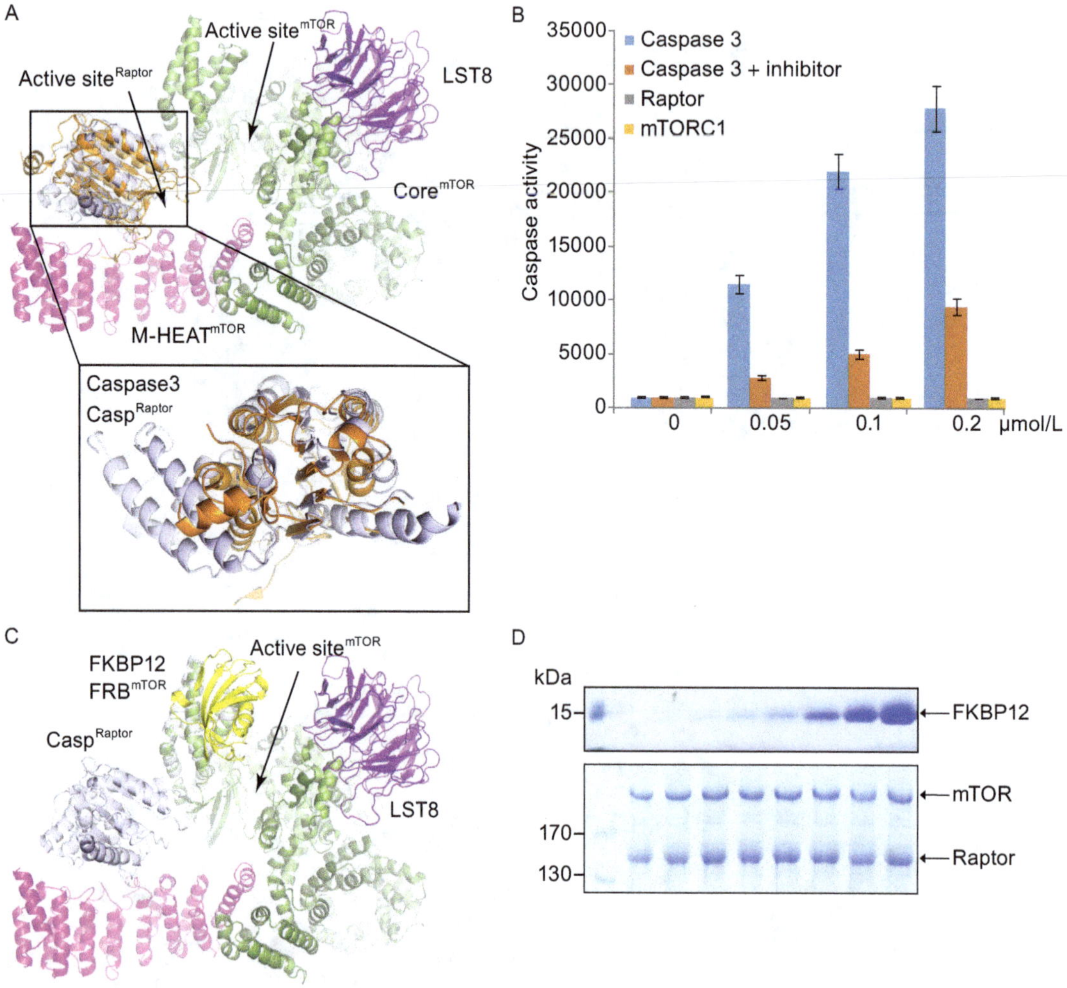

Figure 5. Caspase-like domain of Raptor and FKBP12-mTORC1 interaction. (A) Superimposition of caspase-3 and Raptor in mTORC1 structures. Two structures are shown in ribbon representations with unnecessary regions omitted. Caspase-3 is colored in orange. The active site of kinase domain of mTOR and the active site of caspase-like domain of Raptor are indicated, respectively. Shown below is the closed-up view of the structural comparison between caspase-3 and caspase-like domain of Raptor. (B) Equal amount of purified caspase-3 and Raptor were incubated with 20 μmol/L Ac-DEVD-AMC in the presence or absence of caspase inhibitor Z-VAD-FMK at 37°C for 60 min. The activities were measured using a Spectrafluor Fluorescence Plate Reader with excitation at 400 nm and emission at 505 nm. Error bars, s.d. for triplicate experiments. (C) Superimposition of FRB^{mTOR}-FKBP and mTORC1 structures is shown in ribbon representations. (D) Effect of FKBP12-Rapamycin for mTORC1 assembly. Increased amount of FKBP12-Rapamycin incubated with mTORC1 (Flag-Raptor) immobilized on the Flag resin. Bound proteins were subjected to SDS-PAGE and stained with Coomassie blue.

segments is approximately 20 Å, agreeing with the secondary structural prediction that there are ~10 residues as a flexible loop between the N-HEAT and M-HEAT (Fig. S5). This supports a topology of model I for the mTOR protein in the complex. This model is further supported by crosslinking mass spectrometry analysis of the mTORC1, with the most confident lysine crosslinking pairs between the N-HEAT and the Core region, between the M-HEAT and the Core region, and between the M-HEAT and the caspase-like domain of Raptor (Fig. S6). Notably, this topology (model I) was interpreted differently in the cryo-EM structure of mTORC1 at 5.9 Å resolution (Aylett et al., 2016), but in agreement with the topological interpretation of Tor-Lst8 structure from the thermotolerant yeast *Kluyveromyces marxianus* (Baretic et al., 2016).

Interaction of mTORC1 subunit

We therefore used the model I hereafter for structural analyses of inter- and intramolecular interactions within the mTORC1 (Fig. 4). Within the N-HEAT region, the HEAT repeats 12–13 are involved in the interaction with Raptor and the M-HEAT region. The HEAT repeat 16 binds to three parallel α-helices of FAT domain of the Core region. Other parts of the N-HEAT region have no inter- or intramolecular interactions, suggesting that N-HEAT tends to be dynamic, in consistent with its relatively weak density in the 3D reconstruction. Similar phenomenon has been observed in other proteins such as the flexible HEAT repeats of PP2A complex (Xu et al., 2006). The conformational flexibility of the N-HEAT region may provide significant functional implications, for example, adopting different conformations for the interactions with various mTOR regulators (Kim et al., 2008; Wullschleger et al., 2006; Zoncu et al., 2011).

The M-HEAT region contains 7 HEAT repeats (HEAT repeats 17–23) and adopts extended conformation. The dimerization of mTOR protein in the complex is mediated by the inter-molecular interaction between the N-HEAT region from one mTOR and the M-HEAT region from the other one (Fig. 4A). This interaction is through close proximity between the HEAT repeats 20–23 of the M-HEAT region and the HEAT repeats 12–13 of the N-HEAT region. The formation of this interaction does not seem to need other proteins but is stabilized by Raptor, which binds to the ridge region of the M-HEAT repeats 20–22 in one mTOR and the convex side of the N-HEAT repeats 11–13 in the other mTOR (Fig. 4B). To test Raptor's role in the dimerization of mTOR protein, we purified mTOR-mLST8 complex without Raptor. Gel-filtration analysis indicates that mTOR-mLST8 is sufficient to dimerize (Fig. S7). Thus, mTOR-mLST8 can assemble into dimer of heterodimer, which serves as the core for mTOR holoenzyme (mTORC1 or mTORC2) assembly, as well as a scaffold for association with various binding proteins. We propose that mTOR-mLST8 adopts similar fold in mTORC1 and mTORC2 complexes. The mTOR surface for Raptor interaction may also be involved in Rictor association because Raptor and Rictor are mutually exclusive for mTOR complex formation (Loewith et al., 2002; Sarbassov et al., 2004).

Structure and implication of RNC domain of Raptor

The mTORC1 structure shows that Raptor contains a Raptor N-terminal conserved (RNC) domain, followed by HEAT repeats and a C-terminal WD40 repeats. The RNC domain adopts a compact fold with a six-stranded β-sheet stabilized by three α-helices on each side. In support of previous prediction (Ginalski et al., 2004), the RNC domain shows similar overall structure to that of caspase family proteins (Fig. 5A). Although the catalytic residues (H153, C196) of Raptor are conserved, no caspase activity was detected using Ac-DEVD-AMC as the substrate for either the Raptor alone or mTORC1 (Fig. 5B).

Proposed mechanism of FKBP12-Rapamycin inhibition on mTORC1 activity

It has been well characterized that FKBP12-Rapamycin binds to mTORC1 and inhibits its kinase activity. Structural comparison of mTORC1 with FRB-FKBP12 (Choi et al., 1996) shows that upon binding to FRB, FKBP12 would be in close proximity to kinase domain of mTOR and therefore lead to steric hindrance for substrate entry into the catalytic cavity of mTORC1 (Fig. 5C). Notably, structural superimposition shows that FKBP12 has no other direct contact with mTORC1 besides FKBP-FRB interface. There is still ample space for substrate association, which explains the relatively weak inhibition by FKBP12-Rapamycin compared to the ATP-competitive inhibitor Torin1. To test whether FKBP12 association disrupts interaction between mTOR and Raptor (Kim et al., 2002), we performed an in vitro assay using purified mTORC1 (Flag-Raptor) and increasing amount of FKBP12-Rapamycin (Fig. 5D). In support of the structural analysis, FKBP12-Rapamycin does not disrupt mTORC1 integrity.

DISCUSSION

In this study, we report the cryo-EM structure of mTORC1 at 4.4 Å resolution, the highest resolution for human mTORC1 up to now. According to the biochemical and structural analyses, we proposed a most likely correct topological model for mTOR within the complex. The structure provides a structural basis for understanding the complex assembly of mTORC1 and the regulatory mechanism of mTORC1 by its binding partners and those yet to be identified.

First, we provided four possible models and used several ways to determine the topology of mTOR protein. Our result support the topology of model I, which is in accordance with Tor-Lst8 structure from yeast (Baretic et al., 2016), while is not consistent with mTORC1 at 5.9 Å resolution (Aylett et al. 2016). Although further high-resolution structural analysis should be performed, we here propose that model I is more likely to represent correct topology of mTOR under current understanding (Fig. S4).

Second, from structural analysis, the RNC domain of Raptor bind to the M-HEAT and N-HEAT repeats of mTOR (Fig. 4B). It has been reported that Ser 863, Ser 859 phosphorylation of Raptor by several kinases in response to different conditions, leading to a decrease of mTORC1 activity (Dunlop et al. 2014; Stretton et al., 2015; Yuan et al. 2015) because of reduced interaction between mTOR and Raptor (Stretton et al., 2015). However, structural analysis shows that Ser 859 of Raptor is located within the WD40 repeat domain and would have no direct contact with mTOR. Thus, the structure of mTORC1 also provides a framework to reconsider previous mechanistic explanation of biochemical observations of mTORC1 pathway.

Third, although the RNC domain of Raptor is similar with caspase family proteins from overall structure, we test that Raptor has no caspase activity (Fig. 5B). Previous studies have shown that Raptor binds to and recruits mTORC1 substrate proteins through recognizing TOS motifs, FDIDL in S6K1 and FEMDI in 4EBP1, both containing conserved aspartate residues for caspase cleavage. Therefore, the caspase-like domain may serve as a module to recognize and recruit specific substrates for kinase processing. Consistently, the catalytic site of the caspase-like domain of Raptor faces toward the catalytic cavity of the kinase domain of mTOR.

Fourth, our structural analysis and experiment result show that FKBP12-Rapamycin does not disrupt mTORC1 integrity as previously discussed (Kim et al., 2002). In fact, it provides steric hindrance for substrate entry into the catalytic cavity of mTORC1. In addition, higher resolution structure of mTORC1 is needed to illustrate the detailed information of the complex.

MATERIALS AND METHODS

Reagents

Antibodies against phospho-Thr-389 S6K (9205), 4EBP1 (9652), phos-4EBP1 (9456) and mTOR (2972) were from Cell Signaling Technology; horseradish peroxidase-labeled anti-mouse and anti-rabbit secondary antibodies were from AbMart; FLAG M2-agarose was from Sigma; Ni-NTA resin, Mono Q, Superdex75 (10/300 GL) and Superose 6 (10/300 GL) were from GE Healthcare; Rapamycin and Z-VAD-FMK were from Selleckchem; Torin1 was kindly provided by Dr. Haixin Yuan from IBS, Fudan University; Polyethylenimine (PEI) was from Polysciences (23966); 293F medium was from Sino Biological Inc.

Protein expression and purification

To produce soluble mTORC1 protein, the ORFs of human mTOR, raptor, and mLST8 were sub-cloned into modified pCAG vectors. The three plasmids were co-transfected to 293F cells using PEI. After culture at 37°C for 3 days, cells were collected and lysed in 50 mmol/L HEPES, pH 7.4, 150 mmol/L NaCl, 0.4% CHAPS and 3 mmol/L DTT at 4°C for 30 min, and the insoluble fraction was removed by centrifugation at 15,000 rpm for 30 min. Supernatants were incubated with FLAG-M2 monoclonal antibody-agarose for 1 h and unbound proteins were extensively washed away. The fusion proteins (FLAG-tagged raptor, FLAG-tagged mLST8 and Myc-tagged mTOR) were digested using PreScission protease overnight and the eluted proteins were further purified using ion exchange and gel filtration chromatography. The peak fractions were pooled for biochemical and structural analyses.

In vitro kinase assays

The in vitro kinase assays were performed in the buffer containing 25 mmol/L HEPES, pH 7.4, 100 mmol/L NaCl, 10 mmol/L $MgCl_2$, 2 mmol/L DTT and 0.5 mmol/L ATP for 30 min at 30°C. Reactions were terminated by the addition of SDS sample loading buffer and boiling for 5 min. Samples were subsequently analyzed by SDS-PAGE and immunoblotting.

Caspase activity assay

The purified human caspase-3 and Raptor were respectively incubated with 20 μmol\L Ac-DEVD-AMC with or without caspase inhibitor Z-VAD-FMK at 37°C for 60 min using a Spectrafluor Fluorescence Plate Reader with excitation at 400 nm and emission at 505 nm.

Purification of FKBP12 and in vitro pull-down

FKBP12 protein was purified from *E. coli* BL21(DE3) cells transformed with modified pGEX-6P-1 vector containing the ORF of human FKBP12. GST-FKBP12 was induced at 16 °C overnight with 0.1 mmol/L IPTG. The cells were harvested and disrupted with buffer containing 25 mmol/L Tris, pH 8.0, 150 mmol/L NaCl, 5 mmol/L imidazole. The fusion proteins were purified using Ni-NTA resin and digested by PreScission enzyme. The eluted proteins were purified using ion exchange and gel filtration. Equal molar FKBP12 and rapamycin were incubated with Flag-Raptor/mTOR for 30 min. The unbound proteins were washed away and the bound proteins were subjected to SDS-PAGE and stained by Coomassie blue.

Cross-linking and Mass Spectrometry analysis

The purified mTORC1 (0.8 μg/μL) was cross-linked using disuccinimidyl suberate (DSS) at a 1:150 molar ratio at room temperature for 20 min. The reaction was terminated by 20 mmol/L ammonium bicarbonate. The proteins were precipitated with cooled acetone and lyophilized. The pellet was dissolved in 8 mol/L urea, 100 mmol/L Tris, pH 8.5, followed by TCEP reduction, iodoacetamide alkylation, and trypsin digestion. Trypsin (Promega) digestion was quenched by 5% formic acid. Tryptic peptides were desalted with Pierce C18 spin column (Thermo Fisher) and separated in a Proxeon EASY-nLC liquid chromatography system by applying a step-wise gradient of 0–85% acetonitrile (ACN) in 0.1% formic acid. Peptides eluted from the LC column were directly electrosprayed into the mass spectrometer with a distal 2 kV spray voltage. Data-dependent tandem mass spectrometry (MS/MS) analysis was performed on Thermo Q-Exactive instrument in a 60-minute gradient. Raw data was processed with pLink software (Ellisen et al. 2002).

Cryo-electron microscopy

For cryo-grid preparation, aliquots of 3.5 μL of purified mTOR1 complex at a concentration of ~1.5 mg/mL were applied to glow-discharged holey carbon grids (Quantifoil Cu, R1.2/1.3, 400 mesh). The grids were blotted for 2.5 s and flash-plunged into liquid ethane pre-cooled in liquid nitrogen using an FEI Vitrobot mark IV operated at 22°C and 100% humidity. Data collection was performed on an FEI Titan Krios equipped with Gatan K2 Summit electron counting camera. Images were recorded in the super-resolution mode using UCSF-Image4 (Wang et al., 2015) at a nominal magnification of 22,500, which corresponds to a final pixel size of 1.306 Å by binning 2 of the original micrographs. For each image stack, a total dose of about 50 electrons per $Å^2$ at the specimen were equally fractioned into 32 frames with a total exposure time of 8 s. Defocus values ranged from −1.5 to −2.5 μm.

Image processing

For cryo-EM data sets, beam-induced motion correction was performed as previously described (Li et al., 2013). Micrographs inspection, automatic particle picking, 2D, 3D classification and refinement were performed within RELION 1.4 (Scheres, 2012) and the contrast transfer function parameters were estimated using CTFFIND3 (Mindell and Grigorieff, 2003). About 2000 manually picked particles were used to generate the templates for particle auto-picking. Totally 486,584 particles were auto-picked from 2997 micrographs for further processing. After one round of reference-free 2D classification, 387,806 particles were then subjected to 3D classification using the previously reconstructed 3D map (EMDB:5197) at 20 Å resolution as the initial model. Two rounds of 3D classification were performed to remove the obvious bad particles in the first round and accumulate the most homogeneous population of particles in the second round. A final dataset of 115,039 most homogeneous particles were subjected for final refinement to generate a 4.4 Å resolution map through the process of auto-refinement with C2 symmetry imposed (gold-standard FSC 0.143 criteria) (Chen et al., 2013), particle polishing, and post-processing with auto-mask and the manual-bfactor option (B-factor of −60 Å^2) in RELION. In a final step, a soft mask was imposed on the rigid central portion of the map to improve the density quality by continuing run of the auto-refinement in RELION. This yielded a final 3D map of the central portion at an average of 4.0 Å resolution (corrected gold-standard FSC at 0.143).

Model building into the cryo-EM map

The crystal structure of mTOR (1385–2549aa)/mLST8 (PDB ID: 4JSN) (Yang et al., 2013) was docked into the mTORC1 cryo-EM map using EMfit (Rossmann et al., 2001) while maintaining the two-fold symmetry. The rest parts of mTOR, including N-HEAT and M-HEAT domains, were built by placing the ideal alpha-helices into the density using COOT (Emsley et al., 2010). The caspase-like domain and WD40 domain of Raptor were built by docking the crystal structures of caspase-9 (PDB ID: 2AR9) and WD40 domain (PDB ID: 1J87) into the density. The HEAT repeat domain of Raptor was built manually in COOT (Emsley et al., 2010). The coordinates of the final structure were refined in the real space using phenix.real_space_refine (Adams et al., 2010). Model validation was performed with PROCHECK (Laskowski et al., 1993) and the WHATCHECK routine of WHAT IF (Vriend, 1990).

ACKNOWLEDGEMENTS

We thank staff members Biomedical Core Facility, Fudan University and National Center for Protein Science Shanghai for their help on Mass Spectrometry analyses. We thank the Tsinghua Cryo-EM Facility and High Performance Computation Facility of National Center for Protein Science Beijing for their support in cryo-EM data collection and processing. This work was supported by Grants from the National Natural Science Foundation of China (Grant Nos. U1432242, 31425008, 91419301), Basic Research Project of Shanghai Science and Technology Commission (No. 12JC1402700), the Program of Shanghai Subject Chief Scientist (No. 14XD1400500).

H. Y., J. W., M. L., X. C., Y. X. and H. W. designed the experiments. H. Y. and X. C. purified protein. J. W., M. L., J. W. and H. W. prepared cryo-EM sample, collected the data and determined the structure. H. Y. and X. C. performed biochemical analyses. X. C., M. H., D. T., M. D., H. Y., and C. W. performed crosslinking MS and analyzed the data. H. Y., M. L., J. W., H. W. and Y. X. analyzed the data and wrote the manuscript. Y. X. and H. W. supervised the project.

The electron density map and corresponding atomic coordinates have been deposited in the Protein Data Bank (http://www.rcsb.org/pdb) with PDB ID code: 5H64 and EMDB (http://www.ebi.ac.uk/pdbe/emdb/) with EMDB code: EMD-6668. The authors declare no competing financial interests.

COMPLIANCE WITH ETHICS GUIDELINES

Huirong Yang, Jia Wang, Mengjie Liu, Xizi Chen, Min Huang, Dan Tan, Meng-Qiu Dong, Catherine C. L. Wong, Jiawei Wang, Yanhui Xu and Hong-Wei Wang declare that they have no conflict of interest.

This article does not contain any studies with human or animal subjects performed by the authors.

REFERENCES

Adams PD, Afonine PV, Bunkóczi G, Chen VB, Davis IW, Echols N, Headd JJ, Hung L-W, Kapral GJ, Grosse-Kunstleve RW (2010) PHENIX: a comprehensive Python-based system for macromolecular structure solution. Acta Crystallogr D 66:213–221

Aylett CH, Sauer E, Imseng S, Boehringer D, Hall MN, Ban N, Maier T (2016) Architecture of human mTOR complex 1. Science 351:48–52

Baretic D, Berndt A, Ohashi Y, Johnson CM, Williams RL (2016) Tor forms a dimer through an N-terminal helical solenoid with a complex topology. Nat Commun 7:11016

Benjamin D, Colombi M, Moroni C, Hall MN (2011) Rapamycin passes the torch: a new generation of mTOR inhibitors. Nat Rev Drug Discov 10:868–880

Chen SX, McMullan G, Faruqi AR, Murshudov GN, Short JM, Scheres SHW, Henderson R (2013) High-resolution noise substitution to measure overfitting and validate resolution in 3D structure determination by single particle electron cryomicroscopy. Ultramicroscopy 135:24–35

Choi J, Chen J, Schreiber SL, Clardy J (1996) Structure of the FKBP12-rapamycin complex interacting with the binding domain of human FRAP. Science (NY) 273:239–242

Dazert E, Hall MN (2011) mTOR signaling in disease. Curr Opin Cell Biol 23:744–755

Dunlop EA, Hunt DK, Acosta-Jaquez HA, Fingar DC, Tee AR (2014) ULK1 inhibits mTORC1 signaling, promotes multisite Raptor phosphorylation and hinders substrate binding. Autophagy 7:737–747

Ellisen LW, Ramsayer KD, Johannessen CM, Yang A, Beppu H, Minda K, Oliner JD, McKeon F, Haber DA (2002) REDD1, a developmentally regulated transcriptional target of p63 and p53, links p63 to regulation of reactive oxygen species. Mol Cell 10:995–1005

Emsley P, Lohkamp B, Scott W, Cowtan K (2010) Features and development of COOT. Acta Crystallogr D 66:486–501

Garami A, Zwartkruis FJ, Nobukuni T, Joaquin M, Roccio M, Stocker H, Kozma SC, Hafen E, Bos JL, Thomas G (2003) Insulin activation of Rheb, a mediator of mTOR/S6K/4E-BP signaling, is inhibited by TSC1 and 2. Mol Cell 11:1457–1466

Ginalski K, Zhang H, Grishin NV (2004) Raptor protein contains a caspase-like domain. Trends Biochem Sci 29:522–524

Gingras AC, Gygi SP, Raught B, Polakiewicz RD, Abraham RT, Hoekstra MF, Aebersold R, Sonenberg N (1999) Regulation of 4E-BP1 phosphorylation: a novel two-step mechanism. Genes Dev 13:1422–1437

Holz MK, Blenis J (2005) Identification of S6 kinase 1 as a novel mammalian target of rapamycin (mTOR)-phosphorylating kinase. J Biol Chem 280:26089–26093

Holz MK, Ballif BA, Gygi SP, Blenis J (2005) mTOR and S6K1 mediate assembly of the translation preinitiation complex through dynamic protein interchange and ordered phosphorylation events. Cell 123:569–580

Inoki K, Li Y, Xu T, Guan KL (2003) Rheb GTPase is a direct target of TSC2 GAP activity and regulates mTOR signaling. Genes Dev 17:1829–1834

Inoki K, Corradetti MN, Guan KL (2005) Dysregulation of the TSC-mTOR pathway in human disease. Nat Genet 37:19–24

Kim DH, Sarbassov DD, Ali SM, King JE, Latek RR, Erdjument-Bromage H, Tempst P, Sabatini DM (2002) mTOR interacts with raptor to form a nutrient-sensitive complex that signals to the cell growth machinery. Cell 110:163–175

Kim E, Goraksha-Hicks P, Li L, Neufeld TP, Guan KL (2008) Regulation of TORC1 by Rag GTPases in nutrient response. Nat Cell Biol 10:935–945

Laskowski RA, MacArthur MW, Moss DS, Thornton JM (1993) PROCHECK: a program to check the stereochemical quality of protein structures. J Appl Crystallogr 26:283–291

Li X, Mooney P, Zheng S, Booth CR, Braunfeld MB, Gubbens S, Agard DA, Cheng Y (2013) Electron counting and beam-induced motion correction enable near-atomic-resolution single-particle cryo-EM. Nat Methods 10:584–590

Loewith R, Jacinto E, Wullschleger S, Lorberg A, Crespo JL, Bonenfant D, Oppliger W, Jenoe P, Hall MN (2002) Two TOR complexes, only one of which is rapamycin sensitive, have distinct roles in cell growth control. Mol Cell 10:457–468

Mindell JA, Grigorieff N (2003) Accurate determination of local defocus and specimen tilt in electron microscopy. J Struct Biol 142:334–347

Rossmann MG, Bernal R, Pletnev SV (2001) Combining electron microscopic with X-ray crystallographic structures. J Struct Biol 136:190–200

Sarbassov DD, Ali SM, Kim DH, Guertin DA, Latek RR, Erdjument-Bromage H, Tempst P, Sabatini DM (2004) Rictor, a novel binding partner of mTOR, defines a rapamycin-insensitive and raptor-independent pathway that regulates the cytoskeleton. Curr Biol 14:1296–1302

Scheres SH (2012) RELION: implementation of a Bayesian approach to cryo-EM structure determination. J Struct Biol 180:519–530

Stretton C, Hoffmann TM, Munson MJ, Prescott A, Taylor PM, Ganley IG, Hundal HS (2015) GSK3-mediated raptor phosphorylation supports amino-acid-dependent mTORC1-directed signalling. Biochem J 470:207–221

Tee AR, Blenis J (2005) mTOR, translational control and human disease. Semin Cell Dev Biol 16:29–37

Tee AR, Manning BD, Roux PP, Cantley LC, Blenis J (2003) Tuberous sclerosis complex gene products, Tuberin and Hamartin, control mTOR signaling by acting as a GTPase-activating protein complex toward Rheb. Curr Biol 13:1259–1268

Vriend G (1990) WHAT IF: a molecular modeling and drug design program. J Mol Graph 8:52–56

Wang QS, Yu F, Huang S, Sun B, Zhang KH, Liu K, Wang ZJ, Xu CY, Wang SS, Yang LF et al (2015) The macromolecular crystallography beamline of SSRF. Nucl Sci Technol 26:12–17

Wullschleger S, Loewith R, Hall MN (2006) TOR signaling in growth and metabolism. Cell 124:471–484

Xu Y, Xing Y, Chen Y, Chao Y, Lin Z, Fan E, Yu JW, Strack S, Jeffrey PD, Shi Y (2006) Structure of the protein phosphatase 2A holoenzyme. Cell 127:1239–1251

Yang H, Rudge DG, Koos JD, Vaidialingam B, Yang HJ, Pavletich NP (2013) mTOR kinase structure, mechanism and regulation. Nature 497:217–223

Yip CK, Murata K, Walz T, Sabatini DM, Kang SA (2010) Structure of the human mTOR complex I and its implications for rapamycin inhibition. Mol Cell 38:768–774

Yuan HX, Wang Z, Yu FX, Li F, Russell RC, Jewell JL, Guan KL (2015) NLK phosphorylates Raptor to mediate stress-induced mTORC1 inhibition. Genes Dev 29:2362–2376

Zoncu R, Efeyan A, Sabatini DM (2011) mTOR: from growth signal integration to cancer, diabetes and ageing. Nat Rev Mol Cell Biol 12:21–35

12

E3 ligase UHRF2 stabilizes the acetyltransferase TIP60 and regulates H3K9ac and H3K14ac via RING finger domain

Shengyuan Zeng[1], Yangyang Wang[1], Ting Zhang[2], Lu Bai[3], Yalan Wang[4], Changzhu Duan[1]✉

[1] Department of Cell Biology and Medical Genetics, Molecular Medicine and Cancer Research Center, Chongqing Medical University, Chongqing 400016, China
[2] Department of Obstetrics, The First Affiliated Hospital of Chongqing Medical University, Chongqing 400016, China
[3] Key Laboratory of Medical Molecular Virology, School of Basic Medical Sciences, Shanghai Medical College, Fudan University, Shanghai 200032, China
[4] Department of Pathology, Molecular Medicine and Cancer Research Center, Chongqing Medical University, Chongqing 400016, China
✉ Correspondence: duanchzhu@gmail.com (C. Duan)

ABSTRACT

UHRF2 is a ubiquitin-protein ligase E3 that regulates cell cycle, genomic stability and epigenetics. We conducted a co-immunoprecipitation assay and found that TIP60 and HDAC1 interact with UHRF2. We previously demonstrated that UHRF2 regulated H3K9ac and H3K14ac differentially in normal and cancer cells. However, the accurate signal transduction mechanisms were not clear. In this study, we found that TIP60 acted downstream of UHRF2 to regulate H3K9ac and H3K14ac expression. TIP60 is stabilized in normal cells by UHRF2 ubiquitination. However, TIP60 is destabilized in cancer cells. Depletion or inhibition of TIP60 disrupts the regulatory relationship between UHRF2, H3K9ac and H3K14ac. In summary, the findings suggest that UHRF2 mediated the post-translational modification of histones and the initiation and progression of cancer.

KEYWORDS UHRF2, TIP60, ubiquitination, acetylation, hepatocellular carcinoma

INTRODUCTION

Post-translational modifications (PTMs) of histones regulate gene expression. They are induced by several families of proteins including histone deacetylases (HDACs), histone acetyltransferases (HATs), physphorylases of the aurora kinases family, and histone ubiquitinases of the E3 ligase family (Basu et al., 2009; Pokholok et al., 2005; Tan et al., 2011; Wang et al., 2008). These modifications involving the interaction between histone and DNA regulate gene transcription either synergistically or antagonistically (Arnaudo and Garcia 2013; Su et al., 2016). UHRF2 is a multi-domain E3 ubiquitin ligase, which is related to mouse Np95 and human ICBP90. It comprises UBL, PHD, TTD, SET and RING-associated (SRA/YDG) and RING finger domains (Bronner et al., 2007; Qian et al., 2012; Wang et al., 2012). Several studies have demonstrated that UHRF2 represents a nodal point in the cell cycle network and regulates the epigenetics (Mori et al., 2011). UHRF2 interacts with the DNA methyltransferases (DNMT1, DNMT3a and DNMT3b), HDAC1, G9a and H3K9me2/me3 (Li et al., 2004; Mori et al., 2004; Pichler et al., 2011). Recently we found that UHRF2 influenced the acetylation of histone H3 on lysine 9 (H3K9ac) and lysine 14 (H3K14ac). However, the molecular mechanism was unclear. We investigated the mechanism underlying UHRF2-mediated regulation of H3K9ac and H3K14ac expression and the factors associated with differential regulation in normal and cancer cells.

Histone acetyltransferase TIP60 attracted our interest. It belongs to the MYST family and controls the acetylation of

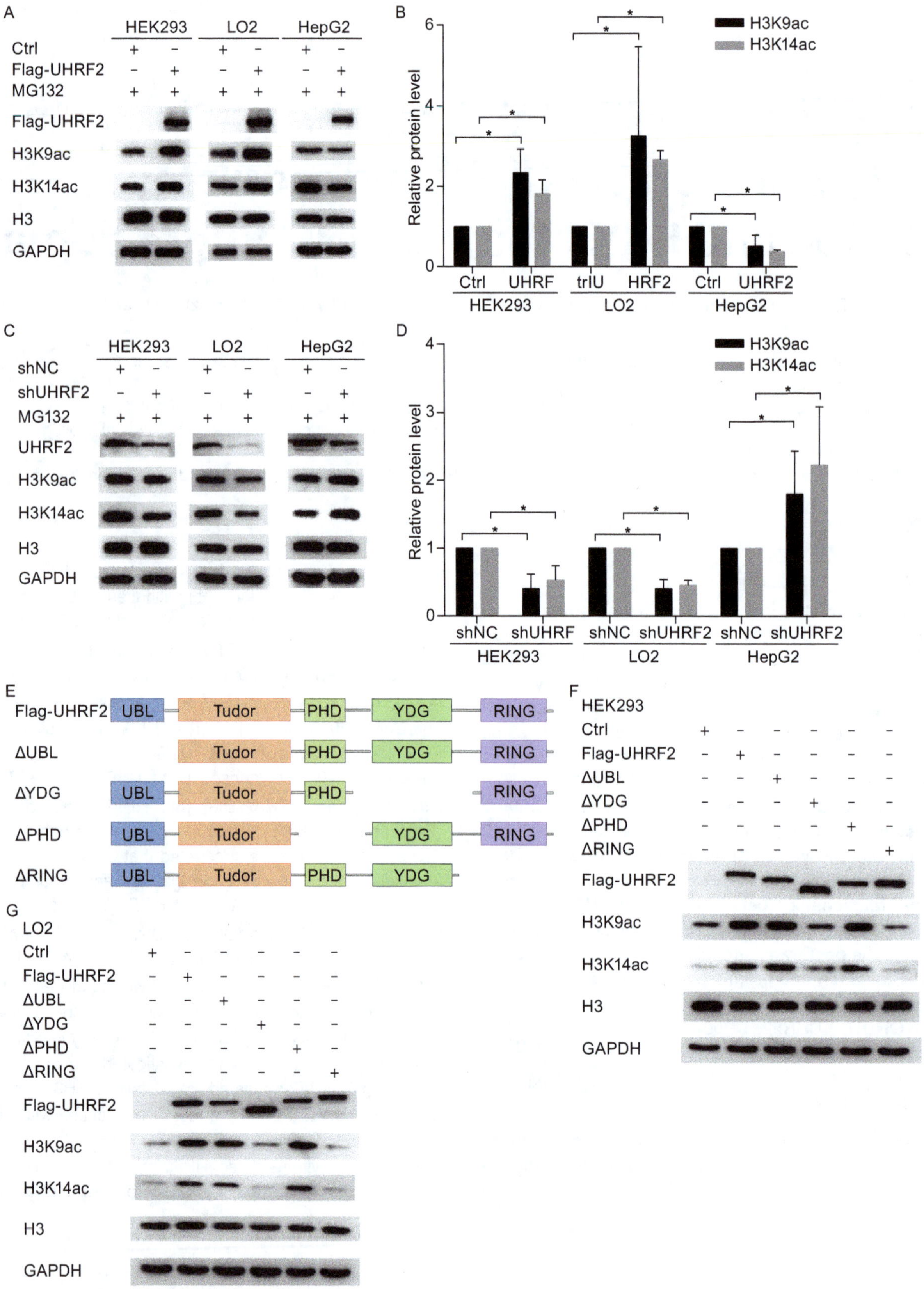
A
HEK293
LO2
HepG2
Ctrl
Flag-UHRF2
MG132
Flag-UHRF2
H3K9ac
H3K14ac
H3
GAPDH
B
Relative protein level
H3K9ac
H3K14ac
Ctrl
UHRF
trlU
HRF2
Ctrl
UHRF2
HEK293
LO2
HepG2
C
HEK293
LO2
HepG2
shNC
shUHRF2
MG132
UHRF2
H3K9ac
H3K14ac
H3
GAPDH
D
Relative protein level
H3K9ac
H3K14ac
shNC
shUHRF
shNC
shUHRF2
shNC
shUHRF2
HEK293
LO2
HepG2
E
Flag-UHRF2
ΔUBL
ΔYDG
ΔPHD
ΔRING
UBL
Tudor
PHD
YDG
RING
F
HEK293
Ctrl
Flag-UHRF2
ΔUBL
ΔYDG
ΔPHD
ΔRING
Flag-UHRF2
H3K9ac
H3K14ac
H3
GAPDH
G
LO2
Ctrl
Flag-UHRF2
ΔUBL
ΔYDG
ΔPHD
ΔRING
Flag-UHRF2
H3K9ac
H3K14ac
H3
GAPDH

◀ **Figure 1. UHRF2 regulates the expression of H3K9ac and H3K14ac.** (A) HEK293, LO2 and HepG2 cells were transfected with the indicated plasmids. Cellular lysates were analyzed by Western blot. UHRF2 was detected using anti-Flag antibodies. H3 and GAPDH were used as loading controls. (B) H3K9ac and H3K14ac levels were normalized to 1 at control groups. Significance was represented as $^*P < 0.05$. (C and D) Endogenous UHRF2 protein levels were decreased by shRNA against UHRF2. The cellular lysates were analyzed by Western blot and quantified using Image J software. (E) Various UHRF2 deletion mutants; (F–H) HEK293, LO2 and HepG2 cells were transiently transfected with indicated plasmids. After 48 h, the cellular lysates were analyzed by Western blot. (I and J) Immunofluorescence analyses were performed using anti-H3K9ac or anti-H3K14ac antibodies.

histone H2A on lysine 5 (H2AK5), histone H3 on lysine 14 (H3K14) and histone H4 (lysines 5, 8, 12, and 16) (Dai et al., 2013; Ikura et al., 2015; Jacquet et al., 2016; Renaud et al., 2016). TIP60 also participates in DNA repair, cellular growth, apoptosis and response to DNA double-strand breakage (Grezy et al., 2016; Jang et al., 2015; Sun et al., 2015). Recently, it was demonstrated that UHRF1 recruits histone acetyltransferase TIP60 and controls its expression and activity (Achour et al., 2009). In this study, the interactions between TIP60 and UHRF2 were detected using co-immunoprecipitation. It is suggested that TIP60 regulates H3K9ac and H3K14ac as downstream signal molecules.

We identified a previously undefined mechanism associated with TIP60-mediated UHRF2-regulation of H3K9ac and H3K14ac expression. In normal cells, UHRF2 overexpression enhances the expression of H3K9ac and H3K14ac, which is reversed in cancer cells. The regulatory mechanism is disrupted following deletion of the YDG or RING finger domains of UHRF2. In HEK293 cells, the interaction between UHRF2, TIP60 and HDAC1 was detected with co-immunoprecipitation, which was mediated via PHD and RING finger domains of UHRF2. Further, we found that TIP60 was a key intermediate in this molecular mechanism. Inhibition of TIP60 expression or activity disrupted the regulatory relationship. We also determined the levels of UHRF2, H3K9ac, H3K14ac, TIP60 and H2AK5ac in human hepatocellular carcinoma (HCC) tissues immunohistochemically. The results showed that all the aforementioned proteins were decreased under the high expression of UHRF2 in HCC tissues.

In summary, we identified a new mechanism underlying UHRF2-TIP60-H3K9ac and H3K14ac signaling axis. UHRF2 may contribute to the initiation and development of primary hepatocellular cancer by TIP60, which warrants further investigation.

RESULTS

UHRF2 regulates the expression of H3K9ac and H3K14ac

To determine the relationship between UHRF2, H3K9ac and H3K14ac, Western blot was performed. In normal cells (HEK293 and LO2 cells), UHRF2 overexpression elevated the expression of H3K9ac and H3K14ac compared with that of the control group. Opposite findings were observed in cancer cells (HepG2 cells) (Fig. 1A and 1B). Depletion of UHRF2 in HEK293 and LO2 cells lowered the expression of H3K9ac and H3K14ac compared with those of the control group. Similarly, HepG2 cells showed an inverse relationship (Fig. 1C and 1D) suggesting totally opposite results of the regulatory mechanism of UHRF2, H3K9ac and H3K14ac in different cells. Furthermore, we found that the YDG and RING finger domains of UHRF2 were required for this regulatory mechanism, based on Western blot and immunofluorescent assays (Fig. 1F–J). UHRF2 is an E3 ligase belonging to the RING-finger family, and the RING finger domain is required for E3 ligase activity. These results suggested that UHRF2 ligase activity was required for the regulatory mechanism.

UHRF2 interacts with TIP60 and HDAC1

Recent studies confirmed that UHRF1 recruits TIP60 and controls its expression and activity (Achour et al., 2009). UHRF2 was implicated in epigenetics via association with DNMTs, G9a, HDAC1, H3K9me2/3 and hemi-methylated DNA (Mori et al., 2011; Pichler et al., 2011). Both UHRF2 and UHRF1 belong to the UHRF family and show 67% sequence similarity (Bronner et al., 2007). Hence, we hypothesized that UHRF2 regulates H3K9ac and H3K14ac expression probably via TIP60 or HDAC1. A co-immunoprecipitation assay was performed to establish the interaction between UHRF2, TIP60 and HDAC1. A significant TIP60 and HDAC1 immunoprecipitation was observed using Flag-UHRF2 (Fig. 2A). In Fig. 2B and 2C, the reciprocal co-immunoprecipitation was performed and the interaction between UHRF2, TIP60 and HDAC1 was verified. In Fig. 2D, we demonstrated that the PHD and RING finger domains of UHRF2 were key domains in the interaction. Deletion of the PHD or RING finger domain of UHRF2 disrupted the interaction between UHRF2 and TIP60. However, the interaction between UHRF2 and HDAC1 was not affected. The transformation was not caused by the mis-localization of the deletion mutants of UHRF2. Double immunofluorescent staining revealed co-localization of UHRF2, TIP60 and HDAC1 proteins in LO2 cells (Fig. 2E and 2F) suggesting that UHRF2 interacted with TIP60 and HDAC1, and the interaction between UHRF2 and TIP60 was induced via the PHD and RING finger domains of UHRF2.

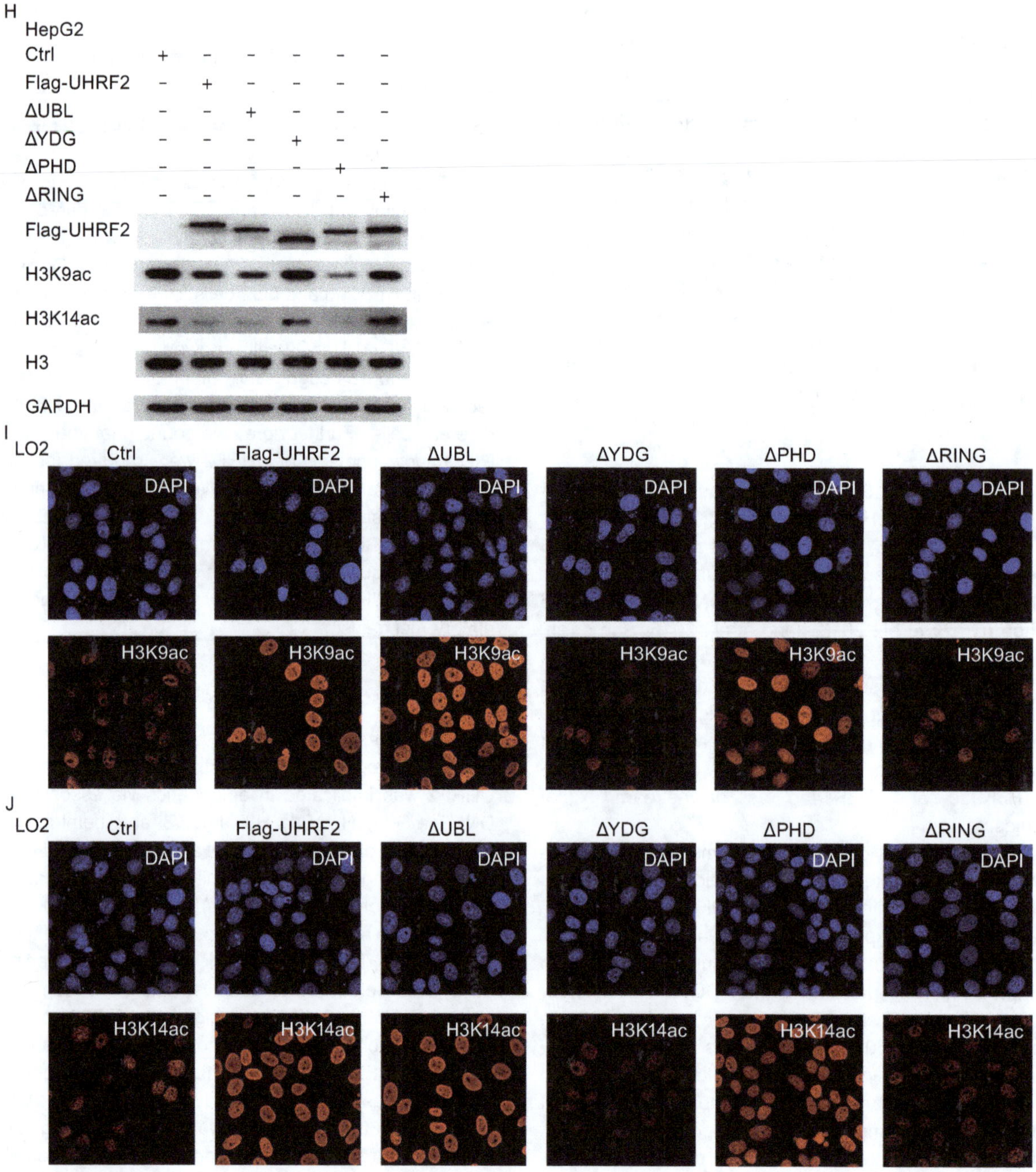

Figure 1. continued.

UHRF2 enhances the expression and activity of TIP60

To address the physiological relevance of these interactions, we tested whether modulation of the cellular levels of UHRF2 affected the TIP60 levels and activity. Recent studies emphasized the role of TIP60 as a unique histone acetyltransferase for the acetylation of H2AK5, which is considered as an endogenous marker of TIP60 activity (Bhaumik et al., 2007). In Fig. 3A, 3C and 3E, UHRF2 overexpression in normal cells induced a significant increase of TIP60 levels and activity, which was reversed in cancer cells. However, no obvious alteration in HDAC1 protein levels was observed. These results suggested that the expression and activity of TIP60 were increased by UHRF2 in normal cells, but decreased in cancer cells. In addition, we employed shRNA to suppress endogenous UHRF2. The results showed that ablation of endogenous UHRF2 decreased TIP60 activity in normal cells, but increased in HepG2 cells. HDAC1 protein levels also showed no

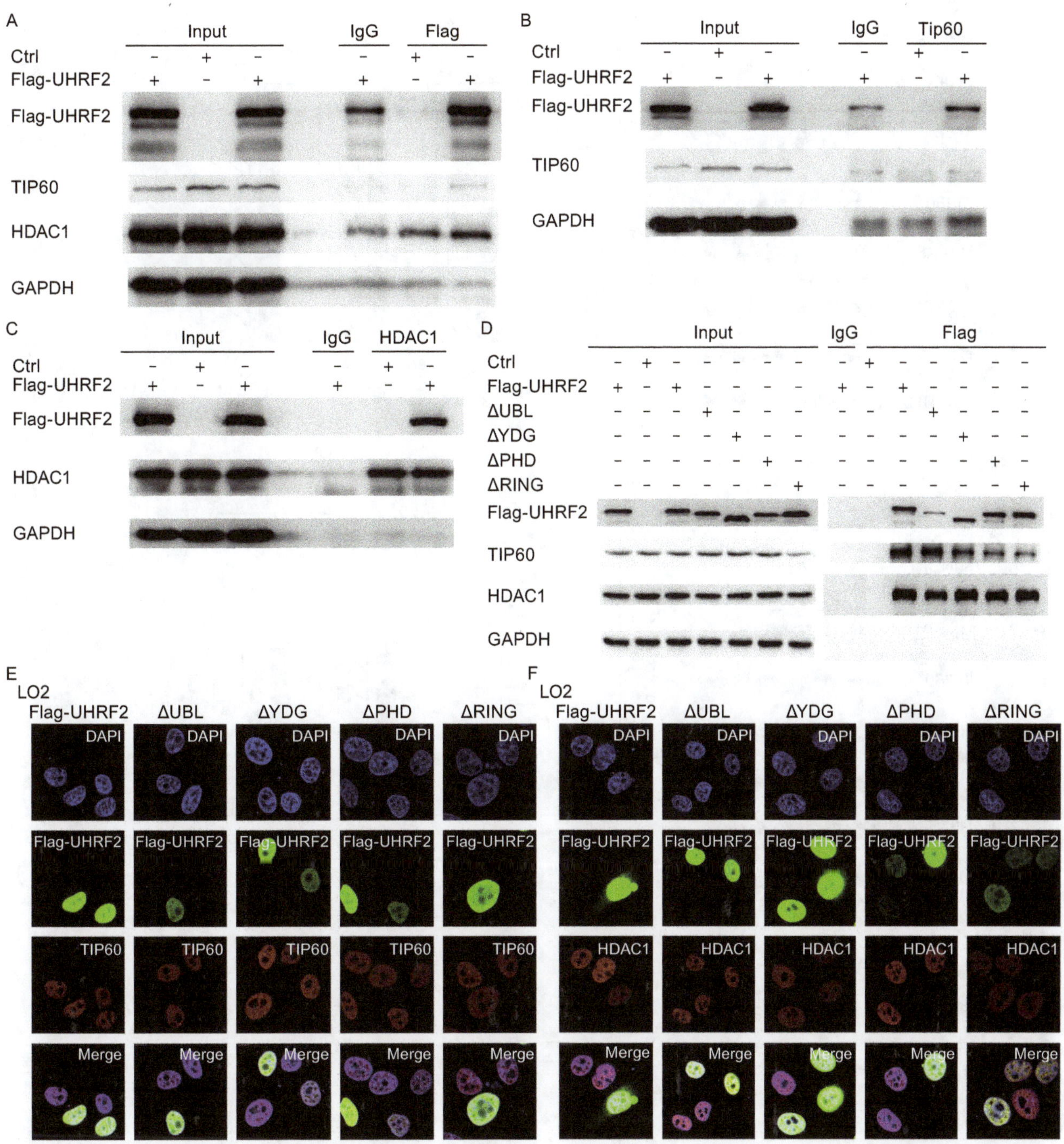

Figure 2. UHRF2 interacts with TIP60 and HDAC1. (A) HEK293 cells were transfected with control or Flag-UHRF2 plasmids, respectively. After 48 h, the total cellular lysates were immunoprecipitated using anti-Flag antibodies. (B) HEK293 cells were transfected with plasmids as shown. Cellular lysates were immunoprecipitated with anti-TIP60 antibodies and analyzed by Western blot. (C) HEK293 cells were transfected with the indicated plasmids. Total cellular lysates were subjected to HDAC1 immunoprecipitation. (D) HEK293 cells were transfected with full-length UHRF2 or various deletion mutant constructs of UHRF2 plasmids. Cellular lysates were immunoprecipitated using anti-Flag antibodies and analyzed by Western blot. (E and F) Double-labeling immunofluorescence combined with CLSM observation revealed co-localization of TIP60 and UHRF2 in LO2 cells. Co-localization of HDAC1 and UHRF2 was also revealed along with nuclear expression of all proteins.

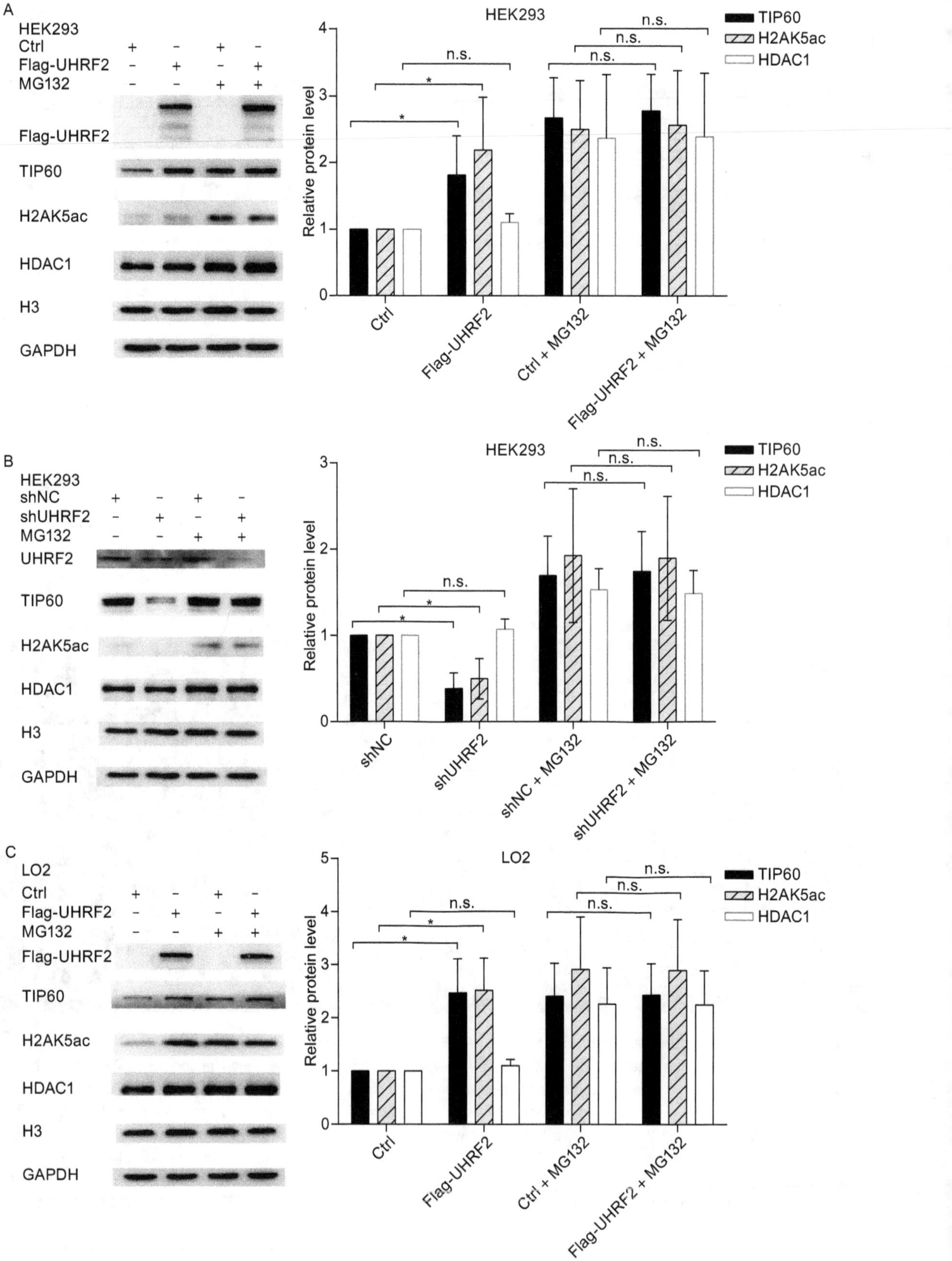
A
HEK293
Ctrl + − + −
Flag-UHRF2 − + − +
MG132 − − + +
Flag-UHRF2
TIP60
H2AK5ac
HDAC1
H3
GAPDH
HEK293
Relative protein level
n.s.
*
TIP60
H2AK5ac
HDAC1
Ctrl
Flag-UHRF2
Ctrl + MG132
Flag-UHRF2 + MG132
B
HEK293
shNC + − + −
shUHRF2 − + − +
MG132 − − + +
UHRF2
TIP60
H2AK5ac
HDAC1
H3
GAPDH
HEK293
Relative protein level
n.s.
*
TIP60
H2AK5ac
HDAC1
shNC
shUHRF2
shNC + MG132
shUHRF2 + MG132
C
LO2
Ctrl + − + −
Flag-UHRF2 − + − +
MG132 − − + +
Flag-UHRF2
TIP60
H2AK5ac
HDAC1
H3
GAPDH
LO2
Relative protein level
n.s.
*
TIP60
H2AK5ac
HDAC1
Ctrl
Flag-UHRF2
Ctrl + MG132
Flag-UHRF2 + MG132

◄ **Figure 3. UHRF2 regulates the expression and activity of TIP60.** (A, C and E) HEK293, LO2 and HepG2 cells were transfected with control or Flag-UHRF2 plasmids. They were exposed to MG132 (10 μg/mL) for 12 h and harvested. The cellular lysates were analyzed by Western blot. (B, D and F) HEK293, LO2 and HepG2 cells were transfected with shNC or shUHRF2. After 48 h, the total cellular lysates were analyzed by Western blot. Data were expressed as means ± SD, $n = 3$. Significance was indicated as $^{*}P < 0.05$, and non-significance as n.s $P > 0.05$. (G–I) HEK293, LO2 and HepG2 cells were transfected with the plasmids as shown. The cellular lysates were analyzed by Western blot. (J and K) Immunofluorescence analyses were performed using anti-TIP60 or anti-HDAC1 antibodies.

significant variation (Fig. 3B, 3D and 3F). Additional investigations showed that deletion of the PHD or RING finger domains of UHRF2 abolished the regulatory relationship (Fig. 3G-I), which suggested that PHD and RING finger domains of UHRF2 were the key domains in this mechanism. Furthermore, immunofluorescence staining showed similar results indicating that the PHD and RING finger domains of UHRF2 were pivotal domains (Fig. 3J and 3K) and that the ligase activity of UHRF2 was required.

UHRF2 ubiquitinates and stabilizes TIP60

To explore whether UHRF2 stabilizes TIP60, a half-life experiment was performed. Under controlled conditions, TIP60 showed a half-life of less than 1 h, which was extended by UHRF2 in normal cells. Deletion of RING finger domain in UHRF2 abrogated the effect. In cancer cells, the degradation of TIP60 was accelerated by UHRF2 (Fig. 4A–C). These results showed that UHRF2 regulated TIP60 stability via RING finger domain differentially in normal and cancer cells. UHRF2 is an E3 ligase and mediates ubiquitination. In this condition, we hypothesized that UHRF2 might ubiquitinate TIP60 *in vivo*. The results showed that TIP60 was immunoprecipitated with anti-HA-Ub antibodies, and was clearly enhanced in the UHRF2-overexpressing group of HEK293 cells (Fig. 4D). Similar results were not observed in HepG2 cells (Fig. 4E) indicating that UHRF2 stabilized TIP60 expression via RING finger domain through ubiquitination.

UHRF2 regulates H3K9ac and H3K14ac expression via interaction with TIP60

Previous studies indicated that TIP60 regulated H3K9ac and H3K14ac expression (Ikura et al., 2015; Renaud et al., 2016; Bassi et al., 2016). In this study, we found similar results in normal and cancer cells (Figure S1). We also suggested that UHRF2 regulated H3K9ac and H3K14ac expression. Hence, we hypothesized that UHRF2 regulates H3K9ac and H3K14ac through TIP60. In Fig. 5A–C, ablation of endogenous TIP60 decreased H3K9ac and H3K14ac protein levels, whereas overexpression of UHRF2 failed to rescue the altered normal cells. In cancer cells, ablation of TIP60 and UHRF2 overexpression further decreased H3K9ac and H3K14ac protein levels. Moreover, co-expression of UHRF2 and TIP60 increased the protein levels of H3K9ac and H3K14ac in normal cells. However, the protein levels of H3K9ac and H3K14ac were not rescued by the co-expression of TIP60 and UHRF2 in cancer cells (Fig. 5D–F). These results clearly showed that TIP60 as a downstream molecule of UHRF2 enhanced the expression of H3K9ac and H3K14ac. We also inhibited TIP60 acetyltransferase activity using a specific inhibitor MG149 that targets acetyl CoA-binding sites. The results showed that inhibition of TIP60 acetyltransferase activity during transfection of full-length UHRF2 or the deletion mutant construct of UHRF2 failed to rescue H3K9ac and H3K14ac expression (Fig. 5G–I). Thus, we concluded that UHRF2 regulated H3K9ac and H3K14ac expression via TIP60 downstream.

Inverse correlation between TIP60 and UHRF2 in HCC tissues

In this study, we found that UHRF2 regulated H3K9ac and H3K14ac expression via interaction with TIP60 in cultured cells. We investigated the regulatory mechanism in human hepatocellular carcinoma tissues using immunohistochemistry. A significant decrease in the expression of TIP60, H3K9ac, H3K14ac and H2AK5ac was found following the elevated expression of UHRF2 in HCC tissues (Fig. 6A). However, in the adjacent non-tumor tissues, a positive correlation between the aforementioned proteins and UHRF2 was observed (Fig. 6B). Similar results were clearly found in human tissues compared with *in vitro* results. Accordingly, we established that UHRF2 regulated the expression of H3K9ac and H3K14ac by stabilizing TIP60 via proteasomal pathway through its RING domain. We developed a model of UHRF2-TIP60-H3K9ac and H3K14ac signaling axis (Fig. 6C).

DISCUSSION

TIP60 is a transcriptional co-factor involved in several essential physiological processes in cells. Its aberrant expression has been reported in several cancers (Feng et al., 2016; Mo et al., 2016). TIP60 plays a major role in histone acetylation to potentially trigger cancer-related gene expression (Sun et al., 2015). A previous study showed that TIP60 co-localizes with the UHRF1/DNMT1 complex, which is involved in ubiquitination as a substrate of E3 ligase (Dai et al., 2013). Our findings strongly suggested that TIP60 was a new substrate for UHRF2. Protein modification via ubiquitination is of great importance in many regulatory processes inside the cells (Hershko et al., 2000). Ubiquitination alters protein turnover via proteasome-mediated degradation of proteins (Harrison et al., 2016; Leithe 2016; Yamano et al.,

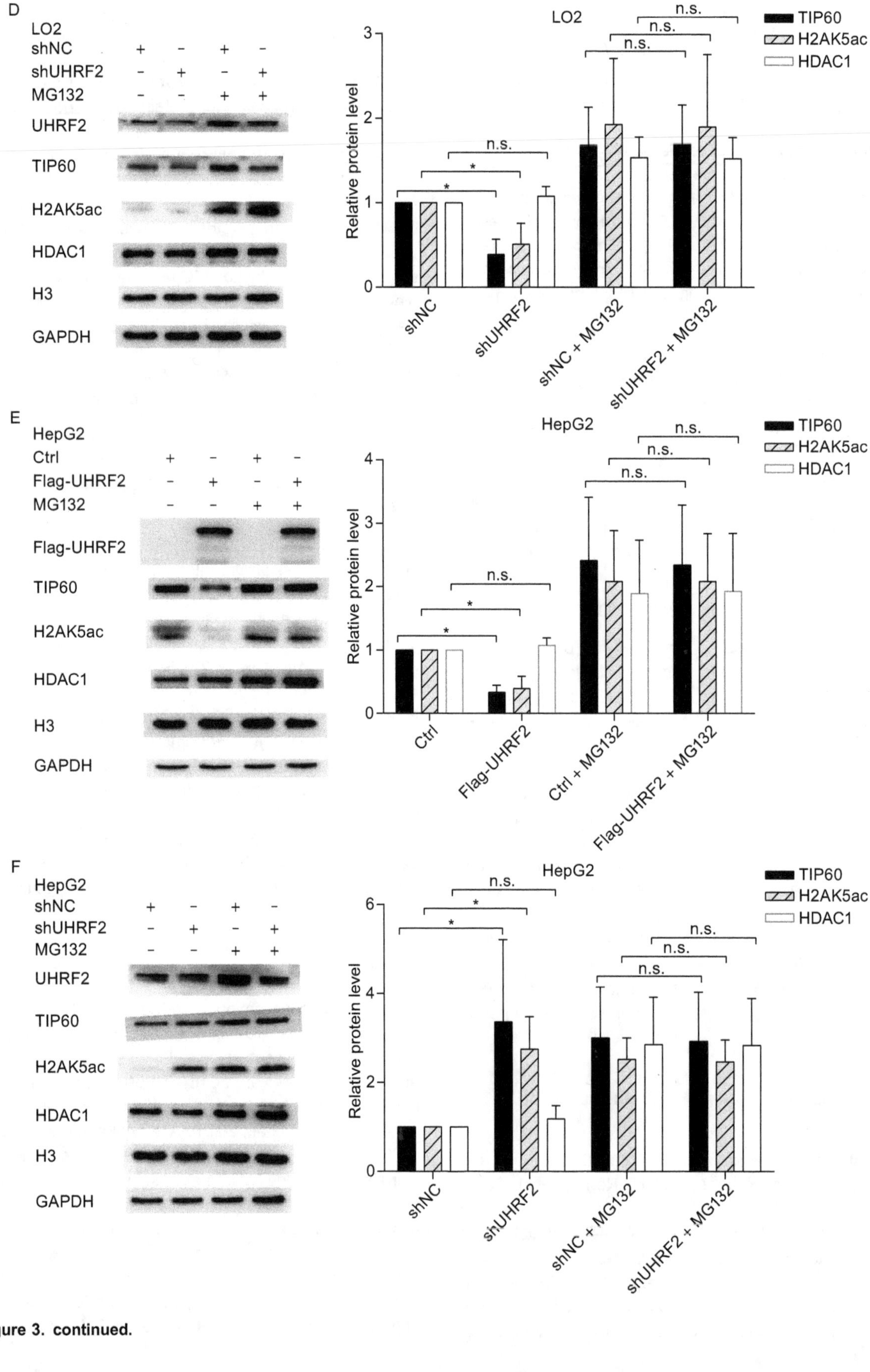

Figure 3. continued.

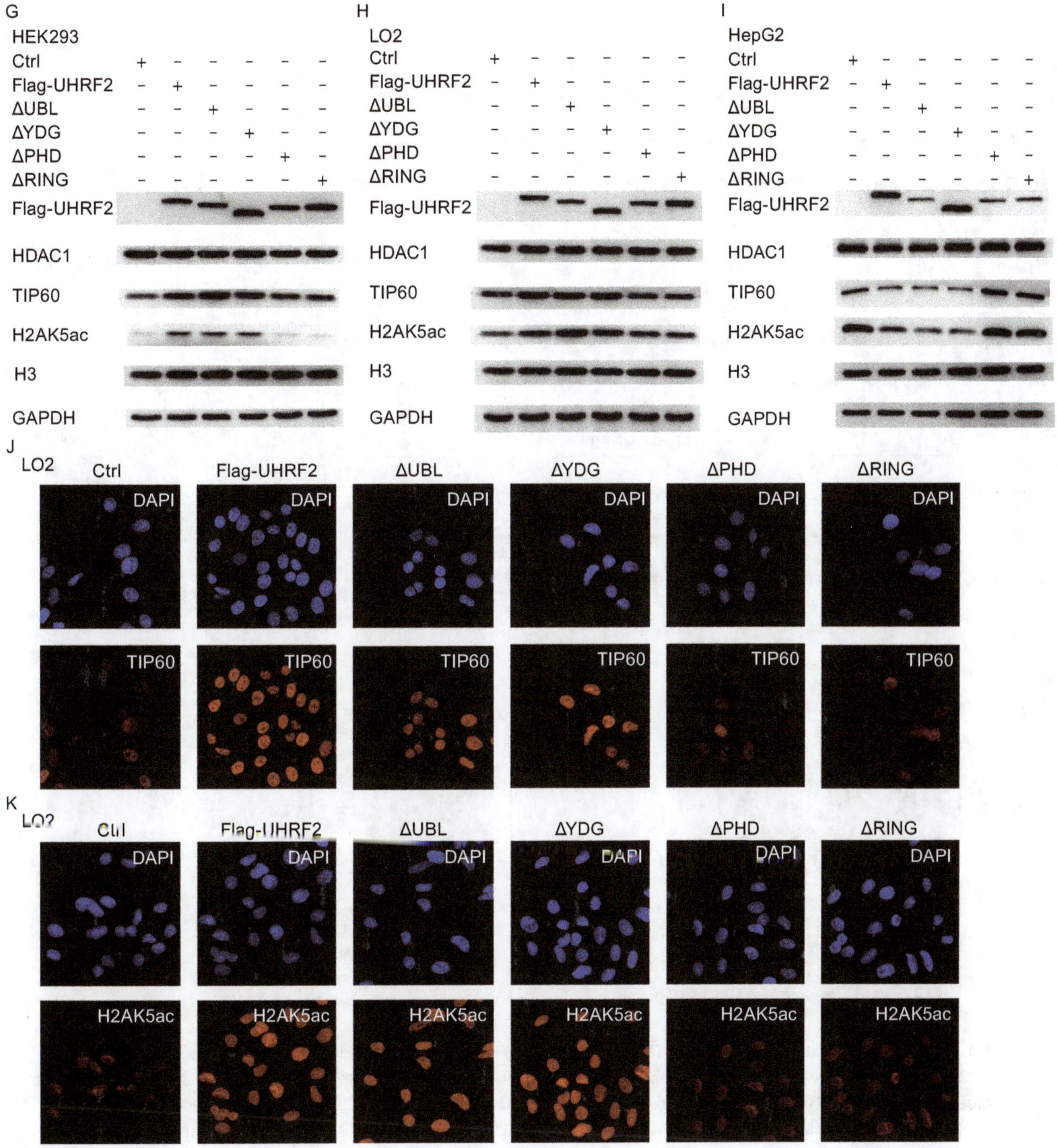

Figure 3. continued.

2015; Lechtenberg et al., 2016). Altered ubiquitination may prolong the half-life of specific proteins, while other proteins are rapidly degraded (Vinther-Jensen et al., 2015). In ubiquitination pathway, the E3 ligase plays an important role in substrate recognition. In this study, we found that E3 ligase UHRF2 ubiquitinated TIP60 and further stabilized its expression and elevated its activity via RING finger domain in normal cells. Conversely, the expression and activity of TIP60 were decreased in tumor cells. These results are of utmost significance, with two diametrically opposite responses in different cells. We conjectured that a series of unknown signal transduction factors mediate this pathway, which remain to be investigated.

PTMs associated with gene expression usually occur at the amino acid level, including methylation, phosphorylation, ubiquitination, SUMOylation, plamitoylation, and acetylation

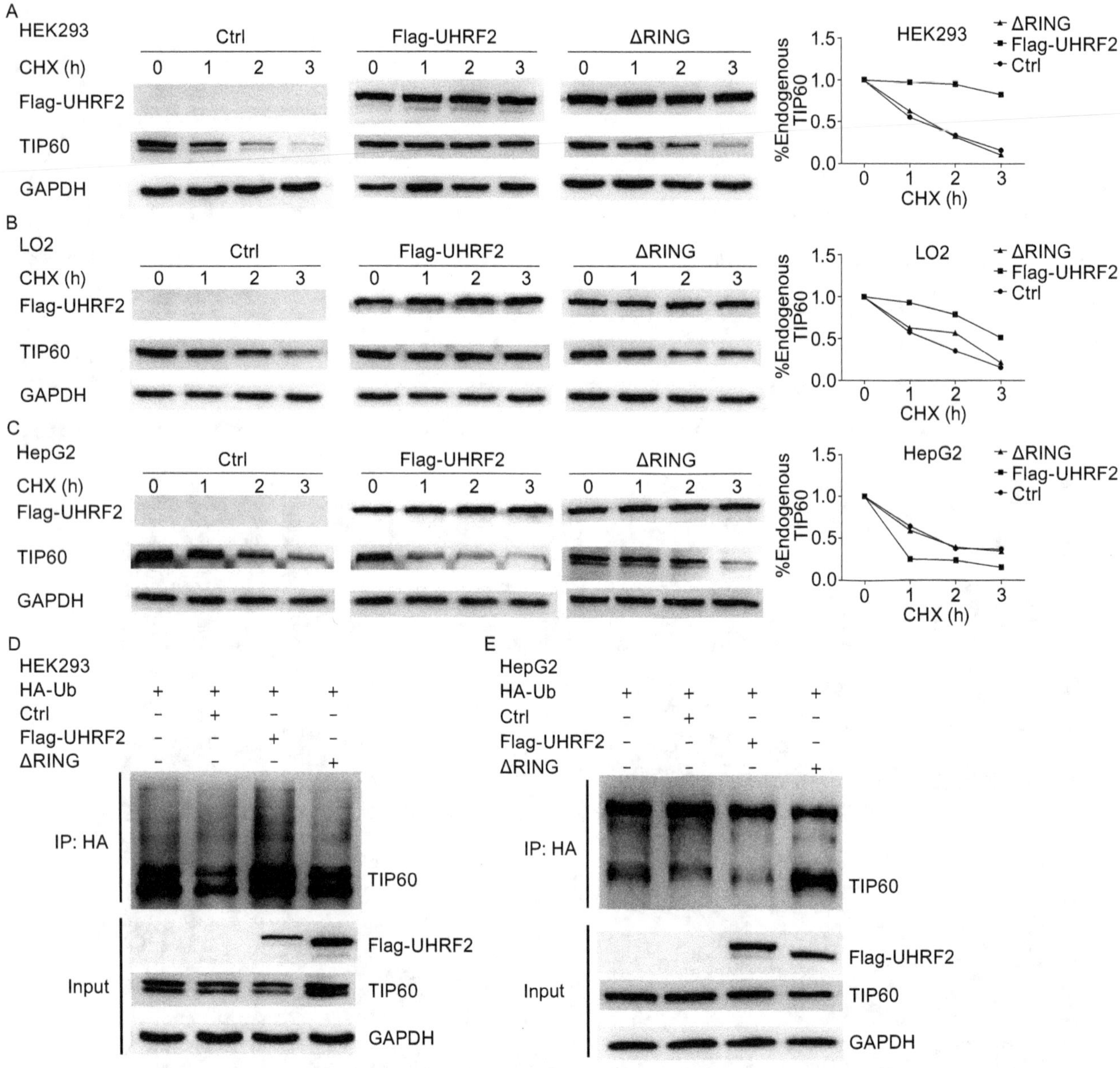

Figure 4. UHRF2 uniquitinates and stabilizes TIP60 *in vivo*. (A–C) HEK293, LO2 and HepG2 cells were transfected with control, Flag-UHRF2 or ΔRING plasmids, respectively. The cellular lysates were analyzed by Western blot. Graph represents collated results from three independent half-life experiments performed in HEK293, LO2 and HepG2 cells. TIP60 levels were normalized to 1 at time zero. (D–E) Various plasmids as shown were transfected into HEK293 and HepG2 cells. The cellular lysates were immunoprecipitated with anti-HA antibodies and immunoblotted with anti-TIP60 antibodies.

(Holt et al., 2015; Kouzarides 2007; Li 2002). Histone acetylation is one of the reversible PTMs catalyzed by HATs and HDACs, which play a key role in the regulation of specific gene expression, chromosome segregation and leukemia (Basu et al., 2009; Wang et al., 2008). Histone acetylation alters the chromosome structure and gene activity, and triggers the initiation and progression of tumors via regulation of cellular proliferation and apoptosis (Pokholok et al., 2005). The proliferation and differentiation levels are balanced under normal conditions through s series of molecular regulatory mechanisms (Tan et al., 2011). Normal cells transform into tumor cells when the balance of histone acetylation and deacetylation is disrupted by physical, chemical and biological factors to trigger abnormal differentiation and chaotic proliferation (Su et al., 2016). Histone H3 is a core component of eukaryotic nucleosome octamer (Karmodiya et al., 2012; Yamada et al., 2013). The two lysine sites 9 and 14 are the focus of intensive study in histone acetylation (Karmodiya et al., 2012). H3K9ac is associated with meiotic recombination hotspots and DNA damage response, and plays a key role in recombination with other factors (Grezy et al., 2016). H3K14ac interacts

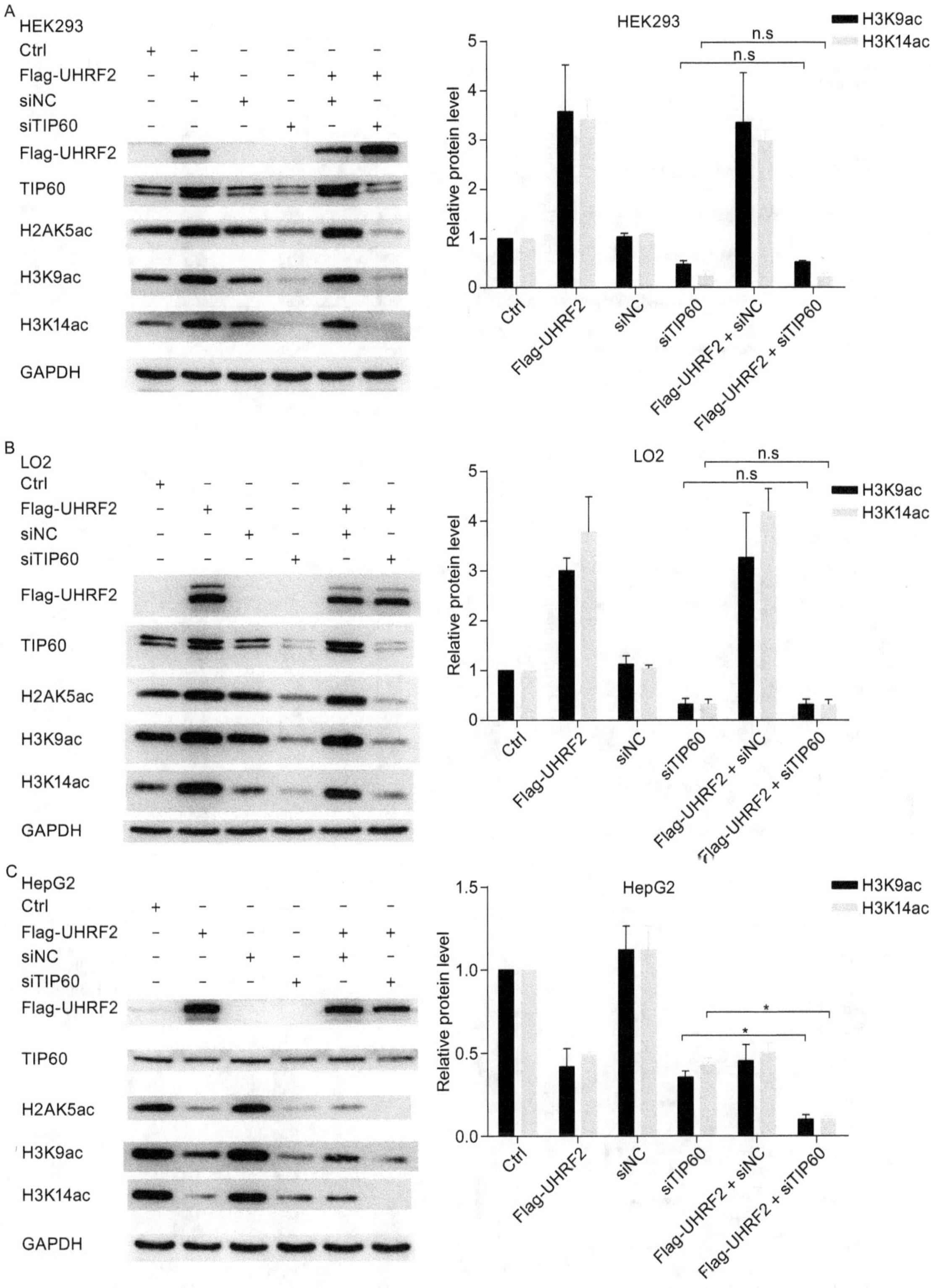

Figure 5. UHRF2 regulates H3K9ac and H3K14ac via TIP60. (A–C) HEK293, LO2 and HepG2 cells were transfected with Flag-UHRF2 plasmid, either alone or in combination with siRNA against TIP60. After 48 h, the cellular lysates were analyzed by Western blot. (D–F) HEK293, LO2 and HepG2 cells were transfected with Flag-UHRF2 plasmid, either alone or in combination with His-TIP60 plasmid. Total cellular lysates were analyzed by Western blot. Data represent means ± SD and were examined with two-sided t-test; *$P < 0.05$; n.s (non-significant) indicated as $P > 0.05$. (G–I) HEK293, LO2 and Hepg2 cells were transfected with various plasmids as shown and treated with MG149 (100 μm) after transfection. Cellular lysates were analyzed by Western blot.

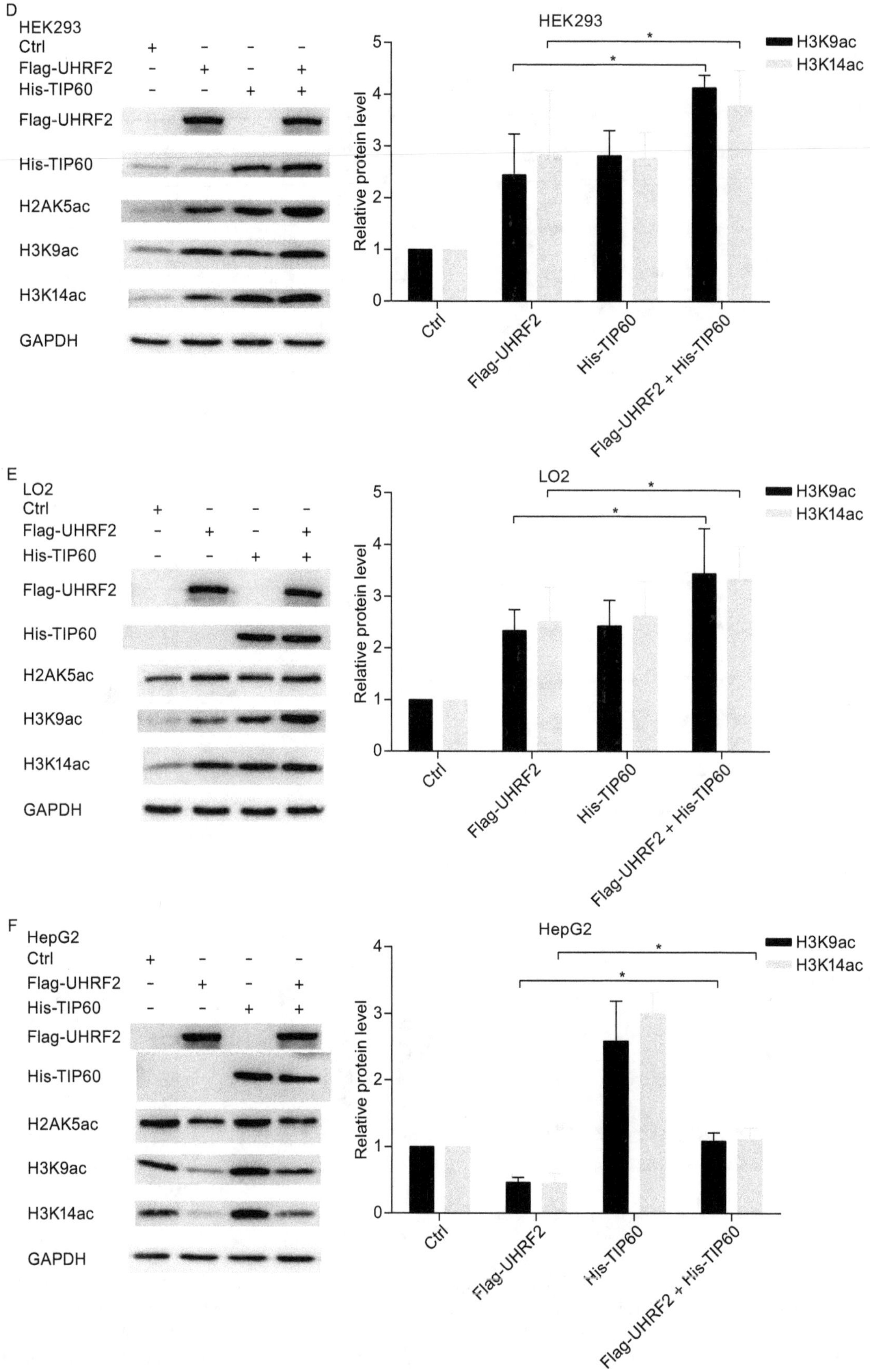

Figure 5. continued.

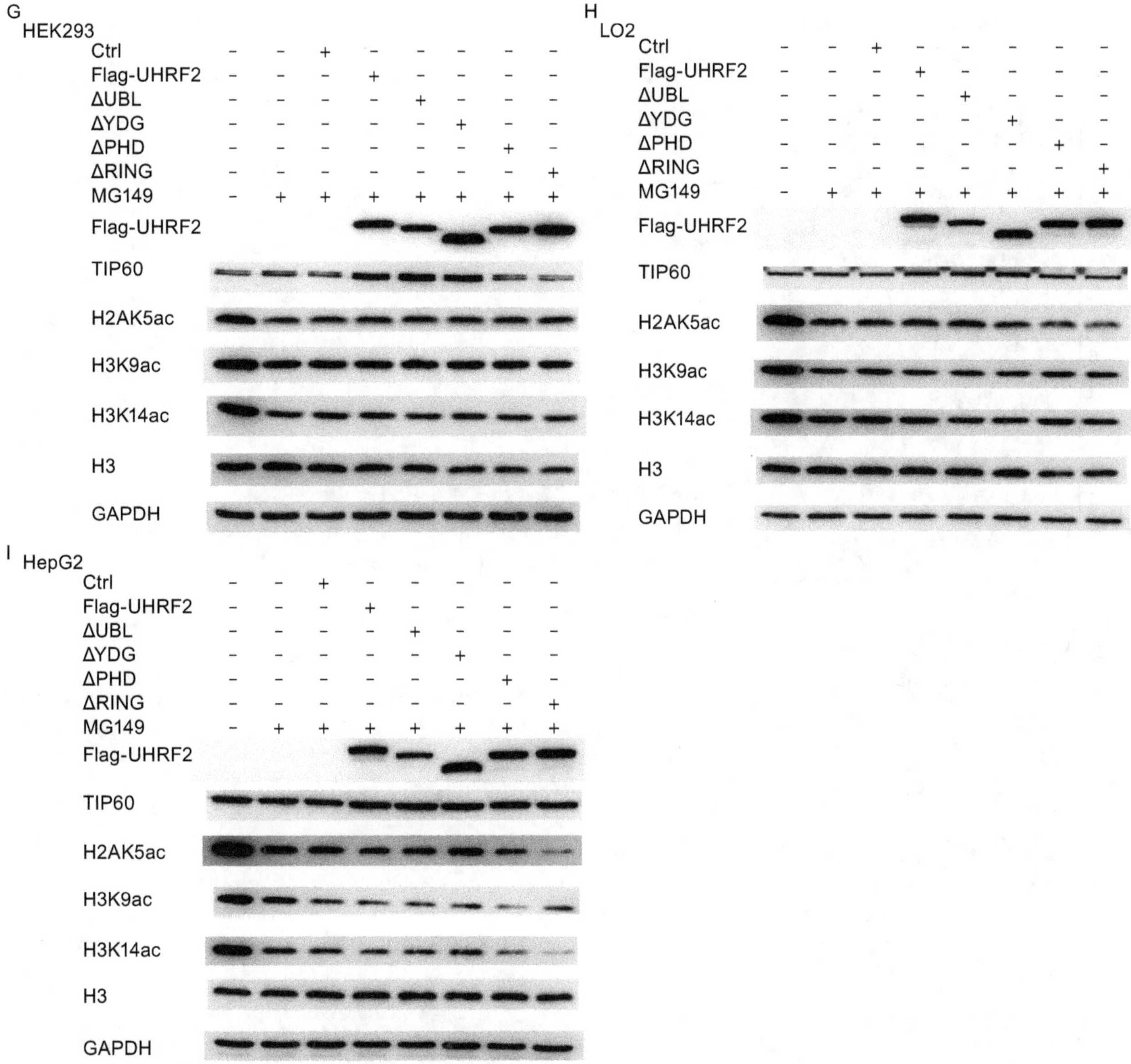

Figure 5. continued.

with transcription factors in tumor initiation and progression (Pichler et al., 2011). In this study, we found that UHRF2 regulated H3K9ac and H3K14ac expression via YDG and RING finger domain, and contrasting results were found in normal and cancer cells. In addition, we found that TIP60 elevated the expression of H3K9ac and H3K14ac. Based on these results, we investigated the UHRF2-mediated regulation of H3K9ac and H3K14ac via interaction with TIP60, and ubiquitination of TIP60 as a new substrate of E3 ligase UHRF2. Interestingly, the differential regulatory mechanism of UHRF2 against TIP60 in normal and cancer cells attracted our attention and investigation.

In summary, we suggest that UHRF2 regulated H3K9ac and H3K14ac expression via the TIP60 downstream signal. It is widely believed that cellular protein-protein interactions are universal. However, in this study, contrasting results were repeatedly found in normal and cancer cells. We used immunohistochemical analysis of HCC tissues and similar results were obtained. The differences in normal and cancer cells were the focus of this study, although we found that an intermediate signaling molecule "TIP60" participated in this mechanism. However, several unknown factors remain to be further investigated. We hypothesized that this mechanism includes specific signal transduction pathways, which trigger carcinogenesis in normal cells, and promote or co-promote malignant transformation of cells.

MATERIALS AND METHODS

Cell lines and cell culture

Human normal liver cells (LO2) were cultured in RPMI-1640 medium (Gibco, USA) with 15% fetal bovine serum (Gibco). Human

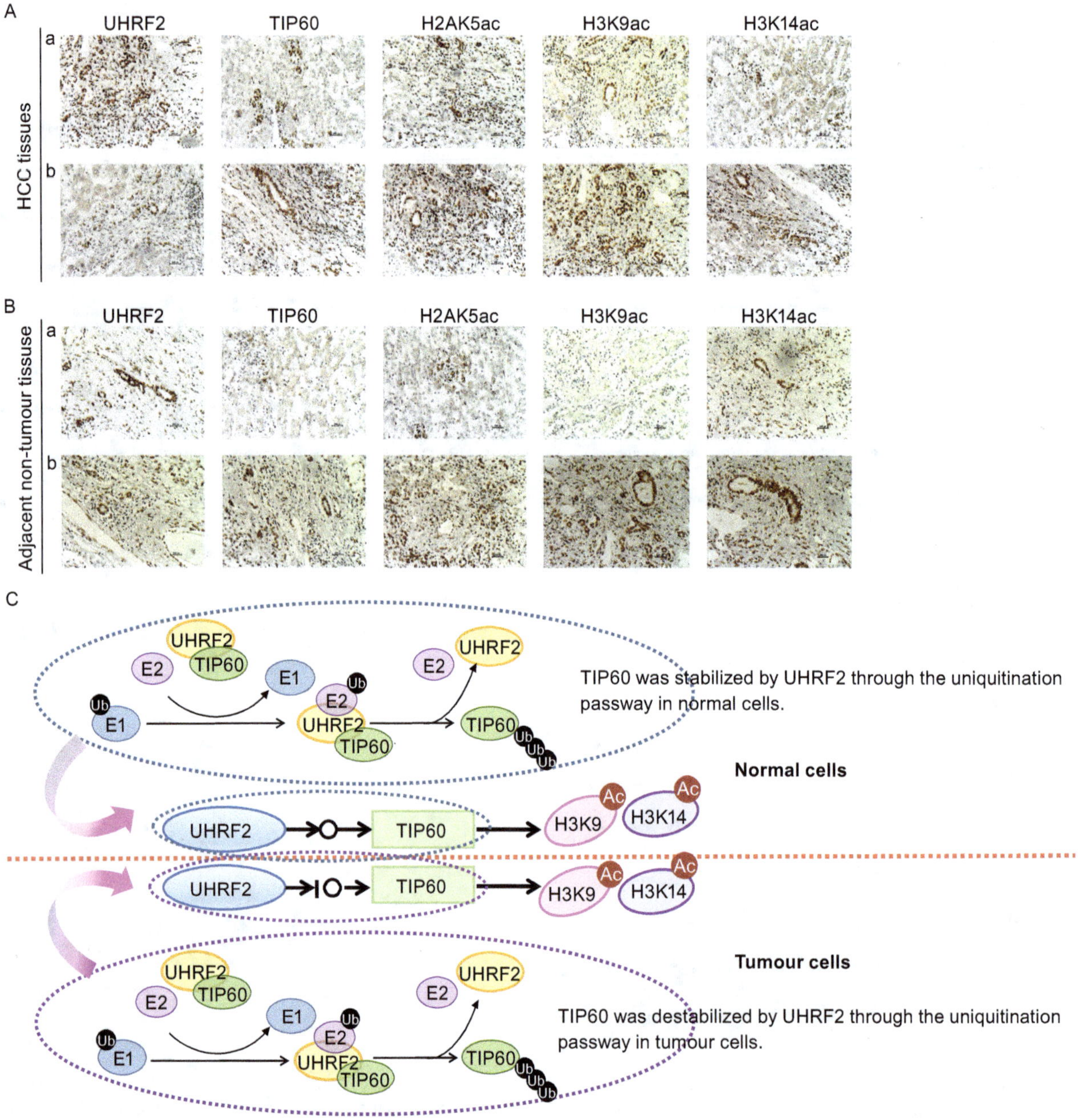

Figure 6. TIP60 and UHRF2 were negatively correlated in HCC tissues. (A) Immunohistochemistry with indicated antibodies in HCC tissues. UHRF2 showed a high expression in group "a" and low expression in group "b". (B) Immunohistochemical analysis using the indicated antibodies in adjacent non-tumor tissues. UHRF2 showed a high expression in group "a" and low expression in group "b". (C) Model of UHRF2-TIP60-H3K9ac and H3K14ac signaling axis.

embryonic kidney 293 cells (HEK293) and human hepatocellular liver carcinoma cells (HepG2) were routinely cultured in DMEM/HIGH glucose medium (Hyclone, USA) with 10% fetal bovine serum (Gibco). In this study, HEK293 and LO2 cells were regarded as normal cells. HepG2 cells were regarded as cancer cells.

Plasmids and siRNAs

Full-length and mutant UHRF2 genes (Flag-UHRF2, ΔUBL, ΔYDG, ΔPHD, ΔRING) were obtained by cloning the PCR fragment corresponding to UHRF2 into the pCMV-3×Flag vector. The pcDNA3.0-HA-Ub was a kind gift obtained from Lu Bai (Fudan University). The pcDNA3.1-His-TIP60, short hairpin RNA (shRNA) against UHRF2 and small interfering RNA (siRNA) against TIP60 were purchased from Invitrogen (USA) and GenePhama (China), respectively. Transfection was performed using Lipofectamine 2000 reagent (Invitrogen) according to the manufacturer's instructions.

Western blot

Total protein was isolated with RIPA lysis buffer (Beyotime Biotechnology, China) supplemented with phenylmethanesulfonyl fluoride (PMSF) (Roche, Germany) and a protease inhibitor cocktail (Roche) after 48 h transfection. Protein concentrations were measured using a BCA protein assay Kit (Beyotime Biotechnology) and boiled in loading buffer containing 1% SDS before infection. Proteins were separated by SDS-PAGE followed by Western blot with the indicated antibodies. The primary and secondary antibodies are listed in Table S1.

Tissue samples and immunohistochemistry

Forty-five cases of HCC tissue samples were collected from patients who underwent surgical resection at the department of the First Affiliated Hospital of Chongqing Medical University. The study was approved by the Ethics Committee of Chongqing Medical University (reference number: 2015018). All subjects signed informed consent. Patients' information will not be disclosed completely.

Immunohistochemistry was performed as described elsewhere (Das et al., 2016).

Co-immunoprecipitation

Cells were lysed with NP40 lysis buffer (Beyotime Biotechnology) supplemented with PMSF and a protease inhibitor cocktail at 48 h after transfection. The whole cell extracts were immunoprecipitated overnight at 4°C with the indicated antibody. After extensive washing with lysis buffer, the immune complexes were boiled in a loading buffer containing 1% SDS to deplete the interference of poly-ubiquitin chain on the other molecules in the same complex, and analyzed by Western blot.

Ubiquitination assay

To detect ubiquitinated TIP60 proteins, cells in a 10 mL cell bottle were transiently transfected with 4 μg HA-ubiquitin-expressing plasmids together with the indicated plasmid. Twelve hours before collection, cells were treated with 20 μmol/L of MG132 (Sigma-Aldrich, USA). Cells were lysed using NP40 lysis buffer supplemented with PMSF and protease inhibitor cocktail. The following procedures were identical to the immunoprecipitation assay except for the use of monoclonal HA antibody.

Half-life experiments

Cells were treated with 100 μg/mL of cycloheximide (CHX) (ABmole Bioscience, USA) for the indicated times, to inhibit protein synthesis. The cellular lysates were analyzed by Western blot.

Immunofluorescence

Cells seeded on the coverslips were fixed in 4% paraformaldehyde, and incubated with primary antibodies and secondary antibodies before staining with DAPI, mounted and observed, as described elsewhere (Liang et al., 2016; Takase et al., 2016).

Statistics analysis

All statistical data are described in the figure legends. Two-sided unpaired Student's *t*-test was used for experimental comparisons, after quantification with Image J software. Significance was represented as $^*P < 0.05$, and non-significance was indicated as n.s $P > 0.05$.

ACKNOWLEDGEMENTS

This work was supported byNational Natural Science Foundation of China (Grant No. 30872248). We acknowledge Lu Bai for providing the HA-Ub plasmid. We thank Yan Sun, Ming Li and Yanyan Xuan for their excellent technical assistance.

ABBREVIATIONS

CHX, cycloheximide; H3K9ac, the acetylation of histone H3 on lysine 9; H3K14ac, the acetylation of histone H3 on lysine 14; H2AK5ac, the acetylation of histone H2A on lysine 5; HATs, histone acetyltransferases; HCC, hepatocellular carcinoma; HDACs, histone deacetylases; PMSF, phenylmethanesulfonyl fluoride; PTMs, post-translational modifications; PVDF, polyvinylidene difluoride; RT, room temperature; h, hour; TBST, Tris-buffered saline Tween-20.

COMPLIANCE WITH ETHICS GUIDELINES

Shengyuan Zeng, Yangyang Wang, Ting Zhang, Lu Bai, Yalan Wang and Changzhu Duan declare that they have no conflict of interest. All procedures followed were in accordance with the ethical standards of the responsible committee on human experimentation (institutional and national) and with the Helsinki Declaration of 1975, as revised in 2000 (5). Informed consent was obtained from all patients for being included in the study

REFERENCES

Achour M, Fuhrmann G, Alhosin M, Ronde P, Chataigneau T, Mousli M, Schini-Kerth VB, Bronner C (2009) UHRF1 recruits the histone acetyltransferase Tip60 and controls its expression and activity. Biochem Biophys Res Commun [illegible]:523–528

Arnaudo AM, Garcia BA (2013) Proteomic characterization of novel histonepost-translational modifications. Epigenetics Chromatin 6:24

Bassi C, Li YT, Khu K, Mateo F, Baniasadi PS, Elia A, Mason J, Stambolic V, Pujana MA, Mak TW, Gorrini C (2016) The acetyltransferase Tip60 contributes to mammary tumorigenesis by modulating DNA repair. Cell Death Differ 23:1198–1208

Basu A, Rose KL, Zhang J, Beavis RC, Ueberheide B, Garcia BA, Chait B, Zhao Y, Hunt DF, Segal E, Allis CD, Hake SB (2009) Proteome-wide prediction of acetylation substrates. Proc Natl Acad Sci USA 106:13785–13790

Bhaumik SR, Smith E, Shilatifard A (2007) Covalent modifications of histones during development and disease pathogenesis. Nat Struct Mol Biol 14:1008–1016

Bronner C, Achour M, Arima Y, Chataigneau T, Saya H, Schini-Kerth VB (2007) The UHRF family: oncogenes that are drugable targets for cancer therapy in the near future? Pharmacol Ther 115:419–434

Dai C, Shi D, Gu W (2013) Negative regulation of the acetyltransferase TIP60-p53 interplay by UHRF1 (ubiquitin-like with PHD and RING finger domains 1). J Biol Chem 288:19581–19592

Das TP, Suman S, Papu John AM, Pal D, Edwards A, Alatassi H, Ankem MK, Damodaran C (2016) Activation of AKT negatively regulates the pro-apoptotic function of death-associated protein kinase 3 (DAPK3) in prostate cancer. Cancer Lett 377:134–139

Feng YL, Xiang JF, Kong N, Cai XJ, Xie AY (2016) Buried territories: heterochromatic response to DNA double-strand breaks. Acta Biochim Biophys Sin 48:594–602

Grezy A, Chevillard-Briet M, Trouche D, Escaffit F (2016) Control of genetic stability by a new heterochromatin compaction pathway involving the Tip60 histone acetyltransferase. Mol Biol Cell 27:599–607

Harrison JS, Jacobs TM, Houlihan K, Van Doorslaer K, Kuhlman B (2016) UbSRD: the ubiquitin structural relational database. J Mol Biol 428:679–687

Hershko A, Ciechanover A, Varshavsky A (2000) The ubiquitin system. Nat Am 10:1073–1081

Holt MT, David Y, Pollock S, Tang Z, Jeon J, Kim J, Roeder RG, Muir TW (2015) Identification of a functional hotspot on ubiquitin required for stimulation of methyltransferase activity on chromatin. Proc Natl Acad Sci USA 112:10365–10370

Ikura M, Furuya K, Matsuda S, Matsuda R, Shima H, Adachi J, Matsuda T, Shiraki T, Ikura T (2015) Acetylation of histone H2AX at Lys 5 by the TIP60 histone acetyltransferase complex is essential for the dynamic binding of NBS1 to damaged chromatin. Mol Cell Biol 35:4147–4157

Jacquet K, Fradet-Turcotte A, Avvakumov N, Lambert JP, Roques C, Pandita RK, Paquet E, Herst P, Gingras AC, Pandita TK, Legube G, Doyon Y, Durocher D, Cote J (2016) The TIP60 complex regulates bivalent chromatin recognition by 53BP1 through direct H4K20me binding and H2AK15 acetylation. Mol Cell 62:409–421

Jang SM, Kim JW, Kim CH, An JH, Johnson A, Song PI, Rhee S, Choi KH (2015) KAT5-mediated SOX4 acetylation orchestrates chromatin remodeling during myoblast differentiation. Cell Death Dis 6:e1857

Karmodiya K, Krebs A, Mustapha OS, Kimura H, Tora L (2012) H3K9 and H3K14 acetylation co-occur at many gene regulatory elements, while H3K14ac marks a subset of inactive inducible promoters in mouse embryonic stemcells. BMC Genom 424:1471–2164

Kouzarides T (2007) Chromatin modifications and their function. Cell 128:693–705

Lechtenberg BC, Rajput A, Sanishvili R, Dobaczewska MK, Ware CF, Mace PD, Riedl SJ (2016) Structure of a HOIP/E2~ubiquitin complex reveals RBR E3 ligase mechanism and regulation. Nature 529:546–550

Leithe E (2016) Regulation of connexins by the ubiquitin system: implications for intercellular communication and cancer. Biochim Biophys Acta 1865:133–146

Li E (2002) Chromatin modification and epigenetic reprogramming in mammalian development. Nat Rev Genetics 3:662–673

LI Y, Mori T, Hata H, Homma Y, Kochi H (2004) NIRF induces G1 arrest and associates with Cdk2. Biochem Biophys Res Commun 319:464–468

Liang BQ, Li DH, Hui HL, Hai YL, Xiao YL, Yong BY (2016) Cataract-causing mutation S228P promotes βB1-crystallin aggregation and degradation by separating two interacting loops in C-terminal domain. Protein Cell 7:501–515

Mo F, Zhuang X, Liu X, Yao PY, Qin B, Su Z, Zang J, Wang Z, Zhang J, Dou Z, Tian C, Teng M, Niu L, Hill DL, Fang G, Ding X, Fu C, Yao X (2016) Acetylation of Aurora B by TIP60 ensures accurate chromosomal segregation. Nat Chem Biol 12:226–232

Mori T, Li Y, Hata H, Kochi H (2004) NIRF is a ubiquitin ligase that is capable of ubiquitinating PCNP, a PEST-containing nuclear protein. FEBS Lett 557:209–214

Mori T, Ikeda DD, Fukushima T, Takenoshita S, Kochi H (2011) NIRF constitutes a nodal point in the cell cycle network and is a candidate tumor suppressor. Cell Cycle 10:3284–3299

Pichler G, Wolf P, Schmidt CS, Meilinger D, Schneider K, Frauer C, Fellinger K, Rottach A, Leonhardt H (2011) Cooperative DNA and histone binding by Uhrf2 links the two major repressive epigenetic pathways. J Cell Biochem 112:2585–2593

Pokholok DK, Harbison CT, Levine S, Cole M, Hannett NM, Lee TI, Bell GW, Walker K, Rolfe PA, Herbolsheimer E, Zeitlinger J, Lewitter F, Gifford DK, Young RA (2005) Genome-wide map of nucleosome acetylation and methylation in yeast. Cell 122:517–527

Qian G, Jin F, Chang L, Yang Y, Peng H, Duan C (2012) NIRF, a novel ubiquitin ligase, interacts with hepatitis B virus core protein and promotes its degradation. Biotechnol Lett 34:29–36

Renaud E, Barascu A, Rosselli F (2016) Impaired TIP60-mediated H4K16 acetylation accounts for the aberrant chromatin accumulation of 53BP1 and RAP80 in Fanconi anemia pathway-deficient cells. Nucl Acids Res 44:648–656

Su J, Wang F, Cai Y, Jin J (2016) The functional analysis of histone acetyltransferase MOF in tumorigenesis. Int J Mol Sci. doi:10.3390/ijms17010099

Sun Y, Sun J, Lungchukiet P, Quarni W, Yang S, Zhang X, Bai W (2015) Fe65 suppresses breast cancer cell migration and invasion through Tip60 mediated cortactin acetylation. Sci Rep 5:11529

Takase N, Koma YI, Urakawa N, Nishio M, Arai N, Akiyama H, Shigeoka M, Kakeji Y, Yokozaki H (2016) NCAM- and FGF-2-mediated FGFR1 signaling in the tumor microenvironment of esophageal cancer regulates the survival and migration of tumor-associated macrophages and cancer cells. Cancer Lett 380:47–58

Tan M, Luo H, Lee S, Jin F, Yang JS, Montellier E, Buchou T, Cheng Z, Rousseaux S, Rajagopal N, Lu Z, Ye Z, Zhu Q, Wysocka J, Ye Y, Khochbin S, Ren B, Zhao Y (2011) Identification of 67 histone marks and histone lysine crotonylation as a new type of histone modification. Cell 146:1016–1028

Vinther-Jensen T, Simonsen AH, Budtz-Jorgensen E, Hjermind LE, Nielsen JE (2015) Ubiquitin: a potential cerebrospinal fluid progression marker in Huntington's disease. Eur J Neurol 22:1378–1384

Wang Z, Zang C, Rosenfeld JA, Schones DE, Barski A, Cuddapah S, Cui K, Roh TY, Peng W, Zhang MQ, Zhao K (2008) Combinatorial patterns of histone acetylations and methylations in the human genome. Nat Genetics 40:897–903

Wang F, Zhang P, Ma Y, Yang J, Moyer MP, Shi C, Peng J, Qin H (2012) NIRF is frequently upregulated in colorectal cancer and its oncogenicity can be suppressed by let-7a microRNA. Cancer Lett 314:223–231

Yamada S, Ohta K, Yamada T (2013) Acetylated histone H3K9 is associated with meiotic recombination hotspots, and plays a role in recombination redundantly with other factors including the H3K4 methylase Set1 in fission yeast. Nucl Acids Res 41:3504–3517

Yamano K, Queliconi BB, Koyano F, Saeki Y, Hirokawa T, Tanaka K, Matsuda N (2015) Site-specific interaction mapping of phosphorylated ubiquitin to uncover parkin activation. J Biol Chem 290:25199–25211

13

The recombinant expression systems for structure determination of eukaryotic membrane proteins

Yuan He[1,2], Kan Wang[3], Nieng Yan[1,2]✉

[1] State Key Laboratory of Bio-membrane and Membrane Biotechnology, Tsinghua university, Beijing 100084, China
[2] Center for Structural Biology, School of Medicine, Tsinghua university, Beijing 100084, China
[3] China-Japan Friendship Hospital, Beijing 100029, China
✉ Correspondence: nyan@tsinghua.edu.cn (N. Yan)

ABSTRACT

Eukaryotic membrane proteins, many of which are key players in various biological processes, constitute more than half of the drug targets and represent important candidates for structural studies. In contrast to their physiological significance, only very limited number of eukaryotic membrane protein structures have been obtained due to the technical challenges in the generation of recombinant proteins. In this review, we examine the major recombinant expression systems for eukaryotic membrane proteins and compare their relative advantages and disadvantages. We also attempted to summarize the recent technical strategies in the advancement of eukaryotic membrane protein purification and crystallization.

KEYWORDS eukaryotic membrane proteins, recombinant expression, structural biology, integral membrane proteins (IMPs), fluorescence detected size exclusion chromatography (FSEC), protein purification and crystallization

INTRODUCTION

It is estimated that approximately 30% of the protein-coding genes are for integral membrane proteins (IMPs) in human (Overington et al., 2006; Murray et al., 2012). IMPs are critical players for many important physiological processes including metabolism, signal transduction, and energy conversion and utilization (Krogh et al., 2001). Aberrant expressions and activities of IMPs are associated with a variety of diseases such as cancer, Alzheimer's disease, and metabolic diseases (Ishikawa et al., 2004; Sanders and Myers, 2004; Overington et al., 2006; Aisenbrey et al., 2008; Bkaily and Al-Khoury, 2014). IMPs constitute more than 50% of the US Food and Drug Administration (FDA)-approved drug targets (Russell and Eggleston, 2000; Yildirim et al., 2007). Structures of eukaryotic membrane proteins are actively pursued for structure-based drug development.

In contrast to their physiological and pathophysiological significance, the progress on the structure biology of IMPs, particularly eukaryotic IMPs, has been relatively slow. By the end of March 2014, in total 466 unique membrane protein structures have been reported (Snider and Stephen, 2014), the majority of which are of prokaryotic origins. With respect to eukaryotic IMPs, more than half of the determined structures are for proteins obtained from endogenous sources (Bill et al., 2011). These proteins, exemplified by the electron transport chain complexes (Tsukihara et al., 1996; Xia et al., 1997; Sun et al., 2005), ATP synthases (Abrahams et al., 1994; Liu et al., 2004; Amunts et al., 2007), and photosystems (Kurisu et al., 2003; Liu et al., 2004; Amunts et al., 2007), usually exist in abundance and are biochemically stable, hence representing ideal candidates for structural analysis. However, the total types of endogenously abundant eukaryotic IMPs are limited. The majority of IMPs exist in low copies in the host species. Therefore, structural determination of most eukaryotic IMPs requires recombinant expression of the target proteins. The first atomic-resolution structure of a eukaryotic IMP obtained through recombinant expression, Kv1.2, was reported in 2005 (Long et al., 2005). Ever since, less than seventy structures have been obtained for eukaryotic IMPs generated through recombinant expression systems (Fig. 1).

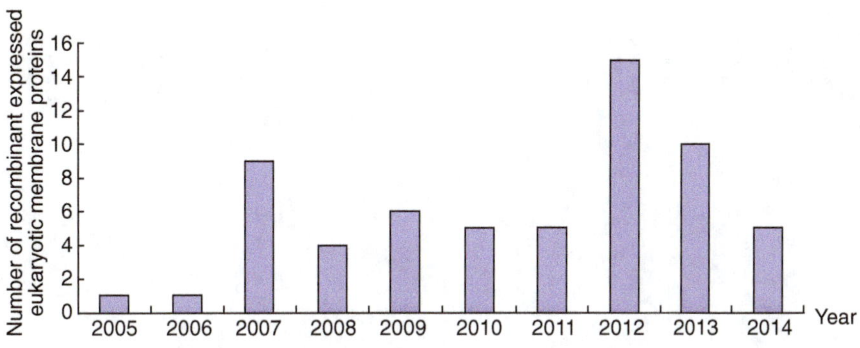

Figure 1. The development trends in recombinant expression eukaryotic membrane proteins. The structure number of eukaryotic membrane protein is limited by some obstacles such as low yield and instability in detergents. Since the first eukaryotic membrane protein structure was determined in 2005, over sixty structures have been emerged until now.

Out of the many challenges facing structural study of eukaryotic IMPs, production of sufficient quantities of well-behaved recombinant proteins represents the real technical bottleneck. Embedded in lipid bilayers, the structural integrity and proper functions of IMPs rely on the interactions with surrounding lipids (Phillips et al., 2009), which stabilize membrane proteins, provide lattice contacts, and in some occasions function as indispensable co-factors (van Meer et al., 2008). Recombinant expression of membrane proteins therefore requires a proper membrane environment. Whereas *Escherichia coli* proved to be the best host for most of prokaryotic IMPs of known structures, eukaryotic IMPs, with very few exceptions, requires eukaryotic expression systems including yeast, baculovirus-infected insect cells, and mammalian cells (Bill et al., 2011; Snider and Stephen, 2014).

In this review, in the hope of extracting some general principles on the expression and crystallization of eukaryotic membrane proteins, we examine the expression systems for the eukaryotic IMPs whose structures are obtained, attempt to summarize and compare the advantages and disadvantages of the representative recombinant expression systems, and delineate the detailed information in eukaryotic membrane protein purification and crystallization (Table 1).

RECOMBINANT EXPRESSION SYSTEMS FOR EUKARYOTIC MEMBRANE PROTEINS

The recombinantly expressed eukaryotic IMPs of known structures were obtained from four systems: *E. coli*, yeasts (*Pichia Pastoris* and *Saccharomyces cerevisiae*), insect cells, and mammalian cells. These expression systems have their respective advantages and disadvantages. The choice of an appropriate expression system remains empirical, largely depending on the biochemical and biological properties of the target proteins (Bernaudat et al., 2011). Among the recombinantly expressed eukaryotic IMPs whose structures have been solved, 4 were expressed in *E. coli*, 20 in yeast, 35 in insect cells, and 3 in mammalian cells. Below we will discuss these four expression systems.

E. coli

As the most frequently exploited recombinant expression system, *E. coli* BL21 (DE3) has the obvious advantage of rapid replication, time-saving operation, inexpensive cost, and mature and easy genetic manipulations (Sahdev et al., 2008). *E. coli* C43 (DE3) and C41 (DE3) strains were developed for over-expression of membrane proteins (Miroux and Walker, 1996; Dumon-Seignovert et al., 2004). Indeed, these *E. coli* strains were employed to over-express the large majority of prokaryotic IMPs whose structures were finally obtained. However, as the prokaryotic expression systems, they may lack the essential lipids, molecular chaperons, and post-translational modifications that are required for the correct membrane insertion, folding, and function of eukaryotic IMPs (Sahdev et al., 2008). As a result, only 4 structures were obtained for eukaryotic IMPs expressed in *E. coli* (Table 2). Despite the challenge to express eukaryotic membrane proteins in *E. coli*, researchers attempted to overcome these hurdles with codon-optimization (Burgess-Brown et al., 2008) and protein fusion with Mistic or GlpF tag to promote protein expression (AegeanSoftware, 2005; Drew et al., 2006; Neophytou et al., 2007), and co-expression of post-translational machineries to facilitate protein folding (Mironova et al., 2005; Mijakovic et al., 2006). Regardless of the effort, *E. coli* may not be an ideal system for eukaryotic IMP expression.

Yeast

Among the many yeast species, *Pichia Pastoris* (*Pichia*) and *Saccharomyces cerevisiae* (*S. cerevisiae*), which have been genetically well characterized, are the major systems to overexpress eukaryotic IMPs (Strausberg and Strausberg, 2001; Bornert et al., 2012). *Schizosaccharomyces pombe* is also employed for overexpression of IMPs, but not as widely

Table 1. Expression, purification, and crystallization information for eukaryotic membrane proteins*

No.	Expression systems		Protein name	Family	Extraction detergent	Purification detergent	Gel filtration detergent	Final concentration (mg/mL)	Temperature (°C)	Methods
1	*E. coli*	Bl21(DE3)	FLAP	MAPEG	DDM	DDM	$C_{12}E_8$ + C_8E_4			Sitting drop
2		C43(DE3)	PfAQP	Water channel	OG	OG	OG	6	18	Hanging drop
3		Bl21(DE3)	Kir3.1-prokaryotic Kir channel chimera	Potassium channel	DDM	DDM	NG	8	20	Sitting drop
4		Bl21(DE3)	Cytochrome b_{561}	Electron transport chain complexes	DM	NM	NG		18	Hanging drop
5	Yeast	*Pichia Pastoris*	Kv1.2 with β subunit	Potassium channel	DDM	DDM	DM	10	20	Hanging drop
6			Kv1.2-Kv2.1paddle	Potassium channel	DDM	DDM	CYMAL6 + CYMAL7	10	20	Hanging drop
7			Kv2.1paddle-Kv1.2 (F233 W)	Potassium channel	DDM	DDM	DDM	10	20	Hanging drop
8			Kir2.2 inward-rectifier	Potassium channel	DM	DM	DM	8	20	Hanging drop
9			GIRK2 (Kir3.2) K^+ channel	Potassium channel	DM	DM	DM	6–7	20	Hanging drop
10			K_{2P}1.1(KWIK-1)	Potassium channel	DDM	DDM	DDM	10	20	Hanging drop
11			K_{2P}4.1(TRAAK)	Potassium channel	DDM	DDM	DM	5	4	Hanging drop
12		*Pichia Pastoris*	Calcium release-activated calcium channel	Calcium channel	DDM	DDM	NM + NG	16	17	Hanging drop
13			SoPIP2;1	Water channel	OG	OG	OG	10	4	Hanging drop
14			HsAQP5	Water channel	NG	NG	NG	10	8	Hanging drop
15	Yeast		HsAQP4	Water channel	OG	OG	OG	30	25	Hanging drop
16			P-Glycoprotein	ABC transporter	Triton	DDM	DDM	10	4	Sitting drop
17			P-Glycoprotein	ABC transporter	DDM	DDM	UDM		4	Hanging drop
18			LTC_4S	MAPEG	DDM	DDM	DDM	6.5	4	Sitting drop
19			Histamine H_1 receptor	GPCR	DDM	DDM	DDM	30–40	20	LCP
20		*S. cerevisiae*	AHA2 (H^+ pump)	Pump	DDM	DDM	C12E8 + CYMAL5	20–30	4	Hanging drop
21			VrH^+-Ppase	M-PPase	DDM	DM	DM	10	20	Hanging drop

Table 1. continued

No.	Expression systems		Protein name	Family	Extraction detergent	Purification detergent	Gel filtration detergent	Final concentration (mg/mL)	Temperature (°C)	Methods
22			NRT1.1	MFS transporter	DDM	DDM	DDM	10	4	Hanging drop
23			CAAX protease Ste24p	Intramembrane protease	DDM	DDM/$C_{12}E_7$	$C_{12}E_7$	7.35	4/17	Hanging drop
24			PiPT	MFS transporter	DDM	DDM	NG	10–15	20	Hanging drop
25	Insect cell	*S. frugiperda*	β_2AR (Fab)	GPCR	DDM	DDM	DDM	8–12	22	Bicelle
26			β_2AR (T4L)	GPCR	DDM	DDM	DDM	Concentrated	22	LCP
27			β_2AR-agonist complex	GPCR	MNG	MNG	MNG (0.1%)	50	20	LCP
28			β_2AR-GS complex	GPCR	MNG	MNG	MNG (0.1%)	90	20	LCP
29			A2A adenosine receptor	GPCR	DDM	DDM	DDM	70	20	LCP
30			CXCR4	GPCR	DDM	DDM	DDM	60–70	20	LCP
31			Dopamine D3 receptor	GPCR	DDM	DDM	DDM	20–30	20	LCP
32			Sphingosine 1-phosphate receptor	GPCR	DDM	DDM	DDM	100	20	LCP
33			M2 muscarinic acetylcholine receptor	GPCR	Digitonin + Na-cholate	DM	MNG	30	20	LCP
34			M3 muscarinic acetylcholine receptor	GPCR	DDM	DDM	MNG	60	20	LCP
35			κ-Opioid receptor	GPCR	DDM	DDM	DDM	40	20	LCP
36			μ-Opioid receptor	GPCR	DDM + CHAPS + CHS	DDM + CHAPS + CHS	MNG + CHS	30	20	LCP
37			δ-Opioid receptor	GPCR	MNG + CHAPS + CHS	MNG + CHAPS + CHS	MNG + CHAPS + CHS	50	20	LCP
38	Insect cell	*S. frugiperda*	N/OFQ receptor	GPCR	DDM + CHS	DDM + CHS	DDM + CHS	40	20	LCP
39			CCR5	GPCR	DDM + CHS	DDM + CHS	DDM + CHS	40–50	20	LCP
40			PAR1	GPCR	DDM + CHS	DDM + CHS	DDM + CHS	40–50	20	LCP
41			5-$HT_{1B/2B}$ serotonin receptor	GPCR	DDM + CHS	DDM + CHS	DDM + CHS	50–80	20	LCP
42			Smoothened receptor	GPCR	DDM + CHS	DDM + CHS	DDM + CHS	50–60	20	LCP
43			Glucagon receptor	GPCR	DDM + CHS	DDM + CHS	DDM + CHS	80	20	LCP
44			Metabotropic Glutamate Receptor 1	GPCR	DDM + CHS	DDM + CHS	DDM + CHS	50–80	20	LCP
45			P2X4	Channel	DDM	DDM	DDM	2	4	Hanging drop

Table 1. continued

No.	Expression systems		Protein name	Family	Extraction detergent	Purification detergent	Gel filtration detergent	Final concentration (mg/mL)	Temperature (°C)	Methods
46			ASIC1	Channel	DDM	DDM	DDM	5	4	Hanging drop
47			GluA2	Channel	DDM	DDM	C_{11}Thio + lipids	2	4	Hanging drop
48			GLuCla	Cys-loop receptor	DDM	DDM	DDM	2	4	Hanging drop
49			CX26	Gap junction	DDM	DDM	UDM	30	4	Hanging drop
50			UT-B	Urea Transporter	DM	DM	OG	8	4	Sitting drop
51			ZMPSTE24	Intramembrane protease	DDM + CHS or OGNG + CHS	DDM + CHS or OGNG + CHS	DDM + CHS or OGNG + CHS	9–11	20	Sitting drop
52	Insect cell		ABCB10	ABC transporter	DDM	DDM	DDM		20	Sitting drop
53			Claudin-15	Tight junction	DDM	LMNG	LMNG	7	20	LCP
54			NRT1.1	MFS transporter	DDM	DDM	DDM	10	4	Hanging drop
55		*Trichoplusiani*	β1 adrenergic receptor	GPCR	DM	Octylthioglucoside	Octylthioglucoside	6	18	Hanging drop
56			NTS1 Neurotensin Receptor	GPCR	MNG + CHS	MNG + CHS	MNG + CHS	20–25	20	LCP
57			CmClC	H^+/Cl^- exchange transporter	DM	DM	DM	10	20	Hanging drop
58			Corticotropin-releasing factor receptor	GPCR	DM	DM	DM	20–30	22.5	LCP
59			GLUT1	MFS transporter	DDM	DDM	NG	10	4	Hanging drop
60		COS-1 cells	Rhodopsin	Rhodopsin	DDM	DDM	C_8E_4			
61	Mammalian cell	HEK293	RhCG	Channel	OG	OG	OG	5	20	Hanging drop
62			Dopamine transporter	Solute carrier transpoter	DDM	DDM CHS POPC: POPE:POPG = 3:1:1	DM CHS POPE	3	4	Hanging drop

* The blank in the table is due to the details in the reported method has not been mentioned.

Table 2. ***E. coli*** **as an expression system for eukaryotic membrane protein**

Expression systems		No.	Protein	Species	PDB code	Reference
E. coli	Bl21(DE3)	1	FLAP	*Homo sapiens*	2Q7 M 2Q7R	Ferguson et al., 2007
	C43 (DE3)	2	PfAQP	*Plasmodium falciparum*	3C02	Newby et al., 2008
	Bl21(DE3)	3	Kir3.1-prokaryotic Kir channel chimera	*Streptomyces lividans*	2QKS	Nishida et al., 2007
	Bl21(DE3)	4	Cytochrome b_{561}	*Arabidopsis thaliana*	4O6Y, 4O79, 4O7G	Lu et al., 2014

Table 3. Yeast as an expression system for eukaryotic membrane protein*

Expression systems		No.	Protein	Species	PDB code	Reference
Yeast	*Pichia Pastoris*	1	Kv1.2 with β subunit	*Drosophila melanogaster*	2A79	Long et al., 2005
		2	Kv1.2-Kv2.1 paddle	*Rattus norvegicus*	2R9R	Long et al., 2007
		3	Kv2.1paddle-Kv1.2 (F233 W)	*Rattus norvegicus*	3LNM	Tao et al., 2010
		4	Kir2.2 Inward-Rectifier	*Gallus gallus*	3JYC	Tao et al., 2009
		5	GIRK2 (Kir3.2) channel	*Mus musculus*	3SYO	Whorton and MacKinnon, 2011
		6	$K_{2P}1.1$ (KWIK-1)	*Homo sapiens*	3UKM	Miller and Long, 2012
		7	$K_{2P}4.1$ (TRAAK)	*Homo sapiens*	3UM7	Brohawn et al., 2012
		8	Calcium release-activated calcium channel	*Drosophila melanogaster*	4HKR	Xiaowei Hou, 2012
		9	SoPIP2;1	*Spinacia oleracea*	1Z98 2B5F	Tornroth-Horsefield et al., 2006
		10	HsAQP5	*Homo sapiens*	3D9S	Horsefield et al., 2008
		11	HsAQP4	*Homo sapiens*	3GD8	Ho et al., 2009
		12	P-Glycoprotein	*M. musculus*	3G5U, 3G60, 3G61	Aller et al., 2009
		13	P-Glycoprotein	*Caenorhabditis elegans*	4F4C	Jin et al., 2012
		14	LTC4S	*Homo sapiens*	2PNO	Ago et al., 2007
		15	Histamine H_1 receptor	*Homo sapiens*	3RZE	Shimamura et al., 2011
	S. cerevisiae	16	AHA2 (H+ pump)	*Arabidopsis thaliana*	3B8C	Pedersen et al., 2007
		17	VrH+-Ppase	*Vigna radiata*	4A01	Lin et al., 2012
		18	NRT1.1	*Arabidopsis thaliana*	4CL4	Parker and Newstead, 2014
		19	CAAX protease Ste24p	*Saccharomyces mikatae*	4IL3	Pryor et al., 2013
		20	PiPT	*Piriformospora indica*	4J05	Pedersen et al., 2013

* For some proteins like GPCR and potassium channel, only the representative ones are listed.

Table 4. Insect cell as an expression system for eukaryotic membrane protein*

Expression systems		No.	Protein	Species	PDB code	Reference
Insect cell	*S. frugiperda*	1	β_2AR (Fab)	*Homo sapiens*	2R4R 2R4S	Rasmussen et al., 2007
		2	β_2AR (T4L)	*Homo sapiens*	2RH1	Cherezov et al., 2007
		3	β_2AR-agonist complex	*Homo sapiens*	3PDS	Rosenbaum et al., 2011
		4	β_2AR-GS complex	*Homo sapiens*	3SN6	Rasmussen et al., 2011a, b
		5	A_{2A} adenosine receptor	*Homo sapiens*	3EML	Jaakola et al., 2008
		6	CXCR4	*Homo sapiens*	3ODU 3OE8	Wu et al., 2010
		7	Dopamine D3 receptor	*Homo sapiens*	3PBL	Chien et al., 2010
		8	Sphingosine 1-phosphate receptor subtype 1	*Homo sapiens*	3V2 W 3V3Y	Hanson et al., 2012
		9	M2 muscarinic acetylcholine receptor	*Homo sapiens*	3UON	Haga et al., 2012
		10	M3 muscarinic acetylcholine receptor	*Rattus norvegicus*	4DAJ	Kruse et al., 2012
		11	κ-Opioid receptor	*Homo sapiens*	4DJH	Wu et al., 2012
		12	μ-Opioid receptor	*Mus musculus*	4DKL	Manglik et al., 2012
		13	δ-Opioid receptor	*Mus musculus*	4EJ4	Granier et al., 2012
		14	N/OFQ receptor	*Homo sapiens*	4EA3	Thompson et al., 2012
		15	CCR5	*Homo sapiens*	4MBS	Tan et al., 2013
		16	PAR1	*Homo sapiens*	3VW7	Zhang et al., 2012
		17	5-$HT_{1B/2B}$ serotonin receptor	*Homo sapiens*	4IAR 4IB4	Wang et al., 2013a, b; Wacker et al., 2013
		18	Smoothened receptor	*Homo sapiens*	4JKV	Wang et al., 2013a, b
		19	Glucagon receptor	*Homo sapiens*	4L6R	Siu et al., 2013
		20	Metabotropic glutamate receptor1	*Homo sapiens*	4OR2	Wu et al., 2014
		21	$P2X_4$	*Danio rerio (Zebra fish)*	3I5D 3H9 V 4DW1	Kawate et al., 2009; Hattori and Gouaux, 2012
		22	ASIC1	*Gallus gallus*	2QTS 3HGC	Jasti et al., 2007; Gonzales et al., 2009
		23	GluA2	*Rat*	3KG2 3KGC	Sobolevsky et al., 2009
		24	GLuClα	*Caenorhabditis elegans*	3RHW, 3RIF, 3RI5 3RIA	Hibbs and Gouaux, 2011
		25	CX26	*Homo sapiens*	2ZW3	Maeda et al., 2009
		26	UT-B	*Bos taurus*	4EZC 4EZD	Levin et al., 2012
		27	ZMPSTE24	*Homo sapiens*	4AW6	Quigley et al., 2013
		28	ABCB10	*Homo sapiens*	4AYT	Shintre et al., 2013
		29	Caludin-15	*Mus Musculus*	4P79	Suzuki et al., 2014
		30	NRT1.1	*Arabidopsis thaliana*	4OH3	Sun et al., 2014
	Trichoplusia ni	31	β1 adrenergic receptor	*Meleagris gallopavo*	2VT4	Warne et al., 2008
		32	NTS1 Neurotensin Receptor	*Rattus norvegicus*	4GRV	White et al., 2012
		33	CmClC	*Cyanidios-chyzonmerolae*	3ORG	Feng et al., 2010
		34	Corticotropin-releasing factor receptor	*Homo sapiens*	4K5Y	Hollenstein et al., 2013
		35	GLUT1	*Homo sapiens*	4PYP	Deng et al., 2014

* For some proteins like GPCR and potassium channel, only the representative ones are listed.

Table 5. Mammalian cell as an expression system for eukaryotic membrane protein

Expression system		No.	Protein	Species	PDB code	Reference
Mammalian	HEK293	1	Rodopsin	*Homo sapiens*	2J4Y	Standfuss et al., 2007
		2	RhCG	*Homo sapiens*	3HD6	Gruswitz et al., 2010
		3	Dopamine transporter	*Homo sapiens*	4M48	Penmatsa et al., 2013

Table 6. Comparison among four expression system

	E. coli	Yeast (*Pichia*)	Insect cell (Sf9)	Mammalian cell (HEK293)
Duration time before cell cultivation (Days)	3–5	6–8	25–30	Transient: 3–5 Stable: at least 30
Cell cultivation time for 1L test (Days)	1–2	3–7	2–4	2–4
Cost for 1L test ($) in China	15–20	20–25	200–250	200–250
Number of available eukaryotic IMP structures	4	20	34	3

as *Pichia* and *S. cerevisiae* (Yang et al., 2009). During the past thirty years, yeast has proved to be a useful expression system: 15 eukaryotic IMP structures have been determined for proteins expressed in *Pichia* expression system and 5 by *S. cerevisiae*. Most of the structurally available eukaryotic channels such as potassium channels and water channels were expressed in yeast, as listed in Table 3.

Pichia is considered the best expression system among yeast species (Cereghino and Cregg, 2000). Several elements contribute to its increasing applications, including the simplicity of genetic manipulation, the high yield of heterologous protein, the cost-effective chemical reagents, as well as the ability of post-translational modifications (Macauley-Patrick et al., 2005). For these reasons, *Pichia* is a more suitable expression system for producing eukaryotic IMP than *E. coli*. *Pichia* shares the advantage of the molecular and genetic manipulation with *S. cerevisiae*, yet it adds extra advantage of 10- to 100- fold biomass out of the same cultural volume comparing with *S. cerevisiae* (Macauley-Patrick et al., 2005).

The improved techniques and the commercial availability together promote the development of *Pichia* (Cereghino and Cregg, 2000). *Pichia* is a methylotrophic yeast, capable of utilizing methanol as its sole carbon source (Yurimoto and Sakai, 2009). A promoter derived from the alcohol oxidase I (*AOXI*), which is the first-step enzyme in the methanol metabolism, strictly controls the expression of the foreign proteins (Macauley-Patrick et al., 2005). The commercial vector pPICZ (or pPICZα) takes advantage of the *AOXI* promoter, being induced by methanol (Li et al., 2007). *AOXI* promoter is prevailing than other promoters like *PMA1* and *GPD1* for its strong and highly inducible ability (Cereghino and Cregg, 2000). After the vector is readily prepared and transformed into the competent cells, the target gene can be inserted into the *Pichia* genome in high efficiency via homologous recombination to generate stable cell lines, and then the colonies with multiple copies that exhibit the highest protein expression level will be screened out through zeocin-spread plates (Daly and Hearn, 2005). This zeocin selective marker for transformation selection is important regarding to the convenience of genetic manipulation in yeast. All the procedure typically takes about 10–15 days for a complete procedure from subcloning to protein expression. A potential disadvantage of the yeast culture concerns the difficulty in cell disruption due to the thick and hard cell walls.

Insect cell

The baculovirus infected insect cell system is undoubtedly the dominant heterologous expression system for obtaining eukaryotic IMPs (Contreras-Gomez et al., 2014). The most common method for generating recombinant baculovirus is based on the site-specific transposition of an expression cassette into a baculovirus shuttle vector (bacmid) that is amplified in *E. coli* (Ciccarone et al., 1998). The process is very convenient: clone the target gene into the pFastBac vector which uses the strong AcMNPV polyhedron (PH) as the promoter for high level protein expression, then transform the pFastBac vector into DH10Bac *E. coli* competent cells. DH10Bac cells possess a baculovirus shuttle vector (bacmid) with a transposon site and a helper plasmid, thus can help the pFastBac vector to have a transposition on the bacmid. Once the transposition occurs and the recombinant bacmid is generated, the bacmid could be isolated and purified for transfection. After the insect cells are cultured into a desired confluence, they are transfected by the purified bacmid DNA to generate a recombinant baculovirus that used for preliminary expression test (Contreras-Gomez et al., 2014). The pFastBac is ampicillin resistance and Bacmid is kanamycin resistance, and these selective markers provide expedience for this baculovirus expression system. It takes approximately 3–4 weeks to complete these procedures for initial protein expression test.

There are two most popular insect cell lines used for IMP expression, *Spodoptera frugiperda* (Sf9) and *Trichoplusia ni* (Hi5). Heterologous proteins have disparate performances on the yield and behavior when expressed in these two cell lines (Unger and Peleg, 2012). Till now, 30 structures were obtained for eukaryotic IMPs from Sf9 expression system and 5 from Hi5 (Table 4).

After the protein IL-2 was first expressed in large scale with the baculovirus-infected insect cells in 1985, this system has been quickly accepted and widely used (Smith et al., 1983; Maeda et al., 1985). Owing to the convenience of scale up, safety and accuracy (Kost et al., 2005), the baculoviral insect cell system has yielded the largest number of eukaryotic IMPs up to date (Table 4). Notably, among the 35 eukaryotic IMP structures, 23 are of G-protein coupled receptors (GPCR) (Table 4). The insect cell system has been the prevailing expression system for eukaryotic IMP. However, the cost for the cultural medium may represent a serious roadblock for most laboratories.

Mammalian cell

Mammalian expression system has become one of the popular recombinant protein production systems for its proper post-translational modification and human protein-like structure assembly (Khan, 2013). HEK (human embryo kidney) and CHO (Chinese hamster ovary) are two broadly used cell lines for recombinant expression. These two cell lines are extensively applied by researchers to do functional assay such as the electrophysiological assay (Kawate et al., 2009). Both these two cell lines can be applied for transient and stable transfections (Zhu, 2012). For the transient transfection approach, it is relative easier to reach to a reasonable protein expression level, but this expression level may vary from batch to batch. On the other hand, although the proteins have higher productivity and less variation in the stable transfection method, it is very time consuming (one month at least) (Condreay et al., 1999; Baldwin et al., 2003). Consequently, it is a balance for scientists to choose between these two transfection methods.

HEK293 is a specific cell line originally derived from HEK cells, while the number "293" comes from Graham's habit of numbering his experiments (Louis et al., 1997). Large scale, transient transfection of HEK293 in suspension culture is a reliable way to generate milligram quantities of recombinant eukaryotic IMPs. When the gene of interest is ligated into the vector pcDNA3 or pCMV5, the complete plasmid is then transfected into the HEK293 cells and the cells are harvested after 48 h (Thomas and Smart, 2005). The whole procedure is more or less similar to that of the insect cell system, only with a couple of exceptions. For example, 5%–10% CO_2 is required for maintaining the HEK293 cells, and the culture temperature is 37°C for HEK293 but not 27°C as for insect cells. The overall process usually requires one to two weeks from initial cloning to small scale test for the transient expression. However, ascribe to the low yield, slow growth rate and higher cost of complex media (Sunley and Butler, 2010), the number of eukaryotic IMP structures generated based on the mammalian cells is very limited. So far, only three eukaryotic IMP structures are from this system, and two of them are obtained from HEK293 cells (Table 5).

The BacMam system has to be mentioned for its safety, reproducibility and efficiency (Dukkipati et al., 2008). The baculoviruses are engineered by inserting a mammalian expression cassette for delivering foreign genes in mammalian cells. Their non-replicating property makes they are safe and well-tolerated by mammalian cells. BacMam system gains widespread use for its safety and rapid manipulation (Reeves et al., 2002; Baconguis and Gouaux, 2012). Depending on the cell type, cell division rate and transduction efficiency, it lasts 5–14 days to detect the gene expression (Dukkipati et al., 2008). The dopamine transporter structure was determined by the BacMam system (Penmatsa et al., 2013).

From the foregoing discussion, it is concluded that every expression system has their distinctive properties for protein expression. We compare their relative merits for an intuitional understanding of each system which can help researchers to make the best choice for their proteins expression (Table 6).

HOMOLOGUE SCREEN

Eukaryotic membrane proteins are very difficult to yield in large quantities, and most of them tend to be unstable in the presence of detergents. As a result, identification of well-expressed proteins is very essential. Homologue screen is widely applied for researchers to discover well-behaved proteins (Kawate et al., 2009; Xiaowei Hou, 2012).

Fluorescence detected size exclusion chromatography (FSEC) is a powerful method for homologue screen (Drew et al., 2006; Newstead et al., 2007). Compared with the common protocols, GFP fusion membrane proteins can be detected by measuring fluorescence in whole cells during the over-expression process. It saves time to help people preclude proteins that have no expression or low expression level. Also, it is much easier to assess the integrity of proteins by detecting the fluorescence in SDS polyacrylamide gels. Moreover, FSEC could be employed to figure out the most stable detergents in initial detergent screen. Considering these benefits, this technology is very widely applied (Jasti et al., 2007; Gonzales et al., 2009; Kawate et al., 2009; Sobolevsky et al., 2009). Taking P2X receptor as an example (Kawate et al., 2009), because of its aggregation and instability problems, researchers applied this method to screen 35 orthologs and finally got one species which was fit for crystallization. FSEC is proven to be one of the most robust methods to facilitate the identification of appropriate candidates for solving the structures of eukaryotic membrane proteins.

OPTIMAL CONSTRUCTS DESIGN

Optimizing constructs is very beneficial for getting the well-packed crystals. One way for optimizing constructs is to "cut off". Limited proteolysis is a conventional method to find the optimal constructs. Besides, it is worth noting that either N-terminal tag or C-terminal tag is removed before crystallization in most crystallization cases (Long et al., 2005; Long

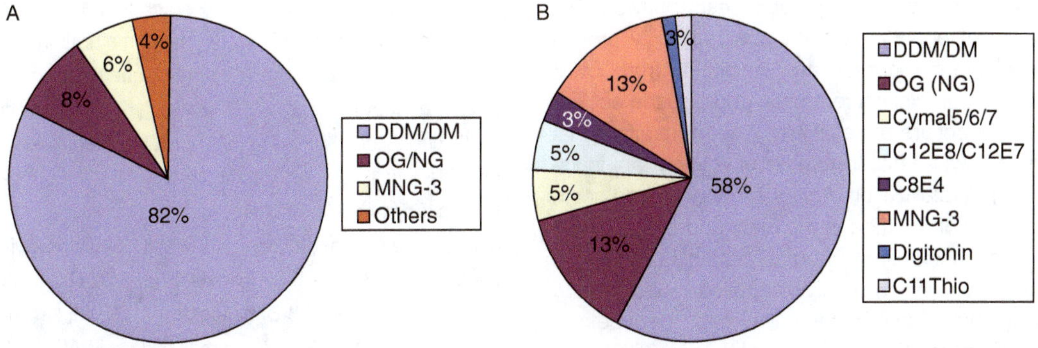

Figure 2. Detergents used for extraction and crystallization of eukaryotic membrane proteins. (A) Detergents for protein extraction and purification. DDM/DM can be applied for most eukaryotic membrane proteins in extraction step. (B) Detergents for protein crystallization. DDM/DM is the major detergent for the crystallization of eukaryotic membrane proteins.

et al., 2007; Gonzales et al., 2009; Maeda et al., 2009; Sobolevsky et al., 2009; Tao et al., 2009). For instance, the desensitized ASIC1 was crystallized by removal of 25 N-terminal and 64 C-terminal residues (Jasti et al., 2007).

The contrary way for optimizing constructs is to "add up". T4 lysozyme (T4L) insertion and Fab/nanobody replacement are applied to produce stable proteins. The T4L fragment is soluble enough to effectively increase the solvent-exposed area, thereby facilitating protein-protein interactions and generating novel crystal packing interfaces (Cherezov et al., 2007). Fab/nanobody, which are generated from monoclonal antibodies, can reduce the protein flexibility and improve the conformational homogeneity (Zhou et al., 2001; Rasmussen et al., 2007). GPCR is one of the most successful cases employing T4L and Fab/nanobody to the ultimate structure determination (Rasmussen et al., 2007, Rasmussen et al., 2011a, b).

Mutagenesis is an alternative way for constructs design. In order to improve the crystallization behavior and stabilize the tetrameric state of the glutamate receptor GluA2, point mutations were introduced, preventing non-specific aggregation and disulphide bond formation (Sobolevsky et al., 2009). And E329Q was introduced in order to stabilize GLUT1 in a certain conformation (Deng et al., 2014). Plus, glycosylation is the most common post-translational modification of eukaryotic membrane proteins and leads to heterogeneity of proteins. Thus, mutating of glycosylation sites or deglycosylation by enzymes is an essential step for crystallization (Deng et al., 2014).

DETERGENTS, LIPIDS AND CRYSTALLIZATION

We have summarized the detergents used for protein purification and crystallization from Table 1. 51 eukaryotic membrane proteins can be extracted from DDM or DM (Fig. 2A), suggesting DDM/DM are the detergents suitable for the extraction process of the majority of eukaryotic membrane proteins. Collaterally, nearly half of the eukaryotic membrane protein crystals are obtained from DDM/DM, indicating DDM/DM are worthy of a trial for crystallization in the first place (Fig. 2B and Table 1). Apart from these conventionally applied detergents, new detergents have also been developed to meet the new requirements. For example, when purifying β_2 adrenergic receptor-Gs protein, the authors stabilized protein complex by exchanging DDM with a newly developed maltose neopentyl glycol detergent MNG-3 (NG310, Anatrace) to prevent the complex dissociated from original detergent DDM (Chae et al., 2010; Rasmussen et al., 2011a, b).

It is worth noting that additional lipids are able to help crystal packing. There are three ways of lipid combinations. The first is mixing lipids with detergent(s) in hanging or sitting drop during crystallization. Take mammalian voltage-dependent *shaker* family potassium channel as an example, the author utilized 0.1 mg/mL 3:1:1 POPC: POPE: POPG throughout purification and crystallization to obtain crystals (Long et al., 2005). The second approach is lipid cubic phase (LCP) method. The lipid cubic phase is a dynamic structure consisting of a highly organized single lipid bilayer pervaded by an inter-connected aqueous channel (Landau and Rosenbusch, 1996). Martin has an elaborate discussion about LCP method which we will not go into details in this review (Caffrey and Cherezov, 2009). The crystal structure of β_2AR-GS complex was determined by the use of 7.7 MAG as the host lipid for crystallization (Rasmussen et al., 2011a, b). The third way is bicelle method, which is regarded as an intermediate approach between the traditional detergent crystallization method and the rigid LCP method. Bicelle can be considered as a lipid bilayer disc that formed by a long chain lipid and a short chain lipid or detergent (Agah and Faham, 2012). The general composition is 3:1 DMPC: CHAPSO. Several protein structures were determined utilizing bicelle method (Rasmussen et al., 2007; Payandeh et al., 2011).

Last but not the least, we will elaborate a few messages for the crystallization of eukaryotic membrane protein drawn from Table 1: (a) Protein concentration: almost all the protein concentration for crystallization is above 5 mg/mL. (b) Crystallization temperature: if we expel the LCP method that is routinely crystallized at 20 ± 2°C, nearly half of the eukaryotic membrane proteins are crystallized at low temperature, especially on 4°C. At cold temperature, for protein with "normal" solubility, protein will be more soluble in high salt and precipitate from lower concentration of the precipitant reagents, and also the equilibrium diffusion rate occurs more slowly. These manifest that crystallization at lower temperature is absolutely an indispensable trial. (c) Crystallization methods: hanging drop or sitting drop crystallization method is the main and conventional approach for most eukaryotic membrane protein. LCP method is an up-rising star which is extensively applied in determining the GPCR's structures which we have mentioned before. Remarkably, LCP method is not only propitious to GPCR, but also is able to be applied for none-GPCR protein structures determination (Suzuki et al., 2014).

CONCLUSION

In this review, we discuss the benefits and drawbacks of different expression systems for eukaryotic membrane protein, and illustrate some general methods of recent advances for eukaryotic membrane protein purification and crystallization. We hope our work can provide help to those who are interested and work on eukaryotic membrane proteins. Although the discussion of eukaryotic membrane protein structure determined by Cryo-EM or NMR is beyond the scope of this review, the general methodologies and technical strategies summarized here also come to an aid in protein yield augment and sample homogeneity improvement for Cryo-EM and NMR. They are very powerful tools to solve structures, for instance, the Cryo-EM was applied to determine TrpV1 structures (Cao et al., 2013; Liao et al., 2013). With the development of advanced technologies, more and more eukaryotic membrane protein structures will emerge to answer the most significant questions in life sciences and provide the novel pharmaceutical targets in drug design.

ACKNOWLEDGEMENTS

We apologize to colleagues whose work could not be cited due to the scope of this review. We would like to thank members in Yan laboratory for discussions. We thank Brendan Lehnert, Xinlei Sheng, Quanxiu Li, Dan Ma and Xinhui Zhou for critical reading. This work was supported by funds from the National Basic Research Program (973 Program) (No. 2011CB910501), the National Natural Science Foundation of China (Grant Nos. 31321062-20131319400, 31125009, and 91017011), and funds from Tsinghua-Peking Center for Life Sciences. The research of N.Y. was supported in part by an International Early Career Scientist grant from the Howard Hughes Medical Institute.

ABBREVIATIONS

β_2AR, human β2 adrenergic G-protein-coupled receptor; ABCB10, ATP-binding cassette (ABC) transporters; AHA2, *Arabidopsis thaliana* auto-inhibited H1-ATPase 2; ASIC1, acid-sensing ion channel 1; CAAX protease Ste24p, C is cysteine redidue, A is an aliphatic residue and X is any residue. It is a zinc metalloprotease catalyzing two proteolytic steps in the maturation of yeast mating pheromone a-factor; C_8E_4, tetraethyleneglycol monooctyl ether; $C_{12}E_7$, dodecylheptaglycol; $C_{12}E_8$, polyoxyethylene dodecyl ether; CHS, cholesteryl hemisuccinate; CmClC, cyanidioschyzon merolae chloride (Cl^-) ions transporter; C_{11}Thio, n-undecyl-β-D-thiomaltopyranoside; CXCR4, human chemokine receptors; CX26, connexin 26 gap junction; CYMAL5, 5-cyclohexyl-1-pentyl-β-D-maltoside; CYMAL6, 6-cyclohexyl-1-hexyl-β-D-maltoside; CYMAL7, 7-cyclohexyl-1-heptyl-β-D-maltoside; DDM, n-dodecyl-β-D-maltoside; DM, n-decyl-β-D-maltoside; FLAP, 5-lipoxygenase-activating protein; GIRK2 (Kir3.2), K^+ channel: G protein-gated K+ channels; GluA2, a-amino-3-hydroxy-5-methyl-4-isoxazole propionic acid (AMPA)-sensitive ionotropic glutamate receptor; GLuClα, caenorhabditis elegans glutamate-gated chloride channel a (GluCl), an inhibitory anion-selective Cys-loop receptor; HsAQP4, human aquaporin 4; HsAQP5, human aquaporin 5; Human BK channel, high-conductance voltage- and Ca21-activated K1 channels; K2P1, two-pore domain potassium (K+) channels; Kv1.2, voltage-dependent shaker family potassium channel; Kv1.2-Kv2.1 paddle, 'paddle-chimaera channel', voltage-sensor paddle has been transferred from Kv2.1 to Kv1.2; LTC4S, cysteinyl leukotrienes; M-Ppase, membrane-integral pyrophosphatases; MAPEG, membrane-associated proteins in eicosanoid and glutathione metabolism; MNG, maltose-neopentyl glycol; NG, n-nonyl-β-D-glucopyranoside; NM, n-nonyl-β-D-maltoside; N/OFQ receptor, nociceptin/orphanin FQ receptor; OG, n-octyl-β-D-glucoside; OGNG, octyl glucose neopentyl glycol; $P2X_4$, cation-selective ion channels gated by extracellular ATP; PAR1, protease-activated receptor 1; PfAQP, *Plasmodium falciparum* aquaglyceroporin; PiPT, a Fungal (*Piriformospora indica*) high-affinity phosphate transporter; POPC, 1-palmitoyl-2-oleoyl-sn-glycero-3-phosphocholine; POPE, 1-palmitoyl-2-oleoyl-sn-glycero-3-phosphoethanolamine; POPG, 1-palmitoyl-2-oleoyl-sn-glycero-3-phosphoglycero; RhCG, rhesus C glycoprotein; SoPIP2, 1, spinach plant plasma membrane aquaporin; TRAAK, TWIK-related arachidonic acid–stimulated K^+ channel; UDTM, n-undecyl-β-maltoside; UT-B, urea transporters-B; VrH^+-Ppase, vigna radiate H1-translocating pyrophosphatases; ZMPSTE24, zinc metallopeptidase STE24.

COMPLIANCE WITH ETHICS GUIDELINES

Yuan He, Kan Wang, and Nieng Yan declare that we have no conflict of interest.

This review does not contain any studies with human or animal subjects performed by the any of the authors.

REFERENCES

Abrahams JP, Leslie AG, Lutter R, Walker JE (1994) Structure at 2.8 A resolution of F1-ATPase from bovine heart mitochondria. Nature 370:621–628

AegeanSoftware (2005) NoteExpress, 2.0 edn. (NoteExpress is a perfect assistant and information manager for researchers, scholars, students, and librarians. NoteExpress is designed to help you organize research notes and bibliographic references, generate bibliographies automatically, search and capture bibliographic data from Internet with efficiency and ease. NoteExpress is well integrated with Microsoft Word. It can format bibliographies in many popular styles)

Agah S, Faham S (2012) Crystallization of membrane proteins in bicelles. Methods Mol Biol 914:3–16

Ago H, Kanaoka Y, Irikura D, Lam BK, Shimamura T, Austen KF, Miyano M (2007) Crystal structure of a human membrane protein involved in cysteinyl leukotriene biosynthesis. Nature 448:609–612

Aisenbrey C, Borowik T, Bystrom R, Bokvist M, Lindstrom F, Misiak H, Sani MA, Grobner G (2008) How is protein aggregation in amyloidogenic diseases modulated by biological membranes? Eur Biophys J 37:247–255

Aller SG, Yu J, Ward A, Weng Y, Chittaboina S, Zhuo R, Harrell PM, Trinh YT, Zhang Q, Urbatsch IL et al (2009) Structure of P-glycoprotein reveals a molecular basis for poly-specific drug binding. Science 323:1718–1722

Amunts A, Drory O, Nelson N (2007) The structure of a plant photosystem I supercomplex at 3.4. A resolution. Nature 447:58–63

Baconguis I, Gouaux E (2012) Structural plasticity and dynamic selectivity of acid-sensing ion channel–spider toxin complexes. Nature 489:400–405

Baldwin SL, Powell TD, Wonderling RS, Keiser KC, Morales T, Hunter S, McDermott M, Radecki SV, Milhausen MJ (2003) Transient and stable transfection of Chinese hamster ovary cells with the recombinant feline erythropoietin gene and expression, purification, and biological activity of feline erythropoietin protein. Am J Vet Res 64:1465–1471

Bernaudat F, Frelet-Barrand A, Pochon N, Dementin S, Hivin P, Boutigny S, Rioux JB, Salvi D, Seigneurin-Berny D, Richaud P et al (2011) Heterologous expression of membrane proteins: choosing the appropriate host. PLoS One 6:e29191

Bill RM, Henderson PJ, Iwata S, Kunji ER, Michel H, Neutze R, Newstead S, Poolman B, Tate CG, Vogel H (2011) Overcoming barriers to membrane protein structure determination. Nat Biotechnol 29:335–340

Bkaily G, Al-Khoury J, Jacques D (2014) Nuclear membranes GPCRs: implication in cardiovascular health and diseases. Curr Vasc Pharmacol 12(2):215–222

Bornert O, Alkhalfioui F, Logez C, Wagner R (2012) Overexpression of membrane proteins using *Pichia pastoris*. Curr Protoc Protein Sci Chapter 29:22–29

Brohawn SG, Del MJ, MacKinnon R (2012) Crystal structure of the human K2P TRAAK, a lipid- and mechano-sensitive K+ ion channel. Science 335:436–441

Burgess-Brown NA, Sharma S, Sobott F, Loenarz C, Oppermann U, Gileadi O (2008) Codon optimization can improve expression of human genes in *Escherichia coli*: a multi-gene study. Protein Exp Purif 59:94–102

Caffrey M, Cherezov V (2009) Crystallizing membrane proteins using lipidic mesophases. Nat Protoc 4:706–731

Cao E, Liao M, Cheng Y, Julius D (2013) TRPV1 structures in distinct conformations reveal activation mechanisms. Nature 504:113–118

Cereghino JL, Cregg JM (2000) Heterologous protein expression in the methylotrophic yeast *Pichia pastoris*. Fems Microbiol Rev 24:45–66

Chae PS, Rasmussen SG, Rana RR, Gotfryd K, Chandra R, Goren MA, Kruse AC, Nurva S, Loland CJ, Pierre Y et al (2010) Maltose-neopentyl glycol (MNG) amphiphiles for solubilization, stabilization and crystallization of membrane proteins. Nat Methods 7:1003–1008

Cherezov V, Rosenbaum DM, Hanson MA, Rasmussen SG, Thian FS, Kobilka TS, Choi HJ, Kuhn P, Weis WI, Kobilka BK et al (2007) High-resolution crystal structure of an engineered human beta2-adrenergic G protein-coupled receptor. Science 318:1258–1265

Chien EY, Liu W, Zhao Q, Katritch V, Han GW, Hanson MA, Shi L, Newman AH, Javitch JA, Cherezov V et al (2010) Structure of the human dopamine D3 receptor in complex with a D2/D3 selective antagonist. Science 330:1091–1095

Ciccarone VC, Polayes DA, Luckow VA (1998) Generation of recombinant baculovirus DNA in *E. coli* using a baculovirus shuttle vector. Methods Mol Med 13:213–235

Condreay JP, Witherspoon SM, Clay WC, Kost TA (1999) Transient and stable gene expression in mammalian cells transduced with a recombinant baculovirus vector. Proc Natl Acad Sci USA 96:127–132

Contreras-Gomez A, Sanchez-Miron A, Garcia-Camacho F, Molina-Grima E, Chisti Y (2014) Protein production using the baculovirus-insect cell expression system. Biotechnol Prog 30:1–18

Daly R, Hearn MT (2005) Expression of heterologous proteins in Pichia pastoris: a useful experimental tool in protein engineering and production. J Mol Recognit 18:119–138

Deng D, Xu C, Sun P, Wu J, Yan C, Hu M, Yan N (2014) Crystal structure of the human glucose transporter GLUT1. Nature 510:121–125

Drew D, Lerch M, Kunji E, Slotboom DJ, de Gier JW (2006) Optimization of membrane protein overexpression and purification using GFP fusions. Nat Methods 3:303–313

Dukkipati A, Park HH, Waghray D, Fischer S, Garcia KC (2008) BacMam system for high-level expression of recombinant soluble and membrane glycoproteins for structural studies. Protein Exp Purif 62:160–170

Dumon-Seignovert L, Cariot G, Vuillard L (2004) The toxicity of recombinant proteins in *Escherichia coli*: a comparison of overexpression in BL21(DE3), C41(DE3), and C43(DE3). Protein Expr Purif 37:203–206

Feng L, Campbell EB, Hsiung Y, MacKinnon R (2010) Structure of a eukaryotic CLC transporter defines an intermediate state in the transport cycle. Science 330:635–641

Ferguson AD, McKeever BM, Xu S, Wisniewski D, Miller DK, Yamin TT, Spencer RH, Chu L, Ujjainwalla F, Cunningham BR et al (2007) Crystal structure of inhibitor-bound human 5-lipoxygenase-activating protein. Science 317:510–512

Gonzales EB, Kawate T, Gouaux E (2009) Pore architecture and ion sites in acid-sensing ion channels and P2X receptors. Nature 460:599–604

Granier S, Manglik A, Kruse AC, Kobilka TS, Thian FS, Weis WI, Kobilka BK (2012) Structure of the delta-opioid receptor bound to naltrindole. Nature 485:400–404

Gruswitz F, Chaudhary S, Ho JD, Schlessinger A, Pezeshki B, Ho CM, Sali A, Westhoff CM, Stroud RM (2010) Function of human Rh based on structure of RhCG at 2.1 A. Proc Natl Acad Sci USA 107:9638–9643

Haga K, Kruse AC, Asada H, Yurugi-Kobayashi T, Shiroishi M, Zhang C, Weis WI, Okada T, Kobilka BK, Haga T et al (2012) Structure of the human M2 muscarinic acetylcholine receptor bound to an antagonist. Nature 482:547–551

Hanson MA, Roth CB, Jo E, Griffith MT, Scott FL, Reinhart G, Desale H, Clemons B, Cahalan SM, Schuerer SC et al (2012) Crystal structure of a lipid G protein-coupled receptor. Science 335:851–855

Hattori M, Gouaux E (2012) Molecular mechanism of ATP binding and ion channel activation in P2X receptors. Nature 485:207–212

Hibbs RE, Gouaux E (2011) Principles of activation and permeation in an anion-selective Cys-loop receptor. Nature 474:54–60

Ho JD, Yeh R, Sandstrom A, Chorny I, Harries WE, Robbins RA, Miercke LJ, Stroud RM (2009) Crystal structure of human aquaporin 4 at 1.8. A and its mechanism of conductance. Proc Natl Acad Sci USA 106:7437–7442

Hollenstein K, Kean J, Bortolato A, Cheng RK, Dore AS, Jazayeri A, Cooke RM, Weir M, Marshall FH (2013) Structure of class B GPCR corticotropin-releasing factor receptor 1. Nature 499:438–443

Horsefield R, Norden K, Fellert M, Backmark A, Tornroth-Horsefield S, Terwisscha VSA, Kvassman J, Kjellbom P, Johanson U, Neutze R (2008) High-resolution x-ray structure of human aquaporin 5. Proc Natl Acad Sci USA 105:13327–13332

Ishikawa Y, Inoue N, Zhenfang Y, Nakae Y (2004) Molecular mechanisms and drug development in aquaporin water channel diseases: the translocation of aquaporin-5 from lipid rafts to the apical plasma membranes of parotid glands of normal rats and the impairment of it in diabetic or aged rats. J Pharmacol Sci 96:271–275

Jaakola VP, Griffith MT, Hanson MA, Cherezov V, Chien EY, Lane JR, Ijzerman AP, Stevens RC (2008) The 2.6 angstrom crystal structure of a human A2A adenosine receptor bound to an antagonist. Science 322:1211–1217

Jasti J, Furukawa H, Gonzales EB, Gouaux E (2007) Structure of acid-sensing ion channel 1 at 1.9. A resolution and low pH. Nature 449:316–323

Jin MS, Oldham ML, Zhang Q, Chen J (2012) Crystal structure of the multidrug transporter P-glycoprotein from Caenorhabditis elegans. Nature 490:566–569

Kawate T, Michel JC, Birdsong WT, Gouaux E (2009) Crystal structure of the ATP-gated P2X(4) ion channel in the closed state. Nature 460:592–598

Khan KH (2013) Gene expression in mammalian cells and its applications. Adv Pharm Bull 3:257–263

Kost TA, Condreay JP, Jarvis DL (2005) Baculovirus as versatile vectors for protein expression in insect and mammalian cells. Nat Biotechnol 23:567–575

Krogh A, Larsson B, von Heijne G, Sonnhammer EL (2001) Predicting transmembrane protein topology with a hidden Markov model: application to complete genomes. J Mol Biol 305:567–580

Kruse AC, Hu J, Pan AC, Arlow DH, Rosenbaum DM, Rosemond E, Green HF, Liu T, Chae PS, Dror RO et al (2012) Structure and dynamics of the M3 muscarinic acetylcholine receptor. Nature 482:552–556

Kurisu G, Zhang H, Smith JL, Cramer WA (2003) Structure of the cytochrome b6f complex of oxygenic photosynthesis: tuning the cavity. Science 302:1009–1014

Landau EM, Rosenbusch JP (1996) Lipidic cubic phases: a novel concept for the crystallization of membrane proteins. Proc Natl Acad Sci USA 93:14532–14535

Levin EJ, Cao Y, Enkavi G, Quick M, Pan Y, Tajkhorshid E, Zhou M (2012) Structure and permeation mechanism of a mammalian urea transporter. Proc Natl Acad Sci USA 109:11194–11199

Li P, Anumanthan A, Gao XG, Ilangovan K, Suzara VV, Duzgunes N, Renugopalakrishnan V (2007) Expression of recombinant proteins in *Pichia pastoris*. Appl Biochem Biotechnol 142:105–124

Liao M, Cao E, Julius D, Cheng Y (2013) Structure of the TRPV1 ion channel determined by electron cryo-microscopy. Nature 504:107–112

Lin SM, Tsai JY, Hsiao CD, Huang YT, Chiu CL, Liu MH, Tung JY, Liu TH, Pan RL, Sun YJ (2012) Crystal structure of a membrane-embedded H+-translocating pyrophosphatase. Nature 484:399–403

Liu Z, Yan H, Wang K, Kuang T, Zhang J, Gui L, An X, Chang W (2004) Crystal structure of spinach major light-harvesting complex at 2.72. A resolution. Nature 428:287–292

Long SB, Campbell EB, Mackinnon R (2005) Crystal structure of a mammalian voltage-dependent Shaker family K+ channel. Science 309:897–903

Long SB, Tao X, Campbell EB, MacKinnon R (2007) Atomic structure of a voltage-dependent K+ channel in a lipid membrane-like environment. Nature 450:376–382

Louis N, Evelegh C, Graham FL (1997) Cloning and sequencing of the cellular-viral junctions from the human adenovirus type 5 transformed 293 cell line. Virology 233:423–429

Lu P, Ma D, Yan C, Gong X, Du M, Shi Y (2014) Structure and mechanism of a eukaryotic transmembrane ascorbate-dependent oxidoreductase. Proc Natl Acad Sci USA 111(5):1813–1818

Macauley-Patrick S, Fazenda ML, McNeil B, Harvey LM (2005) Heterologous protein production using the Pichia pastoris expression system. Yeast 22:249–270

Maeda S, Kawai T, Obinata M, Fujiwara H, Horiuchi T, Saeki Y, Sato Y, Furusawa M (1985) Production of human alpha-interferon in silkworm using a baculovirus vector. Nature 315:592–594

Maeda S, Nakagawa S, Suga M, Yamashita E, Oshima A, Fujiyoshi Y, Tsukihara T (2009) Structure of the connexin 26 gap junction channel at 3.5. A resolution. Nature 458:597–602

Manglik A, Kruse AC, Kobilka TS, Thian FS, Mathiesen JM, Sunahara RK, Pardo L, Weis WI, Kobilka BK, Granier S (2012) Crystal structure of the micro-opioid receptor bound to a morphinan antagonist. Nature 485:321–326

Mijakovic I, Petranovic D, Macek B, Cepo T, Mann M, Davies J, Jensen PR, Vujaklija D (2006) Bacterial single-stranded DNA-binding proteins are phosphorylated on tyrosine. Nucleic Acids Res 34:1588–1596

Miller AN, Long SB (2012) Crystal structure of the human two-pore domain potassium channel K2P1. Science 335:432–436

Mironova R, Niwa T, Handzhiyski Y, Sredovska A, Ivanov I (2005) Evidence for non-enzymatic glycosylation of Escherichia coli chromosomal DNA. Mol Microbiol 55:1801–1811

Miroux B, Walker JE (1996) Over-production of proteins in *Escherichia coli*: mutant hosts that allow synthesis of some membrane proteins and globular proteins at high levels. J Mol Biol 260:289–298

Murray CW, Verdonk ML, Rees DC (2012) Experiences in fragment-based drug discovery. Trends Pharmacol Sci 33:224–232

Neophytou I, Harvey R, Lawrence J, Marsh P, Panaretou B, Barlow D (2007) Eukaryotic integral membrane protein expression utilizing the Escherichia coli glycerol-conducting channel protein (GlpF). Appl Microbiol Biotechnol 77:375–381

Newby ZE, O'Connell JR, Robles-Colmenares Y, Khademi S, Miercke LJ, Stroud RM (2008) Crystal structure of the aquaglyceroporin PfAQP from the malarial parasite Plasmodium falciparum. Nat Struct Mol Biol 15:619–625

Newstead S, Kim H, von Heijne G, Iwata S, Drew D (2007) High-throughput fluorescent-based optimization of eukaryotic membrane protein overexpression and purification in Saccharomyces cerevisiae. Proc Natl Acad Sci USA 104:13936–13941

Nishida M, Cadene M, Chait BT, MacKinnon R (2007) Crystal structure of a Kir3.1-prokaryotic Kir channel chimera. Embo J 26:4005–4015

Overington JP, Al-Lazikani B, Hopkins AL (2006) How many drug targets are there? Nat Rev Drug Discov 5:993–996

Parker JL, Newstead S (2014) Molecular basis of nitrate uptake by the plant nitrate transporter NRT1.1. Nature 507:68–72

Payandeh J, Scheuer T, Zheng N, Catterall WA (2011) The crystal structure of a voltage-gated sodium channel. Nature 475:353–358

Pedersen BP, Buch-Pedersen MJ, Morth JP, Palmgren MG, Nissen P (2007) Crystal structure of the plasma membrane proton pump. Nature 450:1111–1114

Pedersen BP, Kumar H, Waight AB, Risenmay AJ, Roe-Zurz Z, Chau BH, Schlessinger A, Bonomi M, Harries W, Sali A et al (2013) Crystal structure of a eukaryotic phosphate transporter. Nature 496:533–536

Penmatsa A, Wang KH, Gouaux E (2013) X-ray structure of dopamine transporter elucidates antidepressant mechanism. Nature 503:85–90

Phillips R, Ursell T, Wiggins P, Sens P (2009) Emerging roles for lipids in shaping membrane-protein function. Nature 459:379–385

Pryor EJ, Horanyi PS, Clark KM, Fedoriw N, Connelly SM, Koszelak-Rosenblum M, Zhu G, Malkowski MG, Wiener MC, Dumont ME (2013) Structure of the integral membrane protein CAAX protease Ste24p. Science 339:1600–1604

Quigley A, Dong YY, Pike AC, Dong L, Shrestha L, Berridge G, Stansfeld PJ, Sansom MS, Edwards AM, Bountra C et al (2013) The structural basis of ZMPSTE24-dependent laminopathies. Science 339:1604–1607

Rasmussen SG, Choi HJ, Rosenbaum DM, Kobilka TS, Thian FS, Edwards PC, Burghammer M, Ratnala VR, Sanishvili R, Fischetti RF et al (2007) Crystal structure of the human beta2 adrenergic G-protein-coupled receptor. Nature 450:383–387

Rasmussen SG, Choi HJ, Fung JJ, Pardon E, Casarosa P, Chae PS, Devree BT, Rosenbaum DM, Thian FS, Kobilka TS et al (2011a) Structure of a nanobody-stabilized active state of the beta(2) adrenoceptor. Nature 469:175–180

Rasmussen SG, DeVree BT, Zou Y, Kruse AC, Chung KY, Kobilka TS, Thian FS, Chae PS, Pardon E, Calinski D et al (2011b) Crystal structure of the beta2 adrenergic receptor-Gs protein complex. Nature 477:549–555

Reeves PJ, Callewaert N, Contreras R, Khorana HG (2002) Structure and function in rhodopsin: high-level expression of rhodopsin with restricted and homogeneous N-glycosylation by a tetracycline-inducible N-acetylglucosaminyltransferase I-negative HEK293S stable mammalian cell line. Proc Natl Acad Sci USA 99:13419–13424

Rosenbaum DM, Zhang C, Lyons JA, Holl R, Aragao D, Arlow DH, Rasmussen SG, Choi HJ, Devree BT, Sunahara RK et al (2011) Structure and function of an irreversible agonist-beta(2) adrenoceptor complex. Nature 469:236–240

Russell RB, Eggleston DS (2000) New roles for structure in biology and drug discovery. Nat Struct Biol 7(Suppl):928–930

Sahdev S, Khattar SK, Saini KS (2008) Production of active eukaryotic proteins through bacterial expression systems: a review of the existing biotechnology strategies. Mol Cell Biochem 307:249–264

Sanders CR, Myers JK (2004) Disease-related misassembly of membrane proteins. Annu Rev Biophys Biomol Struct 33:25–51

Shimamura T, Shiroishi M, Weyand S, Tsujimoto H, Winter G, Katritch V, Abagyan R, Cherezov V, Liu W, Han GW et al (2011) Structure of the human histamine H1 receptor complex with doxepin. Nature 475:65–70

Shintre CA, Pike AC, Li Q, Kim JI, Barr AJ, Goubin S, Shrestha L, Yang J, Berridge G, Ross J et al (2013) Structures of ABCB10, a human ATP-binding cassette transporter in apo- and nucleotide-bound states. Proc Natl Acad Sci USA 110:9710–9715

Siu FY, He M, de Graaf C, Han GW, Yang D, Zhang Z, Zhou C, Xu Q, Wacker D, Joseph JS et al (2013) Structure of the human glucagon class B G-protein-coupled receptor. Nature 499:444–449

Smith GE, Summers MD, Fraser MJ (1983) Production of human beta interferon in insect cells infected with a baculovirus expression vector. Mol Cell Biol 3:2156–2165

Snider C, Stephen, HW (2014) Membrane proteins of known 3D structure. http://blanco.biomol.uci.edu/mpstruc/

Sobolevsky AI, Rosconi MP, Gouaux E (2009) X-ray structure, symmetry and mechanism of an AMPA-subtype glutamate receptor. Nature 462:745–756

Standfuss J, Xie G, Edwards PC, Burghammer M, Oprian DD, Schertler GF (2007) Crystal structure of a thermally stable rhodopsin mutant. J Mol Biol 372:1179–1188

Strausberg RL, Strausberg SL (2001) Overview of protein expression in saccharomyces cerevisiae. Curr Protoc Protein Sci Chapter 5:t5–t6

Sun F, Huo X, Zhai Y, Wang A, Xu J, Su D, Bartlam M, Rao Z (2005) Crystal structure of mitochondrial respiratory membrane protein complex II. Cell 121:1043–1057

Sun J, Bankston JR, Payandeh J, Hinds TR, Zagotta WN, Zheng N (2014) Crystal structure of the plant dual-affinity nitrate transporter NRT1.1. Nature 507:73–77

Sunley K, Butler M (2010) Strategies for the enhancement of recombinant protein production from mammalian cells by growth arrest. Biotechnol Adv 28:385–394

Suzuki H, Nishizawa T, Tani K, Yamazaki Y, Tamura A, Ishitani R, Dohmae N, Tsukita S, Nureki O, Fujiyoshi Y (2014) Crystal structure of a claudin provides insight into the architecture of tight junctions. Science 344:304–307

Tan Q, Zhu Y, Li J, Chen Z, Han GW, Kufareva I, Li T, Ma L, Fenalti G, Li J et al (2013) Structure of the CCR5 chemokine receptor-HIV entry inhibitor maraviroc complex. Science 341:1387–1390

Tao X, Avalos JL, Chen J, MacKinnon R (2009) Crystal structure of the eukaryotic strong inward-rectifier K+ channel Kir2.2 at 3.1 A resolution. Science 326:1668–1674

Tao X, Lee A, Limapichat W, Dougherty DA, MacKinnon R (2010) A gating charge transfer center in voltage sensors. Science 328:67–73

Thomas P, Smart TG (2005) HEK293 cell line: a vehicle for the expression of recombinant proteins. J Pharmacol Toxicol Methods 51:187–200

Thompson AA, Liu W, Chun E, Katritch V, Wu H, Vardy E, Huang XP, Trapella C, Guerrini R, Calo G et al (2012) Structure of the nociceptin/orphanin FQ receptor in complex with a peptide mimetic. Nature 485:395–399

Tornroth-Horsefield S, Wang Y, Hedfalk K, Johanson U, Karlsson M, Tajkhorshid E, Neutze R, Kjellbom P (2006) Structural mechanism of plant aquaporin gating. Nature 439:688–694

Tsukihara T, Aoyama H, Yamashita E, Tomizaki T, Yamaguchi H, Shinzawa-Itoh K, Nakashima R, Yaono R, Yoshikawa S (1996) The whole structure of the 13-subunit oxidized cytochrome c oxidase at 2.8 A. Science 272:1136–1144

Unger T, Peleg Y (2012) Recombinant protein expression in the baculovirus-infected insect cell system. Methods Mol Biol 800:187–199

van Meer G, Voelker DR, Feigenson GW (2008) Membrane lipids: where they are and how they behave. Nat Rev Mol Cell Biol 9:112–124

Wacker D, Wang C, Katritch V, Han GW, Huang XP, Vardy E, McCorvy JD, Jiang Y, Chu M, Siu FY et al (2013) Structural features for functional selectivity at serotonin receptors. Science 340:615–619

Wang C, Jiang Y, Ma J, Wu H, Wacker D, Katritch V, Han GW, Liu W, Huang XP, Vardy E et al (2013a) Structural basis for molecular recognition at serotonin receptors. Science 340:610–614

Wang C, Wu H, Katritch V, Han GW, Huang XP, Liu W, Siu FY, Roth BL, Cherezov V, Stevens RC (2013b) Structure of the human smoothened receptor bound to an antitumour agent. Nature 497:338–343

Warne T, Serrano-Vega MJ, Baker JG, Moukhametzianov R, Edwards PC, Henderson R, Leslie AG, Tate CG, Schertler GF (2008) Structure of a beta1-adrenergic G-protein-coupled receptor. Nature 454:486–491

White JF, Noinaj N, Shibata Y, Love J, Kloss B, Xu F, Gvozdenovic-Jeremic J, Shah P, Shiloach J, Tate CG et al (2012) Structure of the agonist-bound neurotensin receptor. Nature 490:508–513

Whorton MR, MacKinnon R (2011) Crystal structure of the mammalian GIRK2 K+ channel and gating regulation by G proteins, PIP2, and sodium. Cell 147:199–208

Wu B, Chien EY, Mol CD, Fenalti G, Liu W, Katritch V, Abagyan R, Brooun A, Wells P, Bi FC et al (2010) Structures of the CXCR4 chemokine GPCR with small-molecule and cyclic peptide antagonists. Science 330:1066–1071

Wu H, Wacker D, Mileni M, Katritch V, Han GW, Vardy E, Liu W, Thompson AA, Huang XP, Carroll FI et al (2012) Structure of the human kappa-opioid receptor in complex with JDTic. Nature 485:327–332

Wu H, Wang C, Gregory KJ, Han GW, Cho HP, Xia Y, Niswender CM, Katritch V, Meiler J, Cherezov V et al (2014) Structure of a class C GPCR metabotropic glutamate receptor 1 bound to an allosteric modulator. Science 344:58–64

Xia D, Yu CA, Kim H, Xia JZ, Kachurin AM, Zhang L, Yu L, Deisenhofer J (1997) Crystal structure of the cytochrome bc1 complex from bovine heart mitochondria. Science 277:60–66

Hou X, Pedi L, Diver MM, Long SB (2012) Crystal structure of the calcium release-activated calcium channel orai. Science 338 (6112):1308–1313

Yang Y, Hu Z, Liu Z, Wang Y, Chen X, Chen G (2009) High human GLUT1, GLUT2, and GLUT3 expression in *Schizosaccharomyces pombe*. Biochemistry (Mosc) 74:75–80

Yildirim MA, Goh KI, Cusick ME, Barabasi AL, Vidal M (2007) Drug-target network. Nat Biotechnol. 25:1119–1126

Yurimoto H, Sakai Y (2009) Methanol-inducible gene expression and heterologous protein production in the methylotrophic yeast *Candida boidinii*. Biotechnol Appl Biochem 53:85–92

Zhang C, Srinivasan Y, Arlow DH, Fung JJ, Palmer D, Zheng Y, Green HF, Pandey A, Dror RO, Shaw DE et al (2012) High-resolution crystal structure of human protease-activated receptor 1. Nature 492:387–392

Zhou Y, Morais-Cabral JH, Kaufman A, MacKinnon R (2001) Chemistry of ion coordination and hydration revealed by a K+ channel-Fab complex at 2.0. A resolution. Nature 414:43–48

Zhu J (2012) Mammalian cell protein expression for biopharmaceutical production. Biotechnol Adv 30:1158–1170

14

Arrestins: structural disorder creates rich functionality

Vsevolod V. Gurevich[1✉], **Eugenia V. Gurevich**[1], **Vladimir N. Uversky**[2,3]

[1] Department of Pharmacology, Vanderbilt University, Nashville, TN 37232, USA
[2] Department of Molecular Medicine and USF Health Byrd Alzheimer's Research Institute, Morsani College of Medicine, University of South Florida, Tampa, FL 33612, USA
[3] Institute for Biological Instrumentation, Russian Academy of Sciences, Pushchino, Moscow Region, Russia 142290
✉ Correspondence: vsevolod.gurevich@vanderbilt.edu (V. V. Gurevich)

ABSTRACT

Arrestins are soluble relatively small 44–46 kDa proteins that specifically bind hundreds of active phosphorylated GPCRs and dozens of non-receptor partners. There are binding partners that demonstrate preference for each of the known arrestin conformations: free, receptor-bound, and microtubule-bound. Recent evidence suggests that conformational flexibility in every functional state is the defining characteristic of arrestins. Flexibility, or plasticity, of proteins is often described as structural disorder, in contrast to the fixed conformational order observed in high-resolution crystal structures. However, protein-protein interactions often involve highly flexible elements that can assume many distinct conformations upon binding to different partners. Existing evidence suggests that arrestins are no exception to this rule: their flexibility is necessary for functional versatility. The data on arrestins and many other multi-functional proteins indicate that in many cases, "order" might be artificially imposed by highly non-physiological crystallization conditions and/or crystal packing forces. In contrast, conformational flexibility (and its extreme case, intrinsic disorder) is a more natural state of proteins, representing true biological order that underlies their physiologically relevant functions.

KEYWORDS arrestin, GPCR, crystal structure, NMR, EPR, disorder, protein-protein interactions

INTRODUCTION

Arrestins are globular proteins found throughout the animal kingdom, with four subtypes present in vertebrates (Gurevich and Gurevich, 2006; Indrischek et al., 2017). Structurally, arrestins are elongated two-domain molecules (Granzin et al., 1998; Hirsch et al., 1999; Han et al., 2001; Milano et al., 2002; Sutton et al., 2005; Zhan et al., 2011) (Fig. 1), with relatively few inter-domain interactions (Fig. 1). Specific binding to active phosphorylated receptor was the first arrestin function discovered (Kuhn et al., 1984). Arrestins suppress G protein activation (Wilden et al., 1986) by direct competition (Wilden, 1995; Krupnick et al., 1997). Crystal structures of active rhodopsin in complex with transducin (visual G protein) peptide (Scheerer et al., 2008), of related G protein-coupled receptor (GPCR)β_2-adrenergic receptor in complex with heterotrimeric G_s protein (Rasmussen et al., 2011), of adenosine A_{2A} receptor with an engineered single-subunit G_s protein (Carpenter et al., 2016), and of arrestin-1[1] complex with rhodopsin (Kang et al., 2015; Zhou et al., 2017) reveal the structural basis for this competition: the C-terminus of G protein α-subunit and central "finger loop" of arrestin-1 (Figs. 1 and 2) occupy the same cavity between the cytoplasmic ends of the α-helices that opens upon activation of rhodopsin (Farrens et al., 1996) and other GPCRs (Rasmussen et al., 2011) (Table 1).

Virtually every discovery in the GPCR field was first made in the visual system. The arrestin structure was no exception: the first crystallographic studies revealed three-dimensional structures of visual arrestin-1. Despite their differences in resolution and proposed positions of the N- and C-termini, crystal structures in both studies did not resolve the same elements: the loop between the C-domain and the point of contact that the C-tail makes with the N-domain, as well as distal C-tail (residues 363–404) (Granzin et al., 1998; Hirsch et al., 1999) (Fig. 1). Subsequent structures of arrestin-2 (Han et al., 2001; Milano et al., 2002), arrestin-4 (Sutton et al., 2005), and arrestin-3 (Zhan et al., 2011) also did not reveal these elements. The N-terminus (residues 1–9) is also missing in all structures. Crystal structures indicate disordered elements in several ways. First, some are so flexible that they cannot be even detected in the electron density map: the N-terminus, distal C-terminus, and the connector between the C-domain and the β-strand XX (Figs. 1 and 2). Second, some elements have high B-factors, suggesting that they have alternative positions, but are not disordered enough to become invisible. Third, other elements have distinct conformations in different protomers of crystal oligomer, or in different crystals of the same protein. This applies to many arrestin loops connecting core β-strands in both domains.

The correlation of arrestin structural elements (even those with high structural flexibility) with their functions has been intensively studied. These studies have yielded a significant amount of information on the overall flexibility of these proteins and on the potential roles of such flexibility in arrestin functions. For example, by inserting the six amino-acid motif, C-C-P-G-C-C, into the arrestin-3 (non-visual β-arrestin-2) sequence at sites not directly involved in interactions of this protein with major partners, a set of intramolecular fluorescent arsenical hairpin (FlAsH) bioluminescence resonance energy transfer (BRET) probes (Hoffmann et al., 2005) was created to analyze the effect of various G protein-coupled receptors (GPCRs) on the conformational dynamics and function of this protein (Lee et al., 2016). The analysis using such probes revealed that arrestin-3 structure represents a dynamic conformational ensemble. Different GPCRs are capable of induction of significant structural changes in this protein, generating distinctive arrestin "conformational signatures", information about which is encoded within the population average arrestin-3 conformation (Lee et al., 2016). It was also shown that in living human cells, the GPCR-arrestin-3 interaction rapidly results in the receptor-specific conformational changes leading to activation of arrestin-3. However, this protein is able to keep receptor-specific conformation and remain active after dissociation from receptors (Nuber et al., 2016). Similarly, different phosphorylated GPCRs were shown to induce distinct

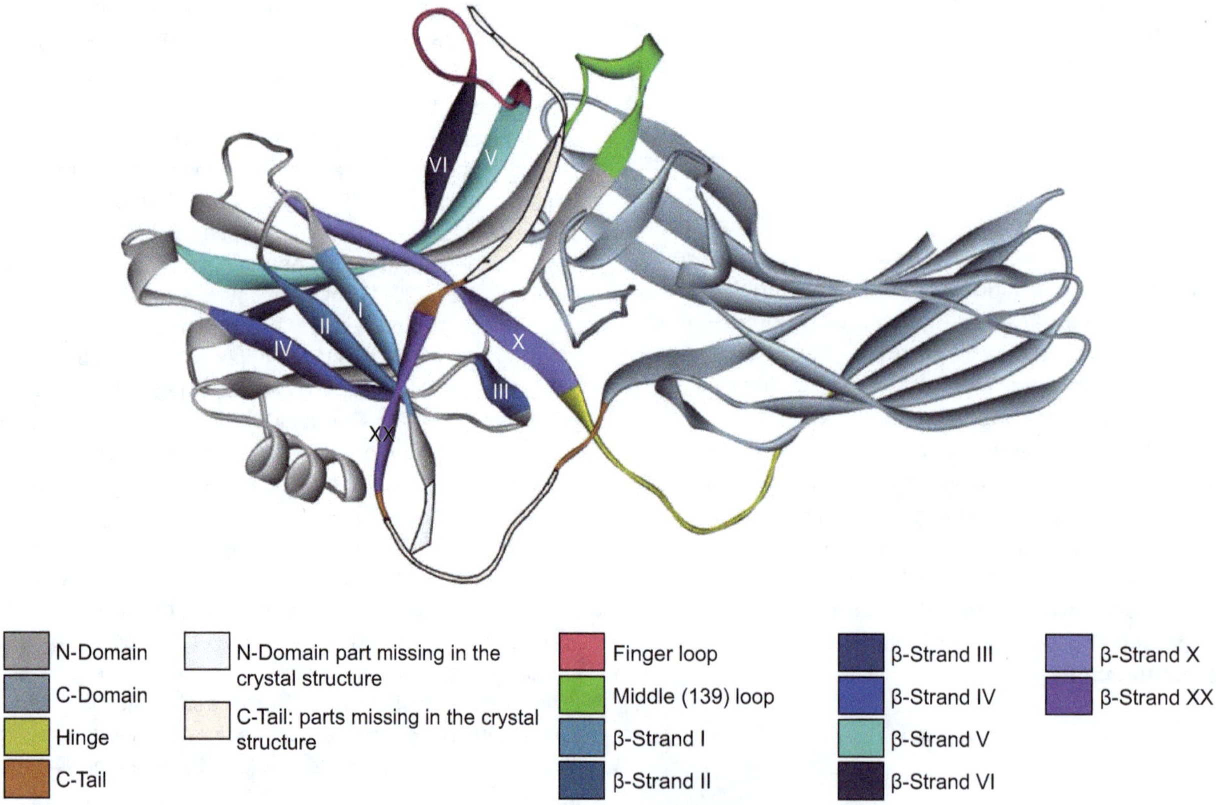

Figure 1. Arrestins structure and structural elements. The structure of arrestin-2 is shown, with main structural elements color-coded, as follows: N-domain, gray; C-domain, teal; inter-domain hinge, yellow; the part of the C-tail resolved in crystal structures, magenta; finger loop, pink; middle loop (termed 139-loop in arrestin-1), green. The part of the N-domain (the N-terminus) and of the C-tail that are not resolved in any crystal structure were added manually and are shown as light grey and light pink, respectively, with black outline. β-strands I, II, III, IV, V, VI, X, and XX are shown in indicated colors and labeled with roman numerals on the structure. Note that β-strand XX, anchored to β-strand I and -α-helix via three-element interaction (see Fig. 2A), includes virtually the whole part of the C-tail resolved in crystal structures.

structural states of arrestin-2 (β-arrestin-1) required for different specific functions of this protein (Yang et al., 2015). The interaction of visual arrestin (arrestin-1) with light-activated, phosphorylated rhodopsin is accompanied by significant movements of the arrestin loop between β-strands V–VI ("finger loop") (Fig. 1), suggesting that the ability of arrestin-1 to adopt a high affinity binding state critically depends on the conformational flexibility of this loop (Sommer et al., 2007). A noticeable difference in the conformational flexibility of arrestins-2 and -3 potentially related to their different receptor selectivity has been recently demonstrated using hydrogen/deuterium exchange mass spectrometry (HDX-MS) (Yun et al., 2015). The basal state of arrestin-3 was characterized by higher flexibility of β-strands II, III, and IV (Fig. 1), whereas high dynamics of the middle loop (called "139-loop" in arrestin-1; Figs. 1 and 2) was characteristic for the basal state of arrestin-2. Although both non-visual arrestins became more flexible as a result of their pre-activation, the scale of such pre-activation-induced increase in structural flexibility was noticeably larger in arrestin-2 (Yun et al., 2015). Based on the solution NMR analysis it was concluded that different sets of arrestin-1 regions are utilized for interaction with different functional states of rhodopsin (such as dark-state phosphorylated rhodopsin (P-Rh), phosphorylated opsin (P-opsin), unphosphorylated light-activated rhodopsin (Rh*), and phosphorylated light-activated rhodopsin (P-Rh*)), and that global conformational changes accompany the binding of arrestin-1 to P-Rh* or P-opsin leading to the formation of a dynamic molten globule-like structure. Since activation of arrestin-1 by P-Rh* is achieved via displacement of the long C-tail of arrestin from its polar core by phosphate groups of receptor (Figs. 1 and 2), and arginine 175 located at the polar core serves as the phosphosensor, mutation of this residue to glutamic acid (R175E) produces a pre-active form of arrestin-1 (Granzin et al., 2015). Analysis of the crystal structure of the R175E mutant revealed the presence of noticeable structural differences with the basal state of wild type arrestin-1, most notably several disordered residues in the receptor-binding finger loop and the C-terminus (residues 361–404) (Granzin et al., 2015) (Figs. 1 and 2). Therefore, arrestins in general are characterized by high conformational dynamics, and some of their regions demonstrate high structural flexibility and might be intrinsically disordered.

What does "disorder" as the opposite of "order" mean in relation to protein functionality? The classic paradigm posits that biologically active proteins should have unique rigid 3D structures; therefore, intrinsically disordered proteins (IDPs) can be ignored. However, the reality is more complex. Well before the IDP concept was introduced at the end of 20th century (Wright and Dyson, 1999; Uversky et al., 2000; Dunker et al., 2001), it became clear that protein molecules are inherently flexible, and this flexibility is crucial for their biological activity. In fact, many globular proteins possess marginal conformational stability under physiological conditions, with $\Delta G_{folding}$ ranging from −5 to −10 kcal/mol (Privalov and Khechinashvili, 1974; Savage et al., 1993; Ruvinov et al., 1997; Vogl et al., 1997; Giver et al., 1998; Taverna and Goldstein, 2002; Williams et al., 2006). Such marginal stability is related to the protein functionality (Tsou, 1998a, b; Zavodszky et al., 1998), ensuring higher flexibility needed for function-related conformational changes (Artymiuk et al., 1979; Frauenfelder et al., 1979; Wagner and Wuthrich, 1979; Wrba et al., 1990; Varley and Pain, 1991; Daniel et al., 1996; Tang and Dill, 1998; Williams et al., 2006). For example, the internal dynamics of enzymes appears to be linked to the mechanism of catalysis (Agarwal, 2005; Eisenmesser et al., 2005).

Although globular proteins are flexible, some parts are more rigid than others. A protein with several stable structural units can form conformational isomers, determined by its overall flexibility and the locations of the more flexible joints (Ma et al., 1999). Whole proteins or various regions can be ordered or disordered to a different degree (Uversky, 2013). Functional proteins contain differently structured regions, some of which are spontaneously folded (foldons), others fold upon interaction with binding partners (inducible foldons), some are always in semi-folded state (semi-foldons), whereas others do not fold at all (non-foldons) or even need to undergo order-disorder transition to become functional (unfoldons) (Uversky, 2013).

IDPs and hybrid proteins containing ordered domains and IDPRs constitute a very important addition to the more "traditional" set of globular, trans-membrane, and fibrous proteins (Wright and Dyson, 1999; Uversky et al., 2000; Dunker et al., 2001; Uversky and Dunker, 2010; Tompa, 2012; Uversky, 2013; Habchi et al., 2014; van der Lee et al., 2014; Uversky, 2016). The functions of IDPs/IDPRs and ordered proteins/domains are complementary (Dunker et al., 2001; Radivojac et al., 2007). Many IDPs bind multiple partners and act as signaling molecules controlling various aspects of cell behavior (Uversky et al., 2000; Dunker et al., 2001; Iakoucheva et al., 2002; Uversky et al., 2005; Uversky, 2013; Habchi et al., 2014; van der Lee et al., 2014; Uversky, 2015). IDPs/IDPRs are tightly regulated via post-translational modifications (Iakoucheva et al., 2004; Pejaver et al., 2014) and alternative splicing (Romero et al., 2006; Buljan et al., 2013). Dysfunctions of IDPs are associated with human diseases, including amyloidosis, cancer, cardiovascular disorders, and neurodegeneration (Uversky et al., 2008; Uversky et al., 2014).

Arrestins are typical hybrid proteins: they consist of two domains with fairly rigid cores and flexible loops connecting the β-strands, along with mostly disordered C-tail and N-terminus (Fig. 1). Upon arrestin activation, several elements significantly change conformation, some becoming disordered in this state. Here we analyze the connections of structural disorder in arrestins with known functions. The emerging picture is intriguing: structural disorder of arrestin elements appears to be linked to their functionality. The propensity of different parts to be disordered is a surprisingly good predictor of their functional importance.

CONFORMATIONAL DYNAMICS OF ARRESTINS

First, let us explore the basal structure of free arrestins and destabilizing mutations that likely enhance disorder. The idea that the conformations of free and receptor-bound arrestin are dramatically different was first expressed on the basis of high arrestin-1 activation energy (Schleicher et al., 1989), long before any structures were solved. Wild type (WT) arrestin-1 has remarkable selectivity for active phosphorylated rhodopsin (P-Rh*), demonstrating many times lower binding to active unphosphorylated (Rh*) and inactive phosphorylated (P-Rh) forms, and virtually no binding to inactive unphosphorylated rhodopsin (Gurevich and Benovic, 1992, 1993). The mutations that "activate" arrestin-1, facilitating its binding to non-preferred forms of rhodopsin, fall into four categories: 1) C-terminal deletions (Gurevich and Benovic, 1992) and alanine substitutions of hydrophobic residues in the C-tail (Gurevich, 1998), which eliminate the C-tail or release it from its basal position; 2) short N-terminal deletions (Gurevich and Benovic, 1993; Gurevich et al., 1994); 3) charge elimination/reversal mutations in the polar core (Hirsch et al., 1999) (Fig. 2A), particularly Arg-175 (Gurevich and Benovic, 1995, 1997) in arrestin-1, or homologous Arg-169 in arrestin-2 (Kovoor et al., 1999; Pan et al., 2003) and Arg-170 in arrestin-3 (Celver et al., 2002); 4) point mutations on the receptor-binding surface (Hanson and Gurevich, 2006; Vishnivetskiy et al., 2013) and small deletions of the "139-loop" in the central crest of the molecule (Figs. 1 and 2B) (Vishnivetskiy et al., 2013). Curiously, the first three categories of mutations affect residues that, according to the crystal structure of the complex (Kang et al., 2015; Zhou et al., 2017), are not directly involved in receptor binding.

Crystal structure of free arrestin-1 revealed two intramolecular interactions that serve as "clasps" holding the molecule in the basal state (Granzin et al., 1998; Hirsch et al., 1999) (Fig. 2A). The first three types of activating mutations disrupt these clasps. One of these clasps, the polar core, is an arrangement of five interacting charged residues between the two arrestin domains (Fig. 2A). These charges are virtually solvent-excluded (Hirsch et al., 1999), which is very unusual for a soluble protein, where charged side chains are normally solvent-exposed. The other clasp is the three-element interaction of β-strand I and α-helix in the N-domain, and β-strand XX in the C-tail (Hirsch et al., 1999) (Figs. 1 and 2A). The C-tail binds to the N-domain via hydrophobic residues, replacement of which with alanines "activates" arrestin-1 (Gurevich, 1998). Thus, for high-affinity binding, active phospho-receptor needs to disrupt key interactions that support basal conformation (Gurevich and Gurevich, 2004), thereby making arrestin more flexible. Receptor-attached phosphates destabilize both the polar core (Vishnivetskiy et al., 1999) and three-element interaction (Vishnivetskiy et al., 2000), enabling transition into high-affinity receptor-binding state. Scanning mutagenesis of the whole arrestin-1 molecule (Ostermaier et al., 2014) confirmed earlier findings regarding the role of the polar core and three-element interaction in maintaining the basal arrestin conformation, as well as the role of Arg18 and Lys20 in phosphate binding (Hanson and Gurevich, 2006). In addition, this comprehensive study revealed the role of Arg29 in phosphate binding that was not explored before (Ostermaier et al., 2014). It also confirmed the role of the finger loop (Fig. 1) in receptor binding proposed earlier (Hanson et al., 2006b; Vishnivetskiy et al., 2011) and suggested that the lariat loop containing two of the three negative charges in the polar core (Fig. 2A) also engages the receptor (Ostermaier et al., 2014). The most unexpected was the finding that the distal tip of the C-domain (part of 344 loop in Fig. 2B) likely engages the membrane (Ostermaier et al., 2014), which was subsequently confirmed by fluorescence quenching experiments (Lally et al., 2017).

Figure 2. Localization of key functional elements in arrestins. (A) Residues of arrestin-2 (PDB ID: 1G4M) involved in critical intra-molecular interactions maintaining the basal conformation are shown as CPK models: in the polar core negatively charged Asp26, Asp290, and Asp297 are colored green, positively charged Arg169 and Arg393 are dark blue; in the three-element interaction the β-strand I residues are salmon, α-helix residues, magenta, and C-tail residues, red. Other structural elements are colored, as follows: inter-domain hinge, yellow; C-tail, brown; main clathrin-binding site, light blue; AP2 binding site, red (note that it includes Arg393, which is also part of the polar core). (B) Parts of arrestin-1 (PDB ID: 1CF1) that move upon receptor binding are colored, as follows: finger loop, salmon; 139-loop, green; 157-loop, dark blue; 344-loop, red. The inter-domain hinge and the C-tail are colored, as in panel A. (C). The position of the JNK3-activating T1A peptide, which includes non-resolved N-terminus and the first two β-strands and the loop between them (magenta), in arrestin-3 (PDB ID: 3P2D) is shown. Note that it is shielded by the C-tail (brown). Inter-domain hinge is shown in yellow. ▶

Several methods, such as hydrogen/deuterium exchange (Carter et al., 2005), measurements of spin label mobility in free and receptor-bound arrestin-1 by EPR (Hanson et al., 2006b), and NMR of isotopically labeled arrestin-1 (Zhuang et al., 2010; Zhuang et al., 2013a), indicate that in both free and receptor-bound states arrestin-1 is flexible, likely exploring a wide range of conformations. This applies to disordered elements that were not resolved in crystal and parts that clearly show up in 3D structures. Several loops demonstrate conformational variability in different protomers within crystallographic tetramer of arrestin-1 (Hirsch et al., 1999), or in different crystal forms of arrestin-2 (Han et al., 2001). Structural variability within the same protein is as

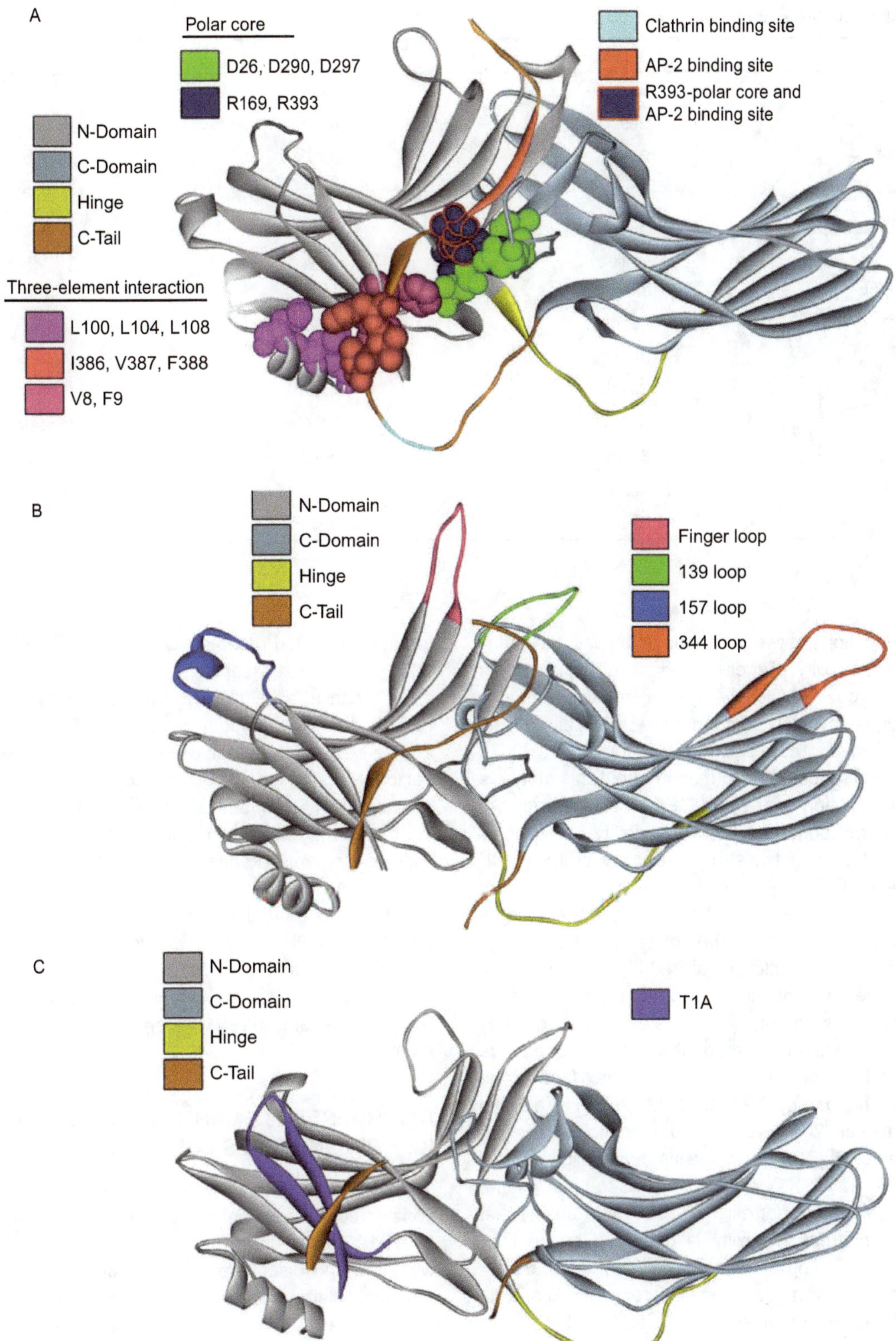
A
Polar core
D26, D290, D297
R169, R393
Clathrin binding site
AP-2 binding site
R393-polar core and AP-2 binding site
N-Domain
C-Domain
Hinge
C-Tail
Three-element interaction
L100, L104, L108
I386, V387, F388
V8, F9
B
N-Domain
C-Domain
Hinge
C-Tail
Finger loop
139 loop
157 loop
344 loop
C
N-Domain
C-Domain
Hinge
C-Tail
T1A

Table 1. Crystal structures of arrestin proteins

Arrestin	State	PDB ID	References
Arrestin-1	Basal	1AYR	(Granzin et al., 1998)
Arrestin-1	Basal	1CF1	(Hirsch et al., 1999)
Arrestin-1, p44	Basal	3UGX	(Granzin et al., 2012)
Arrestin-1, p44	Active	4J2Q	(Kim et al., 2013)
Arrestin-1	Active	4ZWJ	(Kang et al., 2015)
Arrestin-1	Active	5W0P	(Zhou et al., 2017)
Arrestin-1-R175E	Intermediate	4ZRG	(Granzin et al., 2015)
Arrestin-2	Basal	1G4R	(Han et al., 2001)
Arrestin-2	Basal	1JSY	(Milano et al., 2002)
Arrestin-2-IP6	Basal	1ZSH	(Milano et al., 2006)
Arrestin-2	Active	4JQI	(Shukla et al., 2013)
Arrestin-3	Basal	3P2D	(Zhan et al., 2011)
Arrestin-3-IP6	Active	5TV1	(Chen et al., 2017)
Arrestin-4	Basal	1SUJ	(Sutton et al., 2005)

great as the variations between different subtypes, indicating that it reflects flexibility of these regions, rather than their subtype-specific conformations. Although arrestins appear to assume another distinct conformation upon binding to microtubules (Hanson et al., 2006a; Hanson et al., 2007), due to the lack of structural information about this state we will limit the discussion to free and GPCR-bound forms.

To illustrate the range of structural flexibility in the visual arrestin-1, Fig. 3A shows two structures of this protein (PDB IDs: 1AYR and 1CF1) with several loops demonstrating noticeable structural divergence. Whereas in the structures of arrestin-1 (Hirsch et al., 1999), arrestin-2 (Han et al., 2001), and arrestin-3 (Sutton et al., 2005) some of these loops participate in contacts with other protomers, their structural differences do not seem to be explained solely by lattice contacts: the first β-strand on the concave face of the N-domain and the first β-strand on the convex face of the C-domain are engaged by the sister protomers in all arrestin crystals (Zhan et al., 2011), yet their conformation is identical in all protomers. Fig. 3B provides structural alignment of bovine arrestin-1 (1CF1), arrestin-2 (1G4M), arrestin-3 (3P2D), and Salamander arrestin-4 (1SUJ) and shows that the greatest structural variability is observed in the loop regions. Fig. 3C–E show disorder profiles generated for these arrestins by several common disorder predictors. Both N- and C-termini are predicted to be disordered (disorder scores exceed the threshold of 0.5). Several internal regions with disorder scores between 0.2 and 0.5 are expected to have high conformational flexibility. Sequence analysis suggests that arrestin-3 is the most disordered subtype (Fig. 3). Indeed, it is the most promiscuous non-visual arrestin with the lowest preference for phosphorylated over unphosphorylated receptor (Celver et al., 2002; Zhan et al., 2011), the lowest receptor subtype selectivity (Barak et al., 1997; Zhan et al., 2011), and higher affinity for many non-receptor partners (Goodman et al., 1996; Ahmed et al., 2011). Molecular dynamics simulations suggest that the flexibility of wild type arrestin-3 approaches that of the pre-activated R175E mutant of bovine arrestin-1 (Sensoy et al., 2016). The crystal structures confirm conformational flexibility of elements predicted to be disordered by sequence analysis: N- and C-termini are not resolved in the crystal structures, loops in the central crest of the molecule and at the tips of both domains are identifiable in the electron density maps and therefore resolved in most cases, but assume distinct conformations in different protomers in crystal oligomers and in different crystal forms of the same protein.

DISORDERED ELEMENTS OF ARRESTINS CONTAIN KEY BINDING SITES FOR CLATHRIN AND AP2

The binding to active phosphorylated GPCRs was long considered the only function of arrestins. Clathrin (Goodman et al., 1996) and clathrin adaptor AP2 (Laporte et al., 1999) were the first discovered non-receptor binding partners of arrestins. Clathrin binds the loop between β-strands XIX and XX, whereas AP2 interacts with the distal C-tail (Fig. 2A), both disordered elements of arrestin (Fig. 1) not resolved in crystal structures. Longer isoform of arrestin-2 contains an additional clathrin binding site also localized in the disordered part (Kang et al., 2009). Interestingly, when these arrestin elements are co-crystallized with clathrin, both demonstrate highly ordered structures (ter Haar et al., 2000;

Kang et al., 2009), in line with the idea of coupled binding and folding (Shoemaker et al., 2000; Sugase et al., 2007).

Disorder does not mean that these elements are equally accessible in free and receptor-bound arrestins. The release of the C-tail in all arrestins is induced by receptor binding (Palczewski et al., 1991; Hanson et al., 2006b; Vishnivetskiy et al., 2010), and poly-anions heparin (Palczewski et al., 1991; Vishnivetskiy et al., 2002; Zhuang et al., 2010), IP6 (Zhuang et al., 2010; Chen et al., 2017), and multi-phosphorylated C-terminus of rhodopsin (Puig et al., 1995) or vasopressin V2 receptor (Xiao et al., 2004; Nobles et al., 2007). While free arrestin-2 and arrestin-3 bind clathrin (Goodman et al., 1996), the affinity of both arrestins is greatly increased by the release of the C-tail (Xiao et al., 2004; Nobles et al., 2007). Activating mutations destabilizing inter-domain polar core significantly enhance clathrin and AP2 binding (Milano et al., 2002), likely by increasing overall flexibility of the molecule and the probability of the C-tail release in free arrestins. The C-tail release was observed upon a small deletion in 139-loop (Kim et al., 2012), which is also an activating mutation increasing arrestin binding to non-preferred forms of rhodopsin (Vishnivetskiy et al., 2013). Thus, while the range of conformations sampled by the arrestin elements that bind clathrin and AP2 in the basal state is wide enough to prevent their visibility in crystal, it shifts upon arrestin binding to receptors, making the C-tail even more flexible.

However, the order (or lack thereof) is not the only factor affecting the interactions of the C-tail of non-visual arrestins with the trafficking machinery. It has been reported that phosphorylation of arrestin-2 at Ser-412 by ERK1/2 inhibits its interactions with clathrin (and possibly clathrin adaptor AP2, that binds the distal part of the arrestin C-tail (Laporte et al., 1999)) (Fig. 2A) and arrestin-dependent internalization of βAR (Lin et al., 1997). Similar consequences of arrestin-3 phosphorylation at Thr-383 by casein kinase II were also described (Kim et al., 2002; Lin et al., 2002). In both cases receptor binding and consequent release of the C-tail was shown to facilitate dephosphorylation of non-visual arrestins, which enhances clathrin binding just when it is needed to recruit the arrestin-receptor complexes to the coated pit (Lin et al., 1997, 2002; Kim et al., 2002). It is unlikely that phosphorylation/dephosphorylation of these inherently disordered parts of the arrestin C-tail affects their secondary structure. Thus, the presence or absence of phosphates likely directly regulates the interactions of the arrestin C-tail with components of the internalization machinery of the coated pit.

An unexpected effect of PKC phosphorylation of arrestin-2 (β-arrestin1) has been described in the process of T cell activation (Fernández-Arenas et al., 2014). Sustained signaling requires constant supply of antigen-binding T cell receptors to the immunological synapse. To achieve this, T cell needs to internalize peripheral T cell receptors not engaged by the antigen and then deliver them to the immunological synapse, where they can bind the antigen and become activated. Arrestin-2 appears to mediate this process upon its phosphorylation at Ser-163, localized in the cavity of the N-domain (Fig. 1), where the GPCR-attached phosphates bind (Shukla et al., 2013; Zhuo et al., 2014). This phosphorylation apparently activates arrestin-2, with the phosphate attached by PKC (which is activated by T cell receptors) to arrestin itself acting as its activator in lieu of receptor-attached phosphates (Gurevich and Gurevich, 2014). While exact mechanism of triggering the arrestin-2 phosphate sensor by the phosphate attached to Ser-163 was not elucidated, triggering of the phosphate sensor by other means, including polar core mutations and bound polyanions, has been shown to release the C-tail of arrestins (Kim et al., 2002; Zhuang et al., 2010) and increase the affinity of non-visual arrestin subtypes for the components of internalization machinery (Kim et al., 2002; Xiao et al., 2004).

Molecular recognition of IDPs and IDPRs is associated with their ability to undergo a disorder-to-order transition upon binding to their partners (Dunker and Obradovic, 2001; Dyson and Wright, 2002; Uversky et al., 2005). There are elements, known as molecular recognition features (MoRFs), located within longer IDPRs, that are disordered in their unbound state and fold upon interaction with specific partners (Oldfield et al., 2005), mediating IDP/IDPR binding via disorder-to-order-transition-based recognition involved in many biological processes (Disfani et al., 2012). Often, in corresponding disorder profiles, MoRFs are indicated as "dips"; i.e., regions with locally increased order propensity within long IDPRs. Potential MoRFs in arrestins (the regions spanning residues 372–383 in arrestin-1, residues 143–151, 165–170, 343–351, and 375–397 in arrestin-2, residues 143–153 and 167–172 in arrestin-3, and residues 341–348 and 364–371 in arrestin-4) were identified by the ANCHOR algorithm (Dosztanyi et al., 2009). Figure 4 maps known functional elements and arrestin switches (ASw, regions that significantly change conformation upon activation) on arrestin-3 structure. These functional elements include T1A peptide (residues 1–25) needed for the JNK3 activation (see also Fig. 2C) (Zhan et al., 2016), JNK3-binding peptides T3 (residues 172–208) and T6 (residues 287–318) (Zhan et al., 2014), ASwI (residues 89–97), ASwII (residues 176–191), and ASwIII (residues 307–316) (Chen et al., 2017), clathrin-binding element (residues 373–377), and clathrin adaptor AP2 binding site (residues 393–399). Importantly, ASwII and ASwIII are parts or the JNK3-binding peptides T3 and T6, respectively. Also, JNK3-binding peptide T3 and ASwII are located in close proximity of one of the MoRFs (residues 167–172). In addition to two real MoRFs, ANCHOR profile for arrestin-3 contains 9 filtered potential disorder-based regions (which are either short regions with length below 6 residues or regions with an average disorder score below 0.1), residues 80–82 (located in close proximity of ASwI), 116–120, 208–209 (overlaps with C-terminal region of the JNK3-binding peptide T3), 238–247, 271–274, 317–326 (located next to ASwIII), 342, 399 (located within clathrin adaptor AP2 binding site), and 402. Therefore, the

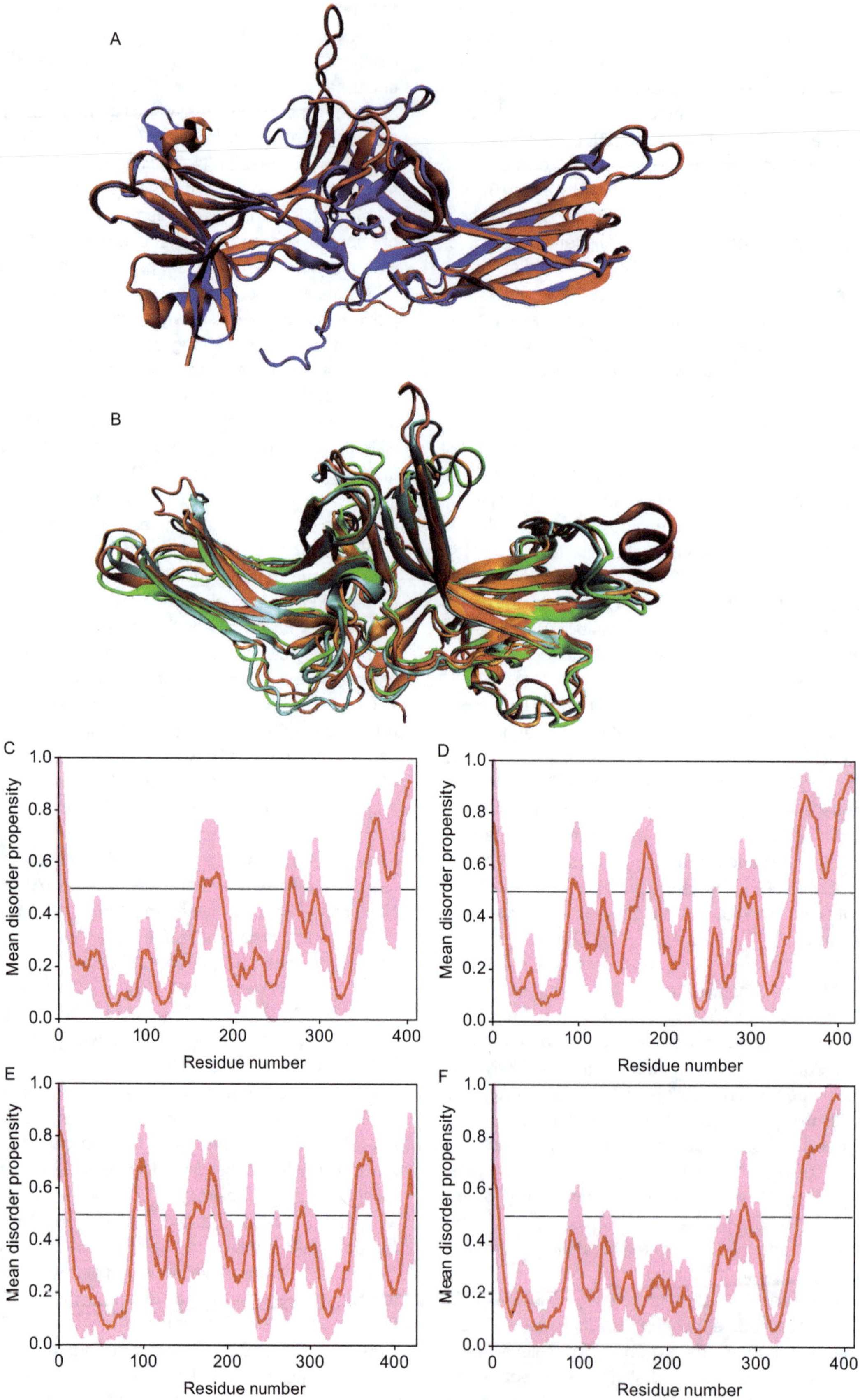
A
B
C
D
E
F
Mean disorder propensity
Residue number
0.0
0.2
0.4
0.6
0.8
1.0
0
100
200
300
400

◄ **Figure 3. Structural disorder in arrestin proteins.** (A) Structural alignment of PDB entries 1AYR (blue ribbon) and 1CF1 (red ribbon) corresponding to the crystal structures of visual arrestin-1 from bovine rod photoreceptors. (B) Multiple structure alignments for bovine arrestin-1 (PDB ID: 1CF1, red ribbon), arrestin-2 (PDB ID: 1G4M, green ribbon), arrestin-3 (PDB ID: 3P2D, cyan ribbon), and arrestin-4 (PDB ID: 1SUJ, orange ribbon). Structures were generated using VMD 1.9.2 (Humphrey et al., 1996). Multiple structure alignment was performed using the MultProt (http://bioinfo3d.cs.tau.ac.il/MultiProt/) (Shatsky et al., 2004). (C–F) Multiparametric computational analysis of the intrinsic disorder predisposition of bovine arrestin-1 (UniProt ID: P08168) (C), arrestin-2 (UniProt ID: P17870) (D) arrestin-3 (UniProt ID: P32120) (E), and arrestin-4 (UniProt ID: Q9PTE7) (F). In each plot, red line shows the mean disorder propensity calculated by averaging disorder profiles generated by six common disorder predictors: PONDR® VLXT (Dunker et al., 2001), PONDR® VSL2 (Peng et al., 2005), PONDR® VL3 (Peng et al., 2006), PONDR® FIT (Xue et al., 2010), and IUPred_short and IUPred_long (Dosztanyi et al., 2005). Light pink shadow around the mean disorder score shows distribution of standard deviations. Predicted intrinsic disorder scores above 0.5 are considered to correspond to the disordered regions, whereas regions with the disorder scores between 0.2 and 0.5 are considered flexible.

comparison of Figs. 3 and 4 shows that functional significance of several MoRFs is already established, whereas the role of other disorder-based binding sites in arrestin interactions with various proteins needs to be tested experimentally.

CAN THE "ORDERED" PART OF ARRESTIN BECOME "DISORDERED"?

The ~350-residue core of all arrestins, from residue ~10 to ~360, is relatively well ordered in all crystal structures of the basal state, with the exception of loops connecting β-strands. C-terminal truncations by alternative splicing (Smith et al., 1994; Pulvermuller et al., 1997) or mutagenesis (Gurevich and Benovic, 1992, 1993) significantly reduce thermal stability of arrestin-1, implying increased flexibility of the remaining part, which lowers energy barrier for the receptor interaction (Pulvermuller et al., 1997; Ahmed et al., 2011) and significantly enhances the arrestin binding to non-preferred forms of GPCRs (Gurevich, 1998; Celver et al., 2002), supporting the idea that less ordered proteins are more inclined to bind their partners. A virtually perfect correlation between increased binding of arrestin-1 mutants to non-preferred forms of rhodopsin and the reduction of their thermal stability was established (Vishnivetskiy et al., 2013). Interestingly, the effects of the C-terminal deletions are not limited to GPCR interactions, facilitating the binding of some non-receptor partners.

A recent study illustrates how nature uses this phenomenon as a gain-of-function change. A common feature of apoptotic cell death is a massive activation of specialized proteases, caspases, that cleave numerous structural and signaling proteins (Lüthi and Martin, 2007), yielding characteristic morphological and biochemical changes. The cleavage of arrestin-2 by caspases at Asp-380 generates a truncated form lacking the C-tail, arrestin-2_{1-380} (Kook et al., 2014). Full-length (FL) arrestin-2 is evenly distributed in the cell (Barak et al., 1997; Song et al., 2006), whereas arrestin-2_{1-380} is recruited to mitochondria, even though it does not have recognizable mitochondrial localization signal. In contrast to FL protein, arrestin-2_{1-380} avidly binds another product of caspase activity, truncated Bid (tBid). Although tBid localizes to mitochondria, arrestin-2_{1-380} moves to this compartment independently, because its mitochondrial localization does not require the presence of tBid or Bid (Kook et al., 2014). Thus, the C-terminal truncation of arrestin-2 confers two new functions: the ability to bind mitochondria and tBid. Both are equally important for the role of arrestin-2_{1-380} in apoptosis, where it enhances tBid-induced cytochrome *c* release from mitochondria, which is the point of no return in the apoptotic cell death (Danial and Korsmeyer, 2004). Apparently, the removal of the C-tail by caspases promotes arrestin-2 interactions with tBid and resident mitochondrial protein(s) by increasing its flexibility, helping arrestin-2_{1-380} fit partners that the more rigid FL protein does not bind. Similarly, C-terminal deletions or C-tail detachment enhance the binding of all arrestin subtypes to microtubules (Hanson et al., 2007).

However, increased flexibility of the core does not indiscriminately enhance arrestin interactions with all possible partners. The C-tail of all arrestins can be forcibly detached by triple alanine substitution of the hydrophobic residues that anchor it to the N-domain (3A) (Gurevich, 1998; Celver et al., 2002; Pan et al., 2003; Song et al., 2009b) (Fig. 2A). These 3A mutations increase the flexibility of the molecule, as revealed by faster H/D exchange (Carter et al., 2005), and reduce activation energy for receptor binding (Gurevich et al., 2011), suggesting that these mutants are in conformation intermediate between basal and receptor-bound. The opposite conformational bias is conferred by seven-residue deletion in the inter-domain hinge (Δ7) (Figs. 1 and 2) that strongly inhibits receptor binding (Vishnivetskiy et al., 2002; Hanson et al., 2007).

Interestingly, the binding of two E3 ubiquitin ligases, Mdm2 and parkin, is inhibited by 3A and enhanced by Δ7 mutations (Song et al., 2006; Ahmed et al., 2011). These data do not necessarily demonstrate that these partners prefer "ordered" arrestins over "disordered". It has been shown that Δ7 mutation also results in the C-tail release (Hanson et al., 2006a), suggesting that it produces a different flavor of disorder, conducive to the binding of Mdm2 and parkin. However, these flavors must have some similarities: the binding of ERK1/2 is facilitated by both 3A and Δ7 mutations in arrestin-2 and -3 (Coffa et al., 2011a). Thus,

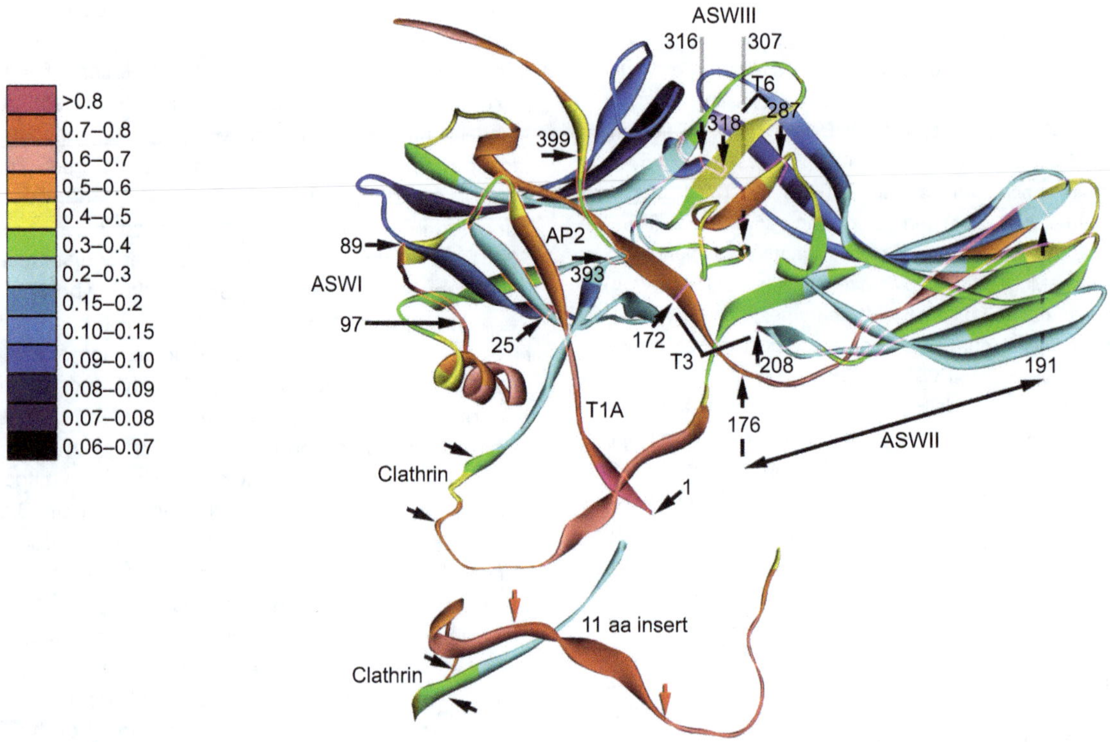

Figure 4. Functional implications of structural disorder. The structure of the basal state of arrestin-3 (PDB ID: 3P2D (Zhan et al., 2011)) is colored according to the propensity for structural disorder (color-code shown on the left). The C-terminus of the long splice variant of arrestin-3 with an 11-residue insert (as compared to the short splice variant (Sterne-Marr et al., 1993)) is shown below. Known functional elements are indicated on the structure: T1A peptide that facilitates the activation of JNK3 in cells (Zhan et al., 2016), residues 1–25; clathrin-binding element (residues 373–377 in the short splice variant), clathrin adaptor AP2 binding site (residues 393–399 in the short splice variant); JNK3-binding peptides T3 (residues 172–208) and T6 (residues 287–318) (Zhan et al., 2014); as well as arrestin switches that significantly change conformation upon activation (Chen et al., 2017): ASwI (residues 89–97), ASwII (residues 176–191), and ASwIII (residues 307–316). Note that JNK3-binding peptides T3 and T6 include ASwII and ASwIII, respectively. Also note that known functional elements are localized in regions of predicted high or moderate disorder. The 11-residue insert in the long splice variant (Sterne-Marr et al., 1993) has very high disorder score. Alternative splicing of arrestin-3 is conserved in mammalian evolution (Gurevich and Gurevich, 2006; Indrischek et al., 2017), suggesting that this insert is likely functionally important, e.g., binds an unknown partner that the short variant does not bind.

different mechanisms of destabilization of the basal arrestin structure appear to increase its flexibility in distinct partially overlapping ways, enhancing the binding of different, or sometimes the same, partners.

FLEXIBILITY OF RECEPTOR-BOUND ARRESTINS

The arrestin-receptor complexes initiate several signaling cascades that free arrestins do not (Peterson and Luttrell, 2017). For example, ERK1/2 binds free arrestin with low affinity (Luttrell et al., 2001; Coffa et al., 2011a), whereas this interaction is greatly enhanced by the arrestin recruitment to the receptor (Luttrell et al., 2001). The arrestin-dependent ERK1/2 activation occurs only upon GPCR stimulation (Luttrell et al., 2001; Coffa et al., 2011b). From structural perspective, the simplest explanation is that receptor-bound arrestin presents a different "face" to the world, where elements engaged by particular partners, such as ERK1/2, become more accessible or better aligned (Gurevich and Gurevich, 2003). However, for a long time the release of the C-tail remained the only conformational rearrangement in arrestins directly ascribed to the receptor binding. Recent study using pulse EPR technique DEER for intra-molecular distance measurements in free and P-Rh*-bound arrestin-1 detected additional rearrangements. These included the movement of the finger loop on the receptor-binding surface (Figs. 1 and 2B) in the direction of the incoming receptor, a dramatic repositioning of the neighboring 139-loop towards the N-domain and to the side, and movements of two loops containing residues 344 and 157 at the tips of the two arrestin domains (Fig. 2B) towards and away from the receptor, respectively (Kim et al., 2012). Interestingly, the

344 loop in the stretch with elevated disorder potential in all arrestins (Fig. 3C–F) directly engages membrane lipids (Lally et al., 2017), consistent with its position in the arrestin-1-rhodopsin complex (Kang et al., 2015).

Earlier findings that deletions in the inter-domain hinge impede the receptor binding (Vishnivetskiy et al., 2002; Hanson et al., 2007) suggested that long hinge is needed for the two arrestin domains to "grab" the receptor like a pincer (Gurevich and Gurevich, 2004). However, DEER studies of arrestin-1 (Kim et al., 2012) and both non-visual subtypes (Zhuo et al., 2014) ruled out such "clam-like" movement of the domains. In an alternative model, the receptor binding is enabled by the two domains rotating relative to each other by ~20° (Modzelewska et al., 2006). Two crystal structures of presumably "active" arrestin-1 and arrestin-2 are consistent with this model. Pre-activated arrestin-2 with C-terminal deletion (Kovoor et al., 1999) was co-crystallized with multi-phosphorylated C-terminus of vasopressin V2 receptor and conformationally selective antibody fragment (Shukla et al., 2013). This structure revealed a large movement of the 139-loop homologue (called "middle loop" in this study), similar to that detected in arrestin-1 by DEER (Kim et al., 2012), rearrangement of the finger loop, and the rotation of N- and C-domains by ~20° (Shukla et al., 2013). The short splice variant of arrestin-1 lacking the C-tail, p44, expressed in some mammalian species, was also shown to be "pre-activated" and bind unphosphorylated Rh* (Pulvermuller et al., 1997). In one crystal structure, this protein looked similar to FL arrestin-1 (Granzin et al., 2012), but the second structure revealed a very different conformation, with characteristic rearrangement of the finger loop, movement of the 139-loop, and ~21° rotation of the two domains (Kim et al., 2013).

These studies have obvious caveats. Both DEER (Kim et al., 2012) and NMR (Zhuang et al., 2013a) were done with arrestin-1 bound to a bona fide receptor, purified rhodopsin, but neither method yields atomic resolution. In higher resolution crystal structures, "active" arrestin-1 (Kim et al., 2013) and arrestin-2 (Shukla et al., 2013) were not bound to a receptor. The recent crystal structure of the arrestin-1-rhodopsin complex, the only structure of the GPCR-bound arrestin so far, solved using innovative method suitable for very small crystals, femtosecond X-ray laser, revealed similar movements of flexible loops and the twisting of the two arrestin-1 domains (Kang et al., 2015; Zhou et al., 2017).

GPCRs are very dynamic proteins (Manglik and Kobilka, 2014; Manglik et al., 2015), which makes them challenging targets for X-ray crystallography. Virtually all structures were solved using heavily engineered receptors, where some elements were deleted and/or replaced with small soluble proteins that provide crystal contacts. The arestin-1-rhodopsin complex was no exception. First, a fusion of human rhodopsin with T4 lysozyme was used. Second, this rhodopsin carried two mutations, M257Y and E113Q, making it fully active even in the absence of retinal, and an additional disulfide bridge between residues 2 and 282 that were replaced with cysteines, which is widely used for rhodopsin stabilization (Standfuss et al., 2007). Third, rhodopsin was not phosphorylated to the level required for tight wild type arrestin-1 binding (Mendez et al., 2000; Vishnivetskiy et al., 2007). Thus, pre-activated mouse arrestin-1-3A mutant that binds Rh* with high affinity was used (Song et al., 2009b; Vishnivetskiy et al., 2013). Finally, although individually expressed rhodopsin and arrestin-1 mutants were able to bind each other, a complex stable enough for crystallization was obtained only when both were expressed as a fusion protein. However, the arrangement of arrestin-1 and rhodopsin in the structure of the complex that was expressed and crystallized as a fusion protein, was confirmed using separately expressed and purified arrestin-1 and rhodopsin by several methods: a) changes in the rate of H/D exchange upon binding; b) disulfide cross-linking between residues in arrestin-1 and rhodopsin predicted to be within range by the crystal structure, and lack of cross-linking between residues predicted to be too far apart; and c) measurements of distances between selected positions in the two proteins using DEER (Kang et al., 2015; Zhou et al., 2017).

Well-diffracting crystals form only when protein conformations are fixed, either naturally, or artificially, by crystallization conditions and/or crystal packing forces. In this crystal, all four arrestin-rhodopsin complexes in the asymmetric unit look very similar (Kang et al., 2015; Zhou et al., 2017), strongly suggesting biological relevance of this structure. In the arrestin-rhodopsin complex three distances were measured by DEER: from rhodopsin residue 74 to arrestin-1 residues 61, 140, and 241 (Kang et al., 2015); four distances were measured to confirm the position of the rhodopsin C-terminus relative to the N-domain of arrestin-1: from rhodopsin positions 335, 337, and 342 to residue 107 in arrestin-1, and from 342 in rhodopsin to 106 in arrestin-1 (Zhou et al., 2017). It is remarkable that the most probable distance in each case matched the crystal structure. However, it is equally remarkable that in every case, multiple additional distances were detected, clearly indicating that the complex explores fairly wide conformational space (Kang et al., 2015; Zhou et al., 2017). This notion is supported by the finding that the distributions of intra-molecular distances measured by DEER in free arrestin-1 are wide, and remain wide upon rhodopsin binding (Kim et al., 2012). The same is true for intra-molecular distances in free and receptor-bound arrestin-2 and -3 (Zhuo et al., 2014). NMR data for ^{13}C-^{15}N-labeled arrestin-1 also suggest its high flexibility in rhodopsin-bound form (Zhuang et al., 2013a). Thus, all evidence indicates that receptor-bound arrestins, which organize a variety of signaling complexes, are very flexible. This appears to be necessary for their functional versatility.

Arrestin-dependent activation of JNKs illustrates one mechanism whereby flexibility translates into function. Out of the four vertebrate arrestins, only arrestin-3 activates JNKs (McDonald et al., 2000; Song et al., 2009a, 2011). Highly homologous arrestin-2 (Attramadal et al., 1992; Sterne-Marr et al., 1993) also binds kinases in the ASK1-MKK4-JNK3 cascade, but does not activate JNKs (Song et al., 2009a, 2011). Although JNK3 and MKK4 binding was discovered first (McDonald et al., 2000), subsequent studies showed that arrestin-3 also binds MKK7 (Zhan et al., 2013) needed for full activation of JNKs (Lawler et al., 1998), as well as ubiquitously expressed JNK1 and JNK2 (Kook et al., 2013). Therefore, the mechanism described below applies to arrestin-3 scaffolding of all JNK activation cascades.

JNK3 engages three arrestin-3 elements (Fig. 4), with the N-terminal 25 residues (T1A peptide) being the most potent binding site (Zhan et al., 2014). This peptide binds all kinases in the pathway and facilitates JNK activation *in vitro* and in cells (Zhan et al., 2016). The concentration dependence of the scaffold effect is usually bell-shaped (Levchenko et al., 2004): low scaffold concentrations facilitate productive interactions, which are suppressed when there is more scaffold molecules than the components of the cascade. The optimal concentration of the scaffold yielding the highest reaction rate depends on scaffold affinities for the proteins it brings together (Zhan et al., 2013). Interestingly, optimal concentrations of T1A peptide for scaffolding of MKK4-JNK3 and MKK7-JNK3 modules are 10 times lower than those of FL arrestin-3 (Zhan et al., 2016). Arrestin-3 structure suggests an explanation for this. Even though free arrestin-3 acts as a scaffold (Song et al., 2009a; Breitman et al., 2012), receptor binding enhances this action (McDonald et al., 2000). In the basal state of arrestin-3, T1A peptide, which includes the N-terminal extension not resolved in crystal and β-strands I and II (Zhan et al., 2014) (Figs. 1 and 4), is partially shielded by the C-tail (Fig. 2C), which is released upon receptor binding (Zhuo et al., 2014), ensuring better exposure of this element. This explains higher effectiveness of the receptor-bound arrestin-3 as JNK3 activator (McDonald et al., 2000), as well as higher apparent affinity of T1A for JNK3, MKK4, and MKK7 (Zhan et al., 2016). Furthermore, expressed as a YFP-fusion, T1A peptide is likely fully exposed and flexible, activating JNK3 more efficiently than full-length arrestin-3 (Zhan et al., 2016). Thus, enhanced flexibility of a protein element facilitates its function.

DOES DISORDER POTENTIAL HAVE PREDICTIVE VALUE?

It is instructive to compare disorder potential of arrestin elements (Fig. 3C–F) with available functional data (Fig. 4). The N-terminus and the C-tail, with high predicted disorder in all arrestins (Fig. 3C–F), are biologically important. The first 25 residues of arrestin-3 (Figs. 2C and 4) are the highest affinity binding site for JNK3 (Zhan et al., 2014), bind all the upstream kinases in JNK3 activation cascade, and serve as a functional scaffold in cells (Zhan et al., 2016). The part of the C-terminus of both arrestin-2 and -3 between the C-domain and β-strand XX (Fig. 1), which is disordered and never resolved in crystal structures, contains both the major (Goodman et al., 1996) and minor (in arrestin-2) (Kang et al., 2009) clathrin-binding sites (Fig. 2A). The distal C-tail (also never resolved in crystal structures) binds clathrin adaptor AP2 (Laporte et al., 1999; Milano et al., 2002; Moaven et al., 2013) (Fig. 2A). Upon receptor binding, when the interactions with clathrin and AP2 come into play, the C-tail of arrestins is released and becomes even more disordered. The comparison of the structure of bovine arrestin-3 activated by IP6 (Chen et al., 2017) with the structures of arrestin-2 activated by receptor phospho-peptide (Shukla et al., 2013), pre-activated arrestin-1 splice variant p44 (Kim et al., 2013), and arrestin-1 in complex with rhodopsin (Kang et al., 2015; Zhou et al., 2017) identified three regions that invariably assume different conformations in basal and active arrestins, termed arrestin switch (ASw) regions (Chen et al., 2017). These include ASwI (one of the long "legs" connecting the α-helix to the body of the N-domain, residues 89–97 in bovine arrestin-3), ASwII (inter-domain hinge and the first β-strand in the C-domain, residues 176–191), and ASwIII (the extension of the lariat loop supplying two of the three negative charges of the polar core, residues 307–316) (Fig. 4). Switches likely serve as docking sites for the non-receptor partners that discriminate between active and inactive arrestins (Chen et al., 2017). Interestingly, all these elements are predicted to have high disorder potential in both non-visual subtypes (Figs. 3D, 3E, and 4) implicated in arrestin-mediated signaling, and appear to be more ordered in visual arrestin-1 and -4 (Fig. 3C and 3F), which lack these functions. Importantly, two other regions implicated in JNK3 binding, peptides T3 (residues 172–208) and T6 (residues 287–318) (Zhan et al., 2014) contain ASwII and ASwIII, respectively (Chen et al., 2017) (Fig. 4). While other protein partners that engage different switch regions still await identification, there is a provocative tendency for the elements with the highest disorder potential to mediate arrestin interactions with other proteins.

CONCLUSIONS

Biologically, the most remarkable characteristic of arrestins is their ability to bind hundreds of different GPCRs and numerous non-receptor partners. Arrestins exist in at least three distinct conformations: free, receptor-bound, and microtubule-bound. Structurally, an important feature of arrestins is high flexibility in every state. X-ray crystallography and biophysical studies clearly show that in both free and receptor-bound state arrestins explore a wide range of conformations. Conceivably, the conformational spaces covered by free, receptor- and microtubule-bound arrestins partially overlap. This flexibility likely underlies remarkable functional versatility of arrestins and represents the structural basis for the ability of arrestins to scaffold signaling pathways. The functionality of free arrestins and their separated elements paves the way for the use of arrestin-based molecular tools to manipulate cellular signaling for experimental and, potentially, therapeutic purposes. The correlation of functional capabilities with high probability of disorder suggests that in search of arrestin elements engaging signaling proteins the field needs to pay closer attention to more disordered parts.

ACKNOWLEDGEMENTS

Funding: NIH RO1 Grants EY011500, GM077561 and GM109955 (the latter two RO1s were replaced by R35 GM122491) (VVG), NS065868 and DA030103 (EVG).

ABBREVIATIONS

ASK, apoptosis signal-regulated kinase; ASw, arrestin switch; DEER, double electron-electron resonance; EPR electron paramagnetic resonance; ERK, extracellular signal-regulated kinase; FL, full length; GPCR, G protein-coupled receptor; IDP, intrinsically disordered protein; IDPR, intrinsically disordered protein region; IP6, inositol hexakisphosphate; JNK, c-Jun N-terminal kinase; MKK, MAP kinase kinase; MoRFs, molecular recognition features; NMR, nuclear magnetic resonance; PKC, protein kinase C; P-Rh, dark (inactive) phosphorylated rhodopsin; P-Rh*, light-activated phosphorylated rhodopsin; Rh*, light-activated rhodopsin; WT, wild type.

FOOTNOTE

[1]Here we use the systematic names of arrestins, where the number after the dash indicates the order of cloning: arrestin-1 (historic names S-antigen, 48 kDa protein, visual or rod arrestin), arrestin-2 (β-arrestin or β-arrestin1), arrestin-3 (β-arrestin2 or hTHY-ARRX), and arrestin-4 (cone or X-arrestin).

COMPLIANCE WITH ETHICS GUIDELINES

Vsevolod V. Gurevich, Eugenia V. Gurevich, and Vladimir N. Uversky declare that they have no conflict of interest.

REFERENCES

Agarwal PK (2005) Role of protein dynamics in reaction rate enhancement by enzymes. J Am Chem Soc 127:15248–15256

Ahmed MR, Zhan X, Song X, Kook S, Gurevich VV, Gurevich EV (2011) Ubiquitin ligase parkin promotes Mdm2-arrestin interaction but inhibits arrestin ubiquitination. Biochemistry 50:3749–3763

Artymiuk PJ, Blake CC, Grace DE, Oatley SJ, Phillips DC, Sternberg MJ (1979) Crystallographic studies of the dynamic properties of lysozyme. Nature 280:563–568

Attramadal H, Arriza JL, Aoki C, Dawson TM, Codina J, Kwatra MM, Snyder SH, Caron MG, Lefkowitz RJ (1992) Beta-arrestin2, a novel member of the arrestin/beta-arrestin gene family. J Biol Chem 267:17882–17890

Barak LS, Ferguson SS, Zhang J, Caron MG (1997) A beta-arrestin/green fluorescent protein biosensor for detecting G protein-coupled receptor activation. J Biol Chem 272:27497–27500

Breitman M, Kook S, Gimenez LE, Lizama BN, Palazzo MC, Gurevich EV, Gurevich VV (2012) Silent scaffolds: inhibition of c-Jun N-terminal kinase 3 activity in the cell by a dominant-negative arrestin-3 mutant. J Biol Chem 287:19653–19664

Buljan M, Chalancon G, Dunker AK, Bateman A, Balaji S, Fuxreiter M, Babu MM (2013) Alternative splicing of intrinsically disordered regions and rewiring of protein interactions. Curr Opin Struct Biol 23:443–450

Carpenter B, Nehmé R, Warne T, Leslie AG, Tate CG (2016) Structure of the adenosine A(2A) receptor bound to an engineered G protein. Nature 536:104–107

Carter JM, Gurevich VV, Prossnitz ER, Engen JR (2005) Conformational differences between arrestin2 and pre-activated mutants as revealed by hydrogen exchange mass spectrometry. J Mol Biol 351:865–878

Celver J, Vishnivetskiy SA, Chavkin C, Gurevich VV (2002) Conservation of the phosphate-sensitive elements in the arrestin family of proteins. J Biol Chem 277:9043–9048

Chen Q, Perry NA, Vishnivetskiy SA, Berndt S, Gilbert NC, Zhuo Y, Singh PK, Tholen J, Ohi MD, Gurevich EV et al (2017) Structural basis of arrestin-3 activation and signaling. Nat Commun 8:1427

Coffa S, Breitman M, Hanson SM, Callaway K, Kook S, Dalby KN, Gurevich VV (2011a) The effect of arrestin conformation on the recruitment of c-Raf1, MEK1, and ERK1/2 activation. PLoS ONE 6:e28723

Coffa S, Breitman M, Spiller BW, Gurevich VV (2011b) A single mutation in arrestin-2 prevents ERK1/2 activation by reducing c-Raf1 binding. Biochemistry 50:6951–6958

Danial NN, Korsmeyer SJ (2004) Cell death: critical control points. Cell 116:205–219

Daniel RM, Dines M, Petach HH (1996) The denaturation and degradation of stable enzymes at high temperatures. Biochem J 317:1–11

Disfani FM, Hsu WL, Mizianty MJ, Oldfield CJ, Xue B, Dunker AK, Uversky VN, Kurgan L (2012) MoRFpred, a computational tool for sequence-based prediction and characterization of short disorder-to-order transitioning binding regions in proteins. Bioinformatics 28:i75–83

Dosztanyi Z, Csizmok V, Tompa P, Simon I (2005) IUPred: web server for the prediction of intrinsically unstructured regions of proteins based on estimated energy content. Bioinformatics 21:3433–3434

Dosztanyi Z, Meszaros B, Simon I (2009) ANCHOR: web server for predicting protein binding regions in disordered proteins. Bioinformatics 25:2745–2746

Dunker AK, Obradovic Z (2001) The protein trinity—linking function and disorder. Nat Biotechnol 19:805–806

Dunker AK, Lawson JD, Brown CJ, Williams RM, Romero P, Oh JS, Oldfield CJ, Campen AM, Ratliff CM, Hipps KW et al (2001) Intrinsically disordered protein. J Mol Graph Model 19:26–59

Dyson HJ, Wright PE (2002) Coupling of folding and binding for unstructured proteins. Curr Opin Struct Biol 12:54–60

Eisenmesser EZ, Millet O, Labeikovsky W, Korzhnev DM, Wolf-Watz M, Bosco DA, Skalicky JJ, Kay LE, Kern D (2005) Intrinsic dynamics of an enzyme underlies catalysis. Nature 438:117–121

Farrens DL, Altenbach C, Yang K, Hubbell WL, Khorana HG (1996) Requirement of rigid-body motion of transmembrane helices for light activation of rhodopsin. Science 274:768–770

Fernández-Arenas E, Calleja E, Martínez-Martín N, Gharbi SI, Navajas R, García-Medel N, Penela P, Alcamí A, Mayor FJ, Albar JP et al (2014) β-Arrestin-1 mediates the TCR-triggered re-routing of distal receptors to the immunological synapse by a PKC-mediated mechanism. EMBO J 33(6):559–577

Frauenfelder H, Petsko GA, Tsernoglou D (1979) Temperature-dependent X-ray diffraction as a probe of protein structural dynamics. Nature 280:558–563

Giver L, Gershenson A, Freskgard PO, Arnold FH (1998) Directed evolution of a thermostable esterase. Proc Natl Acad Sci USA 95:12809–12813

Goodman OB Jr, Krupnick JG, Santini F, Gurevich VV, Penn RB, Gagnon AW, Keen JH, Benovic JL (1996) Beta-arrestin acts as a clathrin adaptor in endocytosis of the beta2-adrenergic receptor. Nature 383:447–450

Granzin J, Wilden U, Choe HW, Labahn J, Krafft B, Buldt G (1998) X-ray crystal structure of arrestin from bovine rod outer segments. Nature 391:918–921

Granzin J, Cousin A, Weirauch M, Schlesinger R, Büldt G, Batra-Safferling R (2012) Crystal structure of p44, a constitutively active splice variant of visual arrestin. J Mol Biol 416:611–618

Granzin J, Stadler A, Cousin A, Schlesinger R, Batra-Safferling R (2015) Structural evidence for the role of polar core residue Arg175 in arrestin activation. Sci Rep 5:15808

Gurevich VV (1998) The selectivity of visual arrestin for light-activated phosphorhodopsin is controlled by multiple nonredundant mechanisms. J Biol Chem 273:15501–15506

Gurevich VV, Benovic JL (1992) Cell-free expression of visual arrestin. Truncation mutagenesis identifies multiple domains involved in rhodopsin interaction. J Biol Chem 267:21919–21923

Gurevich VV, Benovic JL (1993) Visual arrestin interaction with rhodopsin: sequential multisite binding ensures strict selectivity towards light-activated phosphorylated rhodopsin. J Biol Chem 268:11628–11638

Gurevich VV, Benovic JL (1995) Visual arrestin binding to rhodopsin: diverse functional roles of positively charged residues within the phosphorylation-recignition region of arrestin. J Biol Chem 270:6010–6016

Gurevich VV, Benovic JL (1997) Mechanism of phosphorylation-recognition by visual arrestin and the transition of arrestin into a high affinity binding state. Mol Pharmacol 51:161–169

Gurevich VV, Gurevich EV (2003) The new face of active receptor bound arrestin attracts new partners. Structure 11:1037–1042

Gurevich VV, Gurevich EV (2004) The molecular acrobatics of arrestin activation. Trends Pharmacol Sci 25:105–111

Gurevich EV, Gurevich VV (2006) Arrestins are ubiquitous regulators of cellular signaling pathways. Genome Biol 7:236

Gurevich VV, Gurevich EV (2014) Arrestin makes T cells stop and become active. EMBO J 33:531–533

Gurevich VV, Chen C-Y, Kim CM, Benovic JL (1994) Visual arrestin binding to rhodopsin: Intramolecular interaction between the basic N-terminus and acidic C-terminus of arrestin may regulate binding selectivity. J Biol Chem 269:8721–8727

Gurevich VV, Hanson SM, Song X, Vishnivetskiy SA, Gurevich EV (2011) The functional cycle of visual arrestins in photoreceptor cells. Prog Retin Eye Res 30:405–430

Habchi J, Tompa P, Longhi S, Uversky VN (2014) Introducing protein intrinsic disorder. Chem Rev 114:6561–6588

Han M, Gurevich VV, Vishnivetskiy SA, Sigler PB, Schubert C (2001) Crystal structure of beta-arrestin at 1.9 A: possible mechanism of receptor binding and membrane translocation. Structure 9:869–880

Hanson SM, Gurevich VV (2006) The differential engagement of arrestin surface charges by the various functional forms of the receptor. J Biol Chem 281:3458–3462

Hanson SM, Francis DJ, Vishnivetskiy SA, Klug CS, Gurevich VV (2006a) Visual arrestin binding to microtubules involves a distinct conformational change. J Biol Chem 281:9765–9772

Hanson SM, Francis DJ, Vishnivetskiy SA, Kolobova EA, Hubbell WL, Klug CS, Gurevich VV (2006b) Differential interaction of spin-labeled arrestin with inactive and active phosphorhodopsin. Proc Natl Acad Sci USA 103:4900–4905

Hanson SM, Cleghorn WM, Francis DJ, Vishnivetskiy SA, Raman D, Song X, Nair KS, Slepak VZ, Klug CS, Gurevich VV (2007) Arrestin mobilizes signaling proteins to the cytoskeleton and redirects their activity. J Mol Biol 368:375–387

Hirsch JA, Schubert C, Gurevich VV, Sigler PB (1999) The 2.8 A crystal structure of visual arrestin: a model for arrestin's regulation. Cell 97:257–269

Hoffmann C, Gaietta G, Bunemann M, Adams SR, Oberdorff-Maass S, Behr B, Vilardaga JP, Tsien RY, Ellisman MH, Lohse MJ (2005) A FlAsH-based FRET approach to determine G protein-coupled receptor activation in living cells. Nat Methods 2:171–176

Humphrey W, Dalke A, Schulten K (1996) VMD: visual molecular dynamics. J Mol Graph 14(1):33–38

Iakoucheva LM, Brown CJ, Lawson JD, Obradovic Z, Dunker AK (2002) Intrinsic disorder in cell-signaling and cancer-associated proteins. J Mol Biol 323:573–584

Iakoucheva LM, Radivojac P, Brown CJ, O'Connor TR, Sikes JG, Obradovic Z, Dunker AK (2004) The importance of intrinsic disorder for protein phosphorylation. Nucleic Acids Res 32:1037–1049

Indrischek H, Prohaska SJ, Gurevich VV, Gurevich EV, Stadler PF (2017) Uncovering missing pieces: duplication and deletion history of arrestins in deuterostomes. BMC Evol Biol 17:163

Kang DS, Kern RC, Puthenveedu MA, von Zastrow M, Williams JC, Benovic JL (2009) Structure of an arrestin2-clathrin complex reveals a novel clathrin binding domain that modulates receptor trafficking. J Biol Chem 284:29860–29872

Kang Y, Zhou XE, Gao X, He Y, Liu W, Ishchenko A, Barty A, White TA, Yefanov O, Han GW et al (2015) Crystal structure of rhodopsin bound to arrestin determined by femtosecond X-ray laser. Nature 523:561–567

Kim YM, Barak LS, Caron MG, Benovic JL (2002) Regulation of arrestin-3 phosphorylation by casein kinase II. J Biol Chem 277:16837–16846

Kim M, Vishnivetskiy SA, Van Eps N, Alexander NS, Cleghorn WM, Zhan X, Hanson SM, Morizumi T, Ernst OP, Meiler J et al (2012) Conformation of receptor-bound visual arrestin. Proc Natl Acad Sci USA 109:18407–18412

Kim YJ, Hofmann KP, Ernst OP, Scheerer P, Choe HW, Sommer ME (2013) Crystal structure of pre-activated arrestin p44. Nature 497:142–146

Kook S, Zhan X, Kaoud TS, Dalby KN, Gurevich VV, Gurevich EV (2013) Arrestin-3 binds JNK1α1 and JNK2α2 and facilitates the activation of these ubiquitous JNK isoforms in cells via scaffolding. J Biol Chem 288:37332–37342

Kook S, Zhan X, Cleghorn WM, Benovic JL, Gurevich VV, Gurevich EV (2014) Caspase-cleaved arrestin-2 and BID cooperatively facilitate cytochrome C release and cell death. Cell Death Differ 21:172–184

Kovoor A, Celver J, Abdryashitov RI, Chavkin C, Gurevich VV (1999) Targeted construction of phosphorylation-independent b-arrestin mutants with constitutive activity in cells. J Biol Chem 274:6831–6834

Krupnick JG, Gurevich VV, Benovic JL (1997) Mechanism of quenching of phototransduction. Binding competition between arrestin and transducin for phosphorhodopsin. J Biol Chem 272:18125–18131

Kuhn H, Hall SW, Wilden U (1984) Light-induced binding of 48-kDa protein to photoreceptor membranes is highly enhanced by phosphorylation of rhodopsin. FEBS Lett 176:473–478

Lally CC, Bauer B, Selent J, Sommer ME (2017) C-edge loops of arrestin function as a membrane anchor. Nat Commun 8:14258

Laporte SA, Oakley RH, Zhang J, Holt JA, Ferguson SSG, Caron MG, Barak LS (1999) The 2-adrenergic receptor/arrestin complex recruits the clathrin adaptor AP-2 during endocytosis. Proc Natl Acad Sci USA 96:3712–3717

Lawler S, Fleming Y, Goedert M, Cohen P (1998) Synergistic activation of SAPK1/JNK1 by two MAP kinase kinases in vitro. Curr Biol 8:1387–1390

Lee MH, Appleton KM, Strungs EG, Kwon JY, Morinelli TA, Peterson YK, Laporte SA, Luttrell LM (2016) The conformational signature of beta-arrestin2 predicts its trafficking and signalling functions. Nature 531:665–668

Levchenko A, Bruck J, Sternberg PW (2004) Regulatory modules that generate biphasic signal response in biological systems. Syst Biol (Stevenage) 1:139–148

Lin FT, Krueger KM, Kendall HE, Daake Y, Fredericks ZL, Pitcher JA, Lefkowitz RJ (1997) Clathrin-mediated endocytosis of the beta-adrenergic receptor is regulated by phosphorylation/dephosphorylation of beta-arrestin1. J Biol Chem 272:31051–31057

Lin FT, Chen W, Shenoy S, Cong M, Exum ST, Lefkowitz RJ (2002) Phosphorylation of beta-arrestin2 regulates its function in internalization of beta(2)-adrenergic receptors. Biochemistry 41:10692–10699

Lüthi AU, Martin SJ (2007) The CASBAH: a searchable database of caspase substrates. Cell Death Differ 14:641–650

Luttrell LM, Roudabush FL, Choy EW, Miller WE, Field ME, Pierce KL, Lefkowitz RJ (2001) Activation and targeting of extracellular signal-regulated kinases by beta-arrestin scaffolds. Proc Natl Acad Sci USA 98:2449–2454

Ma B, Kumar S, Tsai CJ, Nussinov R (1999) Folding funnels and binding mechanisms. Protein Eng 12:713–720

Manglik A, Kobilka B (2014) The role of protein dynamics in GPCR function: insights from the β2AR and rhodopsin. Curr Opin Cell Biol 27:136–143

Manglik A, Kim TH, Masureel M, Altenbach C, Yang Z, Hilger D, Lerch MT, Kobilka TS, Thian FS, Hubbell WL et al (2015) Structural insights into the dynamic process of beta2-adrenergic receptor signaling. Cell 161:1101–1111

McDonald PH, Chow CW, Miller WE, Laporte SA, Field ME, Lin FT, Davis RJ, Lefkowitz RJ (2000) Beta-arrestin 2: a receptor-regulated MAPK scaffold for the activation of JNK3. Science 290:1574–1577

Mendez A, Burns ME, Roca A, Lem J, Wu LW, Simon MI, Baylor DA, Chen J (2000) Rapid and reproducible deactivation of rhodopsin requires multiple phosphorylation sites. Neuron 28:153–164

Milano SK, Pace HC, Kim YM, Brenner C, Benovic JL (2002) Scaffolding functions of arrestin-2 revealed by crystal structure and mutagenesis. Biochemistry 41:3321–3328

Milano SK, Kim YM, Stefano FP, Benovic JL, Brenner C (2006) Nonvisual arrestin oligomerization and cellular localization are regulated by inositol hexakisphosphate binding. J Biol Chem 281:9812–9823

Moaven H, Koike Y, Jao CC, Gurevich VV, Langen R, Chen J (2013) Visual arrestin interaction with clathrin adaptor AP-2 regulates photoreceptor survival in the vertebrate retina. Proc Natl Acad Sci USA 110:9463–9468

Modzelewska A, Filipek S, Palczewski K, Park PS (2006) Arrestin interaction with rhodopsin: conceptual models. Cell Biochem Biophys 46:1–15

Nobles KN, Guan Z, Xiao K, Oas TG, Lefkowitz RJ (2007) The active conformation of beta-arrestin1: direct evidence for the phosphate sensor in the N-domain and conformational differences in the active states of beta-arrestins1 and -2. J Biol Chem 282:21370–21381

Nuber S, Zabel U, Lorenz K, Nuber A, Milligan G, Tobin AB, Lohse MJ, Hoffmann C (2016) beta-Arrestin biosensors reveal a rapid, receptor-dependent activation/deactivation cycle. Nature 531:661–664

Oldfield CJ, Cheng Y, Cortese MS, Romero P, Uversky VN, Dunker AK (2005) Coupled folding and binding with alpha-helix-forming molecular recognition elements. Biochemistry 44:12454–12470

Ostermaier MK, Peterhans C, Jaussi R, Deupi X, Standfuss J (2014) Functional map of arrestin-1 at single amino acid resolution. Proc Natl Acad Sci USA 111:1825–1830

Palczewski K, Pulvermuller A, Buczylko J, Hofmann KP (1991) Phosphorylated rhodopsin and heparin induce similar conformational changes in arrestin. J Biol Chem 266:18649–18654

Pan L, Gurevich EV, Gurevich VV (2003) The nature of the arrestin x receptor complex determines the ultimate fate of the internalized receptor. J Biol Chem 278:11623–11632

Pejaver V, Hsu WL, Xin F, Dunker AK, Uversky VN, Radivojac P (2014) The structural and functional signatures of proteins that undergo multiple events of post-translational modification. Protein Sci 23:1077–1093

Peng K, Vucetic S, Radivojac P, Brown CJ, Dunker AK, Obradovic Z (2005) Optimizing long intrinsic disorder predictors with protein evolutionary information. J Bioinform Comput Biol 3:35–60

Peng K, Radivojac P, Vucetic S, Dunker AK, Obradovic Z (2006) Length-dependent prediction of protein intrinsic disorder. BMC Bioinform 7:208

Peterson YK, Luttrell LM (2017) The diverse roles of arrestin scaffolds in G protein-coupled receptor signaling. Pharmacol Rev 69:256–297

Privalov PL, Khechinashvili NN (1974) A thermodynamic approach to the problem of stabilization of globular protein structure: a calorimetric study. J Mol Biol 86:665–684

Puig J, Arendt A, Tomson FL, Abdulaeva G, Miller R, Hargrave PA, McDowell JH (1995) Synthetic phosphopeptide from rhodopsin sequence induces retinal arrestin binding to photoactivated unphosphorylated rhodopsin. FEBS Lett 362:185–188

Pulvermuller A, Maretzki D, Rudnicka-Nawrot M, Smith WC, Palczewski K, Hofmann KP (1997) Functional differences in the interaction of arrestin and its splice variant, p44, with rhodopsin. Biochemistry 36:9253–9260

Radivojac P, Iakoucheva LM, Oldfield CJ, Obradovic Z, Uversky VN, Dunker AK (2007) Intrinsic disorder and functional proteomics. Biophys J 92:1439–1456

Rasmussen SG, DeVree BT, Zou Y, Kruse AC, Chung KY, Kobilka TS, Thian FS, Chae PS, Pardon E, Calinski D et al (2011) Crystal structure of the β2 adrenergic receptor-Gs protein complex. Nature 477:549–555

Romero PR, Zaidi S, Fang YY, Uversky VN, Radivojac P, Oldfield CJ, Cortese MS, Sickmeier M, LeGall T, Obradovic Z et al (2006) Alternative splicing in concert with protein intrinsic disorder enables increased functional diversity in multicellular organisms. Proc Natl Acad Sci USA 103:8390–8395

Ruvinov S, Wang L, Ruan B, Almog O, Gilliland GL, Eisenstein E, Bryan PN (1997) Engineering the independent folding of the subtilisin BPN' prodomain: analysis of two-state folding versus protein stability. Biochemistry 36:10414–10421

Savage HJ, Elliott CJ, Freeman CM, Finney JL (1993) Lost hydrogen-bonds and buried surface-area—rationalizing stability in globular-proteins. J Chem Soc Faraday Trans 89:2609–2617

Scheerer P, Park JH, Hildebrand PW, Kim YJ, Krauss N, Choe HW, Hofmann KP, Ernst OP (2008) Crystal structure of opsin in its G-protein-interacting conformation. Nature 455:497–502

Schleicher A, Kuhn H, Hofmann KP (1989) Kinetics, binding constant, and activation energy of the 48-kDa protein-rhodopsin complex by extra-metarhodopsin II. Biochemistry 28:1770–1775

Sensoy O, Moreira IS, Morra G (2016) Understanding the differential selectivity of arrestins toward the phosphorylation state of the receptor. ACS Chem Neurosci 7:1212–1224

Shatsky M, Nussinov R, Wolfson HJ (2004) A method for simultaneous alignment of multiple protein structures. Proteins 56:143–156

Shoemaker BA, Portman JJ, Wolynes PG (2000) Speeding molecular recognition by using the folding funnel: the fly-casting mechanism. Proc Natl Acad Sci USA 97:8868–8873

Shukla AK, Manglik A, Kruse AC, Xiao K, Reis RI, Tseng WC, Staus DP, Hilger D, Uysal S, Huang LY et al (2013) Structure of active β-arrestin-1 bound to a G-protein-coupled receptor phosphopeptide. Nature 497:137–141

Smith WC, Milam AH, Dugger D, Arendt A, Hargrave PA, Palczewski K (1994) A splice variant of arrestin. Molecular cloning and localization in bovine retina. J Biol Chem 269:15407–15410

Sommer ME, Farrens DL, McDowell JH, Weber LA, Smith WC (2007) Dynamics of arrestin-rhodopsin interactions: loop movement is involved in arrestin activation and receptor binding. J Biol Chem 282:25560–25568

Song X, Raman D, Gurevich EV, Vishnivetskiy SA, Gurevich VV (2006) Visual and both non-visual arrestins in their "inactive" conformation bind JNK3 and Mdm2 and relocalize them from the nucleus to the cytoplasm. J Biol Chem 281:21491–21499

Song X, Coffa S, Fu H, Gurevich VV (2009a) How does arrestin assemble MAPKs into a signaling complex? J Biol Chem 284:685–695

Song X, Vishnivetskiy SA, Gross OP, Emelianoff K, Mendez A, Chen J, Gurevich EV, Burns ME, Gurevich VV (2009b) Enhanced arrestin facilitates recovery and protects rod photoreceptors deficient in rhodopsin phosphorylation. Curr Biol 19:700–705

Song X, Vishnivetskiy SA, Seo J, Chen J, Gurevich EV, Gurevich VV (2011) Arrestin-1 expression in rods: balancing functional performance and photoreceptor health. Neuroscience 174:37–49

Standfuss J, Xie G, Edwards PC, Burghammer M, Oprian DD, Schertler GF (2007) Crystal structure of a thermally stable rhodopsin mutant. J Mol Biol 372:1179–1188

Sterne-Marr R, Gurevich VV, Goldsmith P, Bodine RC, Sanders C, Donoso LA, Benovic JL (1993) Polypeptide variants of beta-arrestin and arrestin3. J Biol Chem 268:15640–15648

Sugase K, Dyson HJ, Wright PE (2007) Mechanism of coupled folding and binding of an intrinsically disordered protein. Nature 447:1021–1025

Sutton RB, Vishnivetskiy SA, Robert J, Hanson SM, Raman D, Knox BE, Kono M, Navarro J, Gurevich VV (2005) Crystal structure of cone arrestin at 2.3Å: evolution of receptor specificity. J Mol Biol 354:1069–1080

Tang KES, Dill KA (1998) Native protein fluctuations: The conformational-motion temperature and the inverse correlation of protein flexibility with protein stability. J Biomol Struct Dyn 16:397–411

Taverna DM, Goldstein RA (2002) Why are proteins marginally stable? Proteins 46:105–109

ter Haar E, Harrison SC, Kirchhausen T (2000) Peptide-in-groove interactions link target proteins to the beta-propeller of clathrin. Proc Natl Acad Sci USA 97:1096–1100

Tompa P (2012) Intrinsically disordered proteins: a 10-year recap. Trends Biochem Sci 37:509–516

Tsou CL (1998a) Active site flexibility in enzyme catalysis. Enzyme Eng XIV 864:1–8

Tsou CL (1998b) The role of active site flexibility in enzyme catalysis. Biochemistry 63:253–258

Uversky VN (2013) Unusual biophysics of intrinsically disordered proteins. Biochim Biophys Acta 1834:932–951

Uversky VN (2015) Functional roles of transiently and intrinsically disordered regions within proteins. FEBS J 282:1182–1189

Uversky VN (2016) Dancing protein clouds: the strange biology and chaotic physics of intrinsically disordered proteins. J Biol Chem 291:6681–6688

Uversky VN, Dunker AK (2010) Understanding protein non-folding. Biochim Biophys Acta 1804:1231–1264

Uversky VN, Gillespie JR, Fink AL (2000) Why are "natively unfolded" proteins unstructured under physiologic conditions? Proteins 41:415–427

Uversky VN, Oldfield CJ, Dunker AK (2005) Showing your ID: intrinsic disorder as an ID for recognition, regulation and cell signaling. J Mol Recognit 18:343–384

Uversky VN, Oldfield CJ, Dunker AK (2008) Intrinsically disordered proteins in human diseases: introducing the D2 concept. Annu Rev Biophys 37:215–246

Uversky VN, Dave V, Iakoucheva LM, Malaney P, Metallo SJ, Pathak RR, Joerger AC (2014) Pathological unfoldomics of uncontrolled chaos: intrinsically disordered proteins and human diseases. Chem Rev 114:6844–6879

van der Lee R, Buljan M, Lang B, Weatheritt RJ, Daughdrill GW, Dunker AK, Fuxreiter M, Gough J, Gsponer J, Jones DT et al (2014) Classification of intrinsically disordered regions and proteins. Chem Rev 114:6589–6631

Varley PG, Pain RH (1991) Relation between stability, dynamics and enzyme-activity in 3-phosphoglycerate kinases from yeast and thermus-thermophilus. J Mol Biol 220:531–538

Vishnivetskiy SA, Paz CL, Schubert C, Hirsch JA, Sigler PB, Gurevich VV (1999) How does arrestin respond to the phosphorylated state of rhodopsin? J Biol Chem 274:11451–11454

Vishnivetskiy SA, Schubert C, Climaco GC, Gurevich YV, Velez M-G, Gurevich VV (2000) An additional phosphate-binding element in arrestin molecule: implications for the mechanism of arrestin activation. J Biol Chem 275:41049–41057

Vishnivetskiy SA, Hirsch JA, Velez M-G, Gurevich YV, Gurevich VV (2002) Transition of arrestin in the active receptor-binding state requires an extended interdomain hinge. J Biol Chem 277:43961–43968

Vishnivetskiy SA, Raman D, Wei J, Kennedy MJ, Hurley JB, Gurevich VV (2007) Regulation of arrestin binding by rhodopsin phosphorylation level. J Biol Chem 282:32075–32083

Vishnivetskiy SA, Francis DJ, Van Eps N, Kim M, Hanson SM, Klug CS, Hubbell WL, Gurevich VV (2010) The role of arrestin alpha-helix I in receptor binding. J Mol Biol 395:42–54

Vishnivetskiy SA, Gimenez LE, Francis DJ, Hanson SM, Hubbell WL, Klug CS, Gurevich VV (2011) Few residues within an extensive binding interface drive receptor interaction and determine the specificity of arrestin proteins. J Biol Chem 286:24288–24299

Vishnivetskiy SA, Chen Q, Palazzo MC, Brooks EK, Altenbach C, Iverson TM, Hubbell WL, Gurevich VV (2013) Engineering visual arrestin-1 with special functional characteristics. J Biol Chem 288:11741–11750

Vogl T, Jatzke C, Hinz HJ, Benz J, Huber R (1997) Thermodynamic stability of annexin V E17G: equilibrium parameters from an irreversible unfolding reaction. Biochemistry 36:1657–1668

Wagner G, Wuthrich K (1979) Correlation between the amide proton-exchange rates and the denaturation temperatures in globular-proteins related to the basic pancreatic trypsin-inhibitor. J Mol Biol 130:31–37

Wilden U (1995) Duration and amplitude of the light-induced cGMP hydrolysis in vertebrate photoreceptors are regulated by multiple phosphorylation of rhodopsin and by arrestin binding. Biochemistry 34:1446–1454

Wilden U, Hall SW, Kühn H (1986) Phosphodiesterase activation by photoexcited rhodopsin is quenched when rhodopsin is phosphorylated and binds the intrinsic 48-kDa protein of rod outer segments. Proc Natl Acad Sci USA 83:1174–1178

Williams PD, Pollock DD, Goldstein RA (2006) Functionality and the evolution of marginal stability in proteins: inferences from lattice simulations. Evol Bioinform 2:91–101

Wrba A, Schweiger A, Schultes V, Jaenicke R, Zavodszky P (1990) Extremely thermostable D-glyceraldehyde-3-phosphate dehydrogenase from the eubacterium thermotoga-maritima. Biochemistry 29:7584–7592

Wright PE, Dyson HJ (1999) Intrinsically unstructured proteins: re-assessing the protein structure-function paradigm. J Mol Biol 293:321–331

Xiao K, Shenoy SK, Nobles K, Lefkowitz RJ (2004) Activation-dependent conformational changes in {beta}-arrestin 2. J Biol Chem 279:55744–55753

Xue B, Dunbrack RL, Williams RW, Dunker AK, Uversky VN (2010) PONDR-FIT: a meta-predictor of intrinsically disordered amino acids. Biochim Biophys Acta 1804:996–1010

Yang F, Yu X, Liu C, Qu CX, Gong Z, Liu HD, Li FH, Wang HM, He DF, Yi F et al (2015) Phospho-selective mechanisms of arrestin conformations and functions revealed by unnatural amino acid incorporation and (19)F-NMR. Nat Commun 6:8202

Yun Y, Kim DK, Seo MD, Kim KM, Chung KY (2015) Different conformational dynamics of beta-arrestin1 and beta-arrestin2 analyzed by hydrogen/deuterium exchange mass spectrometry. Biochem Biophys Res Commun 457:50–57

Zavodszky P, Kardos J, Svingor A, Petsko GA (1998) Adjustment of conformational flexibility is a key event in the thermal adaptation of proteins. Proc Natl Acad Sci USA 95:7406–7411

Zhan X, Gimenez LE, Gurevich VV, Spiller BW (2011) Crystal structure of arrestin-3 reveals the basis of the difference in receptor binding between two non-visual arrestins. J Mol Biol 406:467–478

Zhan X, Kaoud TS, Kook S, Dalby KN, Gurevich VV (2013) JNK3 binding to arrestin-3 differentially affects the recruitment of upstream MAP kinase kinases. J Biol Chem 288:28535–28547

Zhan X, Perez A, Gimenez LE, Vishnivetskiy SA, Gurevich VV (2014) Arrestin-3 binds the MAP kinase JNK3α2 via multiple sites on both domains. Cell Signal 26:766–776

Zhan X, Stoy H, Kaoud TS, Perry NA, Chen Q, Vucak G, Perez A, Els-Heindl S, Slagis JV, Iverson TM et al (2016) Peptide mini-scaffold facilitates JNK3 activation in cells. Sci Rep 6:21025

Zhou XE, He Y, de Waal PW, Gao X, Kang Y, Van Eps N, Yin Y, Pal K, Goswami D, White TA et al (2017) Structural identification of phosphorylation codes for arrestin recruitment by G protein-coupled receptors. Cell 170:457–469

Zhuang T, Vishnivetskiy SA, Gurevich VV, Sanders CR (2010) Elucidation of IP6 and heparin interaction sites and conformational changes in arrestin-1 by solution NMR. Biochemistry 49:10473–10485

Zhuang T, Chen Q, Cho M-K, Vishnivetskiy SA, Iverson TM, Gurevich VV, Sanders CR (2013a) Involvement of distinct arrestin-1 elements in binding to different functional forms of rhodopsin. Proc Natl Acad Sci USA 110:942–947

Zhuo Y, Vishnivetskiy SA, Zhan X, Gurevich VV, Klug CS (2014) Identification of receptor binding-induced conformational changes in non-visual arrestins. J Biol Chem 289:20991–21002

15

SIRTain regulators of premature senescence and accelerated aging

Shrestha Ghosh[1], Zhongjun Zhou[1,2]✉

[1] Department of Biochemistry, The University of Hong Kong, Hong Kong, China
[2] Shenzhen Institute of Innovation and Technology, The University of Hong Kong, Hong Kong, China
✉ Correspondence: Zhongjun@hku.hk (Z. Zhou)

ABSTRACT

The sirtuin proteins constitute class III histone deacetylases (HDACs). These evolutionarily conserved NAD$^+$-dependent enzymes form an important component in a variety of cellular and biological processes with highly divergent as well as convergent roles in maintaining metabolic homeostasis, safeguarding genomic integrity, regulating cancer metabolism and also inflammatory responses. Amongst the seven known mammalian sirtuin proteins, SIRT1 has gained much attention due to its widely acknowledged roles in promoting longevity and ameliorating age-associated pathologies. The contributions of other sirtuins in the field of aging are also gradually emerging. Here, we summarize some of the recent discoveries in sirtuins biology which clearly implicate the functions of sirtuin proteins in the regulation of premature cellular senescence and accelerated aging. The roles of sirtuins in various cellular processes have been extrapolated to draw inter-linkage with anti-aging mechanisms. Also, the latest findings on sirtuins which might have potential effects in the process of aging have been reviewed.

KEYWORDS sirtuins, senescence, premature aging, longevity

INTRODUCTION

The phenomenon of premature aging has intrigued researchers all over the globe. Given that the primary motive of nearly all biomedical research is to promote healthspan of individuals or target diseases to improve health of patients, study of the aging process and other pathologies associated with it has ignited huge interest in dissecting the molecular pathways contributing to this process. In this scenario, premature aging disorders in humans and the various animal model systems recapitulating the phenotypes of accelerated aging and cellular senescence constitute a major area of interest in understanding the intricacies of the process of aging. The field of premature aging has garnered immense interest also because of the emerging number of premature aging disorders observed in humans, with a significant proportion of the syndromes originating from mutations in particular genes. For example, laminopathy-based premature aging syndromes (collectively known as laminopathies) originate from mutations in the *LMNA* gene or *ZMPSTE24* gene (Schreiber and Kennedy, 2013). Also, mutation or deletion of genes involved in DNA damage repair or chromatin remodeling, result in premature aging phenotypes (Ghosh and Zhou, 2014). In this regard, the silent information regulator proteins (SIRT) or the sirtuins have displayed intricate roles in promoting anti-aging effects via modulation of a spectrum of biological processes ranging from genomic maintenance, metabolic regulation, tumor suppression, inflammation and others (Fig. 1) (Choi and Mostoslavsky 2014). The sirtuins are evolutionarily conserved NAD$^+$-dependent deacylases and ADP-ribosyltransferases (Saunders and Verdin 2007). In mammals, seven sirtuin proteins (SIRT1–7) have been identified as of now with conserved central catalytic core domains flanked by differing amino and carboxyl termini, which confer individuality to the seven sirtuins in terms of structure, cellular localization and functioning (Saunders and Verdin 2007). In this review, we have analyzed the varying roles of sirtuins with implications in the process of premature cellular senescence and accelerated aging.

SIRT1

SIRT1 is by far the most extensively studied mammalian sirtuin. Its multifaceted roles in a spectrum of pivotal

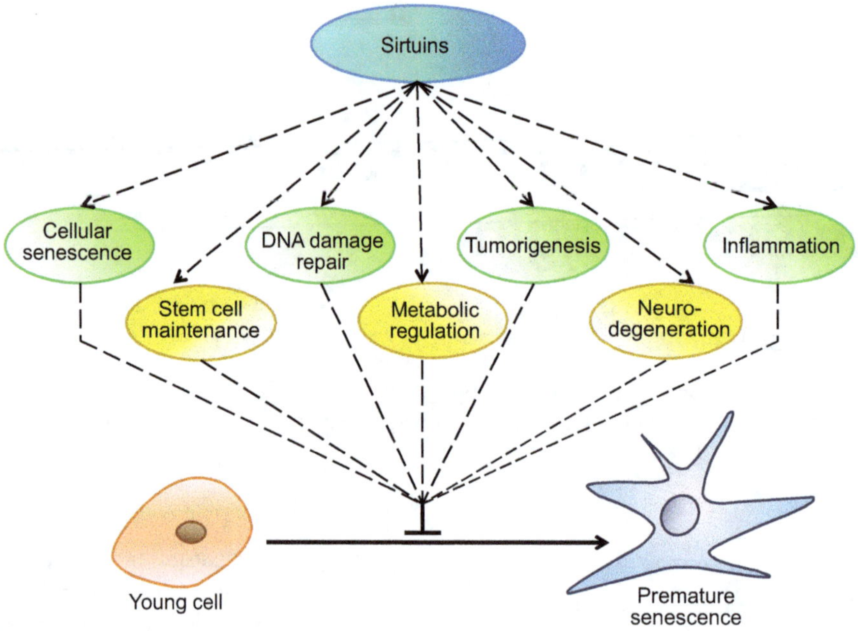

Figure 1. Schematic representation showing the regulation of premature aging by sirtuins via multiple biological processes.

biological functions, such as metabolic regulation, tumor suppression, apoptosis and other stress response signaling pathways, and also aging processes, have brought this sirtuin to limelight (Giblin et al., 2014). SIRT1 has garnered further attention by being involved in longevity expansion and also premature aging via regulation of multiple signaling pathways and cellular processes (Rehan et al., 2014). Some of the recent advancements which solidify the importance of SIRT1 in premature senescence and accelerated aging are discussed below:

CELLULAR SENESCENCE

Mammalian SIRT1, the closest homologue of yeast sirt2 protein, has time and again been linked to premature cellular senescence, one of the critical processes contributing to accelerated aging. The involvement of SIRT1 in modulating cellular senescence was identified with its deacetylation of p53 and eventual attenuation of promyelocytic leukemia protein (PML)-mediated premature cellular senescence (Langley et al., 2002). Loss of SIRT1 level or activity in human umbilical vein endothelial cells (HUVECs) results in premature senescence-associated phenotypes, such as enhanced β-galactosidase (SA-β-gal) activity, persistent growth arrest and also flattened cellular morphology (Ota et al., 2007). This suggests a protective role of SIRT1 in preventing endothelial dysfunction, which is one of the resulting features of cellular senescence. The role of SIRT1 as a guardian against endothelial cellular senescence is further bolstered by the finding that inhibition of SIRT1 expression by miR-217 promotes cellular senescence in the endothelial cells (Menghini et al., 2009). Upregulation of SIRT1 expression by peroxisome proliferator-activated receptor (PPAR)δ also downregulates early senescence in angiotensin (Ang) II-treated human coronary artery endothelial cells (HCAECs) (Kim et al., 2012). It is reported that hyperglycemia promotes premature senescence in vascular endothelial cells by downregulating SIRT1 expression levels (Mortuza et al., 2013). Further, the protective functions of cilostazol (a selective PDE3 inhibitor) in oxidative stress-induced premature senescence in endothelial cells, has been traced back to the upregulation of SIRT1 expression (Ota et al., 2008). On the other hand, the role of persistent insulin-like growth factor (IGF) treatment in triggering premature cellular senescence in a p53-dependent manner is well known (Salvioli et al., 2009). A recent study unraveled that this phenomenon results in attenuation of SIRT1 deacetylase functioning, causing enhanced p53 acetylation and stabilization, thus bringing about premature cellular senescence (Tran et al., 2014). Taken together, SIRT1 stands as a major protector against the progression of cardiovascular diseases which are triggered by endothelial dysfunctioning.

Apart from protecting against endothelial senescence, SIRT1 also plays a key role in preventing stress-induced premature senescence (SIPS) and lung inflammaging, one of the hallmarks of chronic obstructive pulmonary disease (COPD) or emphysema (Yao et al., 2012). The levels of SIRT1 expression and activity are significantly reduced in the lungs of COPD patients. In addition, enhancement of SIRT1 expression or activity inhibits stress-induced premature cellular senescence and also provides protection against cigarette smoke-induced emphysema (Yao et al.,

2012). This SIRT1-mediated protection against emphysema is reported to be catalyzed by FOXO3 transcription factor (Yao et al., 2012).

REPLICATIVE SENESCENCE

Replicative senescence is brought about in primary cells because of telomere shortening after every round of cell cycle, which signals the cells of stress and limits the number of cell divisions. Expression level of SIRT1 is reported to decrease as the primary human fibroblasts enter replicative senescence phase (Michishita et al., 2005). Additionally, SIRT1 overexpression is reported to attenuate oncogene-induced replicative senescence via p53 deacetylation (Langley et al., 2002). However, upon prolonged cell replication in culture, primary *Sirt1*$^{-/-}$ mouse embryonic fibroblasts (MEFs) display resistance towards replicative senescence (Chua et al., 2005). Nevertheless, SIRT1 knockdown in human diploid fibroblasts has been reported to induce cell proliferation (Abdelmohsen et al., 2007). The contribution of SIRT1 in promoting replicative senescence can be further gauged from the finding that it deacetylates histone H1 on lysine 26 (K26), and also histone H4 at K16 and H3 on K9, to endorse facultative heterochromatin formation (Vaquero et al., 2004). It is widely known that loss of heterochromatin is one of the hallmarks for premature cellular senescence and accelerated aging (Shumaker et al., 2006). Hence, given the progressive decline in SIRT1 level with passage, it is plausible that these histone marks get hyperacetylated to further lead to heterochromatin loss, thus resulting in eventual replicative senescence. Moreover, SIRT1 is reported to deacetylate WRN, the protein mutated in Werner syndrome causing telomere attrition (Li et al., 2008). Taken together, SIRT1 plays a key role in blocking replicative senescence via regulation of multiple proteins.

MAINTENANCE OF 'STEMNESS'

SIRT1 expression levels have been reported to be higher in mouse embryonic stem cells (mESCs) as compared to the differentiated cells (Saunders et al., 2010). This phenomenon has been attributed to post-transcriptional down-regulation of SIRT1 by a range of miRNAs, such as miR-9, miR-181a and b, and others (Saunders et al., 2010). Interestingly, the authors also reported an induction in SIRT1 protein levels when the MEFs were reprogrammed into induced pluripotent stem (iPS) cells. Recently, SIRT1 has been implicated in the maintenance of mESCs self-renewal by being directly regulated viamiR-29b in response to reactive oxygen species (ROS) (Xu et al., 2014). In addition, SIRT1 has been reported to be involved in the maintenance of haematopoietic stem cells (HSCs) homeostasis and lineage specification, where loss of SIRT1 in HSCs garnered DNA damage, accumulation of age-linked molecules and other phenotypes of aging HSCs (Rimmelé et al. 2014). Intriguingly, overexpression of SIRT1 or its activation by resveratrol is reported to rescue senescence-associated phenotypes and also angiogenic defects in preterm-endothelial colony forming progenitor cells (PT-ECFCs) (Vassallo et al., 2014). To add to this, overexpression of SIRT1 also delays senescence in the bone marrow-derived mesenchymal stem cells (B-MSCs) (Yuan et al., 2012). Further, cleavage of SIRT1 by cathepsin triggers stress-induced premature senescence (SIPS) in endothelial progenitor cells (Chen et al., 2012). On the whole, SIRT1 stands out as a prominent housekeeper of maintaining stemness by resisting senescence in the cells.

DNA DAMAGE REPAIR AND AGING

Impaired DNA damage repair and accrued genomic instability are some of the chief factors contributing to premature cellular senescence and accelerated aging (Liu et al., 2005). In this arena, SIRT1 is reported to induce DNA damage repair by deacetylating the repair protein Ku70 upon irradiation-induced DNA damage (Jeong et al., 2007). The association of RNA-binding protein HuR, SIRT1 and NBS1 in the maintenance of genomic integrity upon genotoxic stress is also extensively studied (Gorospe and de Cabo, 2008). SIRT1 has been linked in safeguarding against DNA damage and atherosclerosis in human vascular smooth muscle cells (VSMCs) by activating the repair protein Nijmegen breakage syndrome-1 (NBS-1), but not p53 (Gorenne et al., 2013). SIRT1 is also widely acknowledged to deacetylate PARP1, a key player in DNA damage repair and aging, and thus mediate response signaling upon genotoxic stress (Rajamohan et al., 2009). The intricate inter-linkage between SIRT1 and PARP1 has been time and again established to play crucial roles in genotoxic stress and DNA damage responses (Luna et al., 2013) The involvement of SIRT1 in DNA damage repair has been further solidified by the finding that it activates homologous recombination (HR) repair in human cells (Uhl et al., 2010). However, the authors refuted the involvement of Ku70, PARP1 or NBS1 in SIRT1-mediated repair, and advocated the role of WRN protein in this process. This suggests that SIRT1 possibly mediates different repair pathways via modulation of repair proteins in a mutually exclusive way. On the other hand, overexpression of SIRT1 has been reported to negatively regulate Tip60-mediated acetylation of H2AX, an event finely regulated to elicit proper DNA damage response (Yamagata and Kitabayashi, 2009). Thus, SIRT1 possibly represses extensive DNA damage response to fine-tune HR repair process by repressing H2AX over-activation. Recently, SIRT1 has been reported to display prompt recruitment to double strand breaks (DSBs) in post-mitotic neurons, where it deacetylates HDAC1 and stimulates ATM autophosphorylation to mediate non-homologous end joining (NHEJ) repair (Dobbin et al., 2013). Thus, SIRT1 plays a protective role in maintaining genomic integrity not only against genotoxic stress but also to attenuate neurodegeneration.

LIFESPAN EXTENSION

SIR2 was the first identified evolutionarily conserved gene which had been implicated in the expansion of lifespan. Attention was drawn towards the sirtuins when sir2 inactivation in yeasts reduced lifespan by 50%, while introduction of an extra copy of the gene *SIR2* increased yeast lifespan by 30% (Kaeberlein et al., 1999). This finding has been contradicted by an independent research group which failed to replicate longevity expansion in yeast or drosophila by sir2 overexpression (Burnett et al., 2011). However, another group could successfully extend longevity in *C. elegans* by *sir-2.1* overexpression, attributing this occurrence to nicotinamide (NAM) methylation (Schmeisser et al., 2013). Apart from these reports, *Sirt1* transgenic mice did not exhibit any overall lifespan extension (Herranz et al., 2010). However, SIRT1 overexpression in those mice displayed improvement of some age-associated phenotypes, such as wound healing, decreased tumorigenicity, better maintenance of glucose homeostasis and others. Intriguingly, SIRT1 overexpression specifically in brain increases longevity by approximately 11% in both male and female mice, and also reduces incidence of cancer (Satoh et al., 2013). This phenomenon has been largely attributed to SIRT1-mediated upregulation of Orexin type-2 receptor (*Ox2r*) expression specifically in the nuclei of dorsomedial and lateral hypothalamic (DMH and LH) regions of brain. In addition, this organ-specific SIRT1 activation can be explained by the observation that NK2 homeobox 1 (Nkx2-1) specifically colocalizes with SIRT1 in the DMH and LH regions of brain to upregulate Ox2r expression and thus increase longevity. Hence, this study further reinstated the importance of SIRT1 in delaying senescence and premature aging. Nevertheless, loss of sirt1 in mice resulted in severe growth defects causing death of litters in the late prenatal or early postnatal stages, thus implying SIRT1's major contribution in development (Cheng et al., 2003). Given the involvement of SIRT1 in deacetylating and deactivating p53, the authors also observed that p53 was hyperacetylated upon DNA damage in *Sirt1*$^{-/-}$ mice tissues (Cheng et al., 2003). However, mice deficient for both sirt1 and p53 neither displayed any significant differences from *Sirt1*$^{-/-}$ mice, nor any amelioration in the growth and developmental defects (Kamel et al., 2006), thus suggesting that the impairments observed in *Sirt1*$^{-/-}$ mice are not majorly attributable to p53 activation. Recently, SIRT1 activation has been linked to amelioration of skeletal muscle performance in sarcopenia or atrophy during aging, mainly via inhibition of PARP1 (Mohamed et al., 2014). Taken together, SIRT1 can be regarded as a prominent, although much debated, player in promoting longevity and attenuating premature aging.

INVOLVEMENT IN LAMINOPATHY-BASED PREMATURE AGING

Animal models of laminopathy-based premature aging provide excellent model systems for studying the process of accelerated aging and cellular senescence. Although rare, some of the disease phenotypes are highly severe with irreparable mortality in patients in their early teens, such as in Hutchinson-Gilford Progeria Syndrome (HGPS) (Schreiber and Kennedy, 2013). Loss of ZMPSTE24, the metalloproteinase responsible for cleaving the precursor prelamin A into mature lamin A, reiterates progeroid phenotypes in mice (Navarro et al., 2004). Our recent work has further established the role of SIRT1 in delaying premature cellular senescence and aging by being involved in laminopathy-based premature aging (Liu et al., 2012). We reported that lamin A, a nuclear lamina protein, interacts with SIRT1 and also activates it (Liu et al., 2012). This interaction is further induced by resveratrol, thus resulting in ultimate activation of the sirtuin protein. Our study provided a mechanistic explanation to the long debated question of whether or not resveratrol aids in SIRT1 activation (Kulkarni and Cantó 2014). We also observed a prominent decline of adult stem cells (ASCs) in the tissues derived from *Zmpste24*$^{-/-}$ mice, which was primarily attributable to the dissociation of SIRT1 from nuclear matrix in the presence of prelamin A or progerin (the precursor and mutant forms of lamin A respectively). Interestingly, treatment with resveratrol not only rescued ASC attrition, but also extended lifespan in the mutant mice along with reduction of body weight loss and improvement in bone mineral density. This study also reinforced the role played by SIRT1 in the maintenance of stem cells. However, it remains to be seen whether SIRT1 also plays significant roles in delaying senescence or increasing lifespan in other models of laminopathy-based premature aging.

DIVERSE ROLES IN OTHER DISEASES

There are multiple disorders which majorly contribute to the occurrence of premature aging, such as metabolic dysregulation, neurodegenerative diseases, inflammation and also tumorigenesis (Giblin et al., 2014). SIRT1 has been ascribed multifaceted roles in the regulation of proper metabolic homeostasis by maintaining lipogenesis, gluconeogenesis, fatty acid oxidation and others (Chang and Guarente, 2014). Caloric restriction-mediated lifespan extension is also largely attributed to SIRT1 (Chen and Guarente, 2007). So far neurodegenerative disorders are concerned, SIRT1 has been reported to deacetylate tau, the protein causing Alzheimer's disease upon pathogenic aggregation, and trigger its degradation via ubiquitin (Min et al., 2010). Also, overexpression of SIRT1 is reported to inhibit accumulation of α-synuclein, whose pathogenic aggregation in neurons causes Parkinson's disease, and thus leads to longevity (Donmez et al., 2012). Recently, diminished SIRT1 activity has been attributed to be one of the chief reasons for causing Cockayne syndrome, a premature aging disorder characterized by severe neurodegeneration (Scheibye-knudsen et al., 2014). SIRT1 has also been intricately linked to lung inflammaging where it is reported to play significant anti-aging roles (Yao et al., 2012). However, in regard to cancer, SIRT1 plays both tumor suppressive and oncogenic roles (Yuan et al., 2013).

The immense importance of SIRT1 in the process of anti-aging can be further ascertained from the implementation of multiple small molecule drugs which primarily act as SIRT1 activators (Hubbard and Sinclair, 2014). On the whole, SIRT1 can be regarded as a crucial anti-aging protein which mediates its widespread effects in preventing premature senescence and accelerated aging by regulating multiple molecular pathways.

SIRT2

SIRT2 is the second mammalian sirtuin protein which is emerging out as a potential target in preventing age-related disorders. Its possible involvement in the process of aging and longevity can be fathomed from the observation that expression level of SIRT2 was found upregulated in kidney and white-adipose tissues of caloric-restricted mice (Zhu et al., 2012). However, it remains to be seen whether caloric-restriction fails to extend lifespan in mice lacking Sirt2. This will provide direct evidence of the involvement of SIRT2 in this dietary intervention-mediated lifespan extension. SIRT2 was initially identified to deacetylate α-tubulin at lysine 40 and histone H4 at lysine 16 (de Oliveira et al. 2012). Apart from these, SIRT2 is also known to deacetylate forkhead transcription factors of class O, namely FOXO1 and FOXO3 (Wang and Tong, 2009, Wang et al., 2007). Given the widely acknowledged functions of FOXO transcription factors in a spectrum of pathways directly or indirectly linked to aging, such as metabolic regulation, DNA damage repair, apoptosis and others, it is tempting to speculate that SIRT2 may play an intricate role in anti-aging processes. Moreover, an increasing proportion of evidence suggests tumor-suppressive functions of SIRT2 (de Oliveira et al. 2012). Since increasing incidence of cancer is associated with progressive aging, SIRT2 can be considered as a potential anti-aging protein. However, further experimentation is required to confirm this model. Recently, SIRT2 has been implicated in the suppression of glioma formation (Li et al., 2013). In this study, the authors identified p65 (a subunit of NF-κB signaling) as a target of deacetylation by SIRT2 at lysine 310, thereby inhibiting miR-21 transcription via blockage of p65 binding with its promoter region. This report further supported the role of SIRT2 in tumor suppression and opened up new avenues for the involvement of SIRT2 in premature aging via regulation of NF-κB signaling. Additionally, SIRT2 overexpression in BubR1 mice, which harbour hypomorphism for BubR1 and display premature aging phenotypes, has been reported to extend lifespan in the mutant mice (North et al., 2014). Given the many implications of Wnt signaling in tumorigenesis and aging, SIRT2-mediated attenuation of Wnt signaling pathway upon oxidative stress (Nguyen et al., 2014), further insinuates at the involvement of this sirtuin in the regulation of aging via multiple mechanisms. Apart from these, SIRT2 is also ascribed significant roles in safeguarding against neurodegenerative disorders, such as Parkinson's and Huntington's diseases (de Oliveira et al., 2012). Thus, SIRT2 can be considered as a promising target in designing intervention against premature aging.

SIRT3

SIRT3, predominantly located in mitochondria, has been primarily linked to the regulation of a variety of mitochondrial processes, such as β-oxidation, ATP generation, management of reactive oxygen species (ROS) and others (Giblin et al., 2014). However, its roles in aging-associated disorders have begun emerging in the recent years. Although *Sirt3*-knockout mice do not exhibit any severe signs of premature aging or tumorigenesis, the mutant mice do display glucose intolerance, dysregulated ROS production, and heightened white adipose tissue (WAT) development (Jing et al., 2011). Since ROS-induced oxidative stress is one of the critical contributors of premature cellular senescence, it is possible that SIRT3 might have a hand in regulating this process. SIRT3 can be strongly associated with the maintenance of metabolic regulation based on the finding that loss of *Sirt3* in the germline of mice leads to accelerated development of obesity, hepatic steatosis and insulin resistance upon high-fat diet (HFD) feeding (Hirschey et al., 2011). This study also identified reduction in Sirt3 expression level when mice were fed with high-fat diet. Recently, SIRT3 has been implicated in the maintenance of regenerative capacity in the haematopoietic stem cells with progressive age (Brown et al., 2013). Haematopoietic stem cells display high expression levels of SIRT3, where this sirtuin regulates mitochondrial protein acetylation profiles and blocks oxidative stress generation. This groundbreaking discovery not only recognized the significant roles of SIRT3 in maintaining stemness, but also laid out a path for further dissecting its roles in stem cell-based intervention for metabolic disorders resulting in premature aging. In addition, some single nucleotide polymorphisms (SNPs) in *Sirt3* gene have been attributed in the extension of longevity (Giblin et al., 2014). Also, several lines of evidence suggest modulating roles for SIRT3 in response to caloric-restriction (Kincaid and Bossy-Wetzel 2013). Apart from these, a recent study elucidated the involvement of SIRT3 in protection against neurodegeneration in Huntington's disease (Fu et al., 2012). The study identified reduced expression of SIRT3 because of mutant protein huntingtin (Htt), which causes Huntington's disease. Treatment with viniferin, a naturally occurring compound, ameliorated the disease phenotypes by enhancing SIRT3 expression level (Fu et al., 2012). Taken together, these studies clearly reinstate the importance of SIRT3 in the maintenance of metabolic regulation, regeneration in stem cells, and also neuroprotection, all of which contribute directly or indirectly in the acceleration of aging in individuals when dysregulated.

SIRT4

SIRT4 is the second mitochondrial sirtuin protein, with no reported deacetylase activity (Saunders and Verdin 2007). However, this sirtuin possesses ADP-ribosyltransferase activity towards glutamate dehydrogenase (GDH) (Haigis et al., 2006). This study also identified reduction in SIRT4 level upon caloric-restriction, thus suggesting that SIRT4 might antagonize this dietary restriction-mediated effect, unlike SIRT1 and SIRT3. Further, SIRT4 has been implicated in regulating lipid metabolism by attenuating peroxisome proliferator-activated receptor α (PPARα) and thus repress rates of fatty acid oxidation in the hepatocytes (Laurent et al., 2013). SIRT4 has been recently identified to maintain mitochondrial ATP homeostasis (Ho et al., 2013). Although, there has been no direct evidence relating SIRT4 activities with the aging process, the involvement of this sirtuin in mitochondrial biogenesis and ATP production insinuate at a faint possibility that SIRT4 might be linked to the process of aging.

SIRT5

SIRT5 is the third mitochondrial sirtuin protein with very few substrates identified till date. It was first identified to deacetylate carbamoyl phosphate synthetase 1 (CPS1) and thus regulate the urea cycle (Nakagawa and Guarente 2009). It has been assumed that SIRT5 might also play a role in mitochondrial metabolism in response to varying nutritional states (Shih and Donmez, 2013). Lately, SIRT5 has been confided significant desuccinylase activity in mitochondria (Rardin et al., 2013), implicating the role for SIRT5 in the repression of ketogenesis by desuccinylation of the rate-limiting enzyme 3-hydroxy-3-methylglutaryl-CoA synthase 2 (HMGCS2). Since mitochondrial dysregulation plays a key role in bringing about senescence phenotypes, it is not impossible that SIRT5 might play a regulatory role in the process. Furthermore, polymorphisms in the promoter region of *SIRT5* gene have been associated with the development of premature aging in amygdala region of brain (Glorioso et al., 2011), suggesting the possible involvement of SIRT5 promoter polymorphisms in the incidence of Parkinson's disease. Further studies are required in order to clearly differentiate roles (if any) of this sirtuin in the process of accelerated aging.

SIRT6

SIRT6 is predominantly a nuclear protein, whose association with chromatin has time and again been established (Kugel and Mostoslavsky 2012). SIRT6 was primarily identified as a histone deacetylase with specificity towards histone H3 at lysine 9 and 56 (K9 and K56) (Michishita et al., 2008, Michishita et al., 2009). This sirtuin gained prominence when its knockout mouse model developed severe premature aging phenotypes with mortality resulting within a month (Mostoslavsky et al., 2006). Moreover, SIRT6 is the only mammalian sirtuin which displayed clear increase in lifespan when overexpressed in the whole body of mice (Kanfi et al., 2012). Intriguingly, this longevity extension has only been observed in male mice, and the underlying mechanism of this gender-specificity is still under debate. SIRT6 can be regarded as an important anti-aging protein with multifaceted roles in DNA damage repair, metabolic regulation, inflammation and also tumor suppression, as discussed below in details.

MAINTENANCE OF GENOMIC STABILITY

The premature aging phenotypes observed in the *Sirt6*$^{-/-}$ mice were primarily attributed to the defects observed in base excision repair (BER) machinery (Mostoslavsky et al., 2006). This seminal finding ignited interest in the scientific community and multiple roles of SIRT6 in DNA damage response began to be deciphered gradually. For example, SIRT6 is identified to deacetylate CtIP, a DNA end resection protein, and mediate homologous recombination (HR) repair (Kaidi et al., 2010). SIRT6 also mono ADP-ribosylates PARP1 in response to oxidative stress and thus promotes non-homologous end-joining (NHEJ) repair process (Mao et al., 2011). In addition, SIRT6 is one of the prompt recruits to DNA double strand break (DSB) sites, and further helps in the localization of other repair proteins like DNA-PKcs, SNF2H and others (McCord et al., 2009, Toiber et al., 2013). SIRT6 was also observed to progressively decline with passage in human fibroblasts (Mao et al., 2012). In addition, overexpression of this sirtuin alleviated the repression of HR repair observed during replicative senescence. SIRT6 has been also reported to localize in telomeric chromatin where it deacetylates histone H3 at K9, thus stabilizing the interaction of WRN protein and maintaining proper telomeric metabolism (Michishita et al., 2008). Since impairment in DNA damage repair and telomere shortening are some of the hallmarks of premature cellular senescence and are critical contributing factors for accelerated aging, SIRT6 can be considered as a pivotal player in DNA damage repair-mediated anti-aging process.

METABOLIC HOMEOSTASIS

SIRT6 is a major regulator of the maintenance of glucose homeostasis. It has been reported to repress gluconeogenesis in hepatocytes by regulating GCN5-PGC1α pathway, thus preventing diabetic hyperglycemia (Dominy et al., 2012). SIRT6 acts as a co-repressor of the transcription factor Hif1α (which is involved in sensing nutritional state of cells) and also deacetylates H3K9 at the promoters of several glycolytic genes (Zhong et al., 2010). This provides a mechanistic explanation to the lethal hypoglycemia observed in *Sirt6*$^{-/-}$ mice. SIRT6 has been also related in regulation of lipogenesis by repression of lipogenic transcription factors, SREBP1 and SREBP2 via multiple mechanisms (Elhanati et al., 2013). This study identified reduced levels of low-density lipoprotein cholesterol in the *Sirt6* transgenic mice, thus providing another mechanistic explanation to the increased longevity in the mutant mice. The involvement of SIRT6 in fat metabolism is further bolstered by the finding that loss of this sirtuin in the liver of mice results in formation of fatty liver due to upregulated glycolysis and triglyceride synthesis (Kim et al., 2010). SIRT6 is also linked in attenuating insulin-like growth factor (IGF)-AKT signaling and thus preventing cardiac hypertrophy (Sundaresan et al., 2012). On the whole, SIRT6 is a critical housekeeper of metabolic homeostasis, the imbalance of which eventually results in premature aging phenotypes.

TUMOR SUPPRESSION

SIRT6 has been identified as a vital tumor suppressor which critically regulates cancer metabolism (Sebastián et al. 2012). This study also identified downregulation of SIRT6 in several forms of human cancer, such as pancreatic and colorectal cancer. Moreover, overexpression of SIRT6 in non-small cell lung cancer (NSCLC) cells radiosensitizes them which inhibits their proliferation, further supporting the idea of SIRT6 being a tumor suppressor (Cai et al., 2014). However, this concept has been contradicted by an independent research group which reported that depletion of SIRT6 in mouse skin inhibits tumorigenesis (Ming et al., 2014). This is suggestive of particular cancer cell-type specific activities of SIRT6. Nevertheless, the finding that overexpression of SIRT6 triggers heightened apoptotic response specifically in cancerous cells but not in normal cells (Van Meter et al., 2011), further reinstates the role of this sirtuin as a potent tumor suppressor. Also, degradation of SIRT6 by MDM2 has been observed in multiple lines of breast cancer cells (Thirumurthi et al., 2014). All these findings solidify the concept of SIRT6 being a key tumor suppressor, and further strengthen its role in delaying aging, since cancer incidence and progressive aging go hand in hand.

INFLAMMATION

In some recent reports, SIRT6 has been identified to possess anti-inflammatory functions. For example, loss of SIRT6 in human umbilical vein endothelial cells (HUVECs) induced proinflammatory cytokine expression, such as interleukin 6 (IL-6), IL-8 and also IL-1β (Lappas, 2012). Moreover, overexpression of SIRT6 in human rheumatoid arthritis fibroblast-like cells and also in mice with collagen-induced arthritis attenuated proinflammatory cytokine levels (Lee et al., 2013). Given the huge implications of NF-κB in inflammatory responses, the finding that SIRT6 attenuates NF-κB signaling by deacetylating histone H3 at K9 on the promoters of NF-κB target genes (Kawahara et al., 2009), further establishes the role of SIRT6 as a critical anti-inflammatory protein. Also, loss of Sirt6 in immune cells derived from mice display chronic liver inflammation and fibrosis (Xiao et al., 2012). However, another study has advocated that SIRT6 induces the secretion of cytokines and might possibly be involved in pro-inflammatory responses (Bauer et al., 2012). Since inflammatory responses have been linked to the development of premature aging phenotypes, the role of SIRT6 in repressing age-associated pathologies get strengthened more and more.

Taken together, SIRT6 stands out as a key modulator of anti-aging processes which mediates its regulatory roles via multiple pathways to delay cellular senescence and accelerated aging. Hence, identification of SIRT6 activators can have huge therapeutic impact in designing intervention for premature aging disorders and other age-associated pathologies.

SIRT7

SIRT7 is the seventh mammalian sirtuin protein predominantly localized in the nucleolus with an established role in rDNA transcription activation (Kim and Kim, 2013). Increasing evidences are surfacing regarding the functioning of this sirtuin which are gradually hinting its potential role in premature aging. Inactivation of the *Sirt7* gene in mice results in lifespan reduction and the mutant mice also develop cardiac hypertrophy and inflammatory cardiomyopathy (Vakhrusheva et al. 2008a, b). These phenotypic outcomes have been primarily attributed to lack of Sirt7-mediated p53 deacetylation in the mutant mice, thus resulting in p53 hyperactivation and apoptosis. However, no direct evidence has been provided by the authors at this end regarding *in vivo* Western blotting or exact sites of p53 deacetylation by SIRT7. Thus, further experiments need to be performed in order to draw a clear picture. Nevertheless, SIRT7 has been ascribed an anti-proliferative role since its overexpression ceases tumorigenicity in several lines of murine cells (Vakhrusheva et al. 2008a). In contrast to this observation, SIRT7 has been characterized as an oncogene which selectively deacetylates histone H3 at lysine 18 to maintain tumorigenic potential in cancer cells (Paredes et al., 2014). Also, SIRT7 expression level is found heightened in a large number of patients with human hepatocellular carcinoma (HCC), and knockdown of SIRT7 repressed growth in liver cancer cells (Kim et al., 2013). Apart from the recognition of being an oncogene, SIRT7 is also implicated in the aging process since its expression levels were observed to be reduced in senescent cells (Lee et al., 2014). Moreover, the involvement of SIRT7 in attenuation of ER stress and prevention of fatty liver disease formation (Shin et al., 2013), further extend its roles in maintenance of metabolic regulation. SIRT7 is also observed to regulate lipid metabolism in the liver (Yoshizawa et al., 2014). On the whole, it would not be inappropriate to speculate potential roles of SIRT7 in maintaining metabolic homeostasis, cancer metabolism, and thus anti-aging process.

Table 1. Involvement of sirtuins in the pathways contributing to premature aging when dysregulated

Pathways contributing to premature aging	SIRT1	SIRT2	SIRT3	SIRT4	SIRT5	SIRT6	SIRT7
Cellular/replicative senescence	√	?	**	?	?	√	?
Maintenance of stem cells	√	**	√	?	?	√	?
Maintenance of genomic integrity	√	**	?	?	?	√	**
Metabolic regulation	√	√	√	√	√	√	√
Regulation of tumorigenesis	√	√	√	?	?	√	√
Inflammation	√	√	**	**	?	√	√
Neurodegeneration	√	√	√	√	**	?	?

√: Established roles, **: possible roles, ?: No reported roles.

CONCLUSION

Till date, substantial number of evidences has laid foundation to the concept of sirtuins being potent anti-aging proteins. Their multivalent roles in delaying cellular senescence and blocking the development of premature aging phenotypes further mark them as promising targets to design intervention for several age-associated pathologies (Table 1). SIRT1, being the most extensively experimented sirtuin with regards to aging and longevity, has triggered interest in scientific communities to develop small molecule activators or drugs in amelioration of a wide range of aging disorders. Several such activators and drugs for SIRT1 activation are also commercially available for testing in labs (Mellini et al., 2014). For example, the anti-aging effects of resveratrol are primarily attributed to SIRT1 activation. With the emerging reports on potential anti-aging effects of the other sirtuins, specifically SIRT6, it would be important to screen out the activators or modulators of these sirtuins and design drugs to test their effects in the animal models of premature aging. Identification of such potent drugs to stimulate sirtuin functioning can prove to be breakthroughs in ameliorating age-associated disorders. Although development of such small molecule activators of sirtuins or drugs and their inclusion in clinical trials is a long-shot process, the increasing number of findings elaborating the various sirtuins-mediated anti-aging mechanisms, abridge the path.In conclusion, the SIRT proteins can certainly be regarded as essential factors in delaying cellular senescence, repressing premature aging and enhancing longevity and healthspan in the living systems.

ACKNOWLEDGEMENTS

The work in Z ZHOU's lab is supported by grants from Natural Science Foundation of China (Grant No. 81330009), the National Basic Research Program (973 Program) (No. 2011CB964700) and Research Grant Council of Hong Kong (HKU2/CRF/13G).

ABBREVIATIONS

ATM, ataxia telangiectasia mutated; ATP, adenosine triphosphate; CtIP, C-terminal binding protein interacting protein; ER, endoplasmic reticulum; HCC, hepatocellular carcinoma; HDACs, histone deacetylases; HSC, haematopoietic stem cell; IGF, insulin-like growth factor; MEF, mouse embryonic fibroblast; NAD, nicotinamide adenine dinucleotide; PARP1, poly [ADP ribose] polymerase 1; SIPS, stress-induced premature senescence; SIRT, silent information of regulator; ZMPSTE24, zinc metalloproteinase Ste24.

COMPLIANCE WITH ETHICS GUIDELINES

Shrestha Ghosh and Zhongjun Zhou declare that they have no conflict of interest. This article does not contain any studies with human or animal subjects performed by the any of the authors.

REFERENCES

Abdelmohsen K, Pullmann R Jr, Lal A, Kim HH, Galban S, Yang X, Blethrow JD, Walker M, Shubert J, Gillespie DA et al (2007) Phosphorylation of HuR by Chk2 regulates SIRT1 expression. Mol cell 25:543–557

Bauer I, Grozio A, Lasigliè D, Basile G, Sturla L, Magnone M, Sociali G, Soncini D, Caffa I, Poggi A et al (2012) The NAD+-dependent histone deacetylase SIRT6 promotes cytokine production and migration in pancreatic cancer cells by regulating Ca2+ responses. J BiolChem 287:40924–40937

Brown K, Xie S, Qiu X, Mohrin M, Shin J, Liu Y, Zhang D, Scadden DT, Chen D (2013) SIRT3 reverses aging-associated degeneration. Cell reports 3:319–327

Burnett C, Valentini S, Cabreiro F, Goss M, Somogyvári M, Piper MD, Hoddinot M, Sutphin GL, Leko V, Mcelwee JJ et al (2011) Absence of effects of Sir2 overexpression on lifespan in *C. elegans* and Drosophila. Nature 477:482–485

Cai Y, Sheng ZY, Liang SX (2014) Radiosensitization Effect of Overexpression of Adenovirus-mediated SIRT6 on A549 Non-small Cell Lung Cancer Cells. Asian Pacific Journal of Cancer Prevention 15:7297–7301

Chang HC, Guarente L (2014) SIRT1 and other sirtuins in metabolism. Trends in Endocrinology & Metabolism 25:138–145

Chen D, Guarente L (2007) SIR2: a potential target for calorie restriction mimetics. Trends in molecular medicine 13:64–71

Chen J, Xavier S, Moskowitz-Kassai E, Chen R, Lu CY, Sanduski K, Spes A, Turk B, Goligorsky MS (2012) Cathepsin cleavage of sirtuin 1 in endothelial progenitor cells mediates stress-induced premature senescence. The American journal of pathology 180:973–983

Cheng HL, Mostoslavsky R, Saito SI, Manis JP, Gu Y, Patel P, Bronson R, Appella E, Alt FW, Chua KF (2003) Developmental defects and p53 hyperacetylation in Sir2 homolog (SIRT1)-deficient mice. ProcNatlAcadSci 100:10794–10799

Choi JE, Mostoslavsky R (2014) Sirtuins, metabolism, and DNA repair. Current Opinion In Genetics & Development 26:24–32

Chua KF, Mostoslavsky R, Lombard DB, Pang WW, Saito SI, Franco S, Kaushal D, Cheng HL, Fischer MR, Stokes N et al (2005) Mammalian SIRT1 limits replicative life span in response to chronic genotoxic stress. Cell Metab 2:67–76

de Oliveira RM, Sarkander J, Kazantsev AG, Outeiro TF (2012) SIRT2 as a therapeutic target for age-related disorders. Frontiers in pharmacology 3:82

Dobbin MM, Madabhushi R, Pan L, Chen Y, Kim D, Gao J, Ahanonu B, Pao PH, Qiu Y, Zhao Y et al (2013) SIRT1 collaborates with ATM and HDAC1 to maintain genomic stability in neurons. Nature neuroscience 16:1008–1015

Dominy JE Jr, Lee Y, Jedrychowski MP, Chim H, Jurczak MJ, Camporez JP, Ruan HB, Feldman J, Pierce K, Mostoslavsky R et al (2012) The deacetylase Sirt6 activates the acetyltransferase GCN5 and suppresses hepatic gluconeogenesis. Mol cell 48:900–913

Donmez G, Arun A, Chung CY, McLean PJ, Lindquist S, Guarente L (2012) SIRT1 protects against α-synuclein aggregation by activating molecular chaperones. The Journal of Neuroscience 32:124–132

Elhanati S, Kanfi Y, Varvak A, Roichman A, Carmel-Gross I, Barth S, Gibor G, Cohen HY (2013) Multiple regulatory layers of SREBP1/2 by SIRT6. Cell reports 4:905–912

Fu J, Jin J, Cichewicz RH, Hageman SA, Ellis TK, Xiang L, Peng Q, Jiang M, Arbez N, Hotaling K et al (2012) trans-(−)-ϵ-Viniferin increases mitochondrial sirtuin 3 (SIRT3), activates AMP-activated protein kinase (AMPK), and protects cells in models of Huntington Disease. J BiolChem 287:24460–24472

Ghosh S, Zhou Z (2014) Genetics of aging, progeria and lamin disorders. Current opinion in genetics & development 26:41–46

Giblin W, Skinner ME, Lombard DB (2014) Sirtuins: guardians of mammalian healthspan. Trends in Genetics 30:271–286

Glorioso C, Oh S, Douillard GG, Sibille E (2011) Brain molecular aging, promotion of neurological disease and modulation by Sirtuin5 longevity gene polymorphism. Neurobiology of disease 41:279–290

Gorenne I, Kumar S, Gray K, Figg N, Yu H, Mercer J, Bennett M (2013) Vascular smooth muscle cell sirtuin 1 protects against DNA damage and inhibits atherosclerosis. Circulation 127:386–396

Gorospe M, de Cabo R (2008) AsSIRTing the DNA damage response. Trends in cell biology 18:77–83

Haigis MC, Mostoslavsky R, Haigis KM, Fahie K, Christodoulou DC, Murphy AJ, Valenzuela DM, Yancopoulos GD, Karow M, Blander G et al (2006) SIRT4 inhibits glutamate dehydrogenase and opposes the effects of calorie restriction in pancreatic β cells. Cell 126:941–954

Herranz D, Muñoz-Martin M, Cañamero M, Mulero F, Martinez-Pastor B, Fernandez-Capetillo O, Serrano M (2010) Sirt1 improves healthy ageing and protects from metabolic syndrome-associated cancer. Nat Comm 1:3

Hirschey MD, Shimazu T, Jing E, Grueter CA, Collins AM, Aouizerat B, Stancakova A, Goetzmanz E, Lam MM, Schwer B et al (2011) SIRT3 deficiency and mitochondrial protein hyperacetylation accelerate the development of the metabolic syndrome. Mol Cell 44:177–190

Ho L, Titus AS, Banerjee KK, George S, Lin W, Deota S, Saha AK, Nakamura K, Gut P, Verdin E et al (2013) SIRT4 regulates ATP homeostasis and mediates a retrograde signaling via AMPK. Aging (Albany NY) 5:835

Hubbard BP, Sinclair DA (2014) Small molecule SIRT1 activators for the treatment of aging and age-related diseases. Trends in pharmacological sciences 35:146–154

Jeong J, Juhn K, Lee H, Kim S, Min B, Lee K, Cho MH, Park GH, Lee K (2007) SIRT1 promotes DNA repair activity and deacetylation of Ku70. Experimental and Molecular Medicine 39:8

Jing E, Emanuelli B, Hirschey MD, Boucher J, Lee KY, Lombard D, Verdin EM, Kahn CR (2011) Sirtuin-3 (Sirt3) regulates skeletal muscle metabolism and insulin signaling via altered mitochondrial oxidation and reactive oxygen species production. Proc Natl Acad Sci 108:14608–14613

Kaeberlein M, McVey M, Guarente L (1999) The SIR2/3/4 complex and SIR2 alone promote longevity in Saccharomyces cerevisiae by two different mechanisms. Genes & Dev 13:2570–2580

Kaidi A, Weinert BT, Choudhary C, Jackson SP (2010) Human SIRT6 promotes DNA end resection through CtIPdeacetylation. Science 329:1348–1353

Kamel C, Abrol M, Jardine K, He X, McBurney MW (2006) Sirt1 fails to affect p53-mediated biological functions. Aging cell 5:81–88

Kanfi Y, Naiman S, Amir G, Peshti V, Zinman G, Nahum L, Bar-Joseph Z, Cohen HY (2012) The sirtuin SIRT6 regulates lifespan in male mice. Nature 483:218–221

Kawahara TL, Michishita E, Adler AS, Damian M, Berber E, Lin M, McCord RA, Onqaiui KC, Boxer LD, Chang HY et al (2009) SIRT6 links histone H3 lysine 9 deacetylation to NF-κB-dependent gene expression and organismal life span. Cell 136:62–74

Kim W, Kim JE (2013) SIRT7 an emerging sirtuin: deciphering newer rolES. JPP 64:531–534

Kim HS, Xiao C, Wang RH, Lahusen T, Xu X, Vassilopoulos A, Vazqzuez-Ortiz G, Jeong WI, Park O, Ki SH et al (2010) Hepatic-specific disruption of SIRT6 in mice results in fatty liver formation due to enhanced glycolysis and triglyceride synthesis. Cell Metab 12:224–236

Kim MY, Kang EL, Ham SA, Hwang JS, Yoo TS, Lee H, Paek KS, Park C, Lee HT, Kim JH et al (2012) The PPARδ-mediated inhibition of angiotensin II-induced premature senescence in human endothelial cells is SIRT1-dependent. Biochemical pharmacology 84:1627–1634

Kim JK, Noh JH, Jung KH, Eun JW, Bae HJ, Kim MG, Chang YG, Shen Q, Park WS, Lee JY et al (2013) Sirtuin7 oncogenic potential in human hepatocellular carcinoma and its regulation by the tumor suppressors MiR-125a-5p and MiR-125b. Hepatology 57:1055–1067

Kincaid B, Bossy-Wetzel E (2013) Forever young: SIRT3 a shield against mitochondrial meltdown, aging, and neurodegeneration. Frontiers in aging neuroscience 5:48

Kugel S, Mostoslavsky R (2012) Chromatin and beyond: the multitasking roles for SIRT6. Trends in biochemical sciences 39:72–81

Kulkarni SS, Cantó C (2014) The molecular targets of Resveratrol. Biochimicaet Biophysica Acta (BBA)-Molecular Basis of Disease 14(8):557–562

Langley E, Pearson M, Faretta M, Bauer UM, Frye RA, Minucci S, Pelicci PG, Kouzarides T (2002) Human SIR2 deacetylates p53 and antagonizes PML/p53-induced cellular senescence. The EMBO journal 21:2383–2396

Lappas M (2012) Anti-inflammatory properties of sirtuin 6 in human umbilical vein endothelial cells. Mediators of Inflammation 2012:597514

Laurent G, de Boer VC, Finley LW, Sweeney M, Lu H, Schug TT, Cen Y, Jeong SM, Li X, Sauve AA et al (2013) SIRT4 represses peroxisome proliferator-activated receptor α activity to suppress hepatic fat oxidation. Mol Cell Biol 33:4552–4561

Lee HS, Ka SO, Lee SM, Lee SI, Park JW, Park BH (2013) Overexpression of Sirtuin 6 Suppresses Inflammatory Responses and Bone Destruction in Mice With Collagen-Induced Arthritis. Arthritis & Rheumatism 65:1776–1785

Lee N, Kim DK, Kim ES, Park SJ, Kwon JH, Shin J, Park SM, Moon YH, Wang HJ, Gho YS et al (2014) Comparative interactomes of SIRT6 and SIRT7: Implication of functional links to aging. Proteomics 14:1610–1622

Li K, Casta A, Wang R, Lozada E, Fan W, Kane S, Ge Q, Gu W, Orren D, Luo J (2008) Regulation of WRN protein cellular localization and enzymatic activities by SIRT1-mediated deacetylation. J BiolChem 283:7590–7598

Li Y, Dai D, Lu Q, Fei M, Li M, Wu X (2013) Sirt2 suppresses glioma cell growth through targeting NF-κB-miR-21 axis. BiochemBiophys Res Commun 441:661–667

Liu B, Wang J, Chan KM, Tjia WM, Deng W, Guan X, Huang JD, Li KM, Chau PY, Chen DJ et al (2005) Genomic instability in laminopathy-based premature aging. Nat Med 11:780–785

Liu B, Ghosh S, Yang X, Zheng H, Liu X, Wang Z, Jin G, Zheng B, Kennedy BK, Suh Y et al (2012) Resveratrol rescues SIRT1-dependent adult stem cell decline and alleviates progeroid features in laminopathy-based progeria. Cell Metab 16:738–750

Luna A, Aladjem MI, Kohn KW (2013) SIRT1/PARP1 crosstalk: connecting DNA damage and metabolism. Genome integrity 4:6

Mao Z, Hine C, Tian X, Van Meter M, Au M, Vaidya A, Seluanov A, Gorbunova V (2011) SIRT6 promotes DNA repair under stress by activating PARP1. Science 332:1443–1446

Mao Z, Tian X, Van Meter M, Ke Z, Gorbunova V, Seluanov A (2012) Sirtuin 6 (SIRT6) rescues the decline of homologous recombination repair during replicative senescence. Proc Natl Acad Sci USA 109:11800–11805

McCord RA, Michishita E, Hong T, Berber E, Boxer LD, Kusumoto R, Guan S, Shi X, Gozani O, Burlingame AL et al (2009) SIRT6 stabilizes DNA-dependent protein kinase at chromatin for DNA double-strand break repair. Aging (Albany NY) 1:109

Mellini P, Valente S, Mai A (2014) Sirtuin modulators: an updated patent review (2012–2014). Expert Opinion on Therapeutic Patents (0), 1–11

Menghini R, Casagrande V, Cardellini M, Martelli E, Terrinoni A, Amati F, Vasa-Nicotera M, Ippoliti A, Novelli G, Melino G et al (2009) MicroRNA 217 modulates endothelial cell senescence via silent information regulator 1. Circulation 120:1524–1532

Michishita E, Park JY, Burneskis JM, Barrett JC, Horikawa I (2005) Evolutionarily conserved and nonconserved cellular localizations and functions of human SIRT proteins. Mol Biol Cell 16:4623–4635

Michishita E, McCord RA, Berber E, Kioi M, Padilla-Nash H, Damian M, Cheung P, Kusumoto R, Kawahara TL, Barrett JC et al (2008) SIRT6 is a histone H3 lysine 9 deacetylase that modulates telomeric chromatin. Nature 452:492–496

Michishita E, McCord RA, Boxer LD, Barber MF, Hong T, Gozani O, Chua KF (2009) Cell cycle-dependent deacetylation of telomeric histone H3 lysine K56 by human SIRT6. Cell Cycle 8:2664–2666

Min SW, Cho SH, Zhou Y, Schroeder S, Haroutunian V, Seeley WW, Huang EJ, Shen Y, Masliah E, Mukherjee C et al (2010) Acetylation of tau inhibits its degradation and contributes to tauopathy. Neuron 67:953–966

Ming M, Han W, Zhao B, Sundaresan NR, Deng CX, Gupta MP, He YY (2014) SIRT6 promotes COX-2 expression and acts as an oncogene in skin cancer. Cancer research 74:5925–5933

Mohamed JS, Wilson JC, Myers MJ, Sisson KJ, Alway SE (2014) Dysregulation of SIRT-1 in aging mice increases skeletal muscle fatigue by a PARP-1-dependent mechanism. Aging 6:820–834

Mortuza R, Chen S, Feng B, Sen S, Chakrabarti S (2013) High glucose induced alteration of SIRTs in endothelial cells causes rapid aging in a p300 and FOXO regulated pathway. PloS one 8: e54514

Mostoslavsky R, Chua KF, Lombard DB, Pang WW, Fischer MR, Gellon L, Liu P, Mostoslavsky G, Franco S, Murphy MM et al (2006) Genomic instability and aging-like phenotype in the absence of mammalian SIRT6. Cell 124:315–329

Nakagawa T, Guarente L (2009) Urea cycle regulation by mitochondrial sirtuin, SIRT5. Aging 1:578–581

Navarro CL, De Sandre-Giovannoli A, Bernard R, Boccaccio I, Boyer A, Geneviève D, Hadj-Rabia S, Gaudy-Marqueste C, Smitt HS, Vabres P et al (2004) Lamin A and ZMPSTE24 (FACE-1) defects cause nuclear disorganization and identify restrictive dermopathy as a lethal neonatal laminopathy. Human molecular genetics 13:2493–2503

Nguyen P, Lee S, Lorang-Lenis D, Trepel J, Smart DK (2014) SIRT2 interacts with β-catenin to inhibit Wnt signaling output in response to radiation-induced stress. Mol Cancer Res 12:1244–1253

North BJ, Rosenberg MA, Jeganathan KB, Hafner AV, Michan S, Dai J, Baker DJ, Cen Y, Wu LE, Sauve AA et al (2014) SIRT2 induces the checkpoint kinase BubR1 to increase lifespan. The EMBO journal 33:1438–1453

Ota H, Akishita M, Eto M, Iijima K, Kaneki M, Ouchi Y (2007) Sirt1 modulates premature senescence-like phenotype in human endothelial cells. Journal of Molecular and Cellular Cardiology 43:571–579

Ota H, Eto M, Kano MR, Ogawa S, Iijima K, Akishita M, Ouchi Y (2008) Cilostazol inhibits oxidative stress–induced premature senescence via upregulation of Sirt1 in human endothelial cells. Arteriosclerosis, thrombosis, and vascular biology 28:1634–1639

Paredes S, Villanova L, Chua KF (2014) Molecular Pathways: Emerging Roles of Mammalian Sirtuin SIRT7 in Cancer. Clinical Cancer Research 20:1741–1746

Rajamohan SB, Pillai VB, Gupta M, Sundaresan NR, Birukov KG, Samant S, Hottiger MO, Gupta MP (2009) SIRT1 promotes cell survival under stress by deacetylation-dependent deactivation of poly (ADP-ribose) polymerase 1. Mol Cell Biol 29:4116–4129

Rardin MJ, He W, Nishida Y, Newman JC, Carrico C, Danielson SR, Guo A, Gut P, Sahu AK, Li B et al (2013) SIRT5 regulates the mitochondrial lysine succinylome and metabolic networks. Cell Metab 18:920–933

Rehan L, Laszki-Szcząchor K, Sobieszczańska M, Polak-Jonkisz D (2014) SIRT1 and NAD as regulators of ageing. Life sciences 105:1–6

Rimmelé P, Bigarella CL, Liang R, Izac B, Dieguez-Gonzalez R, Barbet G, Donovan M, Brugnara C, Blander JM, Sinclair DA et al (2014) Aging-like Phenotype and Defective Lineage Specification in SIRT1-Deleted Hematopoietic Stem and Progenitor Cells. Stem Cell Reports 3:44–59

Salvioli S, Capri M, Bucci L, Lanni C, Racchi M, Uberti D, Memo M, Mari D, Govoni S, Franceschi C (2009) Why do centenarians escape or postpone cancer? The role of IGF-1, inflammation and p53. Cancer immunology, immunotherapy 58:1909–1917

Satoh A, Brace CS, Rensing N, Cliften P, Wozniak DF, Herzog ED, Yamada KA, Imai SI (2013) Sirt1 extends life span and delays aging in mice through the regulation of Nk2 homeobox 1 in the DMH and LH. Cell Metab 18:416–430

Saunders LR, Verdin E (2007) Sirtuins: critical regulators at the crossroads between cancer and aging. Oncogene 26:5489–5504

Saunders LR, Sharma AD, Tawney J, Nakagawa M, Okita K, Yamanaka S, Willenbring H, Verdin E (2010) miRNAs regulate SIRT1 expression during mouse embryonic stem cell differentiation and in adult mouse tissues. Aging (Albany NY) 2(7):415–431

Scheibye-Knudsen M, Mitchell SJ, Fang EF, Iyama T, Ward T, Wang J, Dunn CA, Singh N, Veith S, Hasan-olive MM et al (2014) A high-fat diet and NAD^+ activate Sirt1 to rescue premature aging in cockayne syndrome. Cell Metab 20:840–855

Schmeisser K, Mansfeld J, Kuhlow D, Weimer S, Priebe S, Heiland I, Birringer M, Groth M, Segref A, Kanfi Y et al (2013) Role of sirtuins in lifespan regulation is linked to methylation of nicotinamide. NatChemBiol 9:693–700

Schreiber KH, Kennedy BK (2013) When lamins go bad: nuclear structure and disease. Cell 152:1365–1375

Sebastián C, Zwaans BM, Silberman DM, Gymrek M, Goren A, Zhong L, Ram O, Truelove J, Guimaraes AR, Toiber D et al (2012) The histone deacetylase SIRT6 is a tumor suppressor that controls cancer metabolism. Cell 151:1185–1199

Shih J, Donmez G (2013) Mitochondrial sirtuins as therapeutic targets for age-related disorders. Genes & cancer 4:91–96

Shin J, He M, Liu Y, Paredes S, Villanova L, Brown K, Qiu X, Nabavi N, Mohrin M, Wojnoonski K et al (2013) SIRT7 represses Myc activity to suppress ER stress and prevent fatty liver disease. Cell reports 5:654–665

Shumaker DK, Dechat T, Kohlmaier A, Adam SA, Bozovsky MR, Erdos MR, Eriksson M, Goldman AE, Khuon S, Collins FS et al (2006) Mutant nuclear laminA leads to progressive alterations of epigenetic control in premature aging. PNAS 103:8703–8708

Sundaresan NR, Vasudevan P, Zhong L, Kim G, Samant S, Parekh V, Pillai VB, Ravindra PV, Gupta M, Jeevanandam V et al (2012) The sirtuin SIRT6 blocks IGF-Akt signaling and development of cardiac hypertrophy by targeting c-Jun. Nat Med 18:1643–1650

Thirumurthi U, Shen J, Xia W, LaBaff AM, Wei Y, Li CW, Chang WC, Chen CH, Lin HK, Yu D et al (2014) MDM2-mediated degradation of SIRT6 phosphorylated by AKT1 promotes tumorigenesis and trastuzumab resistance in breast cancer. Science Signaling 7: ra71

Toiber D, Erdel F, Bouazoune K, Silberman DM, Zhong L, Mulligan P, Sebastian C, Cosentino C, Martinez-Pastor B, Giacosa S et al (2013) SIRT6 recruits SNF2H to DNA break sites, preventing genomic instability through chromatin remodeling. Mol Cell 51:454–468

Tran D, Bergholz J, Zhang H, He H, Wang Y, Zhang Y, Li Q, Kirkland JL, Xiao ZX (2014) Insulin-like growth factor-1 regulates the SIRT1-p53 pathway in cellular senescence. Aging Cell 13:669–678

Uhl M, Csernok A, Aydin S, Kreienberg R, Wiesmüller L, Gatz SA (2010) Role of SIRT1 in homologous recombination. DNA repair 9:383–393

Vakhrusheva O, Smolka C, Gajawada P, Kostin S, Boettger T, Kubin T, Brawn T, Bober E (2008a) Sirt7 increases stress resistance of cardiomyocytes and prevents apoptosis and inflammatory cardiomyopathy in mice. Circulation Research 102:703–710

Vakhrusheva O, Braeuer D, Liu Z, Braun T, Bober E (2008b) Sirt7-dependent inhibition of cell growth and proliferation might be instrumental to mediate tissue integrity during aging. J PhysiolPharmacol 59:201–212

Van Meter M, Mao Z, Gorbunova V, Seluanov A (2011) SIRT6 overexpression induces massive apoptosis in cancer cells but not in normal cells. Cell cycle 10:3153–3158

Vaquero A, Scher M, Lee D, Erdjument-Bromage H, Tempst P, Reinberg D (2004) Human SirT1 interacts with histone H1 and promotes formation of facultative heterochromatin. Mol Cell 16:93–105

Vassallo PF, Simoncini S, Ligi I, Chateau AL, Bachelier R, Robert S, Morere J, Fernandez S, Guillet B, Marcelli M et al (2014) Accelerated senescence of cord blood endothelial progenitor cells in premature neonates is driven by SIRT1 decreased expression. Blood 123:2116–2126

Wang F, Tong Q (2009) SIRT2 suppresses adipocyte differentiation by deacetylating FOXO1 and enhancing FOXO1's repressive interaction with PPARγ. MolBiolCell 20:801–808

Wang F, Nguyen M, Qin F, Tong Q (2007) SIRT2 deacetylates FOXO3a in response to oxidative stress and caloric restriction. Aging cell 6:505–514

Xiao C, Wang RH, Lahusen TJ, Park O, Bertola A, Maruyama T, Reynolds D, Chen Q, Xu X, Young HA et al (2012) Progression of chronic liver inflammation and fibrosis driven by activation of c-JUN signaling in Sirt6 mutant mice. J BiolChem 287:41903–41913

Xu Z, Zhang L, Fei X, Yi X, Li W, Wang Q (2014) The miR-29b–Sirt1 axis regulates self-renewal of mouse embryonic stem cells in response to reactive oxygen species. Cellular signalling 26:1500–1505

Yamagata K, Kitabayashi I (2009) Sirt1 physically interacts with Tip60 and negatively regulates Tip60-mediated acetylation of H2AX. Biochemical and biophysical research communications 390:1355–1360

Yao H, Chung S, Hwang JW, Rajendrasozhan S, Sundar IK, Dean DA, McBurney MW, Guarente L, Gu W, Ronty M et al (2012) SIRT1 protects against emphysema via FOXO3-mediated reduction of premature senescence in mice. J Clin Invest 122:2032–2045

Yoshizawa T, Karim MF, Sato Y, Senokuchi T, Miyata K, Fukuda T, Go S, Tasaki M, Uchimura K, Kadomatsu T et al (2014) SIRT7 Controls Hepatic Lipid Metabolism by Regulating the Ubiquitin-Proteasome Pathway. Cell metabolism 19(4):712–721

Yuan HF, Zhai C, Yan XL, Zhao DD, Wang JX, Zeng Q, Chen L, Nan X, He LJ, Li ST et al (2012) SIRT1 is required for long-term growth of human mesenchymal stem cells. Journal of molecular medicine 90:389–400

Yuan H, Su L, Chen WY (2013) The emerging and diverse roles of sirtuins in cancer: a clinical perspective. OncoTargets and therapy 6:1399

Zhong L, D'Urso A, Toiber D, Sebastian C, Henry RE, Vadysirisack DD, Guimaraes A, Marinelli B, Wikstrom JD, Nir T et al (2010) The histone deacetylase Sirt6 regulates glucose homeostasis via Hif1α. Cell 140:280–293

Zhu H, Zhao L, Wang E, Dimova N, Liu G, Feng Y, Cambi F (2012) The QKI-PLP pathway controls SIRT2 abundance in CNS myelin. Glia 60:69–82

Zhu Y, Yan Y, Principe DR, Zou X, Vassilopoulos A, Gius D (2014) SIRT3 and SIRT4 are mitochondrial tumor suppressor proteins that connect mitochondrial metabolism and carcinogenesis. Cancer & Metabolism 2:1–11

16

Non-catalytic roles for TET1 protein negatively regulating neuronal differentiation through srGAP3 in neuroblastoma cells

Jie Gao[1,2], Yue Ma[2], Hua-Lin Fu[2,3], Qian Luo[1], Zhen Wang[2,4], Yu-Huan Xiao[1,2], Hao Yang[5], Da-Xiang Cui[2,3], Wei-Lin Jin[1,2,3]✉

[1] School of Life Sciences and Biotechnology, Shanghai Jiao Tong University, Shanghai 200240, China
[2] Institute of Nano Biomedicine and Engineering, Department of Instrument Science and Engineering, School of Electronic Information and Electronic Engineering, Shanghai Jiao Tong University, Shanghai 200240, China
[3] National Center for Translational Medicine, Shanghai Jiao Tong University, Shanghai 200240, China
[4] Department of Experimental Surgery, Tangdu Hospital, Fourth Military Medical University, Xi'an 710038, China
[5] Clinical Stem Cell Research Center, Renji Hospital, School of Medicine, Shanghai Jiao Tong University, Shanghai 200127, China
✉ Correspondence: weilinjin@sjtu.edu.cn (W.-L. Jin)

ABSTRACT

The methylcytosine dioxygenases TET proteins (TET1, TET2, and TET3) play important regulatory roles in neural function. In this study, we investigated the role of TET proteins in neuronal differentiation using Neuro2a cells as a model. We observed that knockdown of TET1, TET2 or TET3 promoted neuronal differentiation of Neuro2a cells, and their overexpression inhibited VPA (valproic acid)-induced neuronal differentiation, suggesting all three TET proteins negatively regulate neuronal differentiation of Neuro2a cells. Interestingly, the inducing activity of TET protein is independent of its enzymatic activity. Our previous studies have demonstrated that srGAP3 can negatively regulate neuronal differentiation of Neuro2a cells. Furthermore, we revealed that TET1 could positively regulate srGAP3 expression independent of its catalytic activity, and srGAP3 is required for TET-mediated neuronal differentiation of Neuro2a cells. The results presented here may facilitate better understanding of the role of TET proteins in neuronal differentiation, and provide a possible therapy target for neuroblastoma.

KEYWORDS methylcytosine dioxygenase, TET1, srGAP3, neuronal differentiation, neuroblastoma cells

INTRODUCTION

DNA methylation at the 5-position of cytosine (5-methylcytosine, 5mC) is an important epigenetic mark involved in regulation of gene expression. Ten-eleven translocation proteins (TET1, TET2, and TET3) are a family of Fe (II) and α-ketoglutarate-dependent dioxygenases that are capable of oxidizing 5mC to 5-hydroxymethylcytosine (5hmC) and further to generate 5-formylcytosine (5fC) and 5-carboxylcytosine (5caC) (Pastor et al., 2013; Ito et al., 2011). TET family proteins share highly homologous protein structural features. All TET proteins contain a C-terminal double stranded β-helix catalytic domain, and both TET1 and TET3 have an N-terminal CXXC domain, while TET2 CXXC domain IDAX (also known as CXXC4) is separated from TET2 following chromosomal rearrangement during evolution (Fig. 1A).

The TET family members were identified to possess a catalytic enzymatic activity conversion 5mC to 5hmC in 2009 (Tahiliani et al., 2009). TET proteins had been proved to be critical for numerous biological processes such as embryonic development, stem cell differentiation, immune regulation, and cancer formation (Tan and Shi, 2012; Xu et al., 2012; Li et al., 2015; Zhang et al., 2015; Dawlaty et al., 2013; Cimmino et al., 2015; Neri et al., 2015; Scourzic et al., 2015). Importantly, 5hmC was first to identify highly enriched in

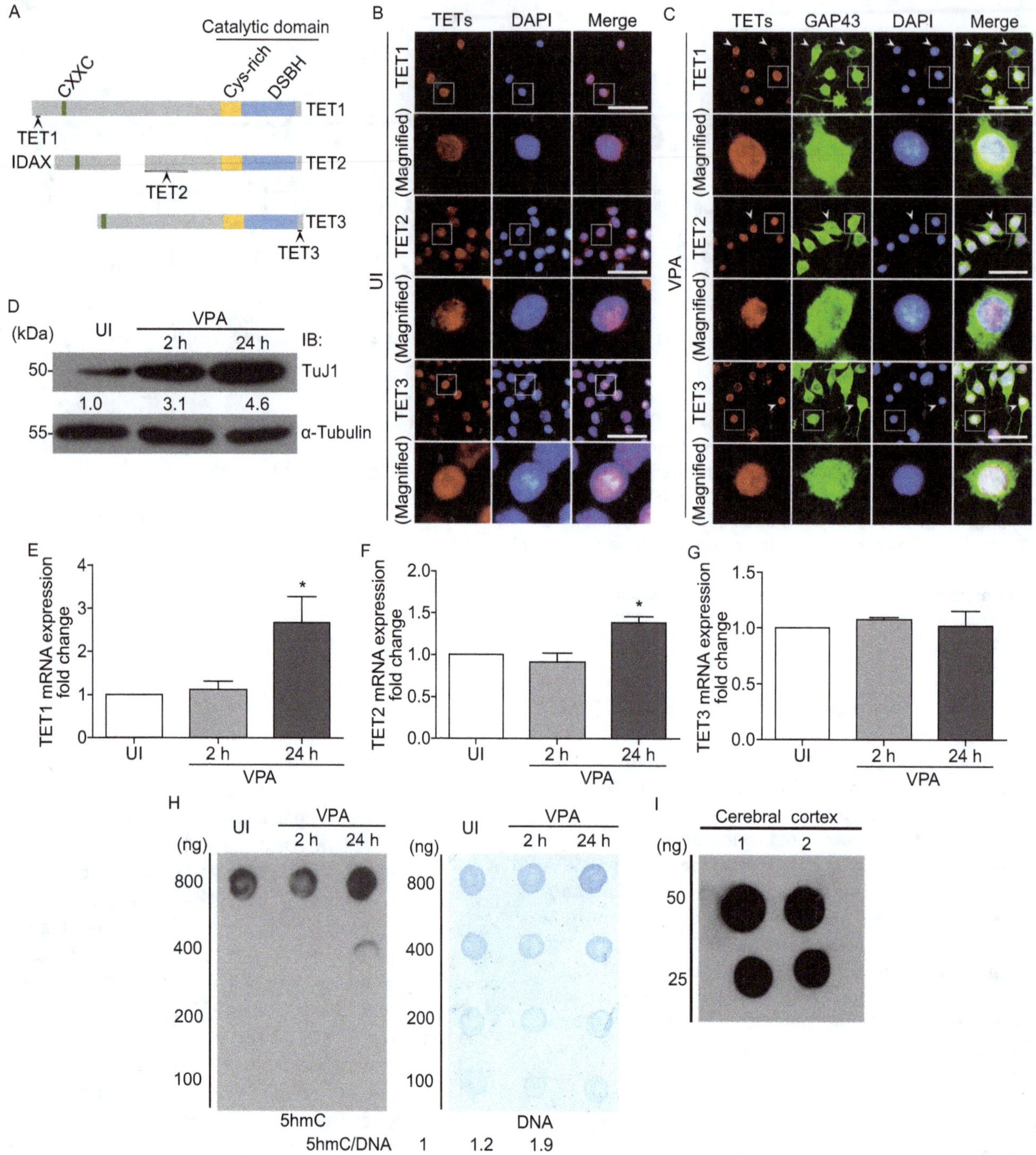

Purkinje cells and the brain tissues, suggesting a potential biological function of 5hmC in these cells and tissues (Kriaucionis and Heintz, 2009). Indeed, accumulating evidence demonstrates that TET proteins play important roles in the regulation of neural function (Yao and Jin, 2014). For example, TET1 deficiency in mice results in impaired neurogenesis and memory (Zhang et al., 2013; Kaas et al., 2013); knockdown of TET2 and TET3 in the mouse cerebral cortex leads to a block in neuronal differentiation (Hahn et al., 2013); TET3 is essential for early eye development in *Xenopus* (Xu et al., 2012). The roles of TET proteins in transcriptional regulation have been extensively investigated (Pastor et al., 2013). In most cases, TET-mediated promoter hypomethylation facilitates gene expression (Ficz et al., 2011; Mariani et al., 2014; Wu et al., 2011) in a dioxygenase activity-dependent manner.

Besides the catalytic domains, the CXXC domains are also involved in TET-mediated gene expression regulation.

◄ **Figure 1. Lower 5hmC level and higher expression of TET proteins in Neuro2a cells.** (A) Schematic presentation of the structure of TET proteins. The antigenic epitopes of the three TET antibodies on the corresponding TET proteins are underlined. (B) The expression and localization of endogenous TET proteins in Neuro2a cells cultured in high glucose DMEM containing 10% FBS (Uninduced, UI). The lower panels show magnifications of the single cell and process outlined by the white box in the upper panels respectively. Scale bar = 100 μm. (C) The expression and localization of endogenous TET proteins in opti-MEM containing 1 mmol/L VPA (VPA-induced, VPA) for 24 h. GAP43 was used as a neuronal differentiation marker. The lower panels show magnifications of the single cell and process outlined by the white box in the upper panels respectively. Scale bar = 100 μm. (D) Western blotting was used to detect the expression level of TuJ1 in uninduced Neuro2a cells (UI), and Neuro2a cells after VPA stimulation (VPA) for 2 h or 24 h. The relative expression level of TuJ1 was calculated by TuJ1/α-Tubulin. (E–G) Quantification of relative mRNA levels of endogenous TET proteins in uninduced Neuro2a cells (UI) and Neuro2a cells after VPA stimulation (VPA) for 2 h or 24 h. GAPDH was used as internal control. (H) Dot blotting was performed to analyze the 5hmC level in uninduced Neuro2a cells (UI), and Neuro2a cells after VPA stimulation (VPA) for 2 h or 24 h. DNA samples were serially two-fold diluted from 800 ng to 100 ng. 5hmC level was detected by using anti-5hmC antibody and the total DNA was stained by methylene blue. The relative level of 5hmC was calculated by 5hmC/DNA. (I) Dot blotting was used to analyze the 5hmC level in P15 mouse cerebral cortex tissues.

The CXXC domains are important for TET proteins binding to specific genomic regions for their action (Xu et al., 2012; Tan and Shi, 2012; Jin et al., 2014), and they can cooperate with the catalytic domain to regulate the key gene expression (Xu et al., 2012; Ko et al., 2013). Interestingly, accumulating evidence suggests that the non-catalytic TET proteins also play important roles in regulating gene expression (Pastor et al., 2013), whereas the regulation mechanisms are far from being fully elucidated.

Neuro2a is a mouse neural crest-derived cell line that has been widely used as an experimental model for neuronal differentiation study. In our previous studies, we used this model to study the role of srGAP3 in neuronal differentiation, and we found srGAP3 negatively regulated valproic acid (VPA)-induced neuronal differentiation of Neuro2a cells (Chen et al., 2011; Ma et al., 2013). In this study, we investigated the role of TET proteins during neuronal differentiation using Neuro2a cells as a model. We found that all three TET proteins could negatively regulate neuronal differentiation of Neuro2a cells. Furthermore, TET1 can negatively modulate neuronal differentiation independent of its catalytic activity and through srGAP3.

RESULTS

The expression of TET proteins is not correlated with 5hmC level in Neuro2a cells

To investigate the roles of TET proteins in neuronal differentiation, we firstly detected TET1–3 expression in Neuro2a cells. Three polyclonal antibodies specific against TET1, TET2, and TET3 protein were applied in the study (Fig. 1A). Immunofluorescence staining was performed to visualize the subcellular localization of endogenous TET proteins (Fig. 1B and 1C). It could be clearly observed that all three TET proteins expressed at detectable levels and localized to the nuclei either in uninduced (UI) or VPA-induced (VPA) Neuro2a cells (Fig. 1B and 1C). TuJ1 was used as a neuronal differentiation marker to indicate the differentiation stages (Fig. 1D). qRT-PCR indicated that the expression levels of TET1 and TET2 but not TET3 were remarkably increased after VPA stimulation for 24 h (Fig. 1E–G).

However, it was reported that 5hmC level is low in Neuro2a cells (Kriaucionis and Heintz, 2009), and this conclusion was confirmed in this study. 5hmC level could be detected by spotting as much as 800 ng DNA in Neuro2a cells (Fig. 1H), compared to only 25 ng DNA in mouse cerebral cortex tissues (Fig. 1I). In addition, 5hmC level increased gradually during VPA-induced Neuro2a cells differentiation (Fig. 1H). Those results indicated Neuro2a cells maintained high level of TET proteins and lower level of 5hmC. The mismatch between TET proteins and 5hmC suggested the catalytic activities of TET proteins might be suppressed in Neuro2a cells.

Knockdown of endogenous TET proteins promote neuronal differentiation of Neuro2a cells

TET proteins play important roles in neuronal development; however, the regulatory mechanisms of TET family proteins remain largely unknown. Here we examined the effects of TET1, TET2, or TET3 depletion on Neuro2a cells by shRNA-based knockdown method. The plasmid pGPU6/GFP/Neo under the control of hU6 promoter and cytomegalovirus immediate-early promoter (Pcmv IE) was used to express shRNA and GFP, respectively (Fig. 2A). The Neuro2a cells transfected with either negative control or shRNA expressing vectors could be recognized by expression of GFP. Cells with neurite processes longer than 1.5 cell bodies were counted as differentiated cells (Fig. 2B). qRT-PCR analysis demonstrated the efficiency of knockdown (Fig. 2C–E). We then examined the effects of TET proteins knockdown on Neuro2a cells differentiation. As shown in Fig. 2F–G, TET proteins depletion promoted neuronal differentiation in Neuro2a cells. The differentiation rate of the two TET1 knockdown groups (TET1 KD1 and TET1 KD2) were 6.7% and 9.6%, respectively, in uninduced Neuro2a cells (UI) compared to the control group (NC) which was 2.9%

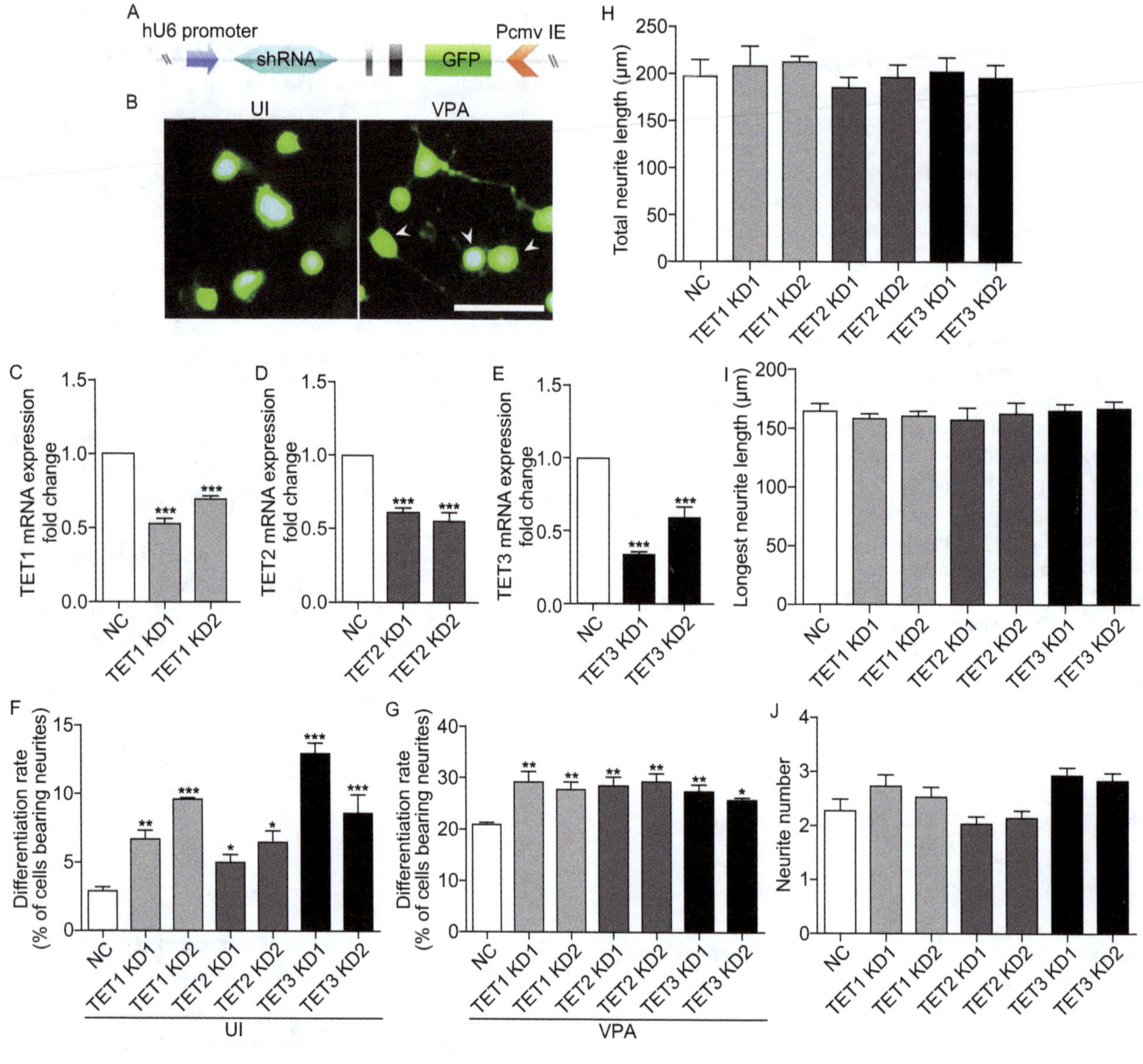

Figure 2. Knockdown of TET proteins facilitate the neuronal differentiation of Neuro2a cells. (A) Schematic presentation of shRNA expression vector. hU6 promoter, human U6 promoter; Pcmv IE, cytomegalovirus immediate-early promoter; GFP, green fluorescent protein. Neuro2a cells transfected with shRNA expression vectors were visualized by GFP. (B) Representative images of uninduced Neuro2a cells (UI) and Neuro2a cells after VPA stimulation (VPA) for 24 h. Arrowheads indicate the differentiated cells. Cells with processes longer than 1.5 cell bodies are considered as differentiated cells. Scale bar = 100 μm. (C–E) shRNA expression vectors of TET1–3 were transfected into Neuro2a cells, and qRT-PCR was used to analyze the effects of TET proteins knockdown. GAPDH was used as internal control. (F and G) Quantifications of the differentiation rates from Neuro2a cells of negative control (NC) and TET1, TET2, or TET3 knockdown in the absence (UI) or presence of VPA stimulation (VPA) for 24 h. Two shRNAs targeted different positions were designed respectively for each gene. (H–J) Assessment of neurite outgrowth of VPA-induced Neuro2a cells after TET1, TET2, or TET3 knockdown. Three endpoints were quantified: total neurite length (H), longest neurite length (I), and neurite number per cell (J). Values are mean ± SEM, *$P < 0.05$, **$P < 0.01$, ***$P < 0.001$.

(Fig. 2F), and were 29.2% and 27.8% in VPA-induced Neuro2a cells (VPA), respectively, compared to the control group, which was 20.9% (Fig. 2G). Additionally, similar effects on neuronal differentiation in Neuro2a cells could be observed after knockdown of TET2 (TET2 KD1 and TET2 KD2) or TET3 (TET3 KD1 and TET3 KD2) (Fig. 2F–G).

Furthermore, we tested whether knockdown of TET proteins could affect the neurite outgrowth of VPA-induced

Neuro2a cells. The results showed that knockdown of TET1, TET2, or TET3 did not affect total neurite length (Fig. 2H), longest neurite length (Fig. 2I), and neurite number (Fig. 2G), suggesting that TET proteins could not alter neurite outgrowth in differentiated Neuro2a cells.

Overexpression of TET proteins inhibit neuronal differentiation of VPA-induced Neuro2a cells

We next tested whether overexpression of TET proteins could affect neuronal differentiation of Neuro2a cells. The plasmid harboring TET1, TET2, or TET3 gene was transfected into Neuro2a cells (Fig. 3A). TET1 overexpression in Neuro2a cells had an approximately 47.0% inhibitory effect on neuronal differentiation by VPA stimulation (Fig. 3B). The differentiation rate was reduced to 10.5% compared to control, which was 19.8% (Fig. 3B). Similarly, overexpression of TET2 or TET3 also leads to a remarkable reduction in differentiation rates (TET2 overexpression: 17.1%; TET3 overexpression: 13.7%) (Fig. 3B). The results demonstrated that TET family proteins are negative regulators in neuronal differentiation of Neuro2a cells.

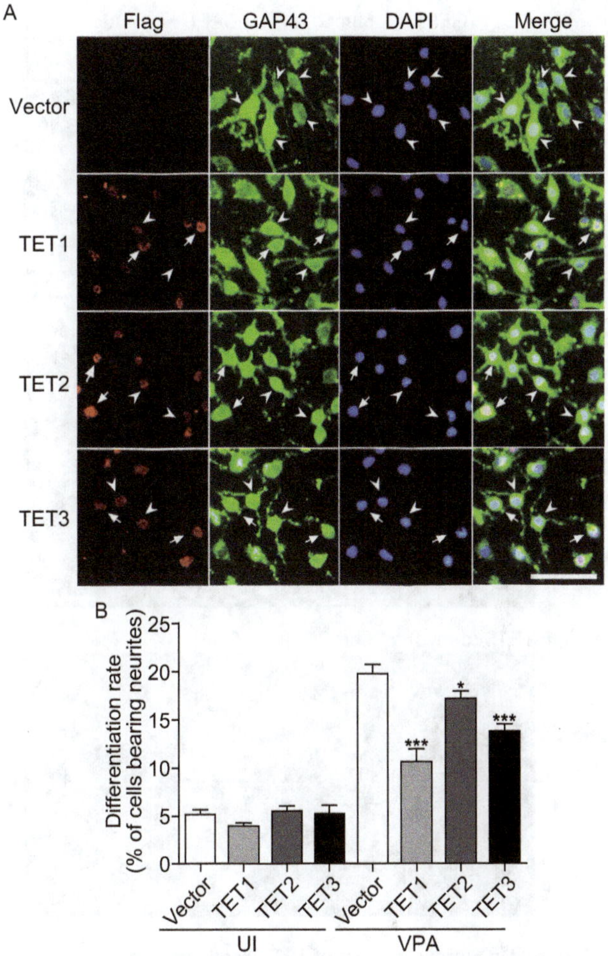

Figure 3. Overexpression of TET proteins inhibits VPA-induced neuronal differentiation of Neuro2a cells. (A) Neuro2a cells transfected with control vector (Vector) or vector harboring TET1, TET2, or TET3 gene were co-immunostained with Flag and GAP43 antibodies. Arrowheads indicate the differentiated cells and arrows indicate the undifferentiated cells. Scale bar = 100 μm. (B) Quantifications of the differentiation rates from Neuro2a cells in the absence (UI) or presence of VPA stimulation (VPA) for 24 h. Values are mean ± SEM, $*P < 0.05$, $***P < 0.001$.

TET proteins enrich 5hmC in Neuro2a cells

It is well known that TET proteins are able to convert 5mC to 5hmC, and enrich 5hmC in the cells. Given a mismatch between TET proteins and 5hmC levels (Fig. 1), we asked whether the overexpressed TET proteins kept their catalytic activities in the Neuro2a cells. As shown in Figure. 4, overexpression of TET1, TET2, or TET3 in Neuro2a cells can slightly increase 5hmC level. Importantly, the 5hmC levels were further increased after VPA stimulation for 24 h. The results suggested their catalytic activities were retained after TET protein overexpression, and the activities might be enhanced by VPA stimulation. However, whether this activity was involved in the regulation of neuronal differentiation of Neuro2a cells was unknown.

Overexpression of non-catalytic TET1 inhibits Neuro2a cells differentiation

As TET1, TET2, and TET3 showed similar effect on Neuro2a differentiation, and all of them shared similar homologous regions, we chose TET1 as a representative for further study. While both dioxygenase and CpG binding activity are critical for TET1 function, it has been reported that TET1 could exert its effect through enzymatic activity-independent manner (Kaas et al., 2013; Wu et al., 2011; Williams et al., 2011). To explore whether the regulation of Neuro2a cells differentiation by TET1 is dependent on its enzymatic activity, we also tested the effects of inactive mutants of TET1 with CpG binding domain CXXC deletion mutant (dCXXC) and catalytic domain mutant (CD mut), respectively (Fig. 5A). The plasmids harboring wild type (WT) TET1, TET1 dCXXC (dCXXC), or TET1 CD mut (CD mut) genes, were transfected into Neuro2a cells. Western blotting confirmed the expression of those proteins (Fig. 5B), and immunocytochemistry confirmed that CD mut lost its catalytic activity, while the dCXXC retained it (Fig. 5C).

Next, we investigated the effects on neuronal differentiation in VPA-induced Neuro2a cells following transfection of these three plasmids. Interestingly, transfection of all three constructs including WT TET1 (TET1), dCXXC, and CD mut significantly inhibited Neuro2a cells differentiation (Fig. 5D and 5E), suggesting that TET1 regulated neuronal

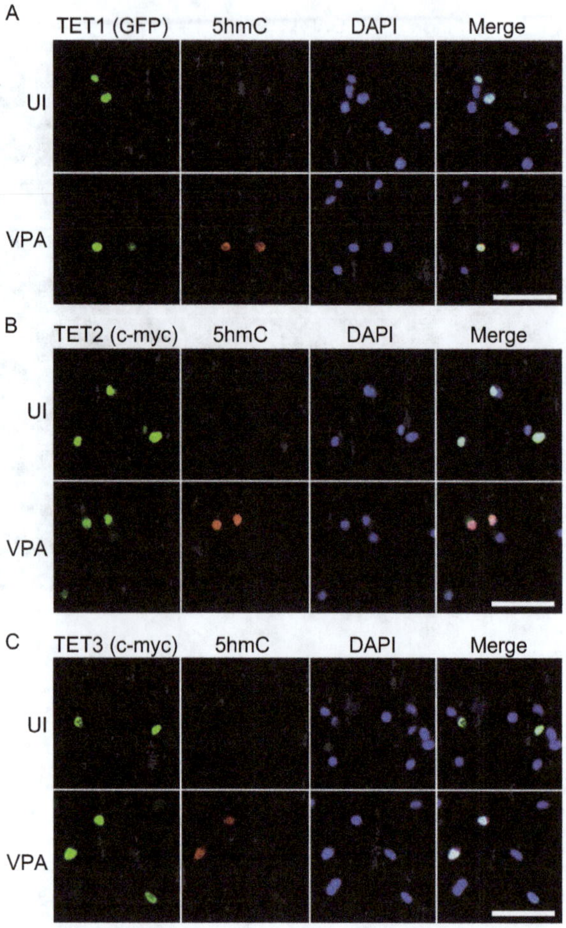

Figure 4. Overexpression of TET proteins increased 5hmC levels in differentiated Neuro2a cells. (A–C) Neuro2a cells were cultured in high glucose DMEM containing 10% FBS (UI) or in opti-MEM containing 1 mmol/L VPA (VPA). 5hmC and the expression of GFP tagged TET1 (A), c-myc tagged TET2 (B), or c-myc tagged TET3 (C) was detected by co-immunofluorescence staining.

differentiation of Neuro2a cells in an enzymatic activity-independent manner.

TET1 positively regulates srGAP3 expression independent of its catalytic domain

Since TET1 has been demonstrated to regulate a series of gene expression, we asked whether TET1 could modulate srGAP3 expression. We examined srGAP3 expression after TET1 knockdown and overexpression by Western blotting. TET1 KD1, one of the two shRNA expressing vectors, which was more effective than the other, was selected for subsequent research. Indeed, the results showed that knockdown of endogenous TET1 by transfection of shRNA against TET1 decreased protein levels of srGAP3 either in uninduced (UI) or VPA-induced (VPA) Neuro2a cells (Fig. 6A), while overexpression of TET1 increased srGAP3 expression (Fig. 6B). In our previous studies, srGAP3 was illustrated to negatively regulate neuronal differentiation of VPA-induced Neuro2a cells in a Rac1-dependent manner (Chen et al., 2011; Ma et al., 2013). Thus, we speculated that TET1 might negatively regulate neuronal differentiation of Neuro2a cells via modulating the expression of srGAP3. Considering that wild type (WT) TET1 can positively regulate the srGAP3 expression, we wonder whether the TET1 mutants can also modulate the srGAP3 expression. We examined the srGAP3 expression level following overexpression of WT TET1 (TET1), dCXXC, and CD mut in Neuro2a cells. As shown in Figure 6C, overexpression of WT TET1 (TET1), dCXXC, and CD mut can upregulate srGAP3 expression.

To confirm if srGAP3 can reverse the neuronal differentiation inducing activity of TET1-depleted neuro2a cells, we then quantified the differentiation rate of Neuro2a cells co-transfected with the knockdown construct for TET1 and srGAP3 overexpression vectors. As expected, overexpression of srGAP3 efficiently attenuated VPA-induced neuronal differentiation of TET1-depleted Neuro2a cells (Fig. 6D and 6E), indicating that TET1 negative regulate neuronal differentiation through srGAP3 expression, independent of its enzymatic activity (Fig. 6F).

DISCUSSION

Accumulating evidence indicates that TET proteins regulate neuronal activity-related gene expression, neuronal survival, neural development, and memory formation (Zhang et al., 2013; Kaas et al., 2013; Rudenko et al., 2013). In this study, we employed a Neuro2a cell model to explore the effect of TET family proteins on neural differentiation *in vitro*. We demonstrated TET proteins could negatively regulate neuronal differentiation of Neuro2a cells, and this regulation is independent of enzymatic activity of TET1. Furthermore, non-catalytic TET1 could negatively regulate neuronal differentiation through upregulating srGAP3.

Both the catalytic domain and CXXC domain are important for TET protein enzymatic activity. Mutations of the TET catalytic domain and CXXC domain can reduce their enzymatic-activity-dependent function (Hsu et al., 2012). Besides functioning in an enzymatic dependent manner, TET proteins can also regulate gene transcription independent of their catalytic activities. For examples, TET1 is found to repress gene transcription through binding to Polycomb group target genes, which is independent of its catalytic activity (Williams et al., 2011; Wu et al., 2011), and it can also upregulate several neuronal memory-associated genes independent of its enzymatic activity (Kaas et al., 2013). Similarly, TET2 was identified to recruit histone deacetylases 2 (Hdac2) to repress IL-6 transcription independent of DNA methylation and hydroxymethylation (Zhang et al., 2015). Additionally, catalytically inactive TET3 mutant can partially rescue the effect on *Xenopus* eye and neural development

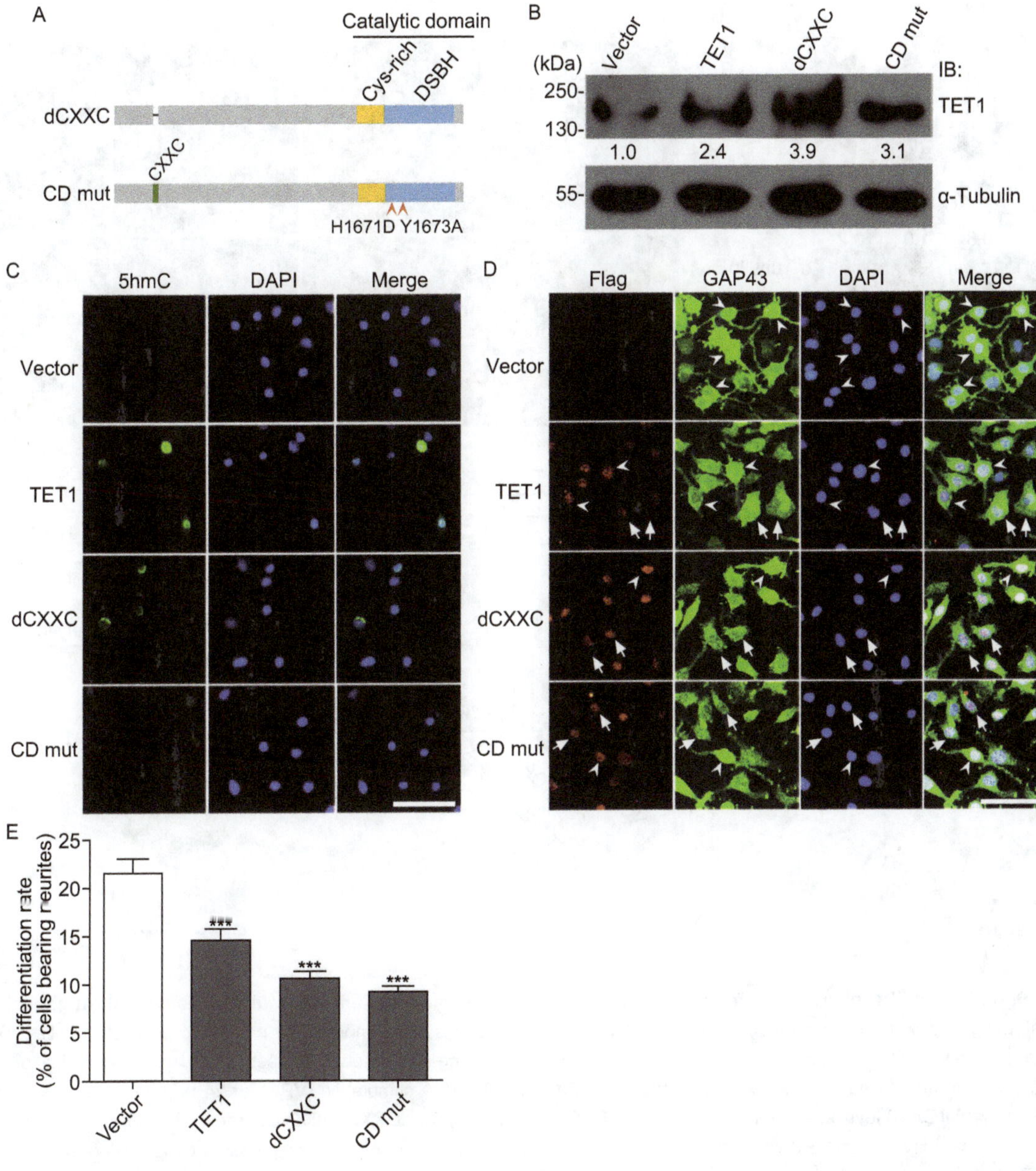

Figure 5. Overexpression of TET1 mutants inhibits neuronal differentiation of Neuro2a cells. (A) Schematic presentation of structure of TET1 mutants. (B) Neuro2a cells transfected with control vector (Vector) or vector harboring WT TET1 (TET1), TET1 dCXXC (dCXXC), and TET1 CD mut (CD mut) were lysed and immunoblotted with TET1 antibody. (C) Immunostaining of 5hmC after overexpression of WT TET1 (TET1), TET1 dCXXC (dCXXC), and TET1 CD mut (CD mut). (D) Neuro2a cells transfected with control vector (Vector) or vector harboring WT TET1 (TET1), TET1 dCXXC (dCXXC), and TET1 CD mut (CD mut) gene were co-immunostained with Flag and GAP43 antibodies. Arrowheads indicate the differentiated cells and arrows indicate undifferentiated cells. Scale bar = 100 μm. (E) Quantification of the differentiation rates from Neuro2a cells after overexpression of WT TET1 (TET1), TET1 dCXXC (dCXXC), and TET1 CD mut (CD mut). Values are mean ± SEM, $^{***}P < 0.001$.

after TET3 depletion, suggesting an enzymatic activity-independent mechanism in gene regulation (Xu et al., 2012). These findings point out a critical role of TET-mediated gene regulation independent of enzymatic activity. In this study, we uncovered that the differentiation rate was significantly decreased after overexpression of both the TET1 dCXXC and TET1 CD mutant, showing that the neuronal differentiation of Neuro2a cells was negatively regulated by TET1 in a dioxygenase activity-independent manner. Furthermore, overexpression of TET1 mutants resulted in a stronger

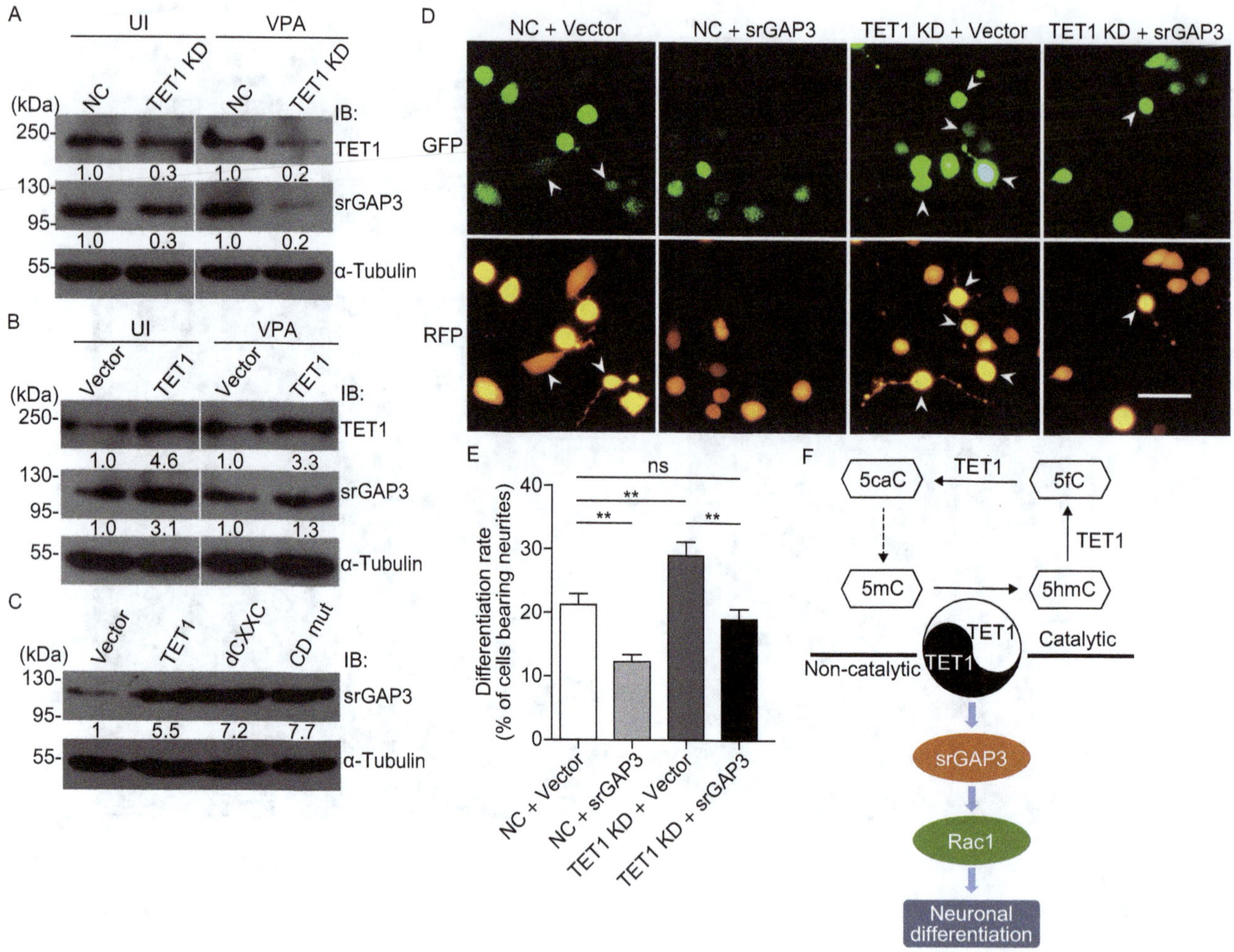

Figure 6. TET1 inhibits neuronal differentiation of Neuro2a cells independent of catalytic activity through srGAP3. (A) Western blotting was used to measure the srGAP3 level after endogenous TET1 knockdown in uninduced Neuro2a cells (UI), and Neuro2a cells after VPA stimulation (VPA) for 24 h. (B) Western blotting was used to measure the srGAP3 level after TET1 overexpression in uninduced Neuro2a cells (UI), and Neuro2a cells after VPA stimulation (VPA) for 24 h. (C) Western blotting was used to analyze srGAP3 level after overexpression of WT TET1 (TET1), TET1 dCXXC (dCXXC), and TET1 CD mut (CD mut). The relative expression level of srGAP3 was calculated by srGAP3/α-Tubulin. (D) SrGAP3 overexpression constructs or control vector and RFP were co-transfected into the normal or TET1-depleted Neuro2a cells, which were induced by VPA for 24 h. Arrowheads indicate the differentiated cells and arrows indicate undifferentiated cells. Scale bar = 100 μm. (E) Cell differentiation rate in (C) was analyzed. (F) A proposed working model for TET1 functions in regulating neuronal differentiation. Catalytic activity of TET1 might be suppressed in Neuro2a cells, and non-catalytic activity of TET1 inhibited neuronal differentiation through srGAP3 and Rac1 pathway. Values are mean ± SEM, $^{**}P < 0.01$.

inhibitory effect on neuronal differentiation even than that of overexpression of WT TET1 (TET1) (Fig. 5E). That might be because WT TET1 may exert an opposite effect on neuronal differentiation in an enzymatic activity-dependent manner.

TET1 can regulate the srGAP3 expression, and this regulation is independent of the catalytic activity of TET1. As srGAP3 can negatively regulate neuronal differentiation of Neuro2a cells, we speculate TET1 may negatively regulate Neuro2a cells differentiation via modulating the expression of srGAP3. Since many proteins have been found to interact with TET proteins to regulate gene expression (Yao and Jin 2014; Qiao et al., 2015; Wang et al., 2015; Perera et al., 2015), the regulation of srGAP3 expression may be due to the interaction between TET1 with some unknown proteins. However, the precise mechanism of neuronal differentiation regulated by TET1 should be further investigated.

In addition, the mRNA levels of TET1 and TET2 were increased after VPA stimulation for 24 h (Fig. 1E and 1F). It has been reported that VPA inhibit histone deacetylases (HDACs) to relieve HDAC-mediated transcriptional repression (Göttlicher et al., 2001). Thus, the increased mRNA of TET1 and TET2 may be due to the increase of histone acetylation level by VPA treatment. Furthermore, 5hmC levels were also increased in VPA-induced Neuro2a cells. These results were consistent with previous observations that levels of 5hmC increased during neuronal differentiation (Hahn et al., 2013). It also suggested the inhibition of catalytic activity of TET proteins might be relieved during neuronal differentiation of Neuro2a cells.

Neuroblastoma is the most common extracranial solid tumor of early childhood. Induction of terminal differentiation is a therapeutic approach for neuroblastoma treatment (Brodeur and Bagatell, 2014). In this study, we found TET proteins could negatively regulate neuronal differentiation of mouse neuroblastoma Neuro2a cells. It suggested inhibition of TET protein expression might be a promising approach for neuroblastoma therapy. Furthermore, in a previous report, it was suggested that hypoxia leads to transcriptional activation of TET1, which facilitated hypoxic gene induction in neuroblastoma (Mariani et al., 2014), showing the important role of TET1 in neuroblastoma development. The results further indicated TET1 might serve as a promising therapeutic target in neuroblastoma.

In summary, we found TET1 could negatively regulate neuronal differentiation of Neuro2a cells independent of its enzymatic activity through srGAP3. This study might provide a better understanding for the role of TET proteins in neuronal differentiation and offer a new potential therapy for neuroblastoma.

MATERIALS AND METHODS

Antibodies

Three homemade TET1 (from 146 to 159 amino acid residues), TET3 (from 1598 to 1610 amino acid residues), and srGAP3 (from 1088 to 1099 amino acid residues) antigen affinity antibodies were used in our previous studies (Zhao et al., 2014; Mi et al., 2015; Fu et al., 2014). These antibodies were purified by antigen affinity column. Rabbit anti-TET2 antibody was purchased from Proteintech Group (Chicago, USA). All of these antibodies could specifically recognize the corresponding proteins in immunofluorescence and Western blotting (Zhao et al., 2014; Mi et al., 2015; Fu et al., 2014). Rabbit anti-5hmC antibody (#39769) was purchased from Active Motif (Carlsbad, USA).

Quantitative RT-PCR

To obtain total RNA, Neuro2a cells were extracted using Trizol Reagent (Invitrogen, Carlsbad USA). The RNA was subsequently treated with RNase-free DNase I (Roche, Mannheim, Germany). The isolated total RNA was reverse-transcribed for two-step quantitative RT-PCR (qRT-PCR) using the PrimeScript RT reagent kit (TAKARA, Kyoto, Japan) according to the manufacturer's instructions. Quantitative RT-PCR was carried out using a Peltier Thermal Cycler (BioRad, Hercules, USA) plus Real time PCR Master Mix (SYBR Green, Toyobo, Japan). Gene-specific primer pairs for PCR are listed in Table S1. GAPDH was served as an internal control.

Plasmids

According to the target sequences of TET1, TET2, and TET3, oligonucleotides encoding shRNA were designed, synthesized, and inserted into pGPU6/GFP/Neo vector (Genepharma, Shanghai, China) to generate recombinant plasmids expressing shRNAs against TET1, TET2, and TET3, respectively. The target sequences for shRNAs are listed in Table S2. The recombinant plasmids expressing full-length TET1, inactive mutant of catalytic mutant (CD mut), and CXXC-deleted mutant (dCXXC) of TET1 were kindly provided by Dr. Li-Jung Juan (Hsu et al., 2012). The recombinant plasmids expressing full-length TET2 and full-length TET3 were obtained from Dr. Ross L. Levine (Abdel-Wahab et al., 2009) and Dr. Toshinobu Nakamura (Nakamura et al., 2012).

Cell culture, transfection, and differentiation assay

Neuro2a cells were cultured in DMEM containing 10% fetal bovine serum (FBS), 1% non-essential amino acid, and 1% penicillin and streptomycin. Neuro2a cells were transfected using X-treme GENE HP DNA Transfection Reagent (Roche, Mannheim, Germany) according to the manufacturer's instructions. To induce neuronal differentiation, Neuro2a cells (about 20% confluences) were transferred to serum-free opti-MEM (Gibco, Grand Island, New York, USA) containing 1 mmol/L VPA for neurite extension.

Immunocytochemistry

Cells on Poly-L-Lysine-coated glass coverslips were fixed with 4% paraformaldehyde for 15 min at room temperature and then permeabilized by treatment with ice-cold methanol for 10 min. For staining 5hmC, DNA was denatured with 2 mol/L HCl for 30 min, and then neutralized with 0.1 mol/L Tris-HCl (pH 8.5) twice (10 min a time). Subsequently, the cells were blocked by 15% normal donkey serum for 30 min, and then were incubated at 4°C overnight with primary antibody. The following antibodies dilutions were used: all TET antibodies were used at 1:200, anti-5hmC (1:5000), anti-GAP43 (1:500), and anti-Flag (PTGlab; 1:200). Then, the cells were rinsed and incubated for 2 h with fluorescence-labeled secondary antibodies. After washing, the coverslips were mounted with PBS containing 50% glycerol and 5 μg/mL DAPI for nuclei staining.

Dot blotting

The genomic DNA of Neuro2a cells were isolated using the EZNA Tissue DNA kit (OMEGA, Norcross, USA) and the DNA concentration was measured by NanoDrop. Dot-blot assays have been described previously (Ito et al., 2011). Briefly, genomic DNA was spotted on nylon membrane, and then baked at 80°C for 2 h. After blocked with 5% skimmed milk for 1 h, nylon membrane was incubated with the anti-5hmC antibody (1:5000) overnight at 4°C, and then incubated with POD-labeled secondary antibodies (1:12500; Roche, Mannheim, Germany). The signals were detected by BM Chemiluminescence Western Blotting kit (Roche, Mannheim, Germany). The densitometry quantification of dot-blot was analyzed by ImageProPlus 6.0.

Western blotting

The procedure had been previously described (Chen et al., 2011). In brief, the cells were lysed and the protein concentrations of the lysates were determined. Then, the lysates were separated by 8% SDS-PAGE and transferred to polyvinylidene fluoride (PVDF) membranes. The membranes were blocked, followed by incubation with primary antibodies and then POD-labeled secondary antibodies (1:12500; Roche, Mannheim, Germany). The primary antibodies were used as follows: TuJ1 (1:1000), TET1 (1:500), srGAP3 (1:1000), and α-Tubulin (Santa Cruz; 1:5000). The signals were detected by BM Chemiluminescence Western Blotting kit (Roche, Mannheim, Germany).

Statistical analysis

For assessing differentiation, cells bearing neurite processes 1.5 times longer than cell bodies were considered to be differentiated. Each group was analyzed by counting at least 300 cells. Assessment of neurite outgrowth was evaluated by counting about 30 to 50 cells per condition. Neurite number and length were quantified using Image Pro-Plus software. In each analysis, the data represent mean ± SEM of at least three experiments. For comparison, statistical significance was tested by one-way ANOVA.

ACKNOWLEDGEMENTS

The authors thank Dr. Heinrich Leonhardt for generously providing the GFP-tagged TET1-FL and TET1-ΔCXXC expression plasmids, Dr. Ross Levine for Flag-TET2, Dr. Toshinobu Nakamura for Flag-TET3, Dr. Li-Jung Juan for Flag-hTET1, Flag-hTET1-CD mut, and Flag-hTET1-ΔCXXC plasmids. This work was supported by the National Natural Science Foundation of China (Grant Nos. 31171033 and 81272801) and Shanghai Jiao Tong University Med-X Fund (No.YG2014MS45).

ABBREVIATIONS

5caC, 5-carboxylcytosine; 5fC, 5-formylcytosine; 5hmC, 5-hydroxymethylcytosine; 5mC, 5-methylcytosine; GAP43, growth associated protein-43; Hdac, histone deacetylases; srGAP3, Slit-Robo Rho GTPase-activating protein 3; TET proteins, ten-eleven translocation proteins; TuJ1, class III β-Tubulin; VPA, valproic acid

COMPLIANCE WITH ETHICS GUIDELINES

Jie Gao, Yue Ma, Hua-Lin Fu, Qian Luo, Zhen Wang, Yu-Huan Xiao, Hao Yang, Da-Xiang Cui, and Wei-Lin Jin declare that they have no conflict of interest.

This article does not contain any studies with human or animal subjects performed by the any of the authors.

REFERENCES

Abdel-Wahab O, Mullally A, Hedvat C, Garcia-Manero G, Patel J, Wadleigh M, Malinge S, Yao J, Kilpivaara O, Bhat R (2009) Genetic characterization of TET1, TET2, and TET3 alterations in myeloid malignancies. Blood 114(1):144–147

Brodeur GM, Bagatell R (2014) Mechanisms of neuroblastoma regression. Nat Rev Clin Oncol 11(12):704–713

Chen K, Mi Y-J, Ma Y, Fu H-L, Jin W-L (2011) The mental retardation associated protein, srGAP3 negatively regulates VPA-induced neuronal differentiation of Neuro2A cells. Cell Mol Neurobiol 31 (5):675–686

Cimmino L, Dawlaty MM, Ndiaye-Lobry D, Yap YS, Bakogianni S, Yu Y, Bhattacharyya S, Shaknovich R, Geng H, Lobry C (2015) TET1 is a tumor suppressor of hematopoietic malignancy. Nat Immunol 16(6):653–662

Dawlaty MM, Breiling A, Le T, Raddatz G, Barrasa MI, Cheng AW, Gao Q, Powell BE, Li Z, Xu M (2013) Combined deficiency of Tet1 and Tet2 causes epigenetic abnormalities but is compatible with postnatal development. Dev Cell 24(3):310–323

Ficz G, Branco MR, Seisenberger S, Santos F, Krueger F, Hore TA, Marques CJ, Andrews S, Reik W (2011) Dynamic regulation of 5-hydroxymethylcytosine in mouse ES cells and during differentiation. Nature 473(7347):398–402

Fu H-L, Ma Y, Lu L-G, Hou P, Li B-J, Jin W-L, Cui D-X (2014) TET1 exerts its tumor suppressor function by interacting with p53-EZH2 pathway in gastric cancer. J Biomed Nanotechnol 10(7):1217–1230

Göttlicher M, Minucci S, Zhu P, Krämer OH, Schimpf A, Giavara S, Sleeman JP, Coco FL, Nervi C, Pelicci PG (2001) Valproic acid defines a novel class of HDAC inhibitors inducing differentiation of transformed cells. EMBO J 20(24):6969–6978

Hahn MA, Qiu R, Wu X, Li AX, Zhang H, Wang J, Jui J, Jin S-G, Jiang Y, Pfeifer GP (2013) Dynamics of 5-hydroxymethylcytosine and chromatin marks in Mammalian neurogenesis. Cell reports 3 (2):291–300

Hsu C-H, Peng K-L, Kang M-L, Chen Y-R, Yang Y-C, Tsai C-H, Chu C-S, Jeng Y-M, Chen Y-T, Lin F-M (2012) TET1 suppresses cancer invasion by activating the tissue inhibitors of metalloproteinases. Cell Rep 2(3):568–579

Ito S, Shen L, Dai Q, Wu SC, Collins LB, Swenberg JA, He C, Zhang Y (2011) Tet proteins can convert 5-methylcytosine to 5-formylcytosine and 5-carboxylcytosine. Science 333(6047):1300–1303

Jin C, Lu Y, Jelinek J, Liang S, Estecio MR, Barton MC, Issa J-PJ (2014) TET1 is a maintenance DNA demethylase that prevents methylation spreading in differentiated cells. Nucleic Acids Res 42(11):6956–6971

Kaas GA, Zhong C, Eason DE, Ross DL, Vachhani RV, Ming G-L, King JR, Song H, Sweatt JD (2013) TET1 controls CNS 5-methylcytosine hydroxylation, active DNA demethylation, gene transcription, and memory formation. Neuron 79(6):1086–1093

Ko M, An J, Bandukwala HS, Chavez L, Äijö T, Pastor WA, Segal MF, Li H, Koh KP, Lähdesmäki H (2013) Modulation of TET2 expression and 5-methylcytosine oxidation by the CXXC domain protein IDAX. Nature 497(7447):122–126

Kriaucionis S, Heintz N (2009) The nuclear DNA base 5-hydroxymethylcytosine is present in Purkinje neurons and the brain. Science 324(5929):929–930

Li T, Yang D, Li J, Tang Y, Yang J, Le W (2015) Critical role of Tet3 in neural progenitor cell maintenance and terminal differentiation. Mol Neurobiol 51(1):142–154

Ma Y, Mi Y-J, Dai Y-K, Fu H-L, Cui D-X, Jin W-L (2013) The inverse F-BAR domain protein srGAP2 acts through srGAP3 to modulate neuronal differentiation and neurite outgrowth of mouse neuroblastoma cells. PLoS One 8(3):e57865

Mariani CJ, Vasanthakumar A, Madzo J, Yesilkanal A, Bhagat T, Yu Y, Bhattacharyya S, Wenger RH, Cohn SL, Nanduri J (2014) TET1-mediated hydroxymethylation facilitates hypoxic gene induction in neuroblastoma. Cell Rep 7(5):1343–1352

Mi Y, Gao X, Dai J, Ma Y, Xu L, Jin W-L (2015) A Novel Function of TET2 in CNS: Sustaining Neuronal Survival. Int J Mol Sci 16 (9):21846–21857

Nakamura T, Liu Y-J, Nakashima H, Umehara H, Inoue K, Matoba S, Tachibana M, Ogura A, Shinkai Y, Nakano T (2012) PGC7 binds histone H3K9me2 to protect against conversion of 5mC to 5hmC in early embryos. Nature 486(7403):415–419

Neri F, Dettori D, Incarnato D, Krepelova A, Rapelli S, Maldotti M, Parlato C, Paliogiannis P, Oliviero S (2015) TET1 is a tumour suppressor that inhibits colon cancer growth by derepressing inhibitors of the WNT pathway. Oncogene 34(32):4168–4176

Pastor WA, Aravind L, Rao A (2013) TETonic shift: biological roles of TET proteins in DNA demethylation and transcription. Nat Rev Mol Cell Biol 14(6):341–356

Perera A, Eisen D, Wagner M, Laube Silvia K, Künzel Andrea F, Koch S, Steinbacher J, Schulze E, Splith V, Mittermeier N, Müller M, Biel M, Carell T, Michalakis S (2015) TET3 is recruited by REST for context-specific hydroxymethylation and induction of gene expression. Cell Rep 11(2):283–294

Qiao Y, Wang X, Wang R, Li Y, Yu F, Yang X, Song L, Xu G, Chin YE, Jing N (2015) AF9 promotes hESC neural differentiation through recruiting TET2 to neurodevelopmental gene loci for methylcytosine hydroxylation. Cell Discov 1:15017

Rudenko A, Dawlaty MM, Seo J, Cheng AW, Meng J, Le T, Faull KF, Jaenisch R, Tsai L-H (2013) Tet1 is critical for neuronal activity-regulated gene expression and memory extinction. Neuron 79 (6):1109–1122

Scourzic L, Mouly E, Bernard OA (2015) TET proteins and the control of cytosine demethylation in cancer. Genome Med 7(1):9

Tahiliani M, Koh KP, Shen Y, Pastor WA, Bandukwala H, Brudno Y, Agarwal S, Iyer LM, Liu DR, Aravind L (2009) Conversion of 5-methylcytosine to 5-hydroxymethylcytosine in mammalian DNA by MLL partner TET1. Science 324(5929):930–935

Tan L, Shi YG (2012) Tet family proteins and 5-hydroxymethylcytosine in development and disease. Development 139(11):1895–1902

Wang Y, Xiao M, Chen X, Chen L, Xu Y, Lv L, Wang P, Yang H, Ma S, Lin H, Jiao B, Ren R, Ye D, Guan KL, Xiong Y (2015) WT1 recruits TET2 to regulate its target gene expression and suppress leukemia cell proliferation. Mol Cell 57(4):662–673

Williams K, Christensen J, Pedersen MT, Johansen JV, Cloos PA, Rappsilber J, Helin K (2011) TET1 and hydroxymethylcytosine in transcription and DNA methylation fidelity. Nature 473(7347): 343–348

Wu H, D'Alessio AC, Ito S, Xia K, Wang Z, Cui K, Zhao K, Sun YE, Zhang Y (2011) Dual functions of Tet1 in transcriptional regulation in mouse embryonic stem cells. Nature 473(7347):389–393

Xu Y, Xu C, Kato A, Tempel W, Abreu JG, Bian C, Hu Y, Hu D, Zhao B, Cerovina T (2012) Tet3 CXXC domain and dioxygenase activity cooperatively regulate key genes for Xenopus eye and neural development. Cell 151(6):1200–1213

Yao B, Jin P (2014) Unlocking epigenetic codes in neurogenesis. Genes Dev 28(12):1253–1271

Zhang R-R, Cui Q-Y, Murai K, Lim YC, Smith ZD, Jin S, Ye P, Rosa L, Lee YK, Wu H-P (2013) Tet1 regulates adult hippocampal neurogenesis and cognition. Cell Stem Cell 13(2):237–245

Zhang Q, Zhao K, Shen Q, Han Y, Gu Y, Li X, Zhao D, Liu Y, Wang C, Zhang X (2015) Tet2 is required to resolve inflammation by recruiting Hdac2 to specifically repress IL-6. Nature 525 (7569):389–393

Zhao X, Dai J, Ma Y, Mi Y, Cui D, Ju G, Macklin WB, Jin W-L (2014) Dynamics of ten-eleven translocation hydroxylase family proteins and 5-hydroxymethylcytosine in oligodendrocyte differentiation. Glia 62(6):914–926

17

MicroRNA-29a modulates axon branching by targeting doublecortin in primary neurons

Hanqin Li[1], Susu Mao[1], Haitao Wang[2], Ke Zen[1✉], Chenyu Zhang[1✉], Liang Li[1✉]

[1] Jiangsu Engineering Research Center for MicroRNA Biology and Biotechnology, State Key Laboratory of Pharmaceutical Biotechnology, Nanjing University School of Life Sciences, Nanjing 210093, China

[2] Department of Pharmacy, 306th Hospital of PLA, Beijing 100101, China

✉ Correspondence: kzen@nju.edu.cn (K. Zen), cyzhang@nju.eud.cn (C. Zhang), lijing@sibs.ac.cn (L. Li)

ABSTRACT

MicroRNAs (miRNAs) are endogenously expressed small, non-coding transcripts that regulate protein expression. Substantial evidences suggest that miRNAs are enriched in central nervous system, where they are hypothesized to play pivotal roles during neural development. In the present study, we analyzed miRNAs expression in mice cerebral cortex and hippocampus at different developmental stages and found miR-29a increased dramatically at postnatal stages. In addition, we provided strong evidences that miR-29a is enriched in mature neurons both *in vitro* and *in vivo*. Further investigation demonstrated that the activation of glutamate receptors induced endogenous miR-29a level in primary neurons. Moreover, we showed that miR-29a directly regulated its target protein Doublecortin (DCX) expression, which further modulated axon branching in primary culture. Together, our results suggested that miR-29a play an important role in neuronal development of mice cerebrum.

KEYWORDS miR-29a, doublecortin, glutamate receptor, mature neurons, axon branching

INTRODUCTION

MiRNAs are approximately 21 nucleotide non-coding transcripts that are derived from hairpin precursors, which regulate protein expression by base pairing to the 3′ untranslated region (UTR) of mRNAs (Chen et al., 2008). Several works show that miRNAs are enriched in central nervous system and emerging evidence have suggested the pivotal roles of miRNAs in neural development (Ji et al., 2013; Shu et al., 2013). Mutations in Drosophila Ago1 cause a severe loss of all types of neurons and glial cells while dicer deficient mice display abnormal neurogenesis and gliogenesis in the developing central nervous system (Kataoka et al., 2001; Kawase-Koga et al., 2009). There are also studies indicating the significant role of specific miRNA in neural development. For example, miR-132 modulates the dopamine neuron differentiation while miR-124 regulates neurite outgrowth during neuronal differentiation (Yang et al., 2012; Yu et al., 2008). Together, these data support the notion that miRNAs modulate cell differentiation, maturation as well as survival in central nervous system.

To investigate the function of miRNAs in the development of mice cerebrum, we screened miRNAs expression in mice cerebrum at different developmental stages. Interestingly, we found miR-29a changed most dramatically among all those miRNAs that increased during cerebrum development. MiR-29 family is consisted of three mature members, miR-29a, miR-29b, and miR-29c, which are encoded by two genome clusters (Wang et al., 2013). Recent studies have demonstrated that miR-29 family is associated with the pathogenesis of sporadic Alzheimer's disease by targeting BACE and NAV3 (Bettens et al., 2009; Shioya et al., 2010). Further investigation indicates that miR-29b has anti-apoptotic effect in mature neuron by targeting BH3-only (Kole et al., 2011). However, the role of miR-29 family during neuronal development remains obscure, which still needs further investigation.

In the present study, we demonstrated that miR-29a, which highly expresses in mature neurons, is up-regulated

Table 1. Copy number of miRNAs that increased significantly during the cerebrum development

	Copy number at different developmental stages				Ratio
	E12.5	E15.5	E18.5	P60	P60/E12.5
miR-22	650.86	753.18	1135	7193.4	11.05
miR-23a	1249.1	1817.3	1482.8	6461.2	5.17
miR-24	3396.8	5393	5354	20931	6.16
miR-29a	61.68	61.58	73.86	5122.3	83.05
miR-31	228.11	245.63	352.49	3118.1	13.67
miR-128	945.99	1333.8	4568.2	7894.6	8.35
miR-132	510.18	901.66	1727.1	16807	32.94
miR-138	1713.5	4395.6	6829.2	23379	13.64
miR-139-5p	359.66	484.26	636.85	4870.7	13.54
miR-143	285	291.6	411.99	2356.9	8.27
miR-145	640.83	638.49	821.15	4777.1	7.45
miR-221	747.36	478.43	901.21	5651.1	7.56
miR-222	463.46	426.43	783.48	9207.9	19.87

both in cortex and hippocampus during cerebrum development especially in postnatal stages. Besides, we provided evidences that activation of glutamate receptors contribute to the increase of miR-29a in primary neurons, suggesting the possible association of neural activity and the increase of miR-29a. Moreover, our data revealed that DCX, a protein that is reported to affect axon branching (Bilimoria et al., 2010), is a direct target of miR-29a. Further functional analysis showed that miR-29a increased axon branching via repressing DCX expression. Taken together, these data indicate that miR-29a plays an important role in neuronal development of mice cerebrum.

RESULTS

Screening miRNAs expression levels in mice cerebral cortex at different developmental stages

To investigate the role of miRNAs in the development of mice cerebrum, we first analyzed the miRNAs levels in the mice cortex at different stages (E12.5, E15.5, E18.5 and P60) by miRNAs array. Here we mainly focused on the increased miRNAs which are selected based on the following two criteria: (1) the minimum copy number of miRNA at P60 is above 2000; (2) the fold change of miRNA between P60 and E12.5 is more than 5. Among all those selected miRNAs, miR-29a increased most significantly in P60 mice cortex (Table 1). Meanwhile, the copy number of miR-29b and miR-29c is less than 300 at the stage of P60 (data not shown). Surprisingly, miR-29a did not change significantly during E12.5–E18.5 stages while it increased drastically (>70 fold) during E18.5–P60 stages. To further understand the correlation between miR-29a and cerebrum development, we analyzed miR-29a expression both in cortex and hippocampus by Q-RTPCR. We found that cortical miR-29a showed a time-dependent increase exclusively in postnatal stages while miR-138 increased gradually during the entire developmental stages analyzed (E12.5–P60), which is consistent with the miRNAs array analysis (Fig. 1A). Similarly, the level of miR-29a in hippocampus began to increase at P7 while the level of miR-138 gradually increased during the whole stages (E18.5–P60) (Fig. 1B). We also compared the expression levels of miR-29 family in cortex and hippocampus of P60 mice by Q-RTPCR and demonstrated that miR-29a concentration is relatively higher than that of the other two family members (Fig. S1).

Expression pattern of miR-29 family in mice cerebrum

To elucidate the expression pattern of miR-29 family in mice cerebrum, we examined three members of miR-29 family including miR-29a, miR-29b and miR-29c by *in situ* hybridization. Our results demonstrated that miR-29a is most intensively expressed in hippocampus neurons while miR-29b and miR-29c showed much weaker signals (Fig. 1C, a–e). Besides, miR-29a is also expressed in cortical neurons, while miR-29b and miR-29c barely showed any positive signals (Fig. 1C, a′–e′). These results are coincident with our Q-RTPCR analysis (Fig. S1). Nevertheless, using lung tissue as positive control, we demonstrated that both miR-29b and miR-29c show relatively strong signals in the mice lung tissue compared with that in cortex and hippocampus (Fig. S2). Importantly, we also analyzed miR-29a expression pattern in dentate gyrus (DG) area at different stages. The signal appeared at P7 and increased at P14 and P60, which is consistent with our Q-RTPCR analysis (Fig. 2A, a–c). Furthermore, the expression of miR-29a in DG area began

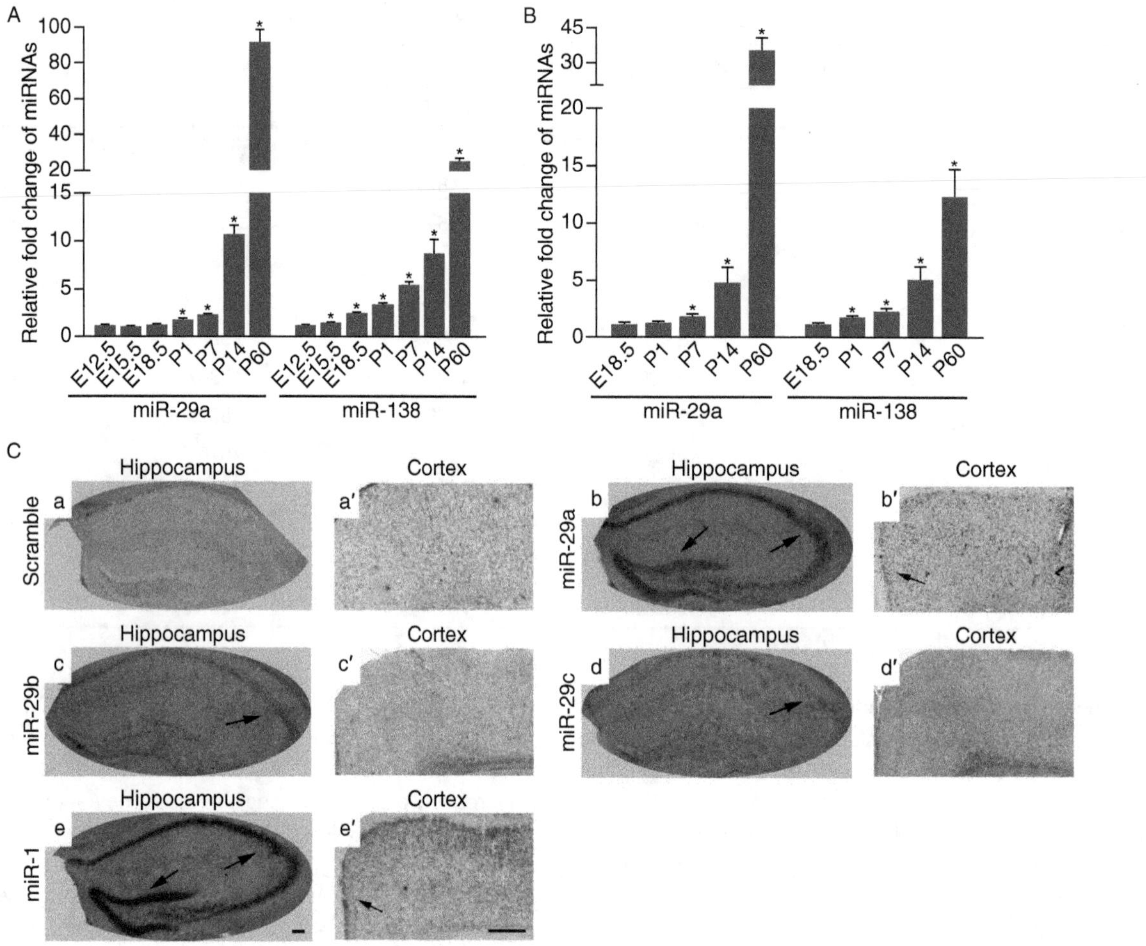

Figure 1. Expression of miR-29a in mice cortex and hippocampus. (A) Relative expression level of miR-29a and miR-138 (U6 is used as internal control) in cortex of different developmental stages, *, $P < 0.01$ compared to the E12.5 stage in each group ($n = 5$); (B) relative expression level of miR-29a and miR-138 in hippocampus of different developmental stages, *, $P < 0.01$ compared to the E18.5 stage in each group ($n = 5$); (C) *in situ* hybridization of different members of miR-29 family in hippocampus (a–e) and cortex (a′–e′) ($n = 3$), miR-124 is used as positive control, arrows indicate neurons with positive signal in cortex or hippocampus, scale bar = 200 μm.

from the outside of granule layer (P7) while gradually spread to the inside (P60), which is similar to the neuron maturation pattern marked by NeuN (mature neuronal marker) (Fig. 2A, d–f) (Snyder et al., 2012). Together, these data provide evidence indicating that miR-29a is highly expressed and functions in the development of cerebrum, especially in mature neurons.

Activation of glutamate receptors increases the level of miR-29a in primary neurons

We have observed the up-regulation of miR-29a during cerebrum development exclusively at postnatal stages (Fig. 1A and 1B). In addition, we also examined the miR-29a level in cultured neuron from cortex or hippocampus at different time points and showed that miR-29a increased as neurons became mature (Fig. 2B). Therefore, it is interesting for us to elucidate the mechanism underlying the regulation of miR-29a expression. We investigated the association between neural activity and miR-29a expression by assessing the level of miR-29a in cortical neurons after glutamate treatment (10 μmol/L, 15 min) and found that miR-29a is significantly increased since 0.5 h after the stimulation. Such increase lasted for 6 h and miR-29a returned to the basal level after 12 h (Fig. 2C). Furthermore, we found that pretreatment of MK-801 (NMDA receptor antagonist) significantly blocked the increase of miR-29a induced by glutamate (Fig. 2D). Similarly, DNQX (AMPA/kainate receptor antagonist) also inhibited such increase and exhibited linear effect when used with MK-801 together (Fig. 2D). These data indicate that the activation of glutamate receptors may contribute to the increase of miR-29a in neurons during cerebrum maturation.

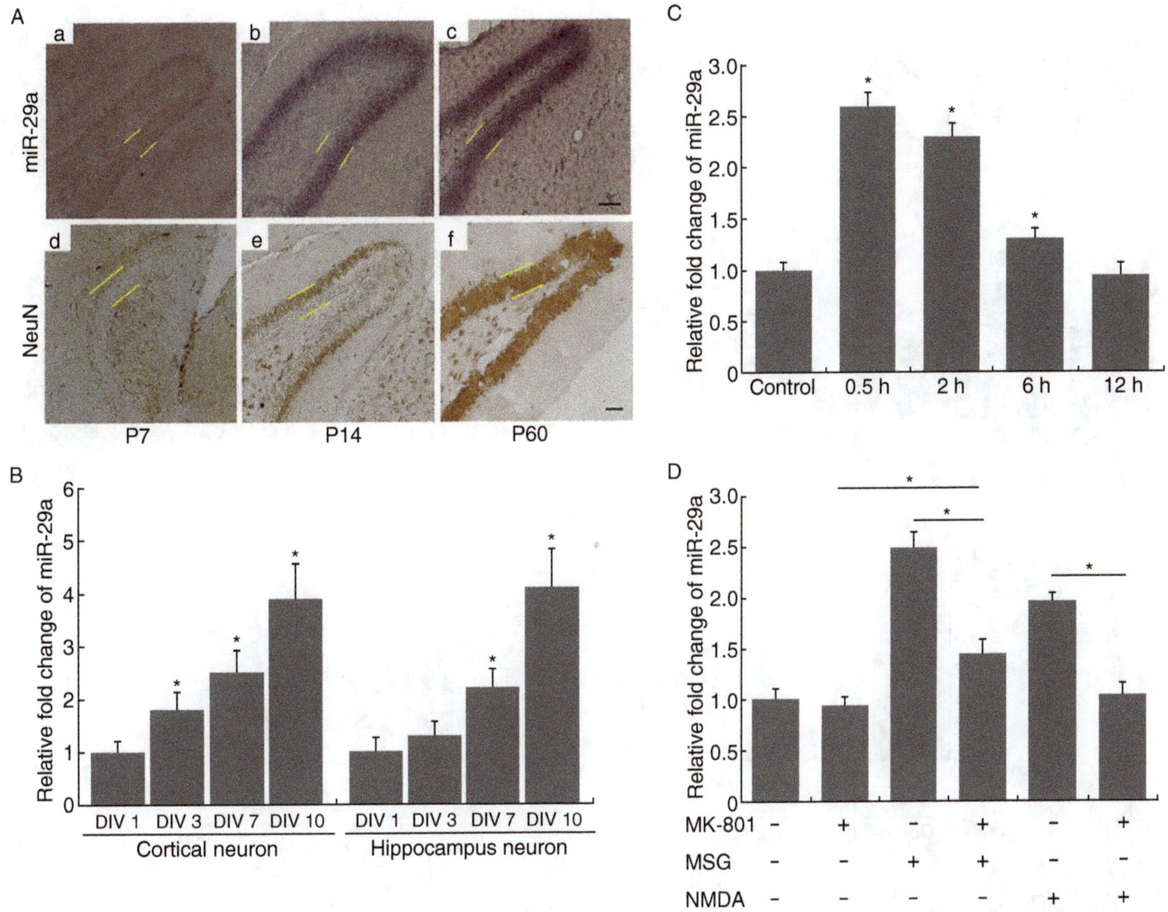

Figure 2. Expression of miR-29a in neurons under different circumstances. (A) Expression pattern of miR-29a (purple) in DG area of hippocampus at different postnatal stages (a–c), expression pattern of NeuN (brown) in DG area of hippocampus at different postnatal stages (d–f), granule cell layers are indicated between two yellow lines (n = 3), scale bar = 50 μm; (B) relative expression level of miR-29a in primary neuron at different cultured stages, *, $P < 0.01$ compared to DIV 1 in each group (n = 5); (C) relative expression level of miR-29a in primary cortical neurons at different time points after the stimulation of 10 μmol/L GSM for 15 min, *, $P < 0.01$ compared to control (n = 3); (D) relative expression level of miR-29a in primary cortical neurons at 2 h after indicated treatments, *, $P < 0.01$ (n = 3).

DCX is a direct target of miR-29a

Through bioinformatics screening, we found that the sequence of miR-29a matched well with 3′ UTR of DCX mRNA (Fig. 3A). DCX is a microtubule-associated protein expressed by neuronal precursor cells and immature neurons, which is reported to modulate axon branching (Bilimoria et al., 2010; Spampanato et al., 2012). In addition, the targeted sequence in DCX mRNA that matched with the seed sequence of miR-29a is highly conservative in many species, indicating the regulation of DCX expression by miR-29a is quite universal. We further evaluate the protein level of DCX in both cortex and hippocampus at different developmental stages. In embryonic stages, DCX level slightly increased in cortex (E18.5) while did not change a lot in hippocampus. However, in postnatal stages, DCX expression significantly decreased both in cortex and hippocampus at P7 (Fig. 3B–D), which is negatively correlated with the change of miR-29a level in cortex and hippocampus (Fig. 1A and 1B). We also demonstrated that the expression patterns of DCX and miR-29a in DG area are mutually complementary (Figs. 3E and 2A). Moreover, we analyzed the expression of DCX during the neuron maturation *in vitro* and found the level of DCX showed a time-course decrease, which is negatively correlated to the level of miR-29a (Fig. 3F and 3G). To prove that DCX is a direct target of miR-29a we conducted luciferase assay and found that miR-29a markedly reduce the translational efficiency of reporter gene tagged with the 3′UTR of DCX mRNA containing the putative miR-29a-binding sites while mutations in these sites significantly

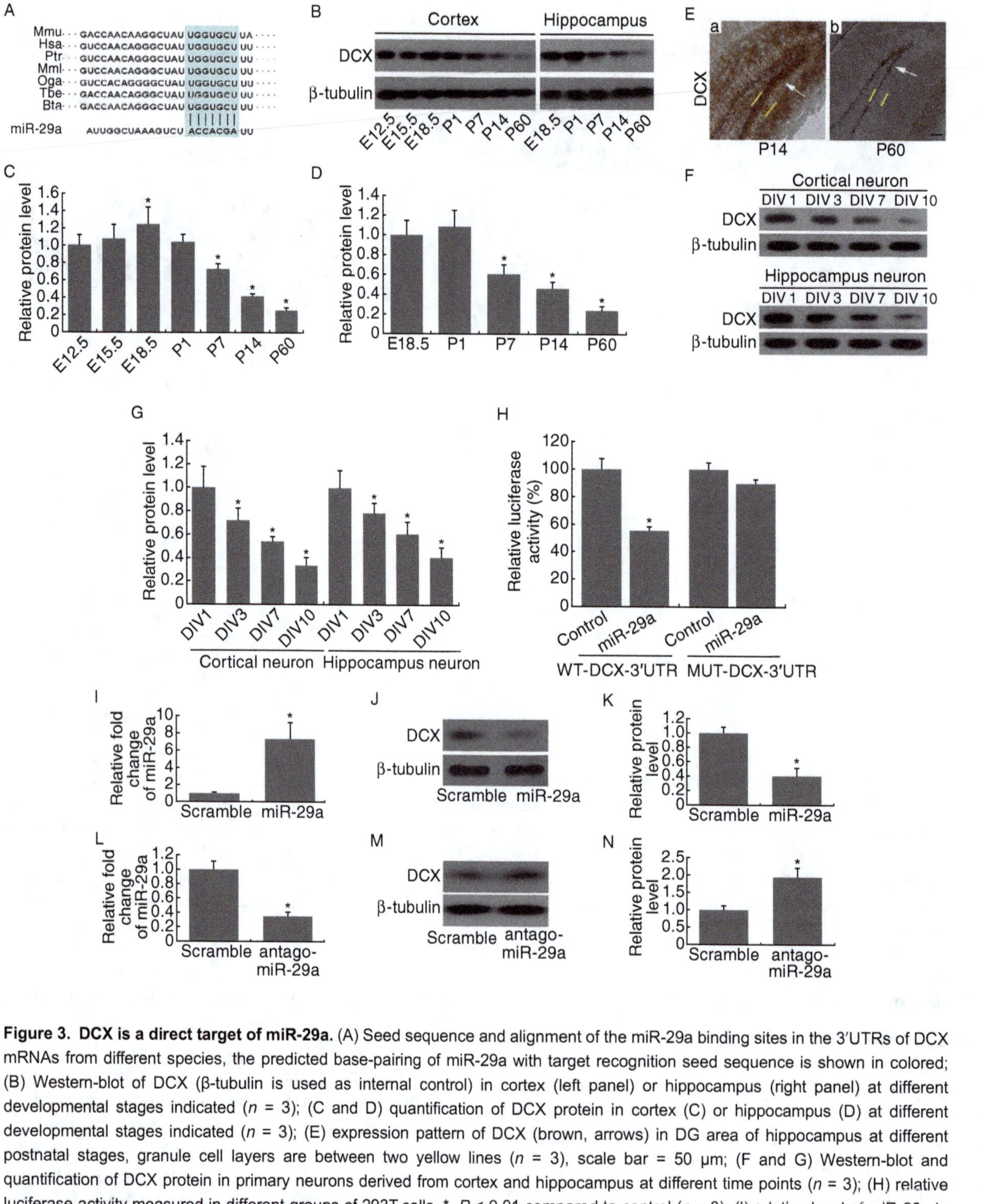

Figure 3. DCX is a direct target of miR-29a. (A) Seed sequence and alignment of the miR-29a binding sites in the 3′UTRs of DCX mRNAs from different species, the predicted base-pairing of miR-29a with target recognition seed sequence is shown in colored; (B) Western-blot of DCX (β-tubulin is used as internal control) in cortex (left panel) or hippocampus (right panel) at different developmental stages indicated (n = 3); (C and D) quantification of DCX protein in cortex (C) or hippocampus (D) at different developmental stages indicated (n = 3); (E) expression pattern of DCX (brown, arrows) in DG area of hippocampus at different postnatal stages, granule cell layers are between two yellow lines (n = 3), scale bar = 50 μm; (F and G) Western-blot and quantification of DCX protein in primary neurons derived from cortex and hippocampus at different time points (n = 3); (H) relative luciferase activity measured in different groups of 293T cells, *, P < 0.01 compared to control (n = 3), (I) relative level of miR-29a in primary cortical neurons transfected with miR-29a mimics, *, P < 0.01 compared to scramble (n = 3); (J and K) Western-blot and quantification of DCX protein in primary cortical neurons treated as in G, *, P < 0.01 compared to scramble (n = 3); (L) relative level of miR-29a in primary cortical neurons transfected with antago-miR-29a, *, P < 0.01 compared to scramble (n = 3); (M, N) Western-blot and quantification of DCX protein in primary cortical neurons treated as in J, *, P < 0.01 compared to scramble (n = 3).

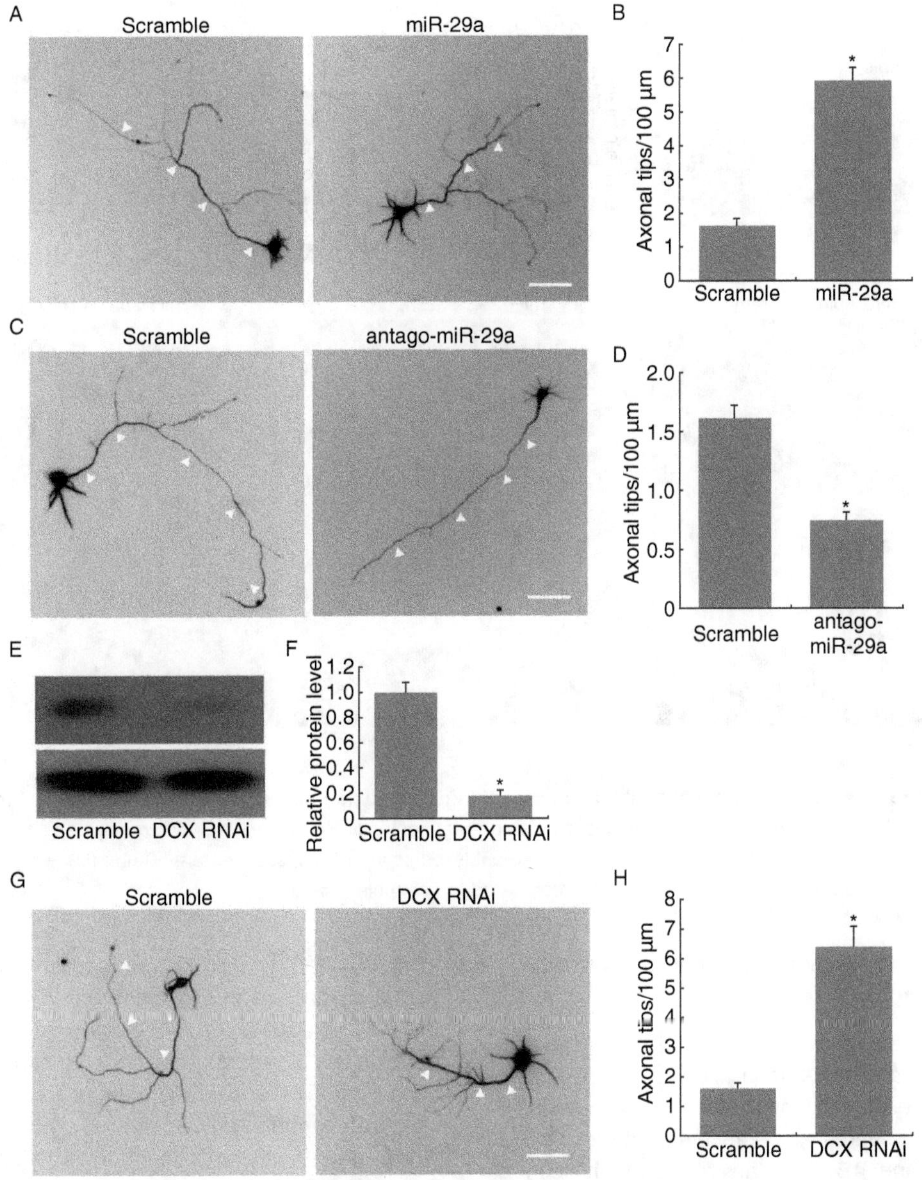

Figure 4. Effects of miR-29a on axon branching of primary cortical neurons. (A) Cortical neurons transfected with scramble or miR-29a mimics together with the eGFP expression plasmid were fixed 4 d after transfection and subjected to immunocytochemistry using the eGFP antibody. In all images of this type, arrowheads indicate primary axons, scale bars = 50 μm; (B) quantification of axon branching of cortical neurons treated in A, *, $P < 0.01$ compared to scramble (n = 20); (C) cortical neurons transfected with scramble or antago-miR-29a together with the eGFP expression plasmid were analyzed as in A, scale bars = 50 μm; (B) quantification of axon branching of cortical neurons treated in C, *, $P < 0.01$ compared to scramble (n = 20); (E, F) Western-blot and quantification of DCX protein in cortical neurons treated with scrambles or DCX RNAi (n = 3); (G) cortical neurons transfected with scrambles or DCX RNAi together with the eGFP expression plasmid were analyzed as in A, scale bars = 50 μm; (H) quantification of axon branching of cortical neurons treated in G, *, $P < 0.01$ compared to scramble (n = 20).

blocked the inhibitory effect (Fig. 3H). Next, through gain- and loss-of-function approaches, we evaluated the effect of miR-29a on DCX expression in primary neurons. Over-expression of miR-29a resulted in a remarkable decrease of DCX protein level in primary neurons (Fig. 3I–K). Meanwhile, antago-miR-29a reduced cellular miR-29a level and increased DCX protein expression in cultured neurons (Fig. 3L–N). Together, these evidences support the notion that miR-29a inhibits DCX expression by directly targeting its 3′UTR of mRNA.

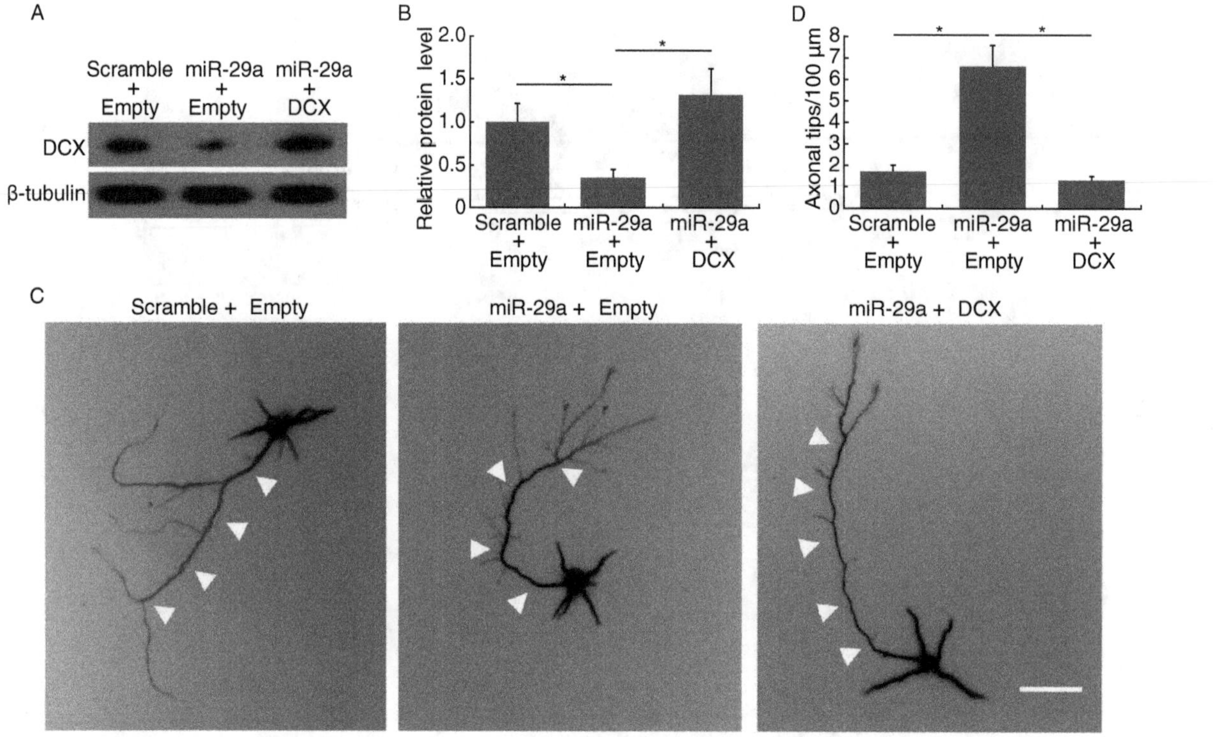

Figure 5. Over-expression of DCX rescues the axon branching phenotype induced by miR-29a. (A and B) Western-blot and quantification of DCX protein in cortical neurons under different treatment indicated ($n = 3$); (C) different groups of cortical neurons transfected together with the eGFP expression plasmid were analyzed as in Fig. 4A, scale bars = 50 μm; (D) quantification of axon branching of cortical neurons treated in C, *, $P < 0.01$ compared to scramble ($n = 20$).

MiR-29a increases axon branching in primary cortical neuron via targeting DCX

Since DCX is reported to restrict axon branching we further investigated the role of miR-29a in axon growth of primary neurons. We over-expressed miR-29a in cortical neurons at DIV 3 and found it triggered exuberant branching of axons at DIV 7 (Fig. 4A and 4B). On the contrary, inhibition of endogenous miR-29a by antago-miR resulted in long axons bearing few or no branches of primary neurons (Fig. 4C and 4D). In addition, to elucidate the role of DCX in miR-29a induced axon branching, we specifically knocked down the endogenous DCX protein by RNAi (Fig. 4E and 4F). In the morphology assay, we found that DCX knockdown also induced excessive axonal branches compared with control neurons (Fig. 4G and 4H). Furthermore, we also over-expressed DCX (without miR-29 binding site) and demonstrated an effective inhibition of the axon branching induced by miR-29a over-expression (Fig. 5). These results indicate that miR-29a may increase axon branching in primary neuron by repressing DCX expression.

DISCUSSION

MiRNAs are small, non-coding transcripts that functioned in many tissues by repressing target gene expression (Chen et al., 2012). Circumstantial evidences suggest that miRNAs play a crucial role in the central nervous system (Ji et al., 2013). To elucidate the role of miRNAs in cerebrum development, we analyzed the miRNAs expression pattern in mice cerebrum and found several miRNAs that increase significantly during different developmental stages, among which miR-29a changed most significantly. The expression level of miR-29a showed no significant differences until postnatal stages when it increased dramatically both in cortex and hippocampus, indicating miR-29a may be associated with the process of neural maturation.

Since it is reported that miR-29 expressed more strongly in cultured astrocyte than in neuron (Smirnova et al., 2005), we investigated the expression pattern of miR-29 *in vivo* by *in situ* hybridization. Our results demonstrated that miR-29a mainly expressed in mature neuron in cerebrum while miR-29b and miR-29c showed relatively weak signals. These discrepancies may be the result of differences between *in vitro* and *in vivo* as well as the techniques with different sensitivity used for detecting miRNAs. In addition, we provided evidence that the expression pattern of miR-29a in DG area is similar to that of NeuN (mature neuronal marker), which suggested that miR-29a is increased during neuron maturation. Furthermore, we showed a time-course increase of miR-29a expression in primary neurons. Together, these data indicate that miR-29a may play an important role in mature neurons.

Since neural stimulation helped to fine-tune the process of neuronal maturation (He et al., 2010), we further investigated the association between miR-29a expression and neural activity. Our results demonstrated that activation of glutamate receptors induced the endogenous level of miR-29a in primary neurons indicating that neural activity may modulate the expression of miR-29a.

Through bioinformatics screening and luciferase assay, we searched and verified one target gene of miR-29a—DCX. DCX initially expressed in neuronal precursor cells and immature neurons (E12.5) while it began to down-regulate around P7, which is coincident with the time that miR-29a aroused in cerebrum (Spampanato et al., 2012). Immunostaining results also revealed that DCX expressed in a complementary pattern compared with the expression of miR-29a in DG area. Besides, protein analysis also demonstrated that miR-29a inhibited DCX translation in primary neurons.

As a target protein of miR-29a, DCX is a microtubule associated protein that stabilized microtubules (Cohen et al., 2008). There is evidence that GSK3β/DCX signaling pathway restricts axon branching (Bilimoria et al., 2010). In addition, neural activity that has been proved to affect miR-29a expression is also reported to regulate axon development (Uesaka et al., 2005). Therefore, it is necessary for us to evaluate the effect of miR-29a on axon growth. Our data suggested that miR-29a triggered excessive axon branching while blocking endogenous miR-29a resulted in the opposite effect. Moreover, DCX RNAi also promoted axon branching of cortical neurons, which is similar to the effect of miR-29a over-expression. MiR-29 has multiple targets that may affect neuronal growth (Govek et al., 2005; Wang et al., 2013). To further confirm the role of DCX in the miR-29a regulated axon branching, we over-expressed DCX and found it effectively rescued the axon branching phenotype induced by miR-29a over-expression. Together, these results suggest that miR-29a modulated axon branching via targeting DCX expression.

Taken together, our data demonstrate that activation of glutamate receptors increased the expression of miR-29a, which further promoted axon branching in the maturation of cerebral neurons by targeting DCX. Whether miR-29a-DCX regulation pathway is involved in the neural activity dependent axon growth *in vivo* needs further investigation.

MATERIAL AND METHODS

Primary culture

Animal care and procedures were performed in accordance with the Laboratory Animal Care Guidelines approved by Model Animal Research Center of Nanjing University. Primary neurons were cultured from cortex of E16.5 mice as described previously (David et al., 2007). Neurons were seeded at 3×10^5 per well in 12 well plate for biochemistry analysis, or at 1×10^4 per well for immunofluorescent staining and morphological observation. Half medium was refreshed every 2 d.

Cultured neurons (DIV 3) were transfected with miR-29a mimics (Genepharma), antago-miR-29a (RiboBio Inc) or DCX RNAi (Genepharma, described in (Ramos et al., 2006)) together with eGFP reporter plamid using lipofectamine 2000 (Invitrogen) according to the manufacturer's protocol. Cells were harvested 24 h later for RNA analysis and 48 h later for protein analysis. For morphological study, cells were fixed 4 d after transfection.

Neurons (DIV 3) were subjected to the stimulation of MSG (10 μmol/L) for 15 min. After that RNA samples were collected at the time indicated. For the inhibition of glutamate receptor, cells were pretreated with MK-801 (10 μmol/L, Sigma) or/and DNQX (10 μmol/L, Sigma) for 2 h before subjecting to MSG stimulation.

RNA and protein quantification

Mice cortex and hippocampus were dissected from C57BL/6 J mouse at different stages indicated. Total RNA was purified from the cultures or tissues with TRIZOL (Invitrogen) according to the manufacturer's instructions. The RNA purity was determined by the OD_{260}/OD_{280} ratio, and the concentration was calculated based on OD_{260}. 50 μg of total RNA of cortexes from different stages (E12.5, E15.5, E18.5 and P60) were sent to the Capitalbio Corporation for miRNAs array analysis.

Q-RTPCR was carried out using TaqMan microRNA probes (Applied Biosystems). Briefly, 2 μg of total RNA was reverse-transcribed to cDNA using AMV reverse transcriptase (TaKaRa) and stem-loop RT primers (Applied Biosystems). Real-time PCR was performed using a TaqMan PCR kit on an Applied Biosystems 7300 Sequence Detection System (Applied Biosystems). All reactions, including the no-template controls, were run in triplicate. After the reactions, the *CT* values were determined using fixed-threshold settings. Upon the evaluation of each miRNA's level, the expression level of U6 was used as internal control.

Cultures or tissues were both lysed in 50 mmol/L Tris, pH 7.5, 150 mmol/L NaCl, 2 mmol/L EDTA, and 1% Triton X-100. The protease inhibitors aprotinin, pepstatin, leupeptin, as well as the reducing agent dithiothreitol were added to this buffer before cell lysis. Lysates were cleared of insoluble material by spinning at maximum speed on a tabletop centrifuge, boiled in sample buffer, and examined using standard SDS-PAGE, followed by Western blotting. The antibody used were DCX (Abcam) and β-tubulin (Abcam).

In situ hybridization

This method was adjusted based on the protocol described by Obernosterer et al (Obernosterer et al., 2007). Brains were fixed in 4% paraformaldehyde overnight at 4°C and cryo protected in 15% sucrose followed by 30% sucrose overnight each. Cryostat section (20 μm) were cut and mounted onto Superfrost Gold Plus slide (Fisher). Then sections were post-fixed in 4% paraformaldehyde followed by washed in PBS for three times and 75% alcohol for one time, dried in RT, and incubated in hybridization buffer (DEPC H_2O with 50% de-ionized formamide (*v/v*), 5× SSC, 0.2 mg/mL yeast tRNA, 0.5 mg/mL herring sperm DNA, 10% dextran sulfate (*w/v*), 1/1000 Tween-20) with a concentration of 10 nmol/L for digoxigenin-labeled LNA probes (Exiqon) overnight at 50°C in a humidified chamber. Following hybridization, slides were washed in 0.1× SSC at 55°C for 3 times, 30 min each, then blocked in 10% FBS in TBS (pH = 7.5) for 1 h. AP-conjuncted-anti-digoxigenin antibody (Roche) was incubated with slides overnight at final concentration of 1:1000. NBT/BCIP kit (Invitrogen) was used to do the color reaction.

Plasmid construction and luciferase assay

Part of mmu-DCX 3′UTR containing the predicted target sequence was obtained from cortex of P1 mouse using RT-PCR with primers as indicated. Forward primer 5′-AAGATGATAGGCTAAATCAAAGCC-3′, reverse primer 5′-CAGGATGAATGCTAGACAATACAC-3′. This PCR product was inserted into pMIR-REPORTTM-Luciferase vector (Ambion) after digested with *Hin*dIII and *Sac*I (Takara) to construct WT vector. Mutant vector was generated in Invitrogen by replacing the predicted target region with its reverse sequence (from AAGCAC CAATAGCCTTGTTGGTC to CTGGTTGTTCCGATAACCACGAA). The oligonucleotides used for the RT-PCR cloning of DCX without miR-29 binding site: forward primer 5′-ggttccaccaaaatATGg (ATG initiator), reverse primer 5′-TCAcatggaatcgccaag (TGA terminator codon).

293T cell was cultured in DMEM (Gibco) containing 10% FBS (Gibco) and seeded in 24 well plate. 12 h after plating, 0.2 μg of firefly luciferase reporter plasmid, 0.2 μg of β-galactosidase expression vector (Ambion), and equal amounts (20 pmol) of miR-29a mimics or scrambled negative control RNA were transfected into cells with Lipofectamine 2000 as manufacturer's instructing. The β-galactosidase vector was used as a transfection control. At 24 h post-transfection, cells were analyzed using a luciferase assay kit (Promega).

Immunochemistry

Primary cortical neurons were briefly washed twice with cold PBS, and then fixed in 4% paraformaldehyde for 10 min at room temperature (RT). Neurons after fixation were washed with cold PBS three times, permeabilized and blocked with 2% BSA and 0.05% Triton X-100 in PBS for 1 h (RT). Next, neurons were incubated with primary antibodies (anti-eGFP, Abcam) in 2% BSA/PBS in a humidified chamber overnight at 4°C and rinsed three times in PBS (3 × 5 min). This was followed by incubation with secondary antibodies (Invitrogen) in 2% BSA/PBS in a light-proof container (2 h at RT).

Slides were washed and blocked with PBS containing 0.3% Triton X-100 and 5% BSA (Sigma), followed with the incubation of primary antibodies (NeuN, Millipore; DCX, Abcam) overnight at 4°C. Sections were washed three times in PBS and then incubated with biotinylated secondary antibody (1:500, Vector Laboratories, Burlingame, California, USA) overnight at 4°C. After that signals were further amplified using peroxidase Vectastain ABC reagent (Vector Laboratories). 3,3′-diaminobenzidine (Sigma) was used for the color reaction.

Morphological analysis

Images were captured on an Olympus microscope and analyzed using Necromantic software. Axon tips were defined as terminal points of the primary axon and all of its branches.

Statistical Analysis

Data are presented as sample means and their standard errors. Two-tailed Student's t test was used for comparison of the means between any two groups. Differences were considered significant when $P < 0.05$.

ACKNOWLEDGEMENTS

This work was supported by grants from the National Natural Science Foundation of China (Grant Nos. 31000478, 31100777 and 31271378).

COMPLIANCE WITH ETHICS GUIDELINES

Hanqin Li, Susu Mao, Haitao Wang, Ke Zen, Chenyu Zhang and Liang Li declare that they have no conflict of interest.

All institutional and national guidelines for the care and use of laboratory animals were followed.

REFERENCES

Bettens K, Brouwers N, Engelborghs S, Van Miegroet H, De Deyn PP, Theuns J, Sleegers K, Van Broeckhoven C (2009) APP and BACE1 miRNA genetic variability has no major role in risk for Alzheimer disease. Hum Mutat 30:1207–1213

Bilimoria PM, de la Torre-Ubieta L, Ikeuchi Y, Becker EB, Reiner O, Bonni A (2010) A JIP3-regulated GSK3beta/DCX signaling pathway restricts axon branching. J Neurosci 30:16766–16776

Chen X, Ba Y, Ma L, Cai X, Yin Y, Wang K, Guo J, Zhang Y, Chen J, Guo X et al (2008) Characterization of microRNAs in serum: a novel class of biomarkers for diagnosis of cancer and other diseases. Cell Res 18:997–1006

Chen X, Liang H, Zhang J, Zen K, Zhang CY (2012) Horizontal transfer of microRNAs: molecular mechanisms and clinical applications. Protein Cell 3:28–37

Cohen D, Segal M, Reiner O (2008) Doublecortin supports the development of dendritic arbors in primary hippocampal neurons. Dev Neurosci 30:187–199

David A, Tiveron MC, Defays A, Beclin C, Camosseto V, Gatti E, Cremer H, Pierre P (2007) BAD-LAMP defines a subset of early endocytic organelles in subpopulations of cortical projection neurons. J Cell Sci 120:353–365

Govek EE, Newey SE, Van Aelst L (2005) The role of the Rho GTPases in neuronal development. Genes Dev 19:1–49

He S, Ma J, Liu N, Yu X (2010) Early enriched environment promotes neonatal GABAergic neurotransmission and accelerates synapse maturation. J Neurosci 30:7910–7916

Ji F, Lv X, Jiao J (2013) The role of microRNAs in neural stem cells and neurogenesis. J Genet Genomics 40:61–66

Kataoka Y, Takeichi M, Uemura T (2001) Developmental roles and molecular characterization of a Drosophila homologue of Arabidopsis Argonaute1, the founder of a novel gene superfamily. Genes Cells 6:313–325

Kawase-Koga Y, Otaegi G, Sun T (2009) Different timings of Dicer deletion affect neurogenesis and gliogenesis in the developing mouse central nervous system. Dev Dyn 238:2800–2812

Kole AJ, Swahari V, Hammond SM, Deshmukh M (2011) miR-29b is activated during neuronal maturation and targets BH3-only genes to restrict apoptosis. Genes Dev 25:125–130

Obernosterer G, Martinez J, Alenius M (2007) Locked nucleic acid-based in situ detection of microRNAs in mouse tissue sections. Nat Protoc 2:1508–1514

Ramos RL, Bai J, LoTurco JJ (2006) Heterotopia formation in rat but not mouse neocortex after RNA interference knockdown of DCX. Cereb Cortex 16:1323–1331

Shioya M, Obayashi S, Tabunoki H, Arima K, Saito Y, Ishida T, Satoh J (2010) Aberrant microRNA expression in the brains of neurodegenerative diseases: miR-29a decreased in Alzheimer disease brains targets neurone navigator 3. Neuropathol Appl Neurobiol 36:320–330

Shu SY, Qing D, Wang B, Zeng QY, Chen YC, Jin Y, Zeng CC, Bao R (2013) Comparison of microRNA expression in hippocampus and the marginal division (MrD) of the neostriatum in rats. J Biomed Sci 20:9

Smirnova L, Grafe A, Seiler A, Schumacher S, Nitsch R, Wulczyn FG (2005) Regulation of miRNA expression during neural cell specification. Eur J Neurosci 21:1469–1477

Snyder JS, Ferrante SC, Cameron HA (2012) Late maturation of adult-born neurons in the temporal dentate gyrus. PLoS ONE 7: e48757

Spampanato J, Sullivan RK, Turpin FR, Bartlett PF, Sah P (2012) Properties of doublecortin expressing neurons in the adult mouse dentate gyrus. PLoS ONE 7:e41029

Uesaka N, Hirai S, Maruyama T, Ruthazer ES, Yamamoto N (2005) Activity dependence of cortical axon branch formation: a morphological and electrophysiological study using organotypic slice cultures. J Neurosci 25:1–9

Wang Y, Zhang X, Li H, Yu J, Ren X (2013) The role of miRNA-29 family in cancer. Eur J Cell Biol 92:123–128

Yang D, Li T, Wang Y, Tang Y, Cui H, Tang Y, Zhang X, Chen D, Shen N, Le W (2012) miR-132 regulates the differentiation of dopamine neurons by directly targeting Nurr1 expression. J Cell Sci 125:1673–1682

Yu JY, Chung KH, Deo M, Thompson RC, Turner DL (2008) MicroRNA miR-124 regulates neurite outgrowth during neuronal differentiation. Exp Cell Res 314:2618–2633

18

Dual phosphorylation of Sin1 at T86 and T398 negatively regulates mTORC2 complex integrity and activity

Pengda Liu, Jianping Guo, Wenjian Gan, Wenyi Wei✉

Department of Pathology, Beth Israel Deaconess Medical Center, Harvard Medical School, Boston, MA 02115, USA
✉ Correspondence: wwei2@bidmc.harvard.edu (W. Wei)

Mammalian target of rapamycin (mTOR) plays essential roles in cell proliferation, survival and metabolism by forming at least two functional distinct multi-protein complexes, mTORC1 and mTORC2. External growth signals can be received and interpreted by mTORC2 and further transduced to mTORC1. On the other hand, mTORC1 can sense inner-cellular physiological cues such as amino acids and energy states and can indirectly suppress mTORC2 activity in part through phosphorylation of its upstream adaptors, IRS-1 or Grb10, under insulin or IGF-1 stimulation conditions. To date, upstream signaling pathways governing mTORC1 activation have been studied extensively, while the mechanisms modulating mTORC2 activity remain largely elusive. We recently reported that Sin1, an essential mTORC2 subunit, was phosphorylated by either Akt or S6K in a cellular context-dependent manner. More importantly, phosphorylation of Sin1 at T86 and T398 led to a dissociation of Sin1 from the functional mTORC2 holo-enzyme, resulting in reduced Akt activity and sensitizing cells to various apoptotic challenges. Notably, an ovarian cancer patient-derived Sin1-R81T mutation abolished Sin1-T86 phosphorylation by disrupting the canonical S6K-phoshorylation motif, thereby bypassing Sin1-phosphorylation-mediated suppression of mTORC2 and leading to sustained Akt signaling to promote tumorigenesis. Our work therefore provided physiological and pathological evidence to reveal the biological significance of Sin1 phosphorylation-mediated suppression of the mTOR/Akt oncogenic signaling, and further suggested that misregulation of this process might contribute to Akt hyper-activation that is frequently observed in human cancers.

mTORC1 AND mTORC2 ARE FUNCTIONALLY DISTINCT mTOR-CONTAINING PROTEIN KINASE COMPLEXES

Mammalian target of rapamycin (mTOR) is a highly conserved protein kinase (van Dam et al., 2011) belonging to the phosphatidylinositol-3-kinase-related protein kinase (PIKK) super-family. Biologically, mTOR is a master regulator of cellular homeostasis, cell growth and proliferation as well as metabolism in a broad range of physiological (Wullschleger et al., 2006; Zoncu et al., 2011b) and pathological settings, including diabetes and cancer (Guertin and Sabatini, 2007; Inoki et al., 2005). In doing so, mTOR serves as a central sensor for cellular physiological cues such as growth signals and nutrients, by functioning as an essential catalytic subunit in two functionally distinct complexes, mTOR complex 1 (mTORC1) (Kim et al., 2002) and mTOR complex 2 (mTORC2) (See (Sabatini, 2006) and (Wullschleger et al., 2006) for review). These two complexes are distinguished by their accessory protein composition, localization and function. Specifically, both mTORC1 and mTORC2 share the common components mTOR and GβL/mLST8, while the unique subunit Raptor defines mTORC1 (Kim et al., 2002) whereas Rictor (Sarbassov et al., 2004) and Sin1 (Frias et al., 2006; Guertin et al., 2006; Jacinto et al., 2006; Wullschleger et al., 2005; Yang et al., 2006) define mTORC2. mTORC1 was reported to localize on the outer membrane of lysosome, which is critical for its activation and physiological function (Bar-Peled and Sabatini, 2012; Thoreen et al., 2012), while mTORC2 association with ribosomes was recently demonstrated to be critical for mTORC2 activity (Zinzalla et al., 2011).

In echoing their unique subcellular localizations, mTORC1 and mTORC2 display distinct cellular functions by targeting different sets of downstream effectors for phosphorylation and function. Notably, mTORC1 promotes cap-dependent mRNA translation (Thoreen et al., 2012) and protein synthesis through direct phosphorylation and activation of its *bona fide* substrates S6K1, 4EBP1 and TFEB1 (Ma and Blenis, 2009; Pena-Llopis et al., 2011), inhibits autophagy through phosphorylation of Ulk1 (Chan, 2009) and establishes Treg-cell function by facilitating cholesterol and lipid metabolism through yet undefined phosphorylation targets (Zeng et al., 2013). On the other hand, mTORC2 was firstly identified to regulate cellular skeletal organization (Jacinto et al., 2004) and later shown to be indispensible in governing cell growth, proliferation, survival and anabolism, which are mainly through direct phosphorylation and activation of its physiological downstream targets including Akt (S473) (Sarbassov et al., 2005), SGK (S422) (Garcia-Martinez and Alessi, 2008) and PKCα (Ikenoue et al., 2008). Moreover, as mTOR serves as an oncogenic pathway to promote cellular growth and survival, deregulation of many components of the mTOR pathway has been implicated in human cancer and metabolic diseases (Weber and Gutmann, 2012; Zoncu et al., 2011b).

mTORC1 AND mTORC2 ARE DIFFERENTIALLY REGULATED IN CELLS

Previous work clearly established that both activation of mTORC1 and mTORC2 are tightly, yet differentially controlled (Laplante and Sabatini, 2012; Weber and Gutmann, 2012). Mechanistically, when stimulated by extra-cellular growth signals, mTORC2 receives activation signals from the Ras/PI3K signaling pathway through undefined mechanisms, and activates mTORC1 by Akt-dependent phosphorylation of TSC2 (Inoki et al., 2002; Manning et al., 2002) or PRAS40 (Vander Haar et al., 2007), releasing their repression on mTORC1. In addition, mTORC1 can also sense cellular energy states and amino acid levels independent of mTORC2. Specifically, when cellular AMP levels are high, which indicates low energy status in cells, the LKB1/AMPK (AMP-activated serine/threonine protein kinase) pathway becomes activated, leading to inhibition of mTORC1 (Shaw, 2009), either by phosphorylation of TSC2(Inoki et al., 2003) or by phosphorylation of the essential mTORC1 component Raptor (Gwinn et al., 2008). Furthermore, mTORC1, but not mTORC2, serves as an inner cellular amino acid sensor (Efeyan et al., 2012) dependent on its recruitment to lysosome surface by Rag (Sancak et al., 2010; Sancak et al., 2008), Ragulator (Bar-Peled et al., 2012) and the lysosome vacuolar H^+-ATPase (Zoncu et al., 2011a) through an inside-out mechanism to control the timely activation of mTORC1.

The differential regulatory mechanisms of these two mTOR-containing complexes extend further to their altered responses to rapamycin. mTORC1 is sensitive to low-dose rapamycin treatment in both cell culture and mouse models, where rapamycin directly binds to FKBP12 and disrupts the interaction between Raptor and mTOR, suppressing mTORC1 assembly and activation (Oshiro et al., 2004; Yip et al., 2010). On the other hand, mTORC2 only responds to prolonged and chronic rapamycin treatment, in part because rapamycin cannot directly interfere with existing mTORC2 complex, but rather only blocks the assembly of mTORC2 from newly synthesized Rictor and mTOR (Sarbassov et al., 2006).

More importantly, mTORC1 could also be activated independent of mTORC2. Mice with deleted *S6K1* display elevated resistance to high-fat diet and weight gain, in part due to deficiency in adipocytes generation (Carnevalli et al., 2010). Adipose specific *Raptor* knockout mice phenocopied the *S6K1* knockout mice with regard to adipocyte generation (Polak et al., 2008), highlighting the critical role of mTORC1 in adipogenesis *in vivo*. However, *Rictor* knockout mice showed no defects in adipogenesis (Kumar et al., 2010), although an mTORC2 substrate, BTSA (a BSD domain-containing protein) has been characterized to be indispensible for adipogenesis (Kumar et al., 2010). Taken together, mTORC1 and mTORC2 could function synergistically or independently to maintain cell physiology.

mTORC1 NEGATIVELY REGULATES mTORC2 ACTIVATION THROUGH INDIRECT MECHANISMS MEDIATED BY PHOSPHORYLATION OF IRS-1 AND GRB10

mTORC1 has also been shown to indirectly suppress mTORC2 signaling that in *TSC2*$^{-/-}$ cells with elevated mTORC1/S6K activity, Akt activation was significantly reduced (Manning et al., 2005). Further studies revealed that elevated mTORC1/S6K could suppress the activation of the PI3K pathway through multiple negative feedback loops (Gual et al., 2005). Specifically, mTORC1 and S6K can phosphorylate IRS-1 to block its interaction with the p85 regulatory subunit of PI3K to negatively regulate the insulin-signaling pathway (Gual et al., 2005). More recently, both the Blennis (Yu et al., 2011) and Sabatini (Hsu et al., 2011) groups independently showed that Grb10 is phosphorylated and stabilized by mTORC1 to block the PI3K signaling. However, in response to certain stimuli such as PDGF and EGF, activation of mTORC2/Akt was not affected by depletion of Grb10, suggesting the presence of an additional mechanism for mTORC1 to suppress mTORC2. As genetic disruption of either *Rictor* or *Sin1* results in the inactivation of mTORC2 (Guertin et al., 2006; Jacinto et al., 2006), we reasoned that mTORC1/S6K-dependent regulation of mTORC2 might occur through these unique mTORC2 components in a direct manner. However, we and others previously showed that AGC kinases-mediated phosphorylation of Rictor at T1135 does not significantly affect mTORC2 complex integrity and its kinase activity (Dibble et al., 2009; Gao et al., 2010), which urged us to further examine whether the other mTORC2 essential component,

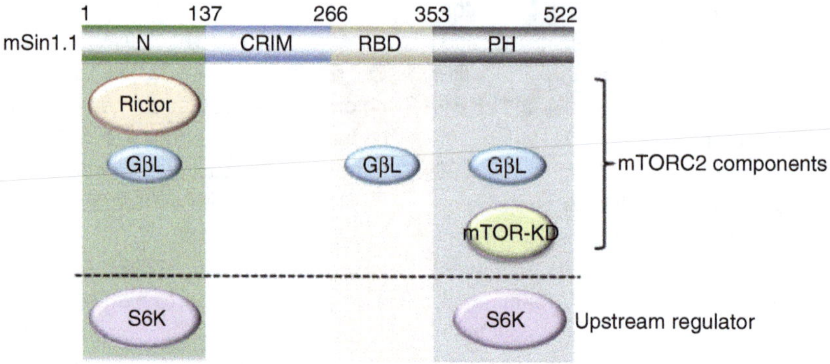

Figure 1. mTORC2 components and S6K interact with different domains of Sin1.

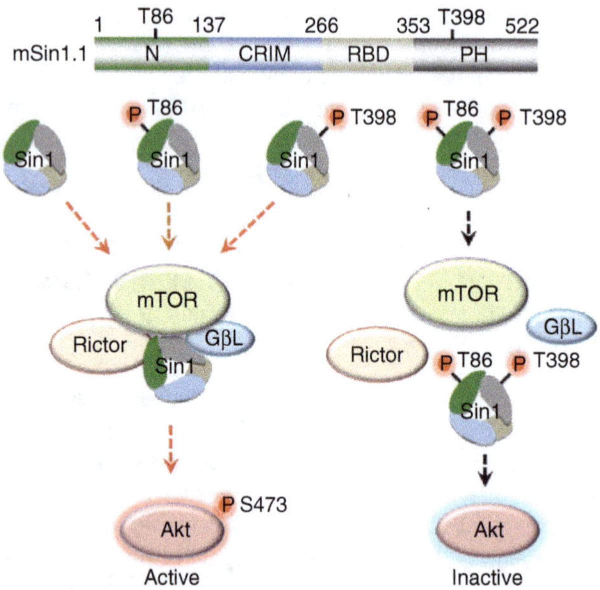

Figure 2. Phosphorylation of Sin1 at both T86 and T398 sites dissociates Sin1 from functional mTORC2 complex and subsequently suppresses Akt activation.

Sin1, is a major target to mediate mTORC1/S6K's negative regulation of mTORC2.

mTORC1 NEGATIVELY REGULATES mTORC2 ACTIVATION THROUGH DIRECT MECHANISMS MEDIATED BY PHOSPHORYLATION OF SIN1

To test if modulation of Sin1 is important for mTORC2 activity, we and others tested if Sin1 could be targeted for phosphorylation by Akt or S6K at either T86 and T398, or both sites, in adipocytes (Humphrey et al., 2013) or epithelial cells (Liu et al., 2013), respectively. Importantly, we found that phosphorylation of Sin1 on both the T86 and T398 sites led to a disassociation of Sin1 from other mTORC2 components, revealing a direct negative regulatory mechanism for mTORC2 governed by mTORC1 (Liu et al., 2013). Mechanistically, we observed that Sin1 utilized various domains to interact with mTORC2 components (Fig. 1), with its N-terminus binding to Rictor and GβL, RBD (Ras binding domain) binding to GβL, and the PH domain interacting with GβL and mTOR-KD (kinase domain). Interestingly, only the N and PH domains were capable of interacting with S6K1 (Fig. 1). Considering that the identified T86 and T398 phosphorylation sites are located in the N- and PH domains, respectively, it is therefore plausible that S6K could directly phosphorylate these two sites to influence mTORC2 integrity. In keeping with this notion, phosphorylation at T86 interfered with Sin1-N-terminus binding to Rictor, while phosphorylation at T398 impaired Sin1-PH domain interaction with the mTOR kinase domain. Importantly, Sin1 dissociation from the mTORC2 complex requires both phosphorylation events (Fig. 2). In addition, we demonstrated that Sin1 phosphorylation at both sites occurred in response to various external cellular stimuli including insulin, IGF1, PDGF and EGF, which significantly blocked the activation of Akt induced by these stimuli, proposing a novel negative feedback loop independent of IRS-1 and Grb10 to suppress mTORC2.

Physiologically, we demonstrated that Sin1 phosphorylation-mediated mTORC2 suppression on Akt led to increased levels of cleaved caspase 3 and subsequently sensitized cells to apoptosis-initiating agents such as etoposide or cisplatin. More importantly, we found that in freshly isolated splenic B cells, rapamycin treatment significantly reduced Sin1-T86 phosphorylation accompanied by an increase in Akt-S473 phosphorylation. A similar inverse correlation between phosphorylation of Sin1-T86 and Akt-S473 was also observed in mouse liver lysates, and rapamycin administration to whole mice led to reduced Sin1-pT86 concomitant with increased Akt-pS473 signals, further confirming the physiological significance of mTORC1-mediated negative regulation on mTORC2 function.

Pathologically, we screened a panel of human ovarian patient clinical samples by IHC and an inverse correlation between Sin1-pT86 and Akt-pS473 was observed in a certain number of cases. However, due to the limited number of

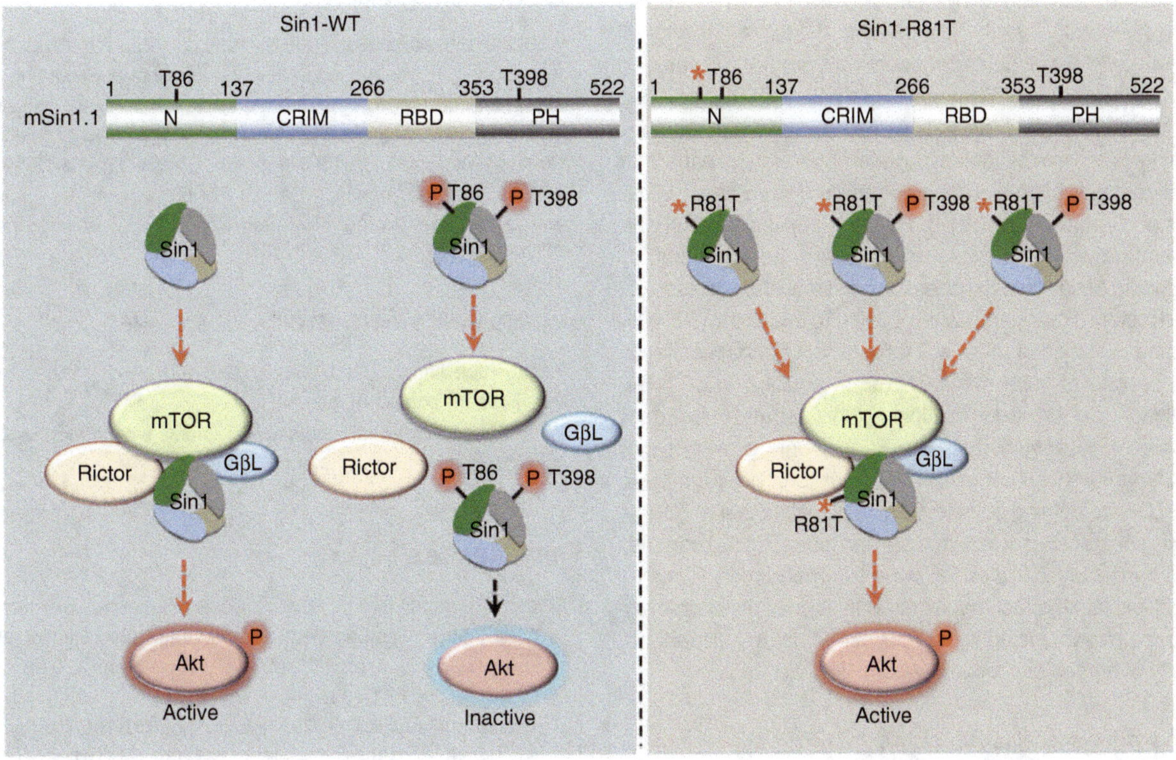

Figure 3. The cancer patient-derived Sin1-R81T mutation leads to stabilized mTORC2 integrity and sustained Akt activation to facilitate tumorigenesis.

patient samples available, the inverse correlation did not reach statistical significance, which warrants further larger-scale examinations. This inverse correlation was also observed in a panel of T-ALL cancer cell lines. Interestingly, two Sin1 mutations (R81T and S84L) were identified in ovarian cancer patients. Using biochemical assays we demonstrated that both of these mutations led to significantly reduced Sin1 phosphorylation at the T86 site, as these mutations impair the canonical AGC kinase consensus recognition motif "RxRxxpS/pT" (Alessi et al., 1996; Manning and Cantley, 2007). As phosphorylation at both T86 and T398 is necessary to dissociate Sin1 from the functional mTORC2 complex, substitutions at R81T or S84L diminished the possibility of co-occurrence of pT86 and pT398, resulting in stabilized mTORC2 integrity and function (Fig. 3). Consistent with this notion, compared to Sin1-WT, relatively sustained Akt phosphorylation was observed in Sin1-R81T expressing cells. At the cellular levels, expressing Sin1-R81T, opposing to the phospho-mimetic Sin1-T86E/T398E mutant, led to comparable levels of cleaved caspase 3 to cells expressing Sin1-WT upon etoposide or cisplatin challenge, subsequently conferring cellular resistance to these agents. More importantly, compared to Sin1-WT, Sin1-R81T expressing cells gained oncogenic ability to promote ovarian cancer cell growth in soft agar as well as in a xenograft mouse model. Together, these data consistently support a model that Sin1 phosphorylation plays critical roles in inhibiting mTORC2 oncogenic function in both physiological and pathological settings.

IMPLICATIONS OF mTORC1-MEDIATED SUPPRESSION OF mTORC2

Compared to the extensively studied regulations of the mTORC1 complex, the upstream regulatory signaling pathways governing mTORC2 activity remained largely to be determined. So far, the only defined upstream regulation of mTORC2 is the requirement for ribosome association for its activation, where the mechanisms remain largely uncharacterized (Zinzalla et al., 2011). As S6K and 4EBP-1, two characterized mTORC1 substrates, are indispensible for translation of capped mRNAs by ribosome, there might be some yet-to-be identified positive regulatory events governed by S6K or 4EBP-1, which are vital for mTORC2 activation on ribosome. Furthermore, given that both mTORC1 and mTOCR2 contain mTOR, it is not surprising that many layers of crosstalks might exist between these two complexes.

Recent studies have begun to reveal a complicated cross-communication between these two mTOR-containing complexes while the exact molecular mechanism(s) remain largely elusive. In this effort, we have defined a novel and independent negative feedback loop through which either S6K1 or Akt directly phosphorylates Sin1 to repress mTORC2 activation in epithelial cells or adipocytes, respectively, providing a possible molecular mechanism for mTORC1 to balance the extra- and intra- cellular signals. Moreover, systematic analyses of many signaling pathways revealed a general oscillation pattern of activation/inactivation signaling dynamics (Purvis et al., 2012; Purvis and Lahav, 2012, 2013). This also applied to mTORC2-mediated growth factor signaling pathways, as suggested by the periodic Akt activation pattern upon stimulation by insulin or growth factors (Purvis and Lahav, 2013). To this end, mTORC1-mediated inhibition of mTORC2 through Sin1 phosphorylation might be one of such mechanisms, in addition to de-phosphorylation of Akt and multiple other negative feedback mechanism, to ensure that mTORC2 is only activated in a "pulse" manner (Chen et al., 2012). Therefore, between these two mTOR-containing complexes, it is plausible that mTORC1 exhibits constant basal activity whereas the mTORC2 complex is only transiently activated following external stimuli. This may partially explain the fact that hyper-activation of the critical signaling pathways including mTORC2/Akt is a hallmark for majority of human cancers (Testa and Tsichlis, 2005).

Finally, the identification of this mTORC1/S6K feedback suppression of mTORC2 expanded the critical role of mTORC1 in regulating and coordinating various growth factor signaling pathways. Compared to the previously identified negative feedback regulatory loops via IRS-1 (Harrington et al., 2004; Shah and Hunter, 2006) or Grb10 (Hsu et al., 2011; Yu et al., 2011), which nicely explained the IGF-1 and insulin but not the PDGF or EGF signaling regulatory pathways, the mTORC1/S6K1-mediated phosphorylation of Sin1 could function to balance the mTORC2/Akt activation in response to a wider range of growth stimuli beyond insulin and IGF-1, including but not limited to PDGF and EGF. Moreover, by targeting the mTORC2-specific component Sin1, instead of IRS-1 or Grb10 that are upstream of both mTORC1 and mTORC2, for phosphorylation and inhibition refines the suppression effects mainly on mTORC2, suggesting that this newly identified mechanism might be a specific mTORC2-targeted negative regulation. Taken together, our study unravels a novel IRS-1/Grb10-independent feedback mechanism of the tightly regulated PI3K/mTORC2/Akt pathway. More significantly, an ovarian cancer patient-derived R81T mutation of Sin1 was demonstrated to gain oncogenic capacity by bypassing this negative regulation due to the lack of Sin1-T86 phosphorylation, providing a molecular mechanism for the elevated mTORC2/Akt activation that could potently promote tumorigenesis at least in this cancer patient. In summary, our work points to the prospect of targeting Sin1 phosphorylation signaling as an effective therapeutic strategy in treating human disorders such as diabetes and cancer.

ACKNOWLEDGEMENTS

We thank Hiroyuki Inuzuka, Kai Xu, Alan Lau, Brian North, Lixin Wan, Xiangpeng Dai, Yueyong Liu, Kohei Ogura, Shavali Shaik and Jifang Zhang for critical reading of the manuscript and helpful discussions. This work was supported by grants from the National Institute of General Medicine, NIH (GM089763, GM094777, CA177910) to W.W. W.W. is an American Cancer Society Scholar and Leukemia and Lymphoma Society Research Scholar. P.L. is supported by 5T32HL007893.

COMPLIANCE WITH ETHICS GUIDELINES

Pengda Liu, Jianping Guo, Wenjian Gan and Wenyi Wei declare that they have no conflict of interest.

This article does not contain any studies with human or animal subjects performed by any of the authors.

REFERENCES

Alessi DR, Caudwell FB, Andjelkovic M, Hemmings BA, Cohen P (1996) Molecular basis for the substrate specificity of protein kinase B; comparison with MAPKAP kinase-1 and p70 S6 kinase. FEBS Lett 399:333–338

Bar-Peled L, Sabatini DM (2012) SnapShot: mTORC1 signaling at the lysosomal surface. Cell 151(1390–1390):e1391

Bar-Peled L, Schweitzer LD, Zoncu R, Sabatini DM (2012) Ragulator is a GEF for the rag GTPases that signal amino acid levels to mTORC1. Cell 150:1196–1208

Carnevalli LS, Masuda K, Frigerio F, Le Bacquer O, Um SH, Gandin V, Topisirovic I, Sonenberg N, Thomas G, Kozma SC (2010) S6K1 plays a critical role in early adipocyte differentiation. Dev Cell 18:763–774

Chan EY (2009) mTORC1 phosphorylates the ULK1–mAtg13–FIP200 autophagy regulatory complex. Sci Signal 2:pe51

Chen JY, Lin JR, Cimprich KA, Meyer T (2012) A two-dimensional ERK–AKT signaling code for an NGF-triggered cell-fate decision. Mol Cell 45:196–209

Dibble CC, Asara JM, Manning BD (2009) Characterization of Rictor phosphorylation sites reveals direct regulation of mTOR complex 2 by S6K1. Mol Cell Biol 29:5657–5670

Efeyan A, Zoncu R, Sabatini DM (2012) Amino acids and mTORC1: from lysosomes to disease. Trends Mol Med 18:524–533

Frias MA, Thoreen CC, Jaffe JD, Schroder W, Sculley T, Carr SA, Sabatini DM (2006) mSin1 is necessary for Akt/PKB phosphorylation, and its isoforms define three distinct mTORC2s. Curr Biol 16:1865–1870

Gao D, Wan L, Inuzuka H, Berg AH, Tseng A, Zhai B, Shaik S, Bennett E, Tron AE, Gasser JA et al (2010) Rictor forms a complex with Cullin-1 to promote SGK1 ubiquitination and destruction. Mol Cell 39:797–808

Garcia-Martinez JM, Alessi DR (2008) mTOR complex 2 (mTORC2) controls hydrophobic motif phosphorylation and activation of serum- and glucocorticoid-induced protein kinase 1 (SGK1). Biochem J 416:375–385

Gual P, Le Marchand-Brustel Y, Tanti JF (2005) Positive and negative regulation of insulin signaling through IRS-1 phosphorylation. Biochimie 87:99–109

Guertin DA, Sabatini DM (2007) Defining the role of mTOR in cancer. Cancer Cell 12:9–22

Guertin DA, Stevens DM, Thoreen CC, Burds AA, Kalaany NY, Moffat J, Brown M, Fitzgerald KJ, Sabatini DM (2006) Ablation in mice of the mTORC components raptor, Rictor, or mLST8 reveals that mTORC2 is required for signaling to Akt–FOXO and PKCalpha, but not S6K1. Dev Cell 11:859–871

Gwinn DM, Shackelford DB, Egan DF, Mihaylova MM, Mery A, Vasquez DS, Turk BE, Shaw RJ (2008) AMPK phosphorylation of raptor mediates a metabolic checkpoint. Mol Cell 30:214–226

Harrington LS, Findlay GM, Gray A, Tolkacheva T, Wigfield S, Rebholz H, Barnett J, Leslie NR, Cheng S, Shepherd PR et al (2004) The TSC1-2 tumor suppressor controls insulin-PI3K signaling via regulation of IRS proteins. J Cell Biol 166:213–223

Hsu PP, Kang SA, Rameseder J, Zhang Y, Ottina KA, Lim D, Peterson TR, Choi Y, Gray NS, Yaffe MB et al (2011) The mTOR-regulated phosphoproteome reveals a mechanism of mTORC1-mediated inhibition of growth factor signaling. Science 332:1317–1322

Humphrey SJ, Yang G, Yang P, Fazakerley DJ, Stockli J, Yang JY, James DE (2013) Dynamic adipocyte phosphoproteome reveals that Akt directly regulates mTORC2. Cell Metab 17:1009–1020

Ikenoue T, Inoki K, Yang Q, Zhou X, Guan KL (2008) Essential function of TORC2 in PKC and Akt turn motif phosphorylation, maturation and signalling. EMBO J 27:1919–1931

Inoki K, Li Y, Zhu T, Wu J, Guan KL (2002) TSC2 is phosphorylated and inhibited by Akt and suppresses mTOR signalling. Nat Cell Biol 4:648–657

Inoki K, Zhu T, Guan KL (2003) TSC2 mediates cellular energy response to control cell growth and survival. Cell 115(5):577–590

Inoki K, Corradetti MN, Guan KL (2005) Dysregulation of the TSC-mTOR pathway in human disease. Nat Genet 37:19–24

Jacinto E, Loewith R, Schmidt A, Lin S, Ruegg MA, Hall A, Hall MN (2004) Mammalian TOR complex 2 controls the actin cytoskeleton and is rapamycin insensitive. Nat Cell Biol 6:1122–1128

Jacinto E, Facchinetti V, Liu D, Soto N, Wei S, Jung SY, Huang Q, Qin J, Su B (2006) SIN1/MIP1 maintains Rictor–mTOR complex integrity and regulates Akt phosphorylation and substrate specificity. Cell 127:125–137

Kim DH, Sarbassov DD, Ali SM, King JE, Latek RR, Erdjument-Bromage H, Tempst P, Sabatini DM (2002) mTOR interacts with raptor to form a nutrient–sensitive complex that signals to the cell growth machinery. Cell 110:163–175

Kumar A, Lawrence JC Jr, Jung DY, Ko HJ, Keller SR, Kim JK, Magnuson MA, Harris TE (2010) Fat cell-specific ablation of Rictor in mice impairs insulin-regulated fat cell and whole-body glucose and lipid metabolism. Diabetes 59:1397–1406

Laplante M, Sabatini DM (2012) mTOR signaling in growth control and disease. Cell 149:274–293

Liu P, Gan W, Inuzuka H, Lazorchak AS, Gao D, Arojo O, Liu D, Wan L, Zhai B, Yu Y et al (2013) Sin1 phosphorylation impairs mTORC2 complex integrity and inhibits downstream Akt signalling to suppress tumorigenesis. Nat Cell Biol 15:1340–1350

Ma XM, Blenis J (2009) Molecular mechanisms of mTOR-mediated translational control. Nat Rev Mol Cell Biol 10:307–318

Manning BD, Cantley LC (2007) AKT/PKB signaling: navigating downstream. Cell 129:1261–1274

Manning BD, Tee AR, Logsdon MN, Blenis J, Cantley LC (2002) Identification of the tuberous sclerosis complex-2 tumor suppressor gene product tuberin as a target of the phosphoinositide 3-kinase/akt pathway. Mol Cell 10:151–162

Manning BD, Logsdon MN, Lipovsky AI, Abbott D, Kwiatkowski DJ, Cantley LC (2005) Feedback inhibition of Akt signaling limits the growth of tumors lacking Tsc2. Genes Dev 19:1773–1778

Oshiro N, Yoshino K, Hidayat S, Tokunaga C, Hara K, Eguchi S, Avruch J, Yonezawa K (2004) Dissociation of raptor from mTOR is a mechanism of rapamycin-induced inhibition of mTOR function. Genes Cells 9:359–366

Pena-Llopis S, Vega-Rubin-de-Celis S, Schwartz JC, Wolff NC, Tran TA, Zou L, Xie XJ, Corey DR, Brugarolas J (2011) Regulation of TFEB and V-ATPases by mTORC1. EMBO J 30:3242–3258

Polak P, Cybulski N, Feige JN, Auwerx J, Ruegg MA, Hall MN (2008) Adipose-specific knockout of raptor results in lean mice with enhanced mitochondrial respiration. Cell Metab 8:399–410

Purvis JE, Lahav G (2012) Decoding the insulin signal. Mol Cell 46:715–716

Purvis JE, Lahav G (2013) Encoding and decoding cellular information through signaling dynamics. Cell 152:945–956

Purvis JE, Karhohs KW, Mock C, Batchelor E, Loewer A, Lahav G (2012) p53 dynamics control cell fate. Science 336:1440–1444

Sabatini DM (2006) mTOR and cancer: insights into a complex relationship. Nat Rev Cancer 6:729–734

Sancak Y, Peterson TR, Shaul YD, Lindquist RA, Thoreen CC, Bar-Peled L, Sabatini DM (2008) The Rag GTPases bind raptor and mediate amino acid signaling to mTORC1. Science 320:1496–1501

Sancak Y, Bar-Peled L, Zoncu R, Markhard AL, Nada S, Sabatini DM (2010) Ragulator–Rag complex targets mTORC1 to the lysosomal surface and is necessary for its activation by amino acids. Cell 141:290–303

Sarbassov DD, Ali SM, Kim DH, Guertin DA, Latek RR, Erdjument-Bromage H, Tempst P, Sabatini DM (2004) Rictor, a novel binding partner of mTOR, defines a rapamycin-insensitive and raptor-independent pathway that regulates the cytoskeleton. Curr Biol 14:1296–1302

Sarbassov DD, Guertin DA, Ali SM, Sabatini DM (2005) Phosphorylation and regulation of Akt/PKB by the Rictor–mTOR complex. Science 307:1098–1101

Sarbassov DD, Ali SM, Sengupta S, Sheen JH, Hsu PP, Bagley AF, Markhard AL, Sabatini DM (2006) Prolonged rapamycin treatment inhibits mTORC2 assembly and Akt/PKB. Mol Cell 22:159–168

Shah OJ, Hunter T (2006) Turnover of the active fraction of IRS1 involves raptor-mTOR- and S6K1-dependent serine phosphorylation in cell culture models of tuberous sclerosis. Mol Cell Biol 26:6425–6434

Shaw RJ (2009) LKB1 and AMP-activated protein kinase control of mTOR signalling and growth. Acta Physiol 196:65–80

Testa JR, Tsichlis PN (2005) AKT signaling in normal and malignant cells. Oncogene 24:7391–7393

Thoreen CC, Chantranupong L, Keys HR, Wang T, Gray NS, Sabatini DM (2012) A unifying model for mTORC1-mediated regulation of mRNA translation. Nature 485:109–113 van Dam TJ, Zwartkruis FJ, Bos JL, Snel B (2011) Evolution of the TOR pathway. J Mol Evol 73:209–220

Vander Haar E, Lee SI, Bandhakavi S, Griffin TJ, Kim DH (2007) Insulin signalling to mTOR mediated by the Akt/PKB substrate PRAS40. Nat Cell Biol 9:316–323

Weber JD, Gutmann DH (2012) Deconvoluting mTOR biology. Cell Cycle 11:236–248

Wullschleger S, Loewith R, Oppliger W, Hall MN (2005) Molecular organization of target of rapamycin complex 2. J Biol Chem 280:30697–30704

Wullschleger S, Loewith R, Hall MN (2006) TOR signaling in growth and metabolism. Cell 124:471–484

Yang Q, Inoki K, Ikenoue T, Guan KL (2006) Identification of Sin1 as an essential TORC2 component required for complex formation and kinase activity. Genes Dev 20:2820–2832

Yip CK, Murata K, Walz T, Sabatini DM, Kang SA (2010) Structure of the human mTOR complex I and its implications for rapamycin inhibition. Mol Cell 38:768–774

Yu Y, Yoon SO, Poulogiannis G, Yang Q, Ma XM, Villen J, Kubica N, Hoffman GR, Cantley LC, Gygi SP et al (2011) Phosphoproteomic analysis identifies Grb10 as an mTORC1 substrate that negatively regulates insulin signaling. Science 332: 1322–1326

Zeng H, Yang K, Cloer C, Neale G, Vogel P, Chi H (2013) mTORC1 couples immune signals and metabolic programming to establish T(reg)-cell function. Nature 499:485–490

Zinzalla V, Stracka D, Oppliger W, Hall MN (2011) Activation of mTORC2 by association with the ribosome. Cell 144: 757–768

Zoncu R, Bar-Peled L, Efeyan A, Wang S, Sancak Y, Sabatini DM (2011a) mTORC1 senses lysosomal amino acids through an inside-out mechanism that requires the vacuolar H(+)-ATPase. Science 334:678–683

Zoncu R, Efeyan A, Sabatini DM (2011b) mTOR: from growth signal integration to cancer, diabetes and ageing. Nat Rev Mol Cell Biol 12:21–35

19

Accelerated regeneration of the skeletal muscle in RNF13-knockout mice is mediated by macrophage-secreted IL-4/IL-6

Jiao Meng, Xiaoting Zou, Rimao Wu, Ran Zhong, Dahai Zhu, Yong Zhang✉

The National Laboratory of Medical Molecular Biology, Institute of Basic Medical Sciences, Chinese Academy of Medical Sciences and School of Basic Medicine, Peking Union Medical College, Beijing 100005, China

✉ Correspondence: dr_zhangyong@126.com (Y. Zhang)

ABSTRACT

RING finger protein 13 (RNF13) is a newly identified E3 ligase reported to be functionally significant in the regulation of cancer development, muscle cell growth, and neuronal development. In this study, the function of RNF13 in cardiotoxin-induced skeletal muscle regeneration was investigated using *RNF13*-knockout mice. *RNF13*$^{-/-}$ mice exhibited enhanced muscle regeneration —characterized by accelerated satellite cell proliferation —compared with wild-type mice. The expression of RNF13 was remarkably induced in macrophages rather than in the satellite cells of wild-type mice at the very early stage of muscle damage. This result indicated that inflammatory cells are important in RNF13-mediated satellite cell functions. The cytokine levels in skeletal muscles were further analyzed and showed that *RNF13*$^{-/-}$ mice produced greater amounts of various cytokines than wild-type mice. Among these, IL-4 and IL-6 levels significantly increased in *RNF*13$^{-/-}$ mice. The accelerated muscle regeneration phenotype was abrogated by inhibiting IL-4/IL-6 action in *RNF13*$^{-/-}$ mice with blocking antibodies. These results indicate that RNF13 deficiency promotes skeletal muscle regeneration via the effects on satellite cell niche mediated by IL-4 and IL-6.

KEYWORDS RNF13, muscle regeneration, satellite cell niche, IL-4/IL-6

INTRODUCTION

Regeneration of skeletal muscles after injury is a coordinated process that progresses through four phases: degeneration; inflammation; regeneration; and remodeling-repair (Carosio et al., 2009; Ten Broek et al., 2010). Injury of myofibers leads to rapid necrosis associated with sarcomere disruption, cell permeabilization, and proteolysis, which induces an acute inflammatory response characterized by the invasion of inflammatory cells (Brunelli and Rovere-Querini, 2008). The regeneration process is characterized by the activation of muscle stem cells (satellite cells), followed by myoblast differentiation and fusion into regenerated myofibers (Hawke, 2001). The final stages of muscle repair include the maturation of regenerated myofibers, the remodeling of extracellular matrix, and the recovery of the contractile capacity of injured muscles (Carosio et al., 2009).

Satellite cells, a quiescent population of myogenic cells that reside between the basal lamina and the plasmalemma, play the central role in skeletal muscle regeneration (Mauro, 1961). Satellite cells can be identified and isolated using a panel of satellite cell-associated markers, such as the paired-box transcription factor Pax7, and some cell surface markers, including M-cadherin, syndecan-3 and -4, CD34, α7-integrin, CXCR4 [chemokine (C-X-C motif) receptor 4], and SM/C2.6 antigen (Vcam) (Beauchamp et al., 2000; Cerletti et al., 2008; Cornelison et al., 2004; Fukada et al., 2007; Gnocchi et al., 2009; Kuang et al., 2008). Among these markers, Pax7 is specifically expressed in quiescent and activated satellite cells and is commonly used as a marker (Halevy et al., 2004; Relaix et al., 2005). MyoD expression, which is remarkably increased in activated satellite cells after injury, is widely regarded as the hallmark of activated satellite cells (Kanisicak et al., 2009; Zammit and

Beauchamp, 2001; Zammit et al., 2002). In general, the immediate environment- or niche-of satellite cells participates in the modulation of satellite cell functions. Numerous factors in this niche influence satellite cell behavior, including adjacent myofibers, basal lamina, microvasculature, and neuromuscular junctions, as well as the secretion of multiple factors by interstitial cells and immune cells (Dhawan and Rando, 2005; Gopinath and Rando, 2008; Kuang et al., 2008; Ten Broek et al., 2010). Among these factors, inflammatory cells infiltrating the injured muscle perform a very important function in satellite cell activation, proliferation, and differentiation.

The innate immune system is activated as an immediate response to acute muscle injury. The first event in this process involves the invasion of damaged areas by inflammatory cells. Neutrophils, the first invasive inflammatory cells, infiltrate the damaged site in 2 h after injury. However, neutrophils die during inflammation by apoptosis and become undetectable in 3–4 days after injury (Tidball and Villalta, 2010). Neutrophils mainly function to promote the removal of cellular debris by secreting cytokines (Tidball, 2005). The next immune cells to invade are macrophages, which are the major cells involved in inflammation and regeneration. As macrophage function is suppressed, the proliferation and differentiation of muscle satellite cells are delayed *in vivo*; as a result, incomplete muscle regeneration and fibrosis occur (Segawa et al., 2008). In previous studies (Brunelli and Rovere-Querini, 2008; Chazaud et al., 2009; Kharraz et al., 2013), macrophages are a heterogeneous population involved in different functions at various stages after damage. Macrophages are classified as M1 and M2, corresponding to classical and alternative activation, respectively. Classically activated M1 macrophages, which are usually found at the early stage, produce high levels of pro-inflammatory cytokines. These macrophages mainly function to phagocytose necrotic fiber debris. M2 macrophages regulate the function of satellite cells by secreting cytokines and chemokines that facilitate vascular and muscle fiber repair. Thus, macrophages are the predominant inflammatory cells and local source of cytokines; playing important roles in the regulation of satellite cell activation, proliferation, and differentiation.

The function of cytokines and chemokines in regulating satellite cell function has been the focus of several studies. The expression and/or secretion of more than 100 cytokines increase by more than fivefold after toxin injection (Hirata et al., 2003). This altered expression and secretion pattern after injury suggests the important function of various cytokines in the regulation of satellite cell activation, proliferation, and differentiation. For instance, tumor necrosis factor (TNF)-α has been shown to increase with muscle injury and regulate muscle regeneration through activation of p38; conversely, TNF-α deficiency impairs muscle regeneration (Chen et al., 2005, 2007; Warren et al., 2002). The interleukin (IL)-6/STAT3 (signal transducer and activator of transcription 3) pathway is essential for macrophage infiltration and myoblast proliferation during muscle regeneration (Zhang et al., 2013a). IL-10 plays an important role in muscle regeneration by regulating the transformation of macrophages from an M1 to an M2 phenotype (Deng et al., 2012). Another study has demonstrated that the activation of IL-4/IL-13 signaling promotes the proliferation of fibro/adipocyte progenitors to support myogenesis while inhibiting their differentiation into adipocytes (Heredia et al., 2013). However, the specific function of these cytokines remains unclear. Under normal conditions, the repertoire and function of cytokines in the satellite cell niche vary, highlighting the importance of investigating the combined effects of cytokines.

We have previously demonstrated that the RING finger ubiquitin ligase RNF13 is a widely expressed membrane-associated E3 ligase evolutionarily conserved among many metazoans (Jin et al., 2011; Zhang et al., 2009; Zhang et al., 2010). RNF13 negatively regulates the proliferation of chicken myoblasts (Zhang et al., 2010). The expression of RNF13, which is positively regulated by myostatin, gradually decreases during skeletal myogenesis. The overexpression of RNF13 inhibits cell proliferation, an effect dependent on the E3 ligase activity of RNF13 (Zhang et al., 2010). We identified snapin, a SNAP25-interacting protein, as a substrate of RNF13, and showed that RNF13 was involved in the regulation of the soluble N-ethylmaleimide-sensitive fusion protein attachment protein receptor (SNARE) complex via snapin ubiquitination (Zhang et al., 2013b).

To further investigate the function of RNF13 in adult muscle development, we established a cardiotoxin (CTX)-induced muscle regeneration model in *RNF13*-knockout (*RNF13*$^{-/-}$) mice (Zhang et al., 2013b). *RNF13*$^{-/-}$ mice exhibited accelerated muscle regeneration after injury. Moreover, RNF13 was significantly induced in inflammatory cells as early as 1 h after CTX injection in wild-type mice. Importantly, RNF13 influenced the concentration of numerous cytokines in damaged areas and modulated muscle regeneration by affecting IL-4/IL-6 function. This study is the first to demonstrate that E3 ligase RNF13 regulates the function of satellite cells by modulating cytokine composition.

RESULTS

Loss of *RNF13* accelerates skeletal muscle regeneration

We have previously demonstrated that *RNF13* expression gradually decreases during skeletal muscle development, and over-expression of *RNF13* inhibits muscle cell proliferation (Zhang et al., 2010). *RNF13*$^{-/-}$ mice display no overt physiological abnormalities during development, as previously described (Zhang et al., 2013b); in the present study, *RNF13*$^{-/-}$ mice were used to assess the significance of RNF13 function in regulating skeletal muscle regeneration. To this end, we established a CTX-induced muscle regeneration model in *RNF13*$^{-/-}$ and *RNF13*$^{+/+}$ mice. *RNF13*$^{-/-}$ and

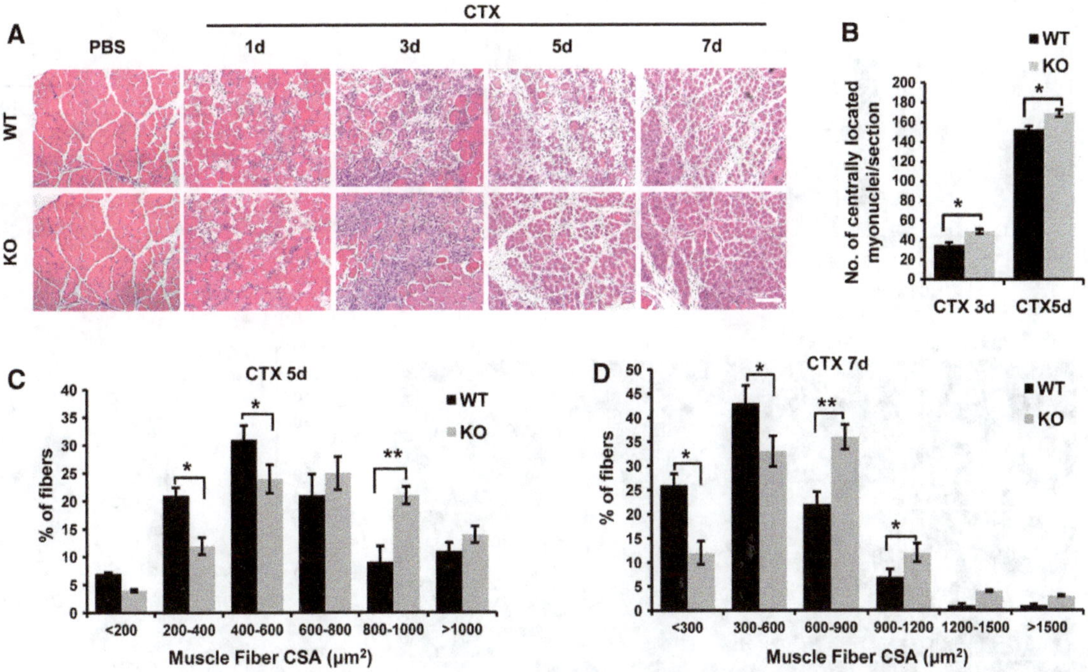

Figure 1. Histological analysis of muscle regeneration in *RNF13*$^{-/-}$ and *RNF13*$^{+/+}$ mice. (A) Sections from the TA muscles of *RNF13*$^{-/-}$ and *RNF13*$^{+/+}$ mice at different times (0, 1, 3, 5, and 7 d) after CTX injection were stained with H&E. Scale bar = 600 m. (B) The percentage of regenerating myofibers, characterized by their centrally located myonuclei, was calculated in H&E-stained sections from *RNF13*$^{-/-}$ and *RNF13*$^{+/+}$ TA muscles at 3 d and 5 d post-injury. (C and D) The cross-sectional areas (CSAs) of regenerating muscle fibers 5 d and 7 d after injury were analyzed in H&E-stained sections. Five pairs of mice were used for each time point, and more than 10 sections for each mouse were analyzed. Data are presented as means SEs (error bars; *P < 0.05, **P < 0.01, ***P < 0.001).

RNF13$^{+/+}$ mice were intramuscularly injected with CTX in the tibialis anterior (TA) muscles and analyzed histologically at 1, 3, 5, and 7 d after injury (Fig. 1A). At 1 d after injury, the muscle cross sections showed that some fibers were damaged and immune cells infiltrated in *RNF13*$^{-/-}$ and *RNF13*$^{+/+}$ mice (Fig. 1A). No phenotypic changes were observed in either of the genotypes. At 3 d and 5 d after injury, H&E staining of the injured TA muscles revealed significantly more regenerating fibers characterized by their centrally located nuclei in *RNF13*$^{-/-}$ mice than in *RNF13*$^{+/+}$ mice (Fig. 1B). The cross-sectional areas (CSAs) of regenerating muscle fibers at 5 d and 7 d after injury were analyzed and showed that the regenerating muscle fibers in *RNF13*$^{-/-}$ mice were larger than those in *RNF13*$^{+/+}$ mice (Fig. 1C and 1D). These results showed that the muscles from *RNF13*$^{-/-}$ mice exhibited an enhanced regeneration compared with those from *RNF13*$^{+/+}$ mice after CTX damage.

Satellite cells proliferate more rapidly in *RNF13*$^{-/-}$ mice than in *RNF13*$^{+/+}$ mice

The improved skeletal muscle regeneration following injury in *RNF13*$^{-/-}$ mice prompted us to investigate the stages of skeletal muscle regeneration in which RNF13 is involved. To accomplish this objective, we initially examined the expression pattern of RNF13 during regeneration in wild-type mice. We found that RNF13 expression was remarkably up-regulated immediately after injury and reached the peak at day 1 (Fig. 2A–C). The significant induction of RNF13 expression at an early stage of muscle regeneration demonstrated the function of RNF13 in regulating satellite cell activation and proliferation, which are important for an efficient muscle regeneration. To test this possibility, we examined whether or not satellite cell activation and proliferation are altered in *RNF13*$^{-/-}$ mice by immunofluorescence staining for Pax7 and MyoD (Fig. 2D). The transcription factor Pax7 is a marker of quiescent satellite cells as well as activated, proliferating satellite cells during muscle regeneration, whereas MyoD is considered as a marker for activated and proliferating satellite cells. Pax7$^+$, MyoD$^+$, and Pax7$^+$/MyoD$^+$ cells were counted and revealed a greater number of activated and proliferating satellite cells in *RNF13*$^{-/-}$ mice at 3 d and 5 d after injury than in *RNF13*$^{+/+}$ mice (Fig. 2E–G). To determine the accelerated proliferation in *RNF13*$^{-/-}$ mice, we performed BrdU and MyoD double staining (Fig. 2H). The statistical analysis results showed that the number of BrdU$^+$/MyoD$^+$ cells was also increased in *RNF13*$^{-/-}$ mice (Fig. 2I). To further confirm this result, we examined Pax7 expression in muscles at different

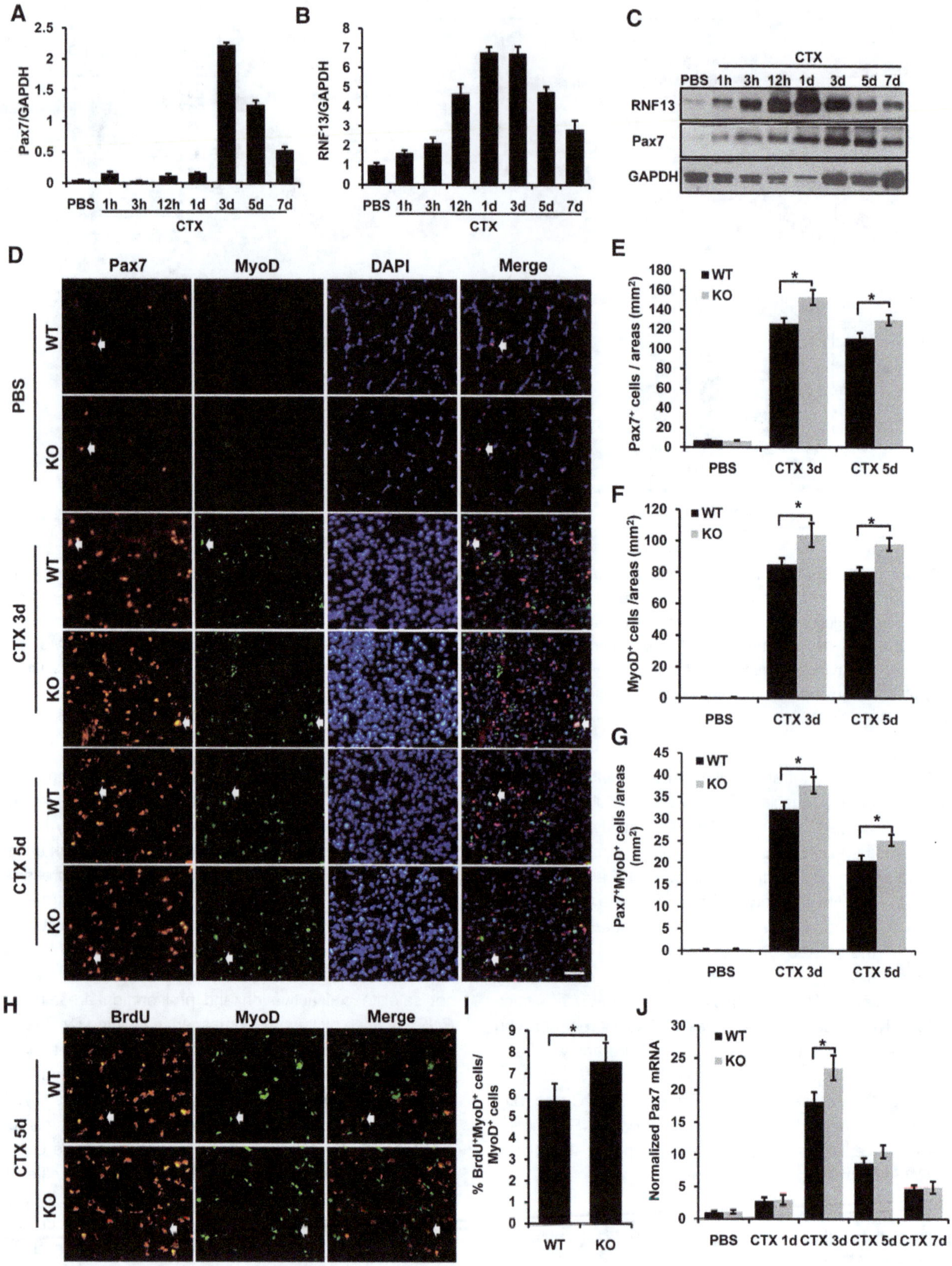
A
Pax7/GAPDH
PBS 1h 3h 12h 1d 3d 5d 7d
CTX
B
RNF13/GAPDH
PBS 1h 3h 12h 1d 3d 5d 7d
CTX
C
CTX
PBS 1h 3h 12h 1d 3d 5d 7d
RNF13
Pax7
GAPDH
D
Pax7
MyoD
DAPI
Merge
PBS
CTX 3d
CTX 5d
WT
KO
E
WT
KO
Pax7⁺ cells / areas (mm²)
PBS CTX 3d CTX 5d
F
WT
KO
MyoD⁺ cells /areas (mm²)
PBS CTX 3d CTX 5d
G
WT
KO
Pax7⁺MyoD⁺ cells /areas (mm²)
PBS CTX 3d CTX 5d
H
BrdU
MyoD
Merge
CTX 5d
WT
KO
I
% BrdU⁺MyoD⁺ cells/ MyoD⁺ cells
WT KO
J
WT
KO
Normalized Pax7 mRNA
PBS CTX 1d CTX 3d CTX 5d CTX 7d

Figure 2. Satellite cell proliferation is accelerated in *RNF13$^{-/-}$* mice. (A–C) Pax7 and RNF13 mRNA and protein expression were determined by qRT-PCR and Western blot, respectively, at different time points after damage. (D) Frozen sections of TA muscles from *RNF13$^{-/-}$* and *RNF13$^{+/+}$* mice damaged for 3 d or 5 d were immunostained for Pax7 (red), MyoD (green), and nuclei (DAPI; blue). The right panel shows a merged image. (E–G) Pax7^{+}, MyoD^{+}, and Pax7^{+}/MyoD^{+} cells in defined areas were counted. More than 50 sections from five individuals were analyzed. (H) Frozen sections of TA muscles from *RNF13$^{-/-}$* and *RNF13$^{+/+}$* mice damaged for 5 d were immunostained for BrdU (red) and MyoD (green). (I) The percentage of BrdU^{+} cells in the MyoD^{+} cell population was determined. More than 50 sections from five individuals were analyzed. (J) Pax7 mRNA levels in uninjured (0 d) and injured muscles at different times after injury (1, 3, 5, and 7 d) from *RNF13$^{-/-}$* and *RNF13$^{+/+}$* mice were determined by qRT-PCR. Scale bar = 200 μm. Data are presented as means ± SEs (error bars; $^{*}P < 0.05$, $^{**}P < 0.01$, $^{***}P < 0.001$).

time points after damage using qRT-PCR and Western blotting analysis (Figs. 2J, S1A, and S1B). The results were consistent with those of immunofluorescence staining, indicating that satellite cells were activated and proliferated more rapidly in *RNF13$^{-/-}$* mice than in *RNF13$^{+/+}$* mice.

Accelerated satellite cell proliferation in *RNF13$^{-/-}$* mice is mainly non-cell-autonomous

Considering that satellite cell activation and proliferation are accelerated in *RNF13$^{-/-}$* mice, we then investigated whether the improvement in satellite cell function is cell autonomous or non-cell autonomous. To test this hypothesis, we assessed the effects of RNF13 on satellite cell activation and proliferation in isolated, cultured single myofibers. In this model, satellite cells initially express Pax7 only; MyoD is subsequently expressed and cells enter the cell cycle. Proliferating cells either down-regulate Pax7, leading to differentiation, or down-regulate MyoD, leading to self-renew (Halevy et al., 2004; Olguin and Olwin, 2004; Zammit et al., 2004). After 24 h or 72 h in culture, the fibers were stained for Pax7 and MyoD (Fig. 3A). Pax7^{+}/MyoD^{+} cells were then counted to assess the satellite cell activation and proliferation. We found no difference in the number of Pax7^{+}/MyoD^{+} cells between the single myofibers isolated from *RNF13$^{-/-}$* mice and *RNF13$^{+/+}$* mice (Fig. 3B). This result suggested that the modulation of satellite cell activation and proliferation by RNF13 ablation is dependent on the niche provided by other cells.

Muscle injury causes an inflammatory immune response in damaged tissues (Tidball, 2005), and inflammatory reaction plays a critical regulatory role in satellite cell activation and proliferation. Considering that RNF13 did not directly affect satellite cell function, we hypothesized that aberrant inflammatory responses contributed to the accelerated regeneration of *RNF13$^{-/-}$* muscles. In addition, RNF13 expression was induced as early as 1 h after CTX-induced muscle damage; this condition occurred earlier than the upregulation of Pax7 (Fig. 2A–C), suggesting that RNF13 exerts its functions before satellite cell activation. To determine the specific cells contributing to RNF13 induction after damage, we separated the cells in damaged TA muscles by flow cytometry (Fig. 3C). Two main subsets of inflammatory cells in damaged muscles can be distinguished on the basis of Ly-6G and CD11b expressions: CD11b^{+}/Ly-6G^{+} cells, which are considered as neutrophils, and CD11b^{+}/Ly-6G^{-} cells, which are considered as macrophages. CD11b^{-}/Ly-6G^{-} cells, which were also collected, are non-inflammatory cells. To assess the purity of the sorted cells, we assessed the expressions of the macrophage marker Mac-1, the neutrophil marker Ly-6G, and the satellite cell marker Pax7 by qRT-PCR analysis (Fig. 3D). Pax7 was highly expressed in the CD11b^{-}/Ly-6G^{-} population, suggesting that this population enriched the satellite cells. The RNF13 expression pattern in these sorted cells was analyzed and the results revealed that RNF13 induced after CTX damage was mainly expressed in inflammatory cells, particularly in macrophages (Fig. 3D). These results indicated that RNF13 regulates satellite cell functions by influencing inflammatory cells. The accelerated satellite cell proliferation observed in *RNF13$^{-/-}$* mice is mainly non-cell autonomous.

Loss of RNF13 induces cytokine production without influencing the recruitment of inflammatory cells

RNF13 is significantly induced in inflammatory cells after injury; thus, RNF13 regulates muscle regeneration primarily by altering muscle inflammation instead of directly affecting satellite cell function. To further confirm this result, we initially investigated whether or not RNF13 deficiency affects inflammatory cell infiltration. We subsequently counted the numbers of intramuscular macrophages and neutrophils in the injured muscles by flow cytometry (Fig. 4A). We found no significant difference in the numbers of intramuscular macrophages or neutrophils between *RNF13$^{-/-}$* mice and *RNF13$^{+/+}$* mice at 1 d or 3 d after damage (Fig. 4B and 4C). Therefore, RNF13 deficiency did not influence the recruitment of inflammatory cells in response to acute injury. We then identified the factor(s) in the niche required for the RNF13 activation of satellite cells at the early stage of muscle regeneration. Considering that cytokines and chemokines are important in muscle regeneration, we determined the presence of 18 different cytokines/chemokines secreted in TA muscles at different time points after CTX-induced damage (0, 4, and 12 h, and 1 d and 3 d) by using a Luminex multiplex platform (Table S2). All of the 18 cytokines/chemokines were stimulated after injury, and the concentrations of nine cytokines, particularly IL-4, IL-6, TNF-α, IL-1α, MIP-1α, and MIP-1β, were greater in *RNF13$^{-/-}$* skeletal muscles than in *RNF13$^{+/+}$* skeletal muscles (Fig. 4D). These

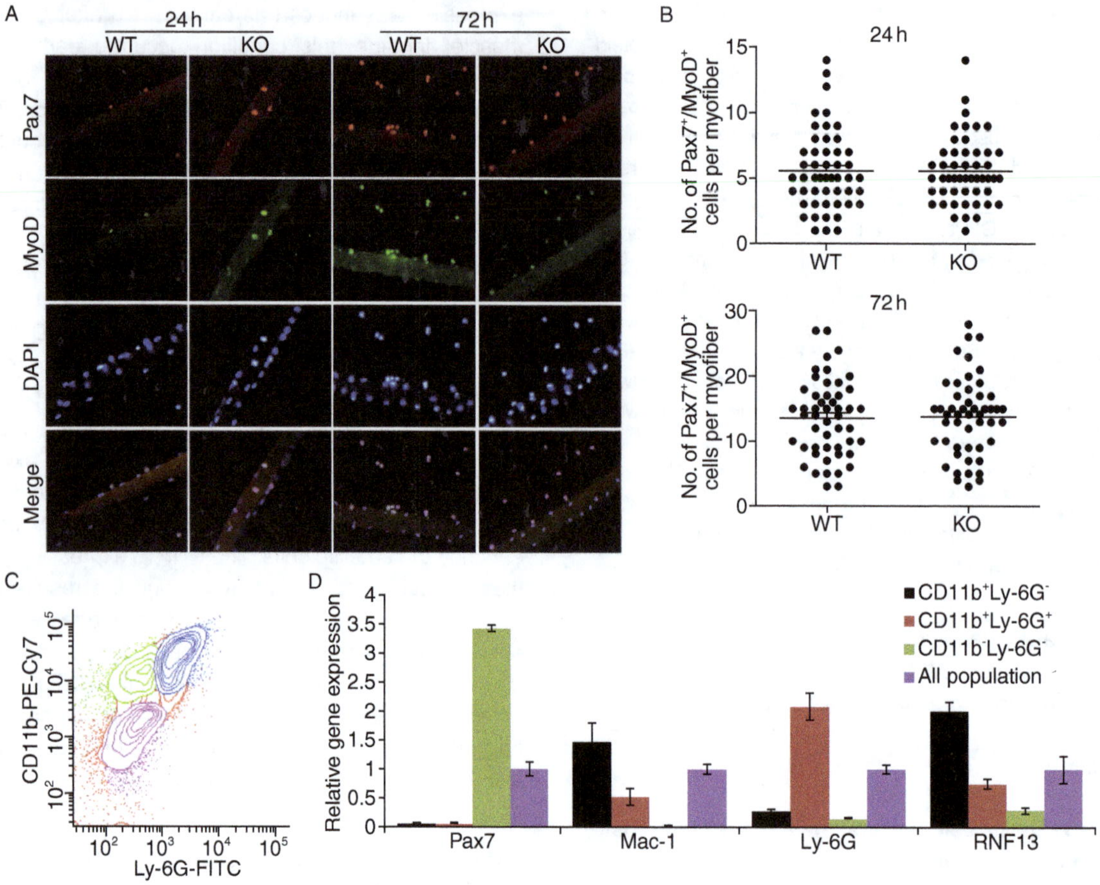

Figure 3. Accelerated satellite cell activation and proliferation in *RNF13*$^{-/-}$ mice is mainly non-cell autonomous. (A) Single myofibers isolated from *RNF13*$^{-/-}$ and *RNF13*$^{+/+}$ mice were cultured for 24 h or 72 h and stained for Pax7 (red), MyoD (green), and DAPI (blue). The bottom panel shows a merged image. (B) Pax7$^+$/MyoD$^+$ cells in each individual fiber were counted after 24 h or 72 h of culture. More than 200 fibers from *RNF13*$^{-/-}$ and *RNF13*$^{+/+}$ mice were analyzed. (C) Inflammatory cells were sorted by FACS. CD11b$^+$/Ly-6G$^+$ cells are considered neutrophils, and CD11b$^+$/Ly-6G$^-$ cells are considered macrophages. (D) The levels of Mac-1 (macrophage marker), Ly-6G (neutrophil marker), Pax7 (satellite cell marker), and RNF13 mRNA in sorted cells were determined by qRT-PCR. Similar results were obtained in three separate experiments.

results suggested that RNF13 deficiency regulates muscle regeneration by affecting the concentration of specific cytokines rather than by influencing the accumulation of inflammatory cells.

RNF13 regulates muscle regeneration by modulating IL-4/IL-6 levels secreted by macrophages

The concentrations of important cytokines and chemokines increased in the damaged muscles of *RNF13*$^{-/-}$ mice. However, this result fails to clarify whether or not this induction can be accounted for the accelerated muscle regeneration in *RNF13*$^{-/-}$ mice. If this condition is possible, the specific cytokines involved in this process remain unknown. Among the nine cytokines that showed greater changes in *RNF13*$^{-/-}$ mice, IL-4 and IL-6 were remarkably and significantly upregulated within 4 h after muscle damage in *RNF13*$^{-/-}$ mice compared with *RNF13*$^{+/+}$ mice (Fig. 4D). These results suggested that IL-4 and IL-6 are possible mediators of RNF13 function. These results are also consistent with those in previous studies, in which IL-4 and IL-6 are important in skeletal muscle regeneration (Heredia et al., 2013; Zhang et al., 2013a). Given the fact that RNF13 is significantly induced in the macrophages of damaged muscles after injury, we hypothesized that the increased levels of IL-4 and IL-6 in such macrophages from *RNF13*$^{-/-}$ mice were due to the loss of RNF13 function in these macrophages. Thus, we initially examined the expressions of IL-4 and IL-6 in different populations of sorted cells from wild-type mice. These two cytokines were also mainly expressed in the macrophages in the damaged muscles (Fig. S2A) and exhibited the similar expression manner with RNF13. This observation was further supported by the significantly increased mRNA levels of IL-4 and IL-6 in sorted macrophages from the damaged

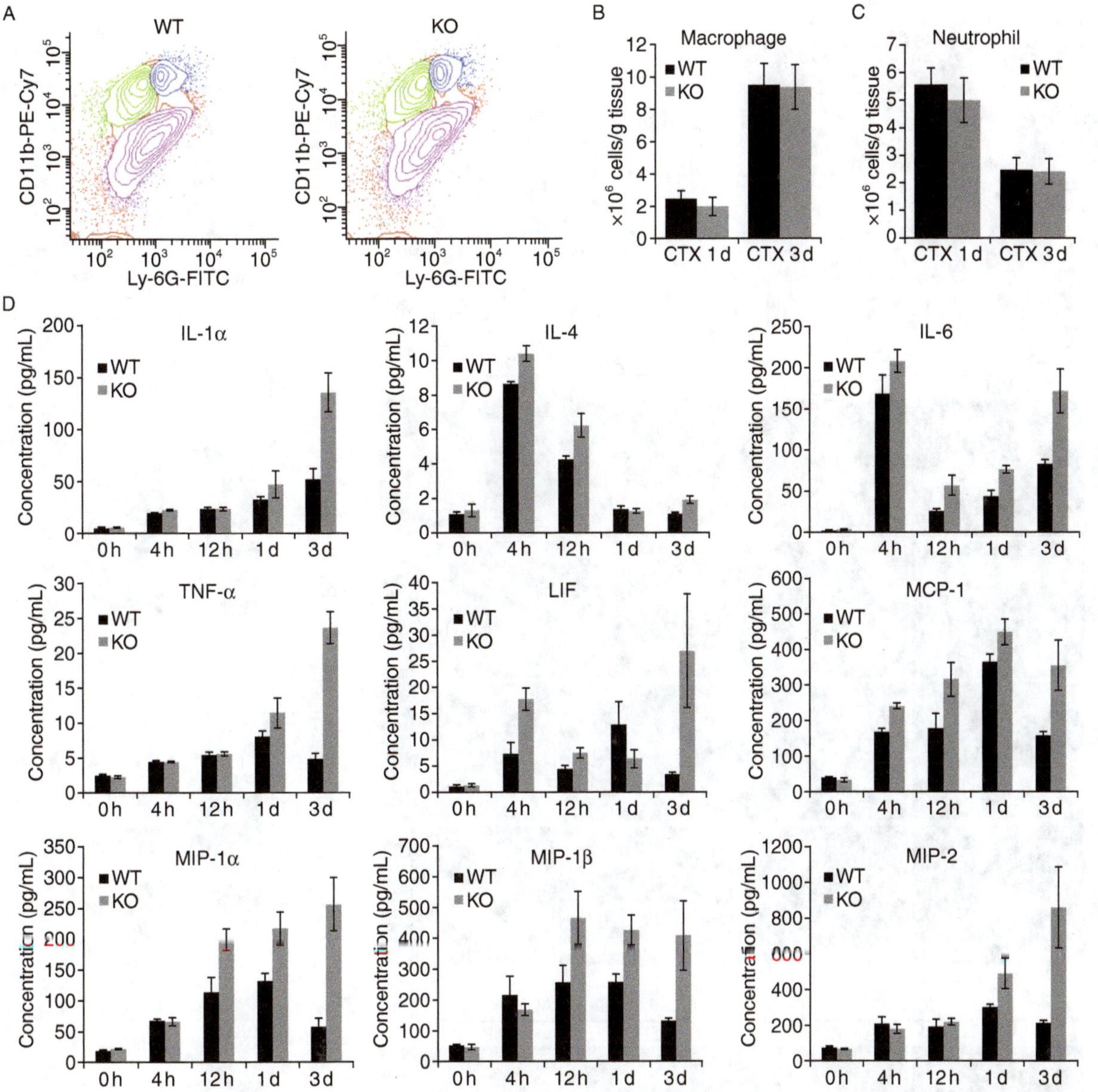

Figure 4. Loss of RNF13 induces the production of cytokines without influencing the recruitment of inflammatory cells. (A) FACS analysis of $CD11b^+/Ly\text{-}6G^+$ cells (macrophages) and $CD11b^+/Ly\text{-}6G^-$ cells (neutrophils) in muscle-derived cells from *RNF13*$^{-/-}$ and *RNF13*$^{+/+}$ mice damaged for 1 d or 3 d. (B and C) The number of intramuscular macrophages and neutrophils at 1 d and 3 d after damage was analyzed. Similar results were obtained in three separate experiments. (D) The concentrations of 18 cytokines and chemokines in damaged TA muscles from *RNF13*$^{-/-}$ and *RNF13*$^{+/+}$ mice were determined at 0, 4, and 12 h, and 1 d and 3 d post-injury using a Luminex multiplex assay. Shown are 9 of 18 cytokines and chemokines whose concentrations in the muscles changed after injury in *RNF13*$^{-/-}$ mice. Data are presented as means ± SEs (error bars; $^*P < 0.05$, $^{**}P < 0.01$, $^{***}P < 0.001$; $n = 5$ mice/time point).

muscles (4 h after injury) of *RNF13*$^{-/-}$ mice compared with *RNF13*$^{+/+}$ mice (Fig. S2B). Finally, we investigated the functional connection between IL-4/IL-6 upregulation and improved muscle regeneration in *RNF13*$^{-/-}$ mice by a neutralization experiment. We injected neutralizing antibodies against IL-4 and IL-6 into the TA muscles of *RNF13*$^{+/+}$ and *RNF13*$^{-/-}$ mice, respectively, before CTX damage. We then analyzed the regeneration after 5 d. Morphological analysis results showed that blocking IL-4 and IL-6 substantially slowed the regeneration process (Fig. 5A). In particular, the analysis results of fiber size showed that the acceleration of muscle regeneration in *RNF13*$^{-/-}$ mice was prevented by blocking IL-4 and IL-6 eliminated (Fig. 5B and 5C; compared with IgG control). Frozen sections were stained concurrently

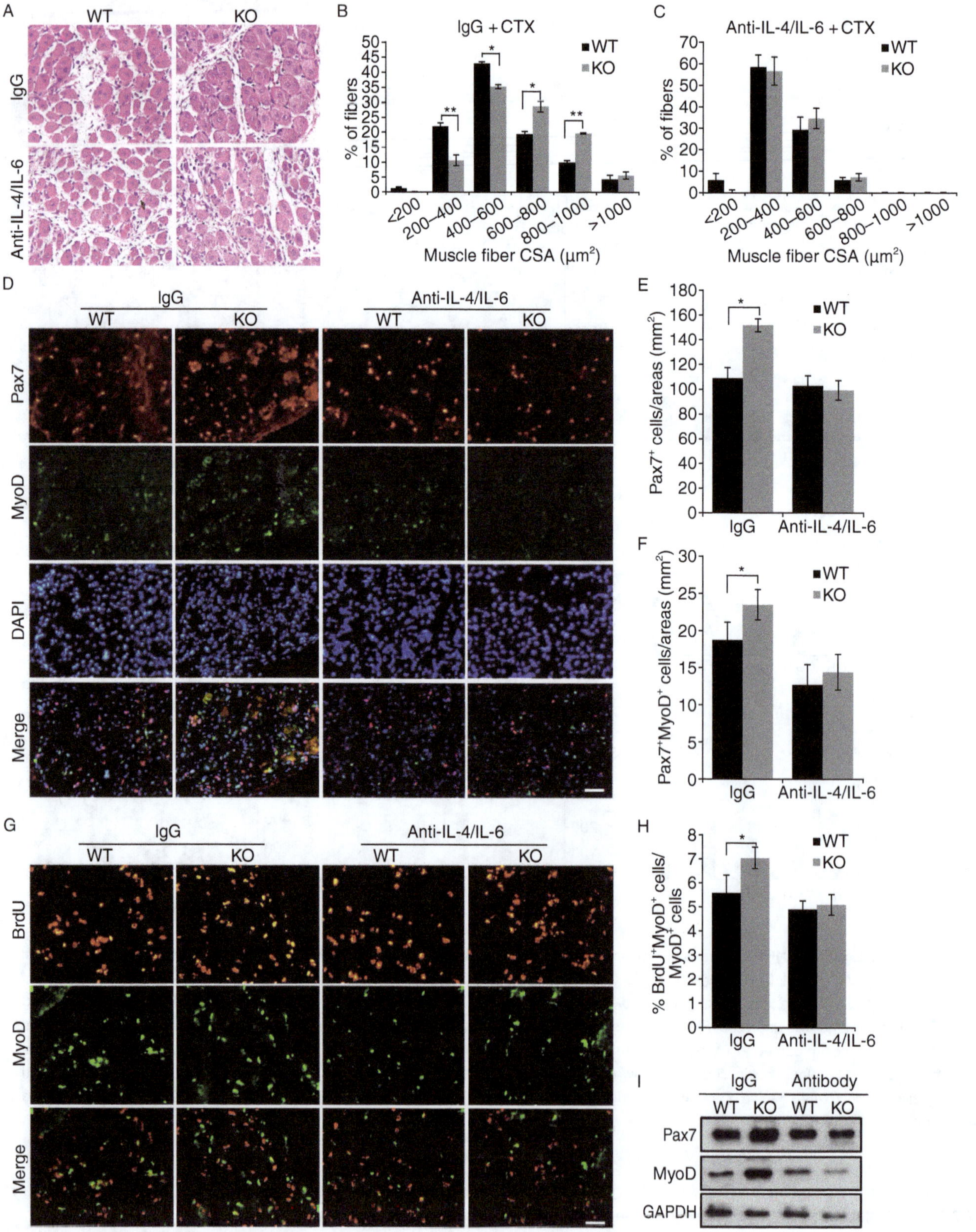
A
WT
KO
IgG
Anti-IL-4/IL-6
B
IgG + CTX
% of fibers
Muscle fiber CSA (μm²)
<200
200–400
400–600
600–800
800–1000
>1000
WT
KO
C
Anti-IL-4/IL-6 + CTX
% of fibers
Muscle fiber CSA (μm²)
D
IgG
Anti-IL-4/IL-6
Pax7
MyoD
DAPI
Merge
E
Pax7⁺ cells/areas (mm²)
F
Pax7⁺MyoD⁺ cells/areas (mm²)
G
BrdU
MyoD
Merge
H
% BrdU⁺MyoD⁺ cells/MyoD⁺ cells
I
IgG
Antibody
Pax7
MyoD
GAPDH

◀ **Figure 5. RNF13 regulates muscle regeneration by enhancing IL-4/IL-6 expression.** (A) Neutralizing antibodies against IL-4 and IL-6 (or control IgG antibodies) were injected into the TA muscles of *RNF13*$^{+/+}$ and *RNF13*$^{-/-}$ mice before CTX damage. Sections from TA muscles damaged for 5 d were stained with H&E. Scale bar = 200 μm. (B and C) The CSAs of regenerating muscle fibers were analyzed in H&E-stained sections. Five pairs of mice were used at each time point, and more than 10 sections from each mouse were analyzed. (D) Frozen sections of TA muscles, injected with anti-IL-4/IL-6 or control antibodies, were immunostained for Pax7 (red), MyoD (green), and nuclei (DAPI; blue). The bottom panel shows a merged image. (E and F) Pax7$^+$ and Pax7$^+$/MyoD$^+$ cells in defined areas were counted. More than 50 sections from five individuals were analyzed. (G) Frozen sections of TA muscles injected with anti-IL-4/IL-6 or control antibodies were immunostained for BrdU (red) and MyoD (green). (H) The percentage of BrdU$^+$ cells in the MyoD$^+$ cell population was determined. More than 50 sections from five individuals were analyzed. (I) Pax7 and MyoD expression in the damaged muscle injected with anti-IL-4/IL-6 or control antibodies was detected by Western blot. Data are presented as means ± SEs (error bars; **P* < 0.05, ***P* < 0.01, ****P* < 0.001).

for Pax7, MyoD, and BrdU to evaluate satellite cell activation and proliferation (Fig. 5D and 5G). Consistent with the morphological data, this immunohistochemical analysis showed that the numbers of Pax7$^+$/MyoD$^+$ cells and BrdU$^+$/MyoD$^+$ cells in RNF13$^{-/-}$ mice decreased after the antibodies were injected (Fig. 5E, 5F and 5H). These results were also confirmed by detection of Pax7 and MyoD expression in the damaged muscle by Western blotting (Fig. 5I). Collectively, these findings indicate that RNF13 regulates satellite cell function and muscle regeneration by modulating IL-4/IL-6 levels in the satellite cell niche.

DISCUSSION

This study provided the first evidence showing that RNF13 deficiency promoted skeletal muscle regeneration after CTX-induced muscle damage. In addition, cytokine levels in the damaged TA muscles of *RNF13*$^{-/-}$ mice, particularly those of IL-4 and IL-6, differed from those of *RNF13*$^{+/+}$ mice. These levels further showed that the accelerated muscle regeneration induced by RNF13 deficiency was partially reversed by blocking IL-4/IL-6 function. Taken together, these results indicated that membrane-associated E3 ligase RNF13 regulated muscle regeneration by modulating IL-4/IL-6 levels.

The Ub-proteasome system (UPS) is a major pathway of protein degradation in mammalian cells. UPS also plays a critical role in the breakdown of muscle proteins during regeneration. Furthermore, UPS selectively degrades target proteins following their ubiquitination by different E3 ligases. In addition to functioning in protein degradation, increasing specific E3 ligases have been found to play a significant role in muscle regeneration. The loss of RNF5, a membrane-associated E3 ligase, is reported to lead to delayed muscle regeneration after CTX damage (Delaunay et al., 2008). Cbl-b, another ubiquitin E3 ligase, functions via RANTES production in macrophages by regulating cytotoxic T-cell infiltration into regenerating muscles. The ablation of Cbl-b gene expression delays muscle regeneration following CTX-induced muscle damage (Kohno et al., 2011). Another recent study demonstrated that the E3 ligase TRIM32 is expressed in the skeletal muscle stem cell lineage of adult mice; myogenic differentiation is disrupted and muscle regeneration is delayed in the absence of TRIM32 (Nicklas et al., 2012). TRAF6, also an ubiquitin E3 ligase, positively regulates myogenic differentiation, thereby promoting myogenic differentiation and muscle regeneration via TAK1/p38 MAPK and Akt pathways (Xiao et al., 2012). The current study defined a novel function of the E3 ligase RNF13. In particular, RNF13 regulates muscle regeneration by mediating the inflammation phase. Considering the findings in the present study and previous studies, we found that E3 ligase-mediated protein modification is important in the regulation of skeletal muscle development and regeneration.

The inflammatory response is a coordinated process that should be finely regulated if an efficient regenerative process is initiated; this progression depends on a balance between excessive and insufficient inflammatory activities (Tidball and Villalta, 2010). In particular, functional inflammation requires well-regulated inflammatory cell infiltration and cytokine secretion. For example, CC chemokine ligand 2 (CCL2) regulates inflammatory responses by affecting the mobilization of macrophages from the bone marrow to the blood and from the blood to injured muscles (Lu et al., 2011). In addition, the local expression of insulin-like growth factor I (IGF-1) rapidly modulates specific inflammatory factors, thereby accelerating regeneration by influencing cytokine secretion after injury (Pelosi et al., 2007).

In our study, RNF13 deficiency did not influence the infiltration of neutrophils or macrophages into the damaged areas. Thus, cytokine concentrations were surprisingly altered in *RNF13*$^{-/-}$ mice. Among the important cytokines upregulated in *RNF13*$^{-/-}$ mice, IL-4 and IL-6 were involved in RNF13 function. In our previous study, snapin, a member of the SNARE complex, was found as a substrate of RNF13. Snapin, expressed ubiquitously in neuronal and non-neuronal cells, interacts with SNAP23 and likely performs a general function in SNARE-mediated vesicle fusion events in non-neuronal cells (BUXTON, 2003). In addition, a subset of t-SNAREs (syntaxin 4/SNAP23/Munc18c) performs a regulatory function in cytokine secretion in macrophages (Pagan, 2003). Therefore, RNF13 likely regulates cytokine secretion in inflammatory cells by modulating t-SNARE assembly, in the same mechanism by which the neuronal SNARE (Vamp2/SNAP25) assembly in the hippocampus is affected (Zhang et al., 2013b). On the basis of this supposition, we confirmed that syntaxin 4, SNAP23, and snapin were selectively expressed in sorted macrophages and neutrophils from damaged muscles by qRT-PCR (Fig. S3A). Considering that RNF13 regulates the association of neuronal SNARE components (Vamp2, SNAP25, and synaptotagmin 1), we also performed co-immunoprecipitation experiments to investigate whether or not RNF13 depletion affects the SNARE complex (syntaxin4 and SNAP23) assembly in sorted macrophages or in the total damaged muscle. Unexpectedly, RNF13 ablation did not regulate the interaction of syntaxin 4 with SNAP23 and snapin (Fig. S3B). Therefore, RNF13 functions in muscle regeneration are not mediated through snapin, suggesting that another substrate of RNF13 is involved. As such, further studies should be conducted to identify the RNF13 substrate(s) involved in the regulation of satellite cell function via macrophage-secreted IL-4/IL-6 in damaged muscles.

Although our results failed to explain the exact mechanism of RNF13 function in detail, the study demonstrated that RNF13 deficiency promotes cytokine production/secretion. Cytokines are a significant part of the satellite cell niche, regulating satellite cell activation, proliferation, and differentiation. The key to normal muscle regeneration is the maintenance of niche homeostasis in response to alterations induced by various physiological stimuli. A deficiency of RNF13 creates a niche that fosters satellite cell function by enhancing IL-4 and IL-6 activities; as a result, muscle regeneration is accelerated. Our finding that RNF13 deficiency promotes muscle regeneration provides new information to develop strategies for the treatment of muscle diseases. This result also emphasizes the importance of targeting specific inflammatory cell populations to optimize therapeutic approaches.

MATERIALS AND METHODS

Mice and skeletal muscle regeneration

The generation of *RNF13*$^{-/-}$ mice has been described previously (Zhang et al., 2013b). *RNF13*$^{-/-}$ and *RNF*13$^{+/+}$ mice were produced by crossing *RNF13*$^{+/-}$ heterozygous mice. The animals were handled and cared for in accordance with the guidelines of the Animal Ethics Committee of Peking Union Medical College. Muscle regeneration was monitored in mice (8–10 weeks old) after inducing muscle injury by injecting 20 μL of 10 μmol/L CTX (Sigma) in phosphate-buffered saline (PBS) into the mid-belly of the right TA muscle; the left TA muscle was injected with PBS (20 μL) as a control. For rescue experiments, IL-4 and IL-6 antibodies (10 μL of 1 μg/μL in PBS; Biolegend) were injected into the TA muscle 2 h before CTX injection.

Muscle histology and hematoxylin and eosin staining

TA muscles were harvested 1, 3, 5, and 7 d following CTX injury. The muscles were fixed in 4% paraformaldehyde (Sigma) in PBS overnight and paraffin embedded for hematoxylin and eosin (H&E) staining. The proportion of fibers with central nuclei (regenerating fibers) was counted in the injured area, and the CSAs of the fibers were measured using Image Pro Plus 5.1 software (Olympus).

Immunofluorescence staining of frozen sections

For Pax7 immunostaining, fresh sections were fixed in 4% paraformaldehyde for 20 min, permeabilized with methanol (-20°C) for 6 min, and blocked first with a solution containing 4% bovine serum albumen (BSA; Jackson) in PBS and then with goat anti-mouse AffiniPure Fab fragment (1:100; Jackson) after antigen retrieval with 100 mmol/L sodium citrate. The sections were then incubated overnight at 4°C with an anti-Pax7 antibody (1:20; Developmental Studies Hybridoma Bank). After washing with PBS, Pax7 signals were visualized by incubating with biotin-conjugated goat anti-mouse IgG_1 (1:1000; Jackson) and Cy3-conjugated streptavidin (Jackson; 1:2500). Nuclei were stained with 4',6-diamidino-2-phenylindole (DAPI; Invitrogen). For MyoD immunostaining, the fresh sections were fixed in 4% paraformaldehyde for 20 min, permeabilized with 0.2% Triton X-100/PBS (PBST) for 10 min, and blocked by incubating with 5% BSA/6% goat serum/0.2% PBST at room temperature for 2 h. Immunostaining with anti-MyoD antibody (1:50; Santa Cruz) was performed by overnight incubation at 4°C. After washing, immunoreactive proteins were visualized by incubating with fluorescein isothiocyanate (FITC)-conjugated goat anti-rabbit IgG. Nuclei were stained with DAPI (Roche). The Pax7^{+} and MyoD^{+} cells in the sections (n = 10) were counted.

BrdU incorporation assay

For *in vivo* BrdU incorporation assays, 125 mg/kg of BrdU (Sigma) was injected into TA muscles 2 h prior to harvesting. Frozen sections were fixed in 4% paraformaldehyde for 20 min, incubated in 2 mol/L HCl at 37°C for 30 min for DNA denaturation, and then immersed twice in 0.1 mol/L borate buffer (5 min each) to neutralize the acid. After three washes with PBS, the sections were blocked with 5% goat serum for 30 min and then incubated with anti-BrdU (Abcam) at 4°C overnight. The sections were then washed with PBS and incubated with FITC-conjugated goat anti-rat IgG to visualize BrdU signals. BrdU-positive cells were quantified in 60 sections from six mice.

Quantitative PCR analysis

Total RNA was extracted using the TRIzol reagent (Life Technologies) and reverse transcribed using RevertAid reverse transcriptase (Thermo Scientific). Quantitative reverse transcription-polymerase chain reaction (qRT-PCR) analyses were performed using a Bio-Rad iQ5 Multicolor Real-Time PCR Detection System (Bio-Rad). The primer sequences are listed in Table S1.

Western blot analysis

Muscle tissues were lysed in a buffer containing 50 mmol/L Tris pH 7.5, 150 mmol/L NaCl, 0.5% Nonidet P-40, and protease and phosphatase inhibitors. Protein lysates were resolved by sodium dodecyl sulfate-polyacrylamide gel electrophoresis and transferred to a polyvinylidene fluoride membrane, followed by immunoblotting with primary antibodies against Pax7 (Developmental Studies

Hybridoma Bank), MyoD (BD Biosciences), and glyceraldehyde 3-phosphate dehydrogenase (GAPDH; Millipore). Membranes were washed for 30 min and incubated with horseradish peroxidase-conjugated secondary antibodies (Zhongshanjinqiao Corp.) for 1 h at room temperature and then washed with PBS for 30 min. Each membrane was then placed into a Detection Solution (Thermo), incubated for 1 min at room temperature, and subsequently exposed to X-ray film. Band densities were quantified using Image J software.

Single fiber isolation and culture

Single myofibers were isolated from the gastrocnemius muscles of eight-week-old mice by collagenase I (Sigma, C-0130) digestion (Rosenblatt et al., 1995). Each gastrocnemius muscle was harvested and incubated in 3 mL of 0.2% collagenase I in Dulbecco's modified Eagle's medium (DMEM; without serum) in a shaking water bath at 35°C for 60–90 min. Digestion was complete when the muscle looked less defined and slightly swollen, with hair-like single fibers appearing to come away from the edge of the muscle. The muscles were placed in a Petri dish, and myofibers were isolated under a microscope. Single fibers were placed in six-well plates precoated with Matrigel (1:3; Becton Dickinson) and allowed to attach for 3 min. Then, 2 mL of fiber medium consisting of DMEM supplemented with 10% horse serum, 0.5% chick embryo extract, and 1% antibiotic/antimycotic was added. The fibers were cultured for 24 h or 72 h at 37°C/5% CO_2, fixed with 4% paraformaldehyde and stained for Pax7 and MyoD, as described above. The numbers of $Pax7^+/MyoD^+$ cells determined from counts of at least 100 single fibers were used for statistical analyses.

Isolation of macrophages and neutrophils by cell sorting

Flow cytometry was performed using muscle single-cell suspensions. TA muscles, damaged for 1 d or 3 d, were harvested from mice; washed in PBS; minced in 2 mL of digestion solution containing 0.2% collagenase II (Sigma), 2.4 U/mL of dispase (Sigma), and 2.5 mmol/L $CaCl_2$; and incubated at 37°C with shaking for 60–90 min. Digestion was stopped by adding an equal volume of 10% FBS in PBS, after which the top liquid layer was filtered through a 40-μm nylon mesh to remove particulates, and centrifuged at 350 *g* for 3 min. The cells were washed twice with PBS and then suspended in PBS. These mononuclear cells were stained with CD11b-PE-CY7 (1:500; BD Biosciences, 561098) and Ly-6G-FITC (1:1000; BD Biosciences, 561105). The cells were sorted using a BD FACS Aria II fluorescence-activated cell sorter. $CD11b^+/Ly\text{-}6G^+$ cells were defined as neutrophils and $CD11b^+/Ly\text{-}6G^-$ cells were defined as macrophages.

Cytokine detection

TA muscles from *RNF13*$^{-/-}$ and *RNF13*$^{+/+}$ mice were harvested at different time points after damage (0, 4 h, and 12 h, and 1 d and 3 d). Muscle tissues were lysed in a buffer containing 50 mmol/L Tris pH 7.5, 150 mmol/L NaCl, 0.5% Nonidet P-40, and protease and phosphatase inhibitors. The concentration of 18 cytokines and chemokines in the damaged muscles was detected using MILLIPLEX MAP Multiplex Assay Kits, according to the manufacturer's instructions (Millipore).

Statistical analysis

Results are presented as means ± SEs. The statistical significance of differences between two means was calculated using Student's *t*-test. The probability level accepted for significance was $P < 0.05$.

ACKNOWLEDGEMENTS

We are grateful to our laboratory staff for providing support throughout this study. This study was supported by the National Basic Research Program (973 Program) (Nos. 2011CBA01104, 2011CB965002 and 2014CB964703) and the National Natural Science Foundation of China (Grant Nos. 31030041, 91019010 and 31271470).

ABBREVIATIONS

CSA, cross-sectional areas; CTX, cardiotoxin; IL-4/6, interleukin 4/6; MyoD, myogenic differentiation antigen; Pax7, paired-box transcription factor; RNF13, RING finger protein 13; SNARE, soluble N-ethylmaleinide-sensitive fusion protein attachment protein receptor; TA, tibialis anterior; UPS, Ub-proteasome system.

COMPLIANCE WITH ETHICS GUIDELINES

Jiao Meng, Xiaoting Zou, Rimao Wu, Ran Zhong, Dahai Zhu and Yong Zhang declare that they have no conflict of interest.

For animal experiments, all institutional and national guidelines for the care and use of laboratory animals were followed.

REFERENCES

Beauchamp JR, Heslop L, Yu DS, Tajbakhsh S, Kelly RG, Wernig A, Buckingham ME, Partridge TA, Zammit PS (2000) Expression of CD34 and Myf5 defines the majority of quiescent adult skeletal muscle satellite cells. J Cell Biol 151:1221–1234

Brunelli S, Rovere-Querini P (2008) The immune system and the repair of skeletal muscle. Pharmacol Res 58:117–121

Buxton P (2003) Identification and characterization of Snapin as a ubiquitously expressed SNARE-binding protein that interacts with SNAP23 in non-neuronal cells. Biochem J 375:433–440

Carosio S, Berardinelli MG, Aucello M, Musaro A (2009) Impact of ageing on muscle cell regeneration. Ageing Res Rev 10:35–42

Cerletti M, Shadrach JL, Jurga S, Sherwood R, Wagers AJ (2008) Regulation and function of skeletal muscle stem cells. Cold Spring Harbor Symp Quant Biol 73:317–322

Chazaud B, Brigitte M, Yacoub-Youssef H, Arnold L, Gherardi R, Sonnet C, Lafuste P, Chretien F (2009) Dual and beneficial roles of macrophages during skeletal muscle regeneration. Exerc Sport Sci Rev 37:18–22

Chen SE, Gerken E, Zhang Y, Zhan M, Mohan RK, Li AS, Reid MB, Li YP (2005) Role of TNF-{alpha} signaling in regeneration of cardiotoxin-injured muscle. Am J Physiol Cell Physiol 289: C1179–C1187

Chen SE, Jin B, Li YP (2007) TNF-alpha regulates myogenesis and muscle regeneration by activating p38 MAPK. Am J Physiol Cell Physiol 292:C1660–C1671

Cornelison DD, Wilcox-Adelman SA, Goetinck PF, Rauvala H, Rapraeger AC, Olwin BB (2004) Essential and separable roles for Syndecan-3 and Syndecan-4 in skeletal muscle development and regeneration. Genes Dev 18:2231–2236

Delaunay A, Bromberg KD, Hayashi Y, Mirabella M, Burch D, Kirkwood B, Serra C, Malicdan MC, Mizisin AP, Morosetti R et al (2008) The ER-bound RING finger protein 5 (RNF5/RMA1) causes degenerative myopathy in transgenic mice and is deregulated in inclusion body myositis. PLoS One 3:e1609

Deng B, Wehling-Henricks M, Villalta SA, Wang Y, Tidball JG (2012) IL-10 triggers changes in macrophage phenotype that promote muscle growth and regeneration. J Immunol 189:3669–3680

Dhawan J, Rando TA (2005) Stem cells in postnatal myogenesis: molecular mechanisms of satellite cell quiescence, activation and replenishment. Trends Cell Biol 15:666–673

Fukada S, Uezumi A, Ikemoto M, Masuda S, Segawa M, Tanimura N, Yamamoto H, Miyagoe-Suzuki Y, Takeda S (2007) Molecular signature of quiescent satellite cells in adult skeletal muscle. Stem Cells 25:2448–2459

Gnocchi VF, White RB, Ono Y, Ellis JA, Zammit PS (2009) Further characterisation of the molecular signature of quiescent and activated mouse muscle satellite cells. PLoS One 4:e5205

Gopinath SD, Rando TA (2008) Stem cell review series: aging of the skeletal muscle stem cell niche. Aging Cell 7:590–598

Halevy O, Piestun Y, Allouh MZ, Rosser BW, Rinkevich Y, Reshef R, Rozenboim I, Wleklinski-Lee M, Yablonka-Reuveni Z (2004) Pattern of Pax7 expression during myogenesis in the posthatch chicken establishes a model for satellite cell differentiation and renewal. Dev Dyn 231:489–502

Hawke TJ (2001) Myogenic satellite cells: physiology to molecular biology. J Appl Physiol 9:534–551

Heredia JE, Mukundan L, Chen FM, Mueller AA, Deo RC, Locksley RM, Rando TA, Chawla A (2013) Type 2 innate signals stimulate fibro/adipogenic progenitors to facilitate muscle regeneration. Cell 153:376–388

Hirata A, Masuda S, Tamura T, Kai K, Ojima K, Fukase A, Motoyoshi K, Kamakura K, Miyagoe-Suzuki Y, Takeda S (2003) Expression profiling of cytokines and related genes in regenerating skeletal muscle after cardiotoxin injection: a role for osteopontin. Am J Pathol 163:203–215

Jin X, Cheng H, Chen J, Zhu D (2011) RNF13: an emerging RING finger ubiquitin ligase important in cell proliferation. FEBS J 278:78–84

Kanisicak O, Mendez JJ, Yamamoto S, Yamamoto M, Goldhamer DJ (2009) Progenitors of skeletal muscle satellite cells express the muscle determination gene, MyoD. Dev Biol 332:131–141

Kharraz Y, Guerra J, Mann CJ, Serrano AL, Munoz-Canoves P (2013) Macrophage plasticity and the role of inflammation in skeletal muscle repair. Mediat Inflamm 2013:491497

Kohno S, Ueji T, Abe T, Nakao R, Hirasaka K, Oarada M, Harada-Sukeno A, Ohno A, Higashibata A, Mukai R et al (2011) Rantes secreted from macrophages disturbs skeletal muscle regeneration after cardiotoxin injection in Cbl-b-deficient mice. Muscle Nerve 43:223–229

Kuang S, Gillespie MA, Rudnicki MA (2008) Niche regulation of muscle satellite cell self-renewal and differentiation. Cell Stem Cell 2:22–31

Lu H, Huang D, Saederup N, Charo IF, Ransohoff RM, Zhou L (2011) Macrophages recruited via CCR2 produce insulin-like growth factor-1 to repair acute skeletal muscle injury. FASEB J 25:358–369

Mauro A (1961) Satellite cell of skeletal muscle fibers. J Biophys Biochem Cytol 9:493–495

Nicklas S, Otto A, Wu X, Miller P, Stelzer S, Wen Y, Kuang S, Wrogemann K, Patel K, Ding H et al (2012) TRIM32 regulates skeletal muscle stem cell differentiation and is necessary for normal adult muscle regeneration. PLoS One 7:e30445

Olguin HC, Olwin BB (2004) Pax-7 up-regulation inhibits myogenesis and cell cycle progression in satellite cells: a potential mechanism for self-renewal. Dev Biol 275:375–388

Pagan JK (2003) The t-SNARE Syntaxin 4 is regulated during macrophage activation to function in membrane traffic and cytokine secretion. Curr Biol 13:156–160

Pelosi L, Giacinti C, Nardis C, Borsellino G, Rizzuto E, Nicoletti C, Wannenes F, Battistini L, Rosenthal N, Molinaro M et al (2007) Local expression of IGF-1 accelerates muscle regeneration by rapidly modulating inflammatory cytokines and chemokines. FASEB J 21:1393–1402

Relaix F, Rocancourt D, Mansouri A, Buckingham M (2005) A Pax3/Pax7-dependent population of skeletal muscle progenitor cells. Nature 435:948–953

Rosenblatt JD, Lunt AI, Parry DJ, Partridge TA (1995) Culturing satellite cells from living single muscle fiber explants. In Vitro Cell Dev Biol Anim 31:773–779

Segawa M, Fukada S, Yamamoto Y, Yahagi H, Kanematsu M, Sato M, Ito T, Uezumi A, Hayashi S, Miyagoe-Suzuki Y et al (2008) Suppression of macrophage functions impairs skeletal muscle regeneration with severe fibrosis. Exp Cell Res 314: 3232–3244

Ten Broek RW, Grefte S, Von den Hoff JW (2010) Regulatory factors and cell populations involved in skeletal muscle regeneration. Journal of cellular physiology 224:7–16

Tidball JG (2005) Inflammatory processes in muscle injury and repair. Am J Physiol Regul Integr Comp Physiol 288:R345–R353

Tidball JG, Villalta SA (2010) Regulatory interactions between muscle and the immune system during muscle regeneration. Am J Physiol Regul Integr Comp Physiol 298:R1173–R1187

Warren GL, Hulderman T, Jensen N, McKinstry M, Mishra M, Luster MI, Simeonova PP (2002) Physiological role of tumor necrosis factor alpha in traumatic muscle injury. FASEB J 16:1630–1632

Xiao F, Wang H, Fu X, Li Y, Wu Z (2012) TRAF6 promotes myogenic differentiation via the TAK1/p38 mitogen-activated protein kinase and Akt pathways. PLoS One 7:e34081

Zammit P, Beauchamp J (2001) The skeletal muscle satellite cell: stem cell or son of stem cell? Differentiation 68:193–204

Zammit PS, Heslop L, Hudon V, Rosenblatt JD, Tajbakhsh S, Buckingham ME, Beauchamp JR, Partridge TA (2002) Kinetics of myoblast proliferation show that resident satellite cells are competent to fully regenerate skeletal muscle fibers. Exp Cell Res 281:39–49

Zammit PS, Golding JP, Nagata Y, Hudon V, Partridge TA, Beauchamp JR (2004) Muscle satellite cells adopt divergent fates: a mechanism for self-renewal? J Cell Biol 166:347–357

Zhang Q, Meng Y, Zhang L, Chen J, Zhu D (2009) RNF13: a novel RING-type ubiquitin ligase over-expressed in pancreatic cancer. Cell Res 19:348–357

Zhang Q, Wang K, Zhang Y, Meng J, Yu F, Chen Y, Zhu D (2010) The myostatin-induced E3 ubiquitin ligase RNF13 negatively regulates the proliferation of chicken myoblasts. FEBS J 277:466–476

Zhang C, Li Y, Wu Y, Wang L, Wang X, Du J (2013a) Interleukin-6/signal transducer and activator of transcription 3 (STAT3) pathway is essential for macrophage infiltration and myoblast proliferation during muscle regeneration. J Biol Chem 288:1489–1499

Zhang Q, Li Y, Zhang L, Yang N, Meng J, Zuo P, Zhang Y, Chen J, Wang L, Gao X et al (2013b) E3 ubiquitin ligase RNF13 involves spatial learning and assembly of the SNARE complex. Cell Mol Life Sci 70:153–165

20

Murine gammaherpesvirus-68 ORF38 encodes a tegument protein and is packaged into virions during secondary envelopment

Sheng Shen[1,2], Haitao Guo[1,2], Hongyu Deng[1]✉

[1] Chinese Academy of Sciences Key Laboratory of Infection and Immunity, Institute of Biophysics, Chinese Academy of Sciences, Beijing 100101, China
[2] University of Chinese Academy of Sciences, Beijing 100049, China
✉ Correspondence: hydeng@moon.ibp.ac.cn (H. Deng)

ABSTRACT

Tegument is the unique structure of a herpesvirion which occupies the space between nucleocapsid and envelope. Accumulating data have indicated that interactions among tegument proteins play a key role in virion morphogenesis. Morphogenesis of gammaherpesviruses including Kaposi's sarcoma-associated herpesvirus (KSHV) and Epstein-Barr virus (EBV) is poorly understood due to the lack of efficient de novo lytic replication in cell culture. Murine gammaherpesvirus-68 (MHV-68) is genetically related to these two human herpesviruses and serves as an effective model to study the lytic replication of gammaherpesviruses. We previously showed that ORF33 of MHV-68 encodes a tegument protein and plays an essential role in virion maturation in the cytoplasm. However, the molecular mechanism of how ORF33 participates in virion morphogenesis has not been elucidated. In this study we demonstrated that ORF38 of MHV-68 is also a tegument protein and is localized to cytoplasmic compartments during both transient transfection and viral infection. Immuno-gold labeling assay showed that ORF38 is only present on virions that have entered the cytoplasmic vesicles, indicating that ORF38 is packaged into virions during secondary envelopment. We further showed that ORF38 co-localizes with ORF33 during viral infection; therefore, the interaction between ORF38 and ORF33 is conserved among herpesviruses. Notably, we found that although ORF33 by itself is distributed in both the nucleus and the cytoplasm, in the presence of ORF38, ORF33 is co-localized to trans-Golgi network (TGN), a site where secondary envelopment takes place.

KEYWORDS gammaherpesvirus, tegument protein, ORF38, ORF33, TGN

INTRODUCTION

The herpesviridae is a family of large DNA viruses. Based on genome structure and biological properties, herpesviruses are classified into three subfamilies, *Alphaherpesvirinae*, *Betaherpesvirinae* and *Gammaherpesvirinae*. Gammaherpesviruses are also called tumor-associated herpesviruses. Epstein-Barr virus (EBV) and Kaposi's sarcoma-associated herpesvirus (KSHV), two human gammaherpesviruses, are causative agents of several malignancies, such as lymphomas, nasopharyngeal carcinoma and Kaposi's sarcoma (Soulier et al., 1995; Ganem, 2006; Kutok and Wang, 2006). Like other herpesviruses, EBV and KSHV are characterized as having two distinct phases in their life cycle: lytic replication and latency (Sunil-Chandra et al., 1992; Decker et al., 1996a, 1996b). Studies have shown that both lytic replication and latency play important roles in tumorigenesis (Bais et al., 1998; Martin et al., 1999; Stine et al., 2000; Blackman and Flano, 2002; Grundhoff and Ganem, 2004; Stevenson and Efstathiou, 2005).

A significant event during lytic replication is the formation of mature virions which are released into extracellular space to initiate new infection cycles in neighboring uninfected cells. A mature herpesvirion consists of a DNA genome, an icosahedral capsid, a thick tegument and an envelope containing lipid bilayer spiked with viral glycoproteins (Dai et al., 2008). Morphogenesis of a herpesvirion is a complicated multiple-step process. Briefly, it begins with the auto-assembly of capsid proteins into a capsid in the nucleus. After encapsidation of virus genome, a nucleocapsid travels to the inner nuclear membrane where it acquires primary envelope and enters into the perinuclear space. Nuclear egress or "de-envelopment" is accomplished by fusion of the primary envelope with the outer nuclear membrane and release of the nucleocapsid into the cytoplasm. The nucleocapsid acquires most tegument proteins and envelope proteins in the cytoplasm and completes its maturation at trans-Golgi network (TGN)-derived vesicles, a process called "secondary envelopment". A mature virion is finally released into the extracellular space through fusion of the vesicles with the cellular membrane (Mettenleiter et al., 2006; Guo et al., 2010).

Until now, studies on morphogenesis of herpesviruses have mostly focused on alphaherpesviruses such as herpes simplex virus-1 (HSV-1) and to a less extent on beta-herpesviruses such as human cytomegalovirus (HCMV) (Mettenleiter et al., 2009). Little is known about the morphogenesis of gammaherpesviruses due to the lack of efficient lytic replication of KSHV and EBV in cell culture. Murine gammaherpesvirus-68 (MHV-68) is genetically related to these two human herpesviruses and replicates robustly in cell culture, presenting an effective model to study the de novo infection process of gammaherpesviruses (Efstathiou et al., 1990; Simas and Efstathiou, 1998). Using multiple approaches, we previously showed that MHV-68 ORF33 is a tegument protein that is essential for virion morphogenesis (Guo et al., 2009). However, the molecular mechanism of ORF33 involved in virion morphogenesis remained unclear. ORF33 is a conserved tegument protein in herpesvirus family. Studies on its homologues in HSV-1 and HCMV (UL16 and UL94, respectively) showed that they both interact with another conserved tegument protein, designated as UL11 in HSV-1 and UL99 in HCMV, and that both interactions are involved in virion maturation in the cytoplasm (Yeh et al., 2008; Phillips et al., 2012). The homologue of UL11 in MHV-68 is ORF38. A previous study on genome-wide protein interaction network of MHV-68 has showed that ORF33 interacts with ORF38 (Lee et al., 2011); however, the role of ORF38 in MHV-68 morphogenesis has not been characterized.

In this study, we first demonstrated that MHV-68 ORF38 is a tegument protein and that it is localized to cytoplasmic compartments during both transient transfection and viral infection. We further showed that ORF38 is packaged into virions within cytoplasmic vesicles during secondary envelopment. In addition, we showed co-localization of ORF38 and ORF33 during viral infection. Notably, we found that although ORF33 by itself is distributed in both the nucleus and the cytoplasm, its interaction with ORF38 makes it localize to TGN, a site where secondary envelopment takes place.

RESULTS

MHV-68 *ORF38* encodes a tegument protein

To facilitate the characterization of MHV-68 ORF38, we first expressed and purified His-tagged ORF38 in *E. coli* and used it to generate a rabbit polyclonal antibody. Lysate from BHK-21 cells infected with MHV-68 was examined by Western blotting analysis to test the specificity of the antibody. A 12-kDa protein was detected in virus infected BHK cells (Fig. 1A, lane 1) but not in mock infected cells (Fig. 1A, lane 2). In contrast, no band was detected from either MHV-68-infected or mock-infected cells by preimmune rabbit serum (data not shown).

The MHV-68 *ORF38* encodes a protein of 75 aa and shares 27% amino acid sequence identity to its homologue in EBV (named BBLF1), which was identified as a component of the virion-associated proteins (Johannsen et al., 2004). The homologues of ORF38 in alpha- and beta-herpesvirus family were also reported to be virion-associated tegument proteins (Kattenhorn et al., 2004; Loret et al., 2008). Therefore, it is very likely that MHV-68 ORF38 is also a virion protein; however, a previous mass spectrometric analysis of purified MHV-68 virions did not report the identification of ORF38 (Bortz et al., 2003). We therefore decided to first determine whether ORF38 is associated with MHV-68 virions.

Mature MHV-68 virions were harvested from supernatant of MHV-68-infected cells and purified through gradient ultracentrifugation. Virions were lysed, resolved on SDS-PAGE, and analyzed by Western blotting, using the anti-ORF38 polyclonal antibody. A 12-kDa protein, consistent with the molecular weight of ORF38 found in MHV-68-infected cell lysate (Fig. 1B, lane 2), was detected in virion lysate (Fig. 1B, lane 1), demonstrating that ORF38 protein is indeed associated with viral particles. We next performed trypsin and detergent treatment experiment, a classical assay to determine the detailed localization of a virion protein within the viral particles (Bortz et al., 2003), and compared the sensitivity of ORF38 to trypsin and/or detergent treatment to that of known capsid protein ORF26, tegument protein ORF33, and envelope protein glycoprotein B (gB). As expected, gB was sensitive to trypsin in both the presence and the absence of detergent (Fig. 1C, bottom panel) whereas capsid protein ORF26 was resistant to trypsin and/or detergent treatment (Fig. 1C, top panel). Similar to ORF33, ORF38 was sensitive to trypsin treatment in the presence of detergent (Fig. 1C, lane 4). However, unlike ORF33, ORF38 was partially sensitive to detergent in the absence of trypsin (Fig. 1C, lane 2). As ORF33 has been identified as a tegument protein of MHV-68 that is associated tightly with the capsid (Guo et al., 2009), these data demonstrated that ORF38 is a tegument protein but binds less strongly to capsid than does ORF33.

MHV-68 ORF38 localizes to trans-Golgi network (TGN)

The fact that ORF38 binds loosely to capsid suggests that ORF38 may be an outer tegument protein and participate in the secondary envelopment of viral particles. As the function of a protein is inevitably linked to its localization within the

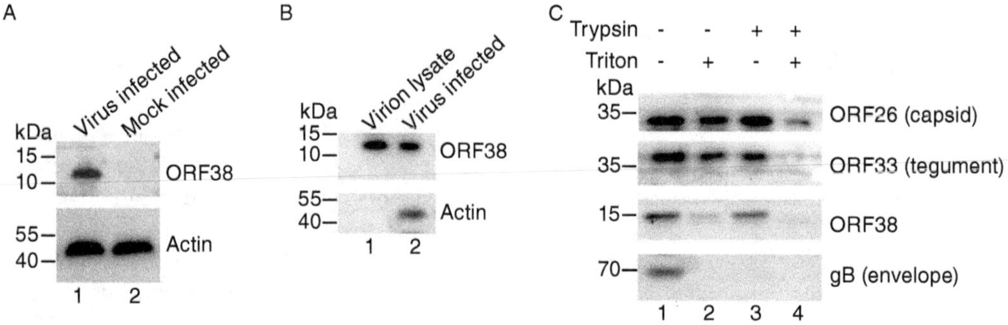

Figure 1. MHV-68 *ORF38* encodes a tegument protein. (A) *ORF38* is expressed during MHV-68 lytic infection. BHK cells were infected with WT MHV-68 at an MOI of 3. At 24 hpi, cells were lysed and subjected to Western blotting analysis with an anti-ORF38 polyclonal antibody or a monoclonal antibody specific for β-actin. Lysate from mock infected cells was used as a negative control. (B) ORF38 is a component of mature MHV-68 virions. Mature virions purified by gradient centrifugation were resolved via SDS-PAGE and analyzed using the anti-ORF38 polyclonal antibody. Lysate from virus infected BHK cells was used as a positive control. (C) Sensitivity of ORF38 to detergent and trypsin treatment. The purified virions were treated with trypsin or detergent or both, as described in MATERIALS AND METHODS. The resulting tegument-nucleocapsid complexes were subjected to Western blotting analysis individually using anti-ORF26, anti-ORF33, anti-ORF38 and anti-gB polyclonal antibodies.

cell, we examined the subcellular localization of ORF38. Cells were infected with MHV-68 and indirect immunofluorescence assay was performed using the anti-ORF38 antibody. As shown in Fig. 2A, ORF38 mainly localized to the cytoplasm, with a predominant distribution in the juxtanuclear space. To further define the subcellular localization of ORF38, we investigated the co-localization of ORF38 with a panel of known protein markers for subcellular compartments. These markers, TGN38 for TGN, EEA1 for early endosome, LAMP1 for late endosome, and VPS37C for multivesicular bodies (MVB), respectively, are all fused to GFP. To facilitate the analysis, we first constructed a plasmid that expresses ORF38 fused to the N-terminus of RFP (named pDsRed-38). We then conducted co-transfection experiment with pDsRed-38 and a plasmid encoding one of the markers. As shown in Fig. 2B, ORF38 displayed partial co-localization with the TGN marker. Significant signal overlap was also observed between ORF38 and the late endosome marker. In contrast, minimal co-localization was observed between ORF38 and the early endosome marker or the MVB marker. As herpesvirus secondary envelopment is believed to occur at the TGN (Mettenleiter et al., 2009), the localization of ORF38 in TGN is consistent with its potential role in secondary envelopment during virion morphogenesis.

ORF38 is packaged into virions during secondary envelopment

To determine the role of ORF38 in MHV-68 virion morphogenesis, we next performed an immuno-gold labeling assay to examine at which step of virion morphogenesis ORF38 is packaged into virions. Cells were infected with MHV-68, fixed and embedded. Samples were processed and ultrathin sections were obtained. The sections were immunolabeled with either the anti-ORF38 rabbit polyclonal antibody or the rabbit IgG as a negative control, followed by a secondary antibody conjugated with 5-nm gold particles. In the negative control sample treated with rabbit IgG, no gold particle was found on virions (Fig. 3A). In the sample treated with the anti-ORF38 antibody, viral particles were readily found in the nucleus, however, no gold particle was detected either on the viral particles or in other areas of the nucleus (Fig. 3B), consistent with the cytoplasmic localization of ORF38 as demonstrated by our immunofluorescence data (Fig. 2A). In contrast, abundant ORF38 was present around or within vesicles in the cytoplasm where virions were undergoing secondary envelopment (Fig. 3C and 3D). Notably, ORF38 was only present on virions that had entered vesicles but not on virions outside the vesicles (Fig. 3D). These data clearly indicated that the tegument protein ORF38 is packaged into virions during secondary envelopment.

ORF38 interacts with ORF33

Tegumentation is an important stage of morphogenesis for a herpesvirus particle and involves a complex and dynamic process of protein-protein interactions. We therefore tested the interaction of ORF38 with other tegument proteins and found out that ORF38 interacts with ORF33, consistent with a previous report (Lee et al., 2011). Briefly, coimmunoprecipitation (co-IP) experiment was performed with cell lysates from 293T cells transiently transfected with either a plasmid expressing HA-tagged ORF33 (HA-33) or a plasmid expressing 3xFLAG-CBP-tagged ORF38 (TAG-38) or both. As shown in Fig. 4A, ORF38 interacted with ORF33 in the absence of other viral proteins. To confirm that the interaction between ORF33 and ORF38 exists in the context of viral infection, we next determined the relative localization of

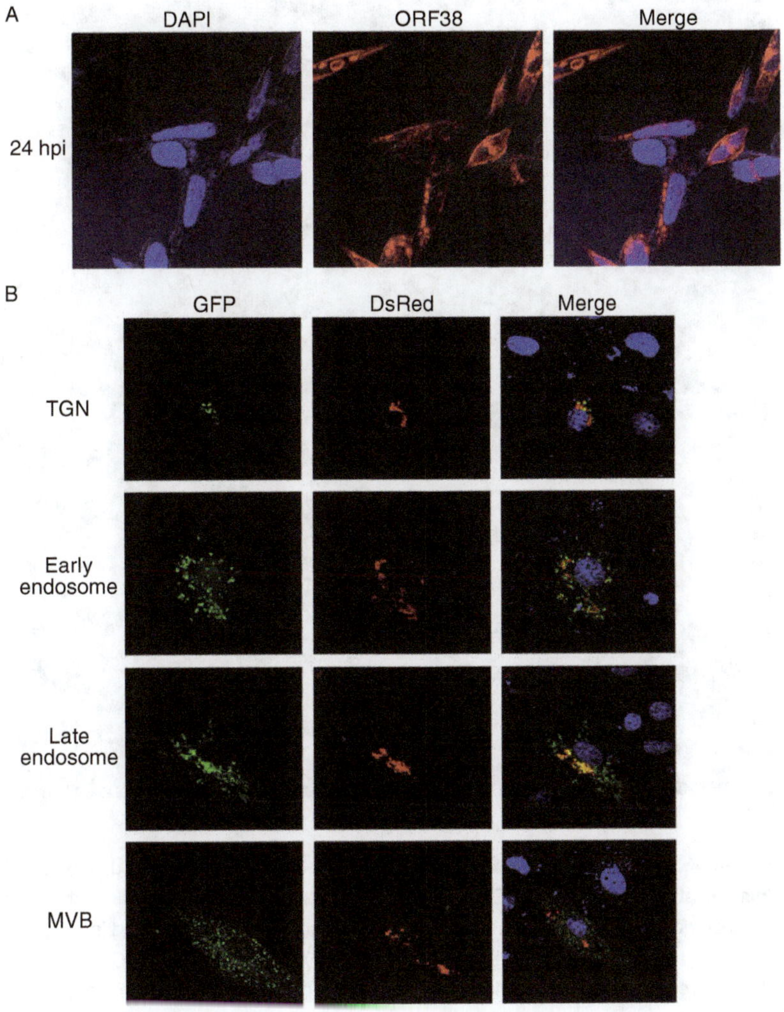

Figure 2. Subcellular localization of MHV-68 ORF38. (A) ORF38 localizes to the cytoplasm. Vero cells were infected with WT MHV-68 at an MOI of 3. At 24 hpi, cells were fixed for indirect immunofluorescence analysis using the anti-ORF38 polyclonal antibody. (B) Co-localization of ORF38 with different cytoplasmic organelles. Vero cells were cotransfected with pDsRed-38 plus pTGN-38 GFP or pEEA1-GFP or pLAMP1-GFP or pVPS37C-GFP. At 24 h posttransfection, cells were fixed, stained with DAPI, and examined with a confocal laser microscope.

these two proteins during viral infection. Cells were infected with MHV-68 and indirect immunofluorescence assay was performed with the anti-ORF38 antibody and an anti-ORF33 antibody we previously prepared. Consistent with our previous observation (Fig. 2), endogenously expressed ORF38 was localized to the cytoplasm (Fig. 4B, column 2). We previously showed that ORF33 was distributed in both the cytoplasm and the nucleus in the absence of other viral proteins (Guo et al., 2009). Here we observed that, in the context of viral infection, endogenously expressed ORF33 accumulated more in the nucleus than in the cytoplasm at 18 h or 24 h postinfection (Fig. 4B, column 3). More interestingly, at these time points, the cytoplasmic portion of ORF33 proteins co-localized with ORF38 (Fig. 4B, column 4).

ORF38 and ORF33 co-localize with TGN

The fact that only the cytoplasmic portion of ORF33 co-localized with ORF38 during viral infection suggested that other viral factors may interact with ORF33 and affect its localization. To focus on the role of ORF38, we further analyzed the interaction of this pair of proteins in the absence of other viral proteins. Cells were transfected with pEGFP-33 or pDsRed-38 or both, stained with DAPI and then observed with a confocal laser microscope. As shown in Fig. 5A, ORF33 by itself localized to both the nucleus and the cytoplasm, whereas ORF38 by itself localized to juxta-nuclear space within the cytoplasm. Notably, in the presence of ORF38, almost all ORF33 accumulated in the cytoplasmic areas where ORF38 localized to. Further analysis using

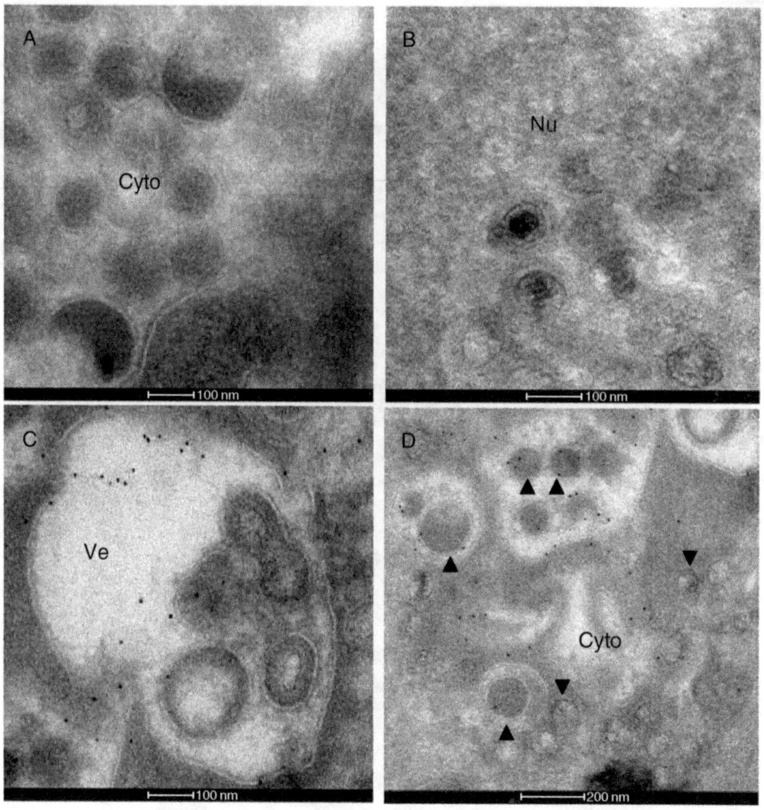

Figure 3. Immuno-gold labeling of ORF38 in MHV-68 infected cells. BHK cells were infected with WT MHV-68 at an MOI of 5. Ultrathin sections were immunolabeled with rabbit IgG as a negative control (A) or anti-ORF38 rabbit polyclonal antibody (B–D), followed by a secondary antibody conjugated with 5-nm gold particles. No gold particle was found in samples treated with rabbit IgG (A). ORF38 was not detected in the nucleus (Nu) of virus infected cells (B). In the cytoplasm (Cy), ORF38 was found around and within the vesicles (C and D) as well as on virions within the vesicles (▲) (D) but not found on virions outside the vesicles (▼)(D). Bars: A, B and C, 100 nm; D, 200 nm.

TGN38, the protein marker for TGN, demonstrated that the cytoplasmic compartment that ORF33 and ORF38 co-localized with is TGN (Fig. 5B).

DISCUSSION

Studies from all herpesvirus subfamilies indicated that tegument, composed mostly of tegument proteins, plays a crucial role during virion morphogenesis. Tegumentation is a complicated process that involves multiple protein-protein interactions (Bortz et al., 2007; Guo et al., 2010; Phillips and Bresnahan, 2012; Wang et al., 2012). However, the molecular mechanisms responsible for gammaherpesvirus tegumentation remained unclear. Here we report that MHV-68 ORF38 is a virion protein (Fig. 1B). Furthermore, our trypsin and detergent sensitivity data indicated that ORF38 is a tegument protein but binds to capsid less tightly than ORF33 does, since ORF38 fell off from virions in the presence of detergent only (Fig. 1C). Therefore ORF38 is very likely a component of the outer tegument and associated with glycoprotein(s) in the virion. This idea is consistent with its localization at TGN (Figs. 2B and 5B), where most viral glycoproteins are located.

Tegument formation is generally thought to originate during primary envelopment, with additional tegument proteins being incorporated in the cytoplasm following nuclear egress and finally at the TGN-derived vesicles. Different tegument proteins are added onto virions at different intracellular locations (Kelly et al., 2009). Using the immuno-gold labeling assay, we previously showed that MHV-68 ORF52, a tegument protein essential for virion maturation in the cytoplasm, is present on both immature virions in the cytoplasm and on virions that have entered vesicles (Wang et al., 2012). Unlike ORF52, ORF38 is only present on virions that have entered the vesicles but not present on immature virions outside the vesicles (Fig. 3). Abundant ORF38 is also present on these vesicles where secondary envelopment takes place. Together, these data indicate that ORF38 is packaged into virions during secondary envelopment.

ORF38 is conserved in all three herpesvirus subfamilies. Its homologues in alpha- and beta-subfamily are UL11 and

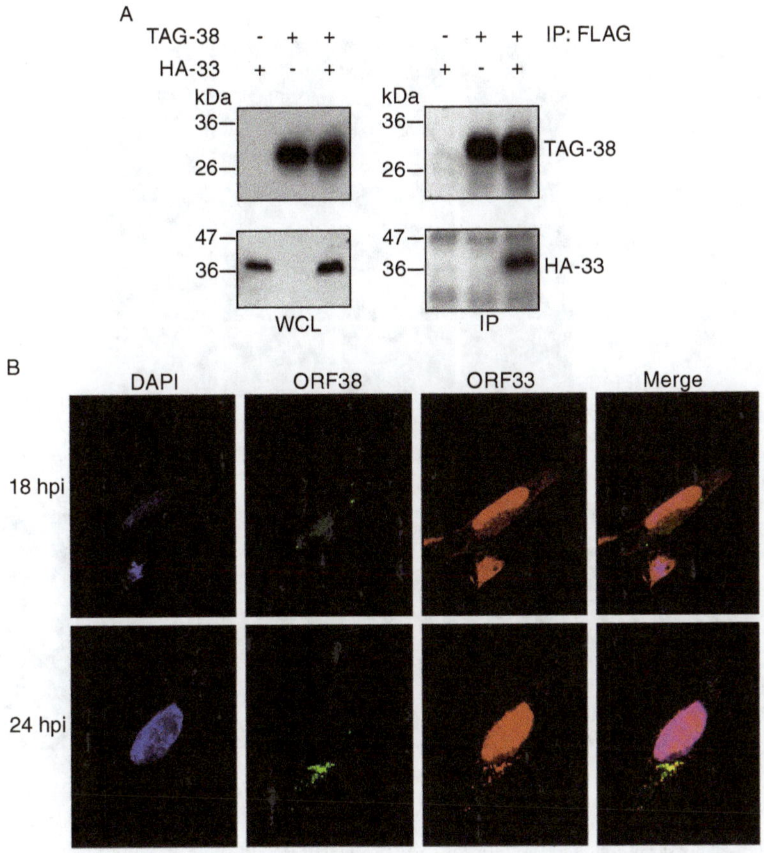

Figure 4. ORF38 interacts with ORF33. (A) ORF38 interacts with ORF33 in transiently transfected cells. 293T cells were transfected with pHA-33 or pTAG-38 or both, and at 48 h posttransfection, whole cell lysates (WCL) were prepared, and the expression of HA-tagged ORF33 and FLAG-tagged ORF38 was examined by Western blotting. The whole cell lysates were further subjected to immunoprecipitation with anti-Flag M2 agarose and analyzed by Western blotting using an anti-HA or an anti-FLAG antibody. (B) Co-localization of ORF38 and ORF33 during virus infection. Vero cells were infected with WT MHV-68 at an MOI of 3. At 24 hpi, cells were fixed for indirect immunofluorescence analysis using an anti-ORF33 antibody and the anti-ORF38 antibody simultaneously.

UL99, respectively. UL11 and UL99 are virion associated tegument proteins. Both are myristoylated and accumulate on the cytoplasmic faces of TGN membrane, a site where secondary envelopment occurs to produce infectious virus particles (Sanchez et al., 2000; Loomis et al., 2001). UL11-null mutant of HSV-1 produced less viral particles and non-enveloped nucleocapsids accumulated in the cytoplasm, providing direct evidence that UL11 is involved in the secondary envelopment of virus (Baines and Roizman, 1992). In HCMV, UL99-null mutation also resulted in the accumulation of non-enveloped viral particles in the cytoplasm (Sanchez et al., 2000). More recent data indicated that the interaction between UL99 and UL94 is essential for virion morphogenesis of HCMV in the cytoplasm (Phillips et al., 2012). The ORF38 homologue in EBV, BBLF1, has been identified as a virion associated protein (Johannsen et al., 2004) and proposed to be involved in virion maturation (Chiu et al., 2012). However, the detailed molecular mechanism of ORF38 involved in virion morphogenesis of gammaherpesviruses has remained unclear. Through co-IP and co-localization studies, we showed in this study that ORF38 interacts with ORF33 in MHV-68 not only during transient transfection but also during viral infection. Therefore, the interaction between ORF38 and ORF33 is a conserved event among herpesviruses.

Virion assembly is thought to be driven by multiple protein-protein interactions (Uetz et al., 2006; Rozen et al., 2008; Mettenleiter et al., 2009). After escape from the nucleus, partially tegumented capsids arrive at TGN for secondary envelopment, depending on a number of protein-protein interactions. The interaction between ORF33 and ORF38 is most likely among these interactions for MHV-68. Our previous study using an ORF33-null mutant of MHV-68 showed that ORF33 is not only essential for the cytoplasmic stage of virion maturation but also plays an important role in virion morphogenesis in the nucleus (Guo et al., 2009).

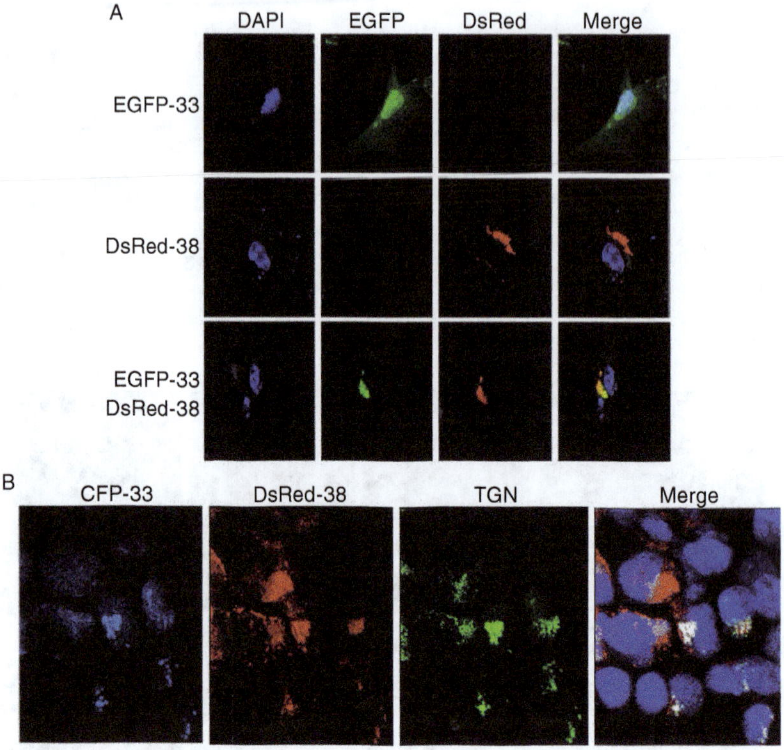

Figure 5. ORF38 and ORF33 co-localize with TGN. (A) Translocation of ORF33 into cytoplasm in transiently transfected cells in the presence of ORF38. HeLa cells were transfected with pEGFP-33 or pDsRed-38 or both. At 24 h posttransfection, cells were stained with DAPI and then observed with a confocal laser microscope. (B) ORF33 and ORF38 co-localize with TGN. 293T cells were co-transfected with pCFP-33, pDsRed-38 and pTGN38-GFP. At 24 h posttransfection, cells were fixed, stained with DAPI, and observed under a confocal laser microscope.

Consistently, ORF33 localized to both the nucleus and the cytoplasm during MHV-68 infection (Fig. 4B). Interestingly, in the absence of other viral proteins, ORF33 and ORF38 together mostly localized to TGN (Fig. 5). These observations indicated that interaction of ORF33 with other virion protein(s) may affect its localization during viral infection. Because ORF33 is likely an inner tegument protein (Guo et al., 2009) and ORF38 likely an outer tegument protein (Fig. 1C), we propose that ORF33 may first bind to capsids within the nucleus and through its interaction with ORF38 may assist with the trafficking of nucleocapsids to TGN for final maturation. Further studies are needed to clarify these sequential events.

MATERIALS AND METHODS

Cells and viruses

BHK-21 cells, Vero cells, HeLa cells and 293T cells were cultured in complete Dulbecco's modified Eagle's medium (DMEM) supplemented with 10% fetal bovine serum. Wide-type (WT) MHV-68 was propagated by infecting BHK-21 cells at a multiplicity of infection (MOI) of 0.03. To infect BHK-21 cells or Vero cells, the viral inoculum in DMEM was incubated with cells for 1 h with occasional swirling. The inoculum was then removed and replaced with fresh DMEM plus 10% fetal bovine serum. The titer of virus was measured by a plaque assay in BHK-21 cells as previously described (Wu et al., 2000).

Plasmids construction

The ORF38 expression plasmid with 3xFLAG-CBP epitope, pTAG-38, is a gift from Dr. Ren Sun (University of California, Los Angeles). pTGN38-GFP, pEEA1-GFP, pLAMP1-GFP and pVPS37C-GFP were kindly provided by Dr. Tao Xu (Institute of Biophysics, Chinese Academy of Sciences). Plasmids pHA-33 and pEGFP-33 were constructed in the lab as described previously (Guo et al., 2009). The full-length MHV-68 ORF38 sequence was cloned into pET-30b (Novagen) via the *Eco*RI and *Xho*I sites to construct pET-30b-38 for expression of His-tagged ORF38 in *E. coli* strain Rosetta (DE3). pDsRed-38 was constructed by inserting the full length ORF38 coding sequences into the *Nhe* I and *Xho* I sites of pDsRed2. The sequences of primers used for plasmid construction are available upon request.

Antibodies, immunoprecipitation and immunoblotting

Purified His-ORF38 of MHV-68 was provided to Experimental Animal Center, Institute of Genetics and Developmental Biology, Chinese Academy of Sciences to generate polyclonal antibodies against ORF38. Polyclonal antibodies for ORF26 and gB were kind gifts from Dr. Ren Sun (University of California, Los Angeles). Polyclonal antibody against ORF33 was described previously (Guo et al., 2009). Mouse monoclonal antibody against the FLAG epitope and the HA epitope were purchased from Sigma. 293T cells seeded onto a 6-cm plate were transfected with 7.5 μg of total DNA by Lipofectamine-2000 (Invitrogen). Forty-eight hours after transfection, cells were washed once with ice-cold phosphate-buffered saline (PBS) and then solubilized in lysis buffer (50 mmol/L Tris-Cl [pH 7.4], 150 mmol/L NaCl, 1% TritonX-100, 1 mmol/L EDTA) with a cocktail of protease inhibitors. Lysates were cleared by centrifugation for 15 min, 13,000 rpm. Ten percent of the supernatant was used as an input control. Soluble proteins were mixed with 20 μL of anti-FLAG M2 agarose (Sigma) and rotated at 4°C for 4 h. Beads were washed three times with lysis buffer prior to use. Immune complexes were washed five times with lysis buffer and supernatant was depleted. Bound proteins were then subjected to immunoblotting analysis. For immunoblotting, samples were heated to 100°C and subjected to SDS-PAGE. Proteins on gels were transferred onto a polyvinylidene difluoride membrane (Millipore) and incubated sequentially with appropriate primary antibody and secondary antibody (anti-rabbit or anti-mouse immunoglobulin G conjugated with horseradish peroxidase), and the proteins were detected by the Enhanced Chemiluminescence system (Millipore).

Detergent and trypsin treatment assay

MHV-68 virions were purified from the supernatant of MHV-68-infected BHK-21 cells by discontinuous sucrose density gradient ultracentrifugation as described previously (Bortz et al., 2003). Briefly, supernatants were cleared of large debris twice by centrifugation at 4°C and 1000 *g* for 15 min. Extracellular viruses were pelleted by ultracentrifugation through a 7.5% sucrose cushion in a P28S rotor (Hitachi, 4°C, 65,000 *g*, 1 h), resuspended in TN buffer (20 mmol/L Tris [pH 7.8], 50 mmol/L NaCl), and purified by 10% to 45% discontinuous sucrose density gradient ultracentrifugation in a P40T rotor (Hitachi, 4°C, 25,000 *g*, 4.5 h) with slow acceleration and braking. Visible bands between 25%–35% starting sucrose positions was saved. Virions were pelleted by ultracentrifugation in a P40T rotor (Hitachi) at 4°C and 65,000 *g* for 1 h and resuspended in 200 μL of TN buffer. For trypsin and detergent treatment, 20 μL of virion stock was diluted into 200 μL of trypsin buffer (50 mmol/L Tris-HCl [pH 7.5], 150 mmol/L NaCl, 1 mmol/L $CaCl_2$) with or without 1% Triton X-100 supplementation, followed by addition of trypsin to a final concentration of 4 μg/mL. After incubation at 37°C for 1 h, the reaction was stopped by adding 0.5 mmol/L phenylmethylsulfonyl fluoride and protease inhibitors, and viral particles were collected by centrifugation at 20,000 *g* for 30 min at 4°C.

Indirect immunofluorescence assay (IFA)

Vero cells were infected with MHV-68 at an MOI of 3. At 18 h or 24 h postinfection, cells were fixed with 4% paraformaldehyde. The fixed cells were permeabilized with phosphate-buffered saline (PBS) plus 0.2% Triton X-100 and processed for immunofluorescence staining, which was performed using a primary antibody against ORF38 or ORF33. Following incubation with fluorescently labeled secondary antibodies, images were captured with a confocal microscope (Olympus).

Immunogold TEM

BHK cells were seeded a day before infection with MHV-68 at 50% confluence in 10-cm plates. The cells were infected at an MOI of 5. Sample preparation and immunostaining were performed as described previously, with some modifications (Wang et al., 2012). Briefly, at 11 h after infection, the cells were fixed with 2% paraformaldehyde plus 0.05% glutaraldehyde in 0.1 mol/L phosphate-buffered saline (PBS) (pH 7.2) for 15 min at 4°C on the plate and embedded in 10% gelatin in 0.1 mol/L PBS. Small blocks were infiltrated in 2.3 mol/L sucrose plus 20% polyvinylpolypyrrolidone (PVP) in PBS overnight at 4°C and quickly plunged into liquid nitrogen. Sections approximately 60-nm thick were cut with a Leica UC6/FC6 ultramicrotome and picked up with a wire loop filled with 2.3 mol/L sucrose. Immunostaining was performed as follows: the sections were washed in BSA buffer (1% BSA and 0.15% glycine in PBS) followed by blocking for 30 min using normal goat serum (1:20 dilution in BSA buffer). Subsequently, the sections were incubated for 1 h at room temperature with the anti-ORF38 polyclonal antibody (or rabbit IgG as a negative control) and then for 1 h at room temperature with goat anti-rabbit IgG conjugated with 5-nm colloidal gold particles (Sigma). After a brief wash in BSA buffer and PBS, the sections were treated with 2.5% glutaraldehyde for 5 min. Finally, the sections were stained by 2% neutral uranyl acetate sealed with methyl cellulose and examined with a 120-KV electron microscope (Tecnai; FEI).

ACKNOWLEDGEMENTS

We thank Drs. Ren Sun and Tao Xu for kindly providing plasmids, Mr. Shufeng Sun at the Center for Biological Imaging, Institute of Biophysics, for help with EM sample preparation and members of the Deng laboratory for helpful discussions. This work was supported by grants from the National Natural Science Foundation of China (Grant No. 81171582) and the National Basic Research Program (973 Program) (No. 2011CB504300).

ABBREVIATION

DMEM, Dulbecco's modified Eagle's medium; EBV, Epstein-Barrvirus; HCMV, human cytomegalovirus; HSV-1, herpes simplex virus-1; IFA, indirect immunofluorescence assay; KSHV, Kaposi's sarcoma-associated herpesvirus; MHV-68, Murine gammaher pesvirus-68; MVB, multivesicular bodies; PBS, phosphate-buffered saline; PVP, polyvinylpolypyrrolidone; TGN, trans-Golgi network; WCL, whole cell lysates; WT, wide-type.

COMPLIANCE WITH ETHICS GUIDELINES

Sheng Shen, Haitao Guo and Hongyu Deng declare that they have no conflict of interest.

This article does not contain any studies with human or animal subjects performed by the any of the authors.

REFERENCES

Baines JD, Roizman B (1992) The UL11 gene of herpes simplex virus 1 encodes a function that facilitates nucleocapsid envelopment and egress from cells. J Virol 66:5168–5174

Bais C, Santomasso B, Coso O, Arvanitakis L, Raaka EG, Gutkind JS, Asch AS, Cesarman E, Gershengorn MC, Mesri EA (1998) G-protein-coupled receptor of Kaposi's sarcoma-associated herpesvirus is a viral oncogene and angiogenesis activator. Nature 391:86–89

Blackman MA, Flano E (2002) Persistent gamma-herpesvirus infections: what can we learn from an experimental mouse model? J Exp Med 195:F29–F32

Bortz E, Whitelegge JP, Jia Q, Zhou ZH, Stewart JP, Wu TT, Sun R (2003) Identification of proteins associated with murine gammaherpesvirus 68 virions. J Virol 77:13425–13432

Bortz E, Wang L, Jia Q, Wu TT, Whitelegge JP, Deng H, Zhou ZH, Sun R (2007) Murine gammaherpesvirus 68 ORF52 encodes a tegument protein required for virion morphogenesis in the cytoplasm. J Virol 81:10137–10150

Chiu YF, Sugden B, Chang PJ, Chen LW, Lin YJ, Lan YC, Lai CH, Liou JY, Liu ST, Hung CH (2012) Characterization and intracellular trafficking of Epstein-Barr virus BBLF1, a protein involved in virion maturation. J Virol 86:9647–9655

Dai W, Jia Q, Bortz E, Shah S, Liu J, Atanasov I, Li X, Taylor KA, Sun R, Zhou ZH (2008) Unique structures in a tumor herpesvirus revealed by cryo-electron tomography and microscopy. J Struct Biol 161:428–438

Decker LL, Klaman LD, Thorley-Lawson DA (1996a) Detection of the latent form of Epstein-Barr virus DNA in the peripheral blood of healthy individuals. J Virol 70:3286–3289

Decker LL, Shankar P, Khan G, Freeman RB, Dezube BJ, Lieberman J, Thorley-Lawson DA (1996b) The Kaposi sarcoma-associated herpesvirus (KSHV) is present as an intact latent genome in KS tissue but replicates in the peripheral blood mononuclear cells of KS patients. J Exp Med 184:283–288

Efstathiou S, Ho YM, Hall S, Styles CJ, Scott SD, Gompels UA (1990) Murine herpesvirus 68 is genetically related to the gammaherpesviruses Epstein-Barr virus and herpesvirus saimiri. J Gen Virol 71(Pt 6):1365–1372

Ganem D (2006) KSHV infection and the pathogenesis of Kaposi's sarcoma. Annu Rev Pathol 1:273–296

Grundhoff A, Ganem D (2004) Inefficient establishment of KSHV latency suggests an additional role for continued lytic replication in Kaposi sarcoma pathogenesis. J Clin Invest 113:124–136

Guo H, Wang L, Peng L, Zhou ZH, Deng H (2009) Open reading frame 33 of a gammaherpesvirus encodes a tegument protein essential for virion morphogenesis and egress. J Virol 83:10582–10595

Guo H, Shen S, Wang L, Deng H (2010) Role of tegument proteins in herpesvirus assembly and egress. Protein Cell 1:987–998

Johannsen E, Luftig M, Chase MR, Weicksel S, Cahir-McFarland E, Illanes D, Sarracino D, Kieff E (2004) Proteins of purified Epstein-Barr virus. Proc Natl Acad Sci USA 101:16286–16291

Kattenhorn LM, Mills R, Wagner M, Lomsadze A, Makeev V, Borodovsky M, Ploegh HL, Kessler BM (2004) Identification of proteins associated with murine cytomegalovirus virions. J Virol 78:11187–11197

Kelly BJ, Fraefel C, Cunningham AL, Diefenbach RJ (2009) Functional roles of the tegument proteins of herpes simplex virus type 1. Virus Res 145:173–186

Kutok JL, Wang F (2006) Spectrum of Epstein-Barr virus-associated diseases. Annu Rev Pathol 1:375–404

Lee S, Salwinski L, Zhang C, Chu D, Sampankanpanich C, Reyes NA, Vangeloff A, Xing F, Li X, Wu TT et al (2011) An integrated approach to elucidate the intra-viral and viral-cellular protein interaction networks of a gamma-herpesvirus. PLoS Pathog 7: e1002297

Loomis JS, Bowzard JB, Courtney RJ, Wills JW (2001) Intracellular trafficking of the UL11 tegument protein of herpes simplex virus type 1. J Virol 75:12209–12219

Loret S, Guay G, Lippe R (2008) Comprehensive characterization of extracellular herpes simplex virus type 1 virions. J Virol 82:8605–8618

Martin DF, Kuppermann BD, Wolitz RA, Palestine AG, Li H, Robinson CA (1999) Oral ganciclovir for patients with cytomegalovirus retinitis treated with a ganciclovir implant. Roche Ganciclovir Study Group. N Engl J Med 340:1063–1070

Mettenleiter TC, Klupp BG, Granzow H (2006) Herpesvirus assembly: a tale of two membranes. Curr Opin Microbiol 9:423–429

Mettenleiter TC, Klupp BG, Granzow H (2009) Herpesvirus assembly: an update. Virus Res 143:222–234

Phillips SL, Bresnahan WA (2012) The human cytomegalovirus (HCMV) tegument protein UL94 is essential for secondary envelopment of HCMV virions. J Virol 86:2523–2532

Phillips SL, Cygnar D, Thomas A, Bresnahan WA (2012) Interaction between the human cytomegalovirus tegument proteins UL94 and UL99 is essential for virus replication. J Virol 86:9995–10005

Rozen R, Sathish N, Li Y, Yuan Y (2008) Virion-wide protein interactions of Kaposi's sarcoma-associated herpesvirus. J Virol 82:4742–4750

Sanchez V, Sztul E, Britt WJ (2000) Human cytomegalovirus pp28 (UL99) localizes to a cytoplasmic compartment which overlaps the endoplasmic reticulum-golgi-intermediate compartment. J Virol 74:3842–3851

Simas JP, Efstathiou S (1998) Murine gammaherpesvirus 68: a model for the study of gammaherpesvirus pathogenesis. Trends Microbiol 6:276–282

Soulier J, Grollet L, Oksenhendler E, Cacoub P, Cazals-Hatem D, Babinet P, d'Agay MF, Clauvel JP, Raphael M, Degos L et al (1995) Kaposi's sarcoma-associated herpesvirus-like DNA sequences in multicentric Castleman's disease. Blood 86:1276–1280

Stevenson PG, Efstathiou S (2005) Immune mechanisms in murine gammaherpesvirus-68 infection. Viral Immunol 18:445–456

Stine JT, Wood C, Hill M, Epp A, Raport CJ, Schweickart VL, Endo Y, Sasaki T, Simmons G, Boshoff C et al (2000) KSHV-encoded CC chemokine vMIP-III is a CCR4 agonist, stimulates angiogenesis, and selectively chemoattracts TH2 cells. Blood 95:1151–1157

Sunil-Chandra NP, Efstathiou S, Nash AA (1992) Murine gammaherpesvirus 68 establishes a latent infection in mouse B lymphocytes in vivo. J Gen Virol 73(Pt 12):3275–3279

Uetz P, Dong YA, Zeretzke C, Atzler C, Baiker A, Berger B, Rajagopala SV, Roupelieva M, Rose D, Fossum E et al (2006) Herpesviral protein networks and their interaction with the human proteome. Science 311:239–242

Wang L, Guo H, Reyes N, Lee S, Bortz E, Guo F, Sun R, Tong L, Deng H (2012) Distinct domains in ORF52 tegument protein mediate essential functions in murine gammaherpesvirus 68 virion tegumentation and secondary envelopment. J Virol 86:1348–1357

Wu TT, Usherwood EJ, Stewart JP, Nash AA, Sun R (2000) Rta of murine gammaherpesvirus 68 reactivates the complete lytic cycle from latency. J Virol 74:3659–3667

Yeh PC, Meckes DG Jr, Wills JW (2008) Analysis of the interaction between the UL11 and UL16 tegument proteins of herpes simplex virus. J Virol 82:10693–10700

PERMISSIONS

All chapters in this book were first published in P&C, by Springer; hereby published with permission under the Creative Commons Attribution License or equivalent. Every chapter published in this book has been scrutinized by our experts. Their significance has been extensively debated. The topics covered herein carry significant findings which will fuel the growth of the discipline. They may even be implemented as practical applications or may be referred to as a beginning point for another development.

The contributors of this book come from diverse backgrounds, making this book a truly international effort. This book will bring forth new frontiers with its revolutionizing research information and detailed analysis of the nascent developments around the world.

We would like to thank all the contributing authors for lending their expertise to make the book truly unique. They have played a crucial role in the development of this book. Without their invaluable contributions this book wouldn't have been possible. They have made vital efforts to compile up to date information on the varied aspects of this subject to make this book a valuable addition to the collection of many professionals and students.

This book was conceptualized with the vision of imparting up-to-date information and advanced data in this field. To ensure the same, a matchless editorial board was set up. Every individual on the board went through rigorous rounds of assessment to prove their worth. After which they invested a large part of their time researching and compiling the most relevant data for our readers.

The editorial board has been involved in producing this book since its inception. They have spent rigorous hours researching and exploring the diverse topics which have resulted in the successful publishing of this book. They have passed on their knowledge of decades through this book. To expedite this challenging task, the publisher supported the team at every step. A small team of assistant editors was also appointed to further simplify the editing procedure and attain best results for the readers.

Apart from the editorial board, the designing team has also invested a significant amount of their time in understanding the subject and creating the most relevant covers. They scrutinized every image to scout for the most suitable representation of the subject and create an appropriate cover for the book.

The publishing team has been an ardent support to the editorial, designing and production team. Their endless efforts to recruit the best for this project, has resulted in the accomplishment of this book. They are a veteran in the field of academics and their pool of knowledge is as vast as their experience in printing. Their expertise and guidance has proved useful at every step. Their uncompromising quality standards have made this book an exceptional effort. Their encouragement from time to time has been an inspiration for everyone.

The publisher and the editorial board hope that this book will prove to be a valuable piece of knowledge for researchers, students, practitioners and scholars across the globe.

LIST OF CONTRIBUTORS

Xiaojuan Chen, Yang Zheng, Yudong Yang and Yaling Xing
Division of Infection and Immunity, Department of Electromagnetic and Laser Biology, Beijing Institute of Radiation Medicine, Beijing 100850, China

Xingxing Yang and Zhongbin Chen
Division of Infection and Immunity, Department of Electromagnetic and Laser Biology, Beijing Institute of Radiation Medicine, Beijing 100850, China
Anhui Medical University, Hefei 230032, China

Yan Jiang, Yanping Zhu
National Laboratory of Biomacromolecules, Institute of Biophysics, Chinese Academy of Sciences, Beijing 100101, China

Zhi-Jie Liu
National Laboratory of Biomacromolecules, Institute of Biophysics, Chinese Academy of Sciences, Beijing 100101, China
iHuman Institute, Shanghai Tech University, Shanghai 201210, China

Songying Ouyang
National Laboratory of Biomacromolecules, Institute of Biophysics, Chinese Academy of Sciences, Beijing 100101, China
Department of Microbiology, Immunology and Molecular Genetics, University of California Los Angeles, Los Angeles, CA 90095, USA

Yumin Hao, Hong Chen, Zengqiang Yuan and Jinbo Cheng
State Key Laboratory of Brain and Cognitive Sciences, Institute of Biophysics, Chinese Academy of Sciences, Beijing 100101, China

Yajin Liao and Qing He
State Key Laboratory of Brain and Cognitive Sciences, Institute of Biophysics, Chinese Academy of Sciences, Beijing 100101, China
College of Life Sciences, University of Chinese Academy of Sciences, Beijing 100049, China

Lu Sun, Yu Zhang, Bao Zhao, Mengmeng Deng, Jun Liu, Xin Li, Junwei Hou, Shuijun Zhang, George F. Gao and Songdong Meng
CAS Key Laboratory of Pathogenic Microbiology and Immunology, Institute of Microbiology, Chinese Academy of Sciences (CAS), Beijing 100101, China

Xiaodong Li
Beijing Institute of Infectious Diseases, Beijing 302 Hospital, Beijing 100039, China

Mingming Gui
Xinjiang Agricultural University, Ürümqi 830052, China

Máté Virágh, Liliána Tóth, Csaba Vágvölgyi and László Galgóczy
Department of Microbiology, Faculty of Science and Informatics, University of Szeged, Szeged, Hungary

Annamária Marton and Csaba Vizler
Institute of Biochemistry, Biological Research Centre, Hungarian Academy of Sciences, Szeged, Hungary

Florentine Marx
Division of Molecular Biology, Biocenter, Innsbruck Medical University, Innsbruck, Austria

Yujia Zhai, Xiaoyun Pang and Zhe Sun
National Key Laboratory of Biomacromolecules, Institute of Biophysics, Chinese Academy of Sciences, Beijing 100101, China

Shengliu Wang and Tongxin Niu
National Key Laboratory of Biomacromolecules, Institute of Biophysics, Chinese Academy of Sciences, Beijing 100101, China
University of Chinese Academy of Sciences, Beijing 100049, China

Fei Sun
National Key Laboratory of Biomacromolecules, Institute of Biophysics, Chinese Academy of Sciences, Beijing 100101, China
Center for Biological Imaging, Institute of Biophysics, Chinese Academy of Sciences, Beijing 100101, China
University of Chinese Academy of Sciences, Beijing 100049, China

Yue-He Ding and Meng-Qiu Dong
National Institute of Biological Sciences, Beijing, Beijing 102206, China

Victor W. Hsu
Department of Medicine, Harvard Medical School, Brigham and Women's Hospital, Boston, MA 02115, USA

Yu-Shi Wang, He Li, Yang Li, Hongyan Zhu and Ying-Hua Jin
Key Laboratory for Molecular Enzymology and Engineering of the Ministry of Education, College of Life Science, Jilin University, Changchun 130012, China

Xinhua Wang, Mary Mathieu and Randall J. Brezski
Genentech, Antibody Engineering, South San Francisco, CA 94080, USA

Junyi Wang, Zongyu Li and Wanli Liu
MOE Key Laboratory of Protein Sciences, Collaborative Innovation Center for Diagnosis and Treatment of Infectious Diseases, School of Life Sciences, Tsinghua University, Beijing 100084, China

Liling Xu
Ragon Institute of Massachusetts General Hospital, Massachusetts Institute of Technology, Harvard University, 400 Technology Square, Cambridge, MA 02139, USA

Hengwen Yang
The First Affiliate Hospital, Biomedical Translational Research Institute, Guangdong Province Key Laboratory of Molecular Immunology and Antibody Engineering, Jinan University, Guangzhou 510632, China

Jing-Xiang Wu, Yunlu Kang and Lei Chen
State Key Laboratory of Membrane Biology, Institute of Molecular Medicine, Peking University, Beijing Key Laboratory of Cardiometabolic Molecular Medicine, Beijing 100871, China
Peking-Tsinghua Center for Life Sciences, Peking University, Beijing 100871, China

Dian Ding, Mengmeng Wang and Xin Zeng
State Key Laboratory of Membrane Biology, Institute of Molecular Medicine, Peking University, Beijing Key Laboratory of Cardiometabolic Molecular Medicine, Beijing 100871, China
Peking-Tsinghua Center for Life Sciences, Peking University, Beijing 100871, China
Academy for Advanced Interdisciplinary Studies, Peking University, Beijing 100871, China

Huirong Yang, Mengjie Liu, Xizi Chen and Yanhui Xu
Fudan University Shanghai Cancer Center, Institute of Biomedical Sciences, Shanghai Medical College of Fudan University, Shanghai 200032, China
Key Laboratory of Molecular Medicine, Ministry of Education, Department of Systems Biology for Medicine, School of Basic Medical Sciences, Shanghai Medical College of Fudan University, Shanghai 200032, China
State Key Laboratory of Genetic Engineering, Collaborative Innovation Center of Genetics and Development, School of Life Sciences, Fudan University, Shanghai 200433, China

Jia Wang, Jiawei Wang and Hong-Wei Wang
Ministry of Education Key Laboratory of Protein Sciences, Tsinghua-Peking Joint Center for Life Sciences, Beijing Advanced Innovation Center for Structural Biology, School of Life Sciences, Tsinghua University, Beijing 100084, China

Min Huang and Catherine C. L. Wong
National Center for Protein Science, Shanghai Institute of Biochemistry and Cell Biology, Shanghai Institutes of Biological Sciences, Chinese Academy of Sciences, Shanghai 200031, China

Dan Tan and Meng-Qiu Dong
National Institute of Biological Sciences, Beijing 102206, China

Shengyuan Zeng, Yangyang Wang and Changzhu Duan
Department of Cell Biology and Medical Genetics, Molecular Medicine and Cancer Research Center, Chongqing Medical University, Chongqing 400016, China

Ting Zhang
Department of Obstetrics, The First Affiliated Hospital of Chongqing Medical University, Chongqing 400016, China

Lu Bai
Key Laboratory of Medical Molecular Virology, School of Basic Medical Sciences, Shanghai Medical College, Fudan University, Shanghai 200032, China

Yalan Wang
Department of Pathology, Molecular Medicine and Cancer Research Center, Chongqing Medical University, Chongqing 400016, China

Yuan He and Nieng Yan
State Key Laboratory of Bio-membrane and Membrane Biotechnology, Tsinghua University, Beijing 100084, China
Center for Structural Biology, School of Medicine, Tsinghua University, Beijing 100084, China

Kan Wang
China-Japan Friendship Hospital, Beijing 100029, China

Vsevolod V. Gurevich and Eugenia V. Gurevich
Department of Pharmacology, Vanderbilt University, Nashville, TN 37232, USA

Vladimir N. Uversky
Department of Molecular Medicine and USF Health Byrd Alzheimer's Research Institute, Morsani College of Medicine, University of South Florida, Tampa, FL 33612, USA
Institute for Biological Instrumentation, Russian Academy of Sciences, Pushchino, Moscow Region, Russia 142290

Shrestha Ghosh
Department of Biochemistry, The University of Hong Kong, Hong Kong, China

Zhongjun Zhou
Department of Biochemistry, The University of Hong Kong, Hong Kong, China
Shenzhen Institute of Innovation and Technology, The University of Hong Kong, Hong Kong, China

Qian Luo
School of Life Sciences and Biotechnology, Shanghai Jiao Tong University, Shanghai 200240, China

Jie Gao and Yu-Huan Xiao
School of Life Sciences and Biotechnology, Shanghai Jiao Tong University, Shanghai 200240, China
Institute of Nano Biomedicine and Engineering, Department of Instrument Science and Engineering, School of Electronic Information and Electronic Engineering, Shanghai Jiao Tong University, Shanghai 200240, China

Wei-Lin Jin
School of Life Sciences and Biotechnology, Shanghai Jiao Tong University, Shanghai 200240, China
Institute of Nano Biomedicine and Engineering, Department of Instrument Science and Engineering, School of Electronic Information and Electronic Engineering, Shanghai Jiao Tong University, Shanghai 200240, China
National Center for Translational Medicine, Shanghai Jiao Tong University, Shanghai 200240, China

Yue Ma
Institute of Nano Biomedicine and Engineering, Department of Instrument Science and Engineering, School of Electronic Information and Electronic Engineering, Shanghai Jiao Tong University, Shanghai 200240, China

Hua-Lin Fu and Da-Xiang Cui
Institute of Nano Biomedicine and Engineering, Department of Instrument Science and Engineering, School of Electronic Information and Electronic Engineering, Shanghai Jiao Tong University, Shanghai 200240, China
National Center for Translational Medicine, Shanghai Jiao Tong University, Shanghai 200240, China

Zhen Wang
Institute of Nano Biomedicine and Engineering, Department of Instrument Science and Engineering, School of Electronic Information and Electronic Engineering, Shanghai Jiao Tong University, Shanghai 200240, China
Department of Experimental Surgery, Tangdu Hospital, Fourth Military Medical University, Xi'an 710038, China

Hao Yang
Clinical Stem Cell Research Center, Renji Hospital, School of Medicine, Shanghai Jiao Tong University, Shanghai 200127, China

Hanqin Li, Susu Mao, Ke Zen, Chenyu Zhang and Liang Li
Jiangsu Engineering Research Center for MicroRNA Biology and Biotechnology, State Key Laboratory of Pharmaceutical Biotechnology, Nanjing University School of Life Sciences, Nanjing 210093, China

Haitao Wang
Department of Pharmacy, 306th Hospital of PLA, Beijing 100101, China

Pengda Liu, Jianping Guo, Wenjian Gan and Wenyi Wei
Department of Pathology, Beth Israel Deaconess Medical Center, Harvard Medical School, Boston, MA 02115, USA

Jiao Meng, Xiaoting Zou, Rimao Wu, Ran Zhong, Dahai Zhu and Yong Zhang
The National Laboratory of Medical Molecular Biology, Institute of Basic Medical Sciences, Chinese Academy of Medical Sciences and School of Basic Medicine, Peking Union Medical College, Beijing 100005, China

Hongyu Deng
Chinese Academy of Sciences Key Laboratory of Infection and Immunity, Institute of Biophysics, Chinese Academy of Sciences, Beijing 100101, China

Sheng Shen and Haitao Guo
University of Chinese Academy of Sciences, Beijing 100049, China

Xiaofei Yang
Key Laboratory of Cognitive Science, Laboratory of Membrane Ion Channels and Medicine, College of Biomedical Engineering, South-Central University for Nationalities, Wuhan 430074, China

Chen Zhang
State Key Laboratory of Biomembrane and Membrane Biotechnology, College of Life Sciences, Peking University, Beijing 100871, China
PKU-IDG/McGovern Institute for Brain Research, Peking University, Beijing 100871, China

Dongmei Hou and Wei Jiang
Key Laboratory of Cognitive Science, Laboratory of Membrane Ion Channels and Medicine, College of Biomedical Engineering, South-Central University for Nationalities, Wuhan 430074, China
State Key Laboratory of Biomembrane and Membrane Biotechnology, College of Life Sciences, Peking University, Beijing 100871, China
PKU-IDG/McGovern Institute for Brain Research, Peking University, Beijing 100871, China

Index

www.ingramcontent.com/pod-product-compliance
Lightning Source LLC
LaVergne TN
LVHW081021210926
848485LV00004B/915